Ottoman Wars, 1700–1870

MODERN WARS IN PERSPECTIVE

General Editors: H. M. Scott and B. W. Collins

This ambitious series offers wide-ranging studies of specific wars, and distinct phases of warfare, from the close of the Middle Ages to the present day. It aims to advance the current integration of military history into the academic mainstream. To that end, the books are not merely traditional campaign narratives, but examine the causes, course and consequences of major conflicts, in their full international political, social and ideological contexts.

ALSO IN THIS SERIES

THE ANGLO–DUTCH WARS OF THE SEVENTEENTH CENTURY
J. R. Jones

THE WARS OF LOUIS XIV
John A. Lynn

THE WARS OF FREDERICK THE GREAT
Dennis Showalter

THE WAR OF THE AUSTRIAN SUCCESSION, 1740–1748
M. S. Anderson

THE WARS OF NAPOLEON
Charles J. Esdaile

THE SPANISH–AMERICAN WAR: CONFLICT IN THE CARIBBEAN AND THE PACIFIC 1895–1902
Joseph Smith

CHINA AT WAR, 1901–1949
Edward L. Dreyer

THE WARS OF FRENCH DECOLONIZATION
Anthony Clayton

A NOBLE CAUSE? AMERICA AND THE VIETNAM WAR
Gerard J. DeGroot

THE NORTHERN WARS
Robert I. Frost

ENGLAND'S COLONIAL WARS 1550–1688: CONFLICTS, EMPIRE AND NATIONAL IDENTITY
Bruce Lenman

BRITAIN'S COLONIAL WARS 1688–1783
Bruce Lenman

THE GREAT WAR
Ian F. W. Beckett

AUSTRIA'S WARS OF EMERGENCE
Michael Hochedlinger

THE FRENCH CIVIL WARS
R. J. Knecht

RUSSIA'S WARS OF EMERGENCE, 1460–1730
Carol B. Stevens

THE ENGLISH REVOLUTION AND THE WARS OF THE THREE KINGDOMS, 1638–1652
Ian Gentles

Ottoman Wars 1700–1870: An Empire Besieged

Virginia H. Aksan

PEARSON
Longman

Harlow, England • London • New York • Boston • San Francisco • Toronto
Sydney • Tokyo • Singapore • Hong Kong • Seoul • Taipei • New Delhi
Cape Town • Madrid • Mexico City • Amsterdam • Munich • Paris • Milan

PEARSON EDUCATION LIMITED

Edinburgh Gate
Harlow CM20 2JE
United Kingdom
Tel: +44 (0)1279 623623
Fax: +44 (0)1279 431059
Website: www.pearsoned.co.uk

First edition published in Great Britain in 2007

ISBN: 978-0-582-30807-7

British Library Cataloguing in Publication Data
A CIP catalogue record for this book can be obtained from the British Library

Library of Congress Cataloging in Publication Data
A CIP catalog record for this book can be obtained from the Library of Congress

10 9 8 7 6 5 4
11

Set by 35 in 10/13.5pt Sabon
Printed in Malaysia (CTP-VVP)

Contents

List of plates x
Acknowledgements xi
A note on place names and foreign terms xiv
Chronology xv
Maps xviii

Introduction 1

Comparative empires 5
The northern defensive line 6
Military reform 7
The outline of the narrative 8
Eastern questions and new Ottomans 11

1 The political map of Central and Eastern Europe circa 1700 **18**

The geopolitics of the Danube 19
Members of the Holy League 23
The Principalities 28
Ottomans and Tatars 30
Eastern border strategies 31
Imperial aims: Ottoman, Habsburg, Romanov 33
Revolts in Istanbul circa 1700 36
Janissary loyalties and ethos 38

2 The Ottoman system **45**

Rethinking imperial military history 46
The nature and composition of the Janissaries 48
Janissary command and discipline 53
Provincial forces: sipahis 54
Frontier defence systems 56

Survival strategies 59
The northern frontier 60
The southern frontier: Egypt and the Arab Provinces 62
Auxiliary forces: the religious factor 62
Realities of Danube warfare 65
Logistics of warfare on the Danube: the Belgrade–Azov fortress line 67
Campaign financing 72

3 From Prut to Belgrade: illusions of success and failure **83**

The political setting 1700–50 83
The Ottoman case 84
Austria's position and ambitions 86
Russian expansion and the pre-modern impasse 87
Centralisation of the control of violence 88
The consolidation of Austrian arms 89
Origins of the Prut campaign 90
On the march – Prut 1711 95
The recovery of the Morea 98
Austria enters the war 101
Renewal of Ottoman–Habsburg conflict on the Danube, 1736–39: stalemate 102
Uneasy alliances 102
Ochakov and the Crimea, 1737 107
The Bosnian army under Governor Ali Pasha 110
Ottoman victory at Orsova, 1738 112
Ottoman military strengths and weaknesses 117

4 The Austrian–Russian–Ottoman Danube waltz, 1768–92 **129**

Overview 129
1768–92 and the Ottoman Empire 130
1768–92 and the Russians 135
1768–92 and the Austrians 137
1768–74: Russian miracle, Ottoman debacle, and Austrian test 138
The centre of the conflict: the lower Danube River 140
Declaration and preparations for war, 1768 142
The collapse of the Ottoman supply system 147
The year of disasters: Çeşme and Kartal (1770) 151
Peace negotiations and the Crimea 154

Austro-Russian–Ottoman confrontations 1787–92 160
Peace – lessons learned 167

5 Selim III and the new order (*Nizâm-ı Cedid*) 180

Introduction: crisis and reform 180
The problem 181
Economic reforms 184
New military organisations 185
Ottoman military reform prior to Selim III 186
Military reforms under Selim III 192
Initial reforms 193
Gunpowder and arms 194
Creation of a new army, the Nizâm-ı Cedid 195
The new order expands beyond Istanbul 196
Foreign advisers and reform of the Artillery Corps 198

6 The internal and external challenges to Selim III 214

Introduction: provincial allies and colonial powers 214
Local Ottoman elites: provincial loyalty and betrayal 215
Caniklizades in North Anatolia 216
Vidin and Osman Pasvanoğlu 219
Radical diplomacy 224
Pre-1793 diplomacy 225
First embassies to Europe 226
Napoleon's invasion of Egypt 229
Pre-Napoleonic Egypt 231
French evacuation of Egypt 235
Imperial impressions 237
The final challenge to Selim III 1806–07 241
The Rusçuk Committee and the end of an era 247

7 Mahmud II and the new absolutism 259

Sened-i İttifak – the Deed of Agreement 261
The state of the Russian military 265
Ottoman campaign preparations 269
The Battle for Rusçuk and the Treaty of Bucharest, 1811–12 270
Confronting rebellion: the Serbs 282
Confronting rebellion: Ali Pasha of Iannina 285
The Greek Revolt 1821–27 288
Revolt in the Peloponnesus and Ionian Islands 291
Missolonghi and Athens 294

8 The Auspicious Occasion and its consequences 306

Mehmed Ali Pasha and the transformation of Egypt 306
The Wahhabi challenge and Mehmed Ali's new army 308
The Auspicious Occasion, 1826 313
Formation of the Eşkinci Corps, May 1826 317
Elimination of the Bektashis 322
The realities of reconstruction 325
Asakir-i Mansure *regulations 7 July 1826* 328
Further regulations 1826–27 329

9 An Empire returns to war 343

Russo-Ottoman conflict resumes on the Danube, 1828–29 343
Standoff at Şumnu and the siege of Varna 349
The campaign of 1829 and Ottoman collapse 352
Assessing the new army 356
The Treaty of Adrianople, 1829 361
The Syrian question 1831–33 363
İbrahim Pasha and the siege of Acre 367
The European powers and the Treaty of Hünkar İskelesi 374
Ottoman reforms 1834–39 376
A new national reserve system 380
The status quo and negotiations 383
Redifs and Kurds 385
The battle of Nizib, 1839 388

10 From Tanzimat to Crimean War 399

New sultan, new advisers 402
Taming Mehmed Ali 404
Further military reforms: divisional armies 408
Further economic reforms: tax farming 414
Provincial resistance: post-1841 Syria 416
Rebellion in Aleppo 418
Seeds of Russo-Ottoman conflict on the Danube 422
1848 revolutions and the Ottoman northern frontier 424
Provincial resistance: Vidin in 1850 427
Provincial resistance: Bosnia and Montenegro 429
Austrian and Russian armies revisited circa 1850 432
Origins of the Crimean War 437

11 Completing the northern arc **449**

On the Danube once again: Vidin and Silistre in 1853–54 450
Crimea and Sevastopol 457
The Caucasus and the Ottomans pre-1853 460
The siege of Kars: the British and the barbarians 466
Reform and the Concert of Europe 476
A turn to the Prussian – 1869 conscription reforms 477

Conclusion 483
Guide to further reading 490
Appendix: Excerpts from William Wittman, *Travels in Turkey, Asia Minor, Syria and Egypt* 500
Glossary of places 512
Glossary of terms 517
Bibliography 522

Index 559

List of plates

1	The Ottoman northern arc, Vidin to Kars, 1750–1830	xviii
2	The southern tier of the empire, 1800–50	xix
3	Greece, the Danube and the Caucasus, 1820–56	xx
4	Nineteenth-century lithograph of the Upper Danube	21
5	Members of various pre-reform Ottoman forces	51
6	Examples of Ottoman transport circa 1700 (Marsigli)	70
7	Map of Lower Danube at time of Habsburg-Ottoman campaigns, 1737–39	104–5
8	Map of Crimea and northern Black Sea Coast, 1737–39	108–9
9	The Danube basin (a map of the Russian dominions in Europe)	132–33
10	Cartoon satirising the French military mission in Istanbul in 1783	201
11	The Ottoman interior of the Baba Vida fortress in Vidin	219
12	Egyptian mamluks training in Cairo	233
13	Nineteenth-century print of Rusçuk	275
14	Mehmed Ali's reformed army uniforms	311
15	Sultan Mahmud II, 1829	314
16	Uniforms of the *Asakir-i Mansure* of Mahmud II after the adoption of trousers	335
17	Column commemorating Mahmud II's visit to Tutrekan in 1837	362
18	Campaigns of Mehmed Ali and his son İbrahim in Greece, 1824–33	368–69
19	A *Redif* muster roll from Nov–Dec 1836	381
20	Sketch of *başıbozuk* by James Robertson	411
21	Sketch of Silistre fortress system and Mecidiye *tabya* still standing	451
22	The Mecidiye *tabya* at Silistre	455
23	Map of fortress of Kars, 1854–55	468–69

Acknowledgements

It is customary for an author to thank profusely a vast array of institutions, granting agencies, colleagues, anonymous readers, friends, neighbours and of course family for the patience with which they have endured the research journey just finished. I assume that as a given here, as I am very fortunate to be situated in a happy place as regards all of the above. Without the well-grounded sense and support of my partner in crime, Oktay Aksan, this book would never have been finished. I am also particularly blessed with the collegiality of a core group of Ottomanists, remarkable scholars all, who tolerate my naive rantings and ramblings with great good humour and sage advice. They know who they are.

I agreed to write the history of Ottoman wars without the least understanding of the journey that would entail, and how long it would take. The project has stretched in time through the seismic shifts in world politics engendered by the war on terror. To my complete horror, I found myself writing about the British bombarding Beirut in 1840, and the folly of the Crimean War, just as Washington made the benighted decision to make war on Iraq. That historians are trapped in their own eras may well be a truism, but finding oneself reflecting on the repeated global idiocy regarding the Middle East was emotionally wrenching.

Ottoman military historians writing in English are few in number: Caroline Finkel, Victor Ostapchuk and Gábor Ágoston have encouraged me to continue this madness. Suraiya Faroqhi, in spite of her well-known dislike of the subject of warfare, has read various pieces of the manuscript, for which I thank her. I am much indebted to Rossitsa Gradeva at the American University of Bulgaria, and her colleagues Fionera Filipova, Director of the Historical Museum in Vidin, and Theodora Bakardjieva at the Historical Museum in Rusçuk, for the particularly eye-opening trip from Vidin to Silistre to tour surviving Ottoman fortresses. Theodora has been instrumental in preserving the columns commemorating the tour of Mahmud II to Bulgaria in the 1830s, one of which is pictured in the book.

What came as the greatest surprise to me was the serendipity of the internet. I have corresponded with a number of scholars, most of whom I have yet to meet, who pointed me to new or otherwise unavailable sources, and even supplied them on occasion. This I wish to acknowledge with heartfelt thanks. They include Avigdor Levy, who sent me a copy of his unpublished PhD thesis which is so important to the discussion of the reforms of Mahmud II; Alexander Mikaberidze, who supplied me with his translation of Danilevsky's history of the Russo-Ottoman War of 1806–12; Mesut Uyar, of Ankara, who is writing a history of the Ottoman military himself, and whose advice has been invaluable; Veysel Şimşek, and Kahraman Şakul, students both, whose own writings and references have enriched this manuscript immensely. I would also like to thank Peter Harrington, Curator of the Anne S.K. Brown Military Collection at Brown University, who sent along a number of possible illustrations, one of which adorns the cover.

I need to thank my colleagues at McMaster University, particularly John Weaver, Stefania Miller and Richard Rempel, for helping me steer the course by reading various parts of this manuscript.

To Hamish Scott I owe a debt of gratitude for taking me on as an untried military historian, and for his insistence that the Ottoman military post-1700 be included in Longman's series *Modern Wars in Perspective*. I am also grateful for his seeing the work into press, and especially thank Christina Wipf Perry and Melanie Carter at Pearson Education for making it happen.

That I chose to situate the Ottomans among their neighbours, and write a comparative imperial history about war and society, should be laid at my door. The book has ambitions to educate a general audience as well as suggest to experts the inherent narrative possibilities of texts and documents. When I began the task such work was just beginning. As I finish, a generation of international scholars is taking up the challenge of comparative Russian–Ottoman history from the point of view of Ottoman as well as Russian sources.

I do not pretend to have dealt with the subject of Ottoman warfare from 1700 to 1870 with anything like completeness. Much of the material lends itself to snapshots of a particular moment, or a particular crisis. The Ottoman archives are immensely rich, vast and intimidating on the post-1826 period, but also require time to digest, share and interpret. At one point I had to abandon the hope of using the archives for all the wars, but remain optimistic that this introduction to the topic will encourage others to take up the cudgel. As they used to say in Turkish when a young child was surrendered to his first teacher at school: 'Eti senin; kemiği benim.' 'His flesh is yours, his bones are mine.'

Virginia H. Aksan
McMaster University
October 2006

Publisher's acknowledgements

We are grateful to the following for permission to reproduce copyright material:

Plate 9 courtesy of McMaster University Library; plate 19 Cevdet Askeriye 635, from Prime Ministry Ottoman Archives, Istanbul, with thanks to Veysel Şimşek, graduate student, Bilkent University, Ankara, Turkey; plate 21 courtesy of the editors of the *International Journal of Turkish Studies*, Machiel Kiel, 'Urban development in Bulgaria', *IJTS* 4 (1989), 79–159.

In some instances we have been unable to trace the owners of copyright material, and we would appreciate any information that would enable us to do so.

A note on foreign terms and place names

In the interests of accessibility, I have kept foreign terms to a minimum. Some, such as pasha, agha, shar'ia, kadi (court judge), ulema, or jihad, have entered the English language. For others, I have used modern Turkish, based on the spelling in the *New Redhouse Turkish-English Dictionary of 1968*. A small glossary with minimal definitions is included at the back of the volume for those terms that are consistently used in Turkish throughout the text.

In the matter of place names, those that are commonly known in English are used that way (Istanbul instead of Constantinople; Izmir instead of Smyrna; Ochakov instead of Özü, for example). I have been a bit stubborn, and used the Ottoman names for other places that have long since disappeared or been renamed (Rusçuk for Ruse, Bulgaria, for example). To help the reader, I have included a glossary of some of the names with variants, and an occasional location if it seems called for. I have also tried to locate the important ones on the maps that follow these pages, as a way of pinpointing particular wars and campaigns.

Chronology

Date	Sultan	Events
1683–99	Mehmed IV (1648–87); Süleyman II (1687–91); Ahmed II (1691–95); Mustafa II (1695–1703)	Habsburg–Ottoman War and Venice, Poland and Russia (The Holy League)
1683		2nd siege of Vienna
1686		Russia joins the Holy League (Austria, Poland, Venice)
1688		Belgrade falls to Habsburgs
1690		Belgrade retaken by Ottomans
1696		**Peter the Great (1696–1725) as sole ruler**
1699		Treaty of Karlowitz
1703	Ahmed III (1703–1730)	'Edirne Incident', deposition of Mustafa II
1711		Battle of Prut – War with Russia
1709–14		Charles XII of Sweden (1697–1718) takes refuge with Ottomans
1715–18		War with Venice and Austria
1717		Belgrade falls to Austria
1718		Treaty of Passarowitz
1722		Collapse of Safavids (Iran)
1724–46		War with Iran; Russo-Ottoman partition of north-west Iran
1727		First Arabic script printing press in Istanbul
1730	Mahmud I (1730–54)	'Patrona Halil' revolt in Istanbul
1736–39		Russo-Austrian Ottoman War
1737		Russian Capture of Ochakov
1739		Treaty of Belgrade; Ottomans regain Belgrade
1740		Ottoman–French capitulations **Frederick the Great of Prussia** (1740–86) **Maria Theresa of Austria** (1740–80)
1753	Osman III (1754–7)	
1756–63	Mustafa III (1757–74)	Seven Years War in Europe and first global conflict
1756		Diplomatic revolution. France and Austria allies
1762		**Catherine the Great** (1762–92)

Date	Sultan	Events
1764		Stanislaus-Augustus Poniatowski (last King of Poland, 1764–95)
1765		**Joseph II of Austria** (1780–90) **with Maria Theresa** (1765–80)
1768–74		War with Russia
1770		Ottoman naval defeat at Çeşme
		Rebellion in Peloponnesus
1772		First partition of Poland
1773–5		Pughachev Revolt
1774	**Abdülhamid I** (1774–89)	Treaty of Küçük Kaynarca
1775–1804		Cezzar Ahmed Pasha in Acre and Syria
1783		Russia annexes Crimea
1787–92	**Selim III** (1789–1807)	War with Russia and Austria (1788–91)
1787–1822		Ali Pasha of Iannina appointed Governor
1790		**Leopold II in Austria** (1790–92)
1791		Treaty of Sistova (with Austria)
1792		Treaty of Jassy (with Russia)
		Franz II 1792–1835 in Austria
		Osman Pasvanoğlu seizes Vidin
1793		First permanent Ottoman embassies in Europe
		Nizâm-ı Cedid 'New Order' era inaugurated
1798		Napoleon invades Egypt
1801–02		British and Ottomans defeat French in Egypt
		Alexander I in Russia (1801–25)
1804		Serbian revolt
1805–6		Wahhabis attack and hold Mecca and Medina
1805		Mehmed Ali Pasha appointed Governor of Egypt
1806–12		War with Russia
1806		Resistance to New Order recruitment in Edirne
		Selim III recognizes Napoleon as emperor
1807		British warships force the Dardanelles
1807	**Mustafa IV** (1807–08)	Revolt against New Order in Istanbul – Selim III deposed
		Treaty of Tilsit: Russia and France agree to divide up Ottoman Balkans
1808	**Mahmud II** (1808–39)	Alemdar Mustafa Pasha enthrones Mahmud II. Selim III murdered
		Sened-i Ittifak
		Massive revolt in Istanbul
1811–18		Mehmed Ali Pasha of Egypt and son Ibrahim subdue Wahhabis
1812		Treaty of Bucharest
1814		Philike Hetairia established in Odessa
1821		Greek rebellion
1822		Ali Pasha of Iannina murdered
1825		**Nicholas I in Russia** (1825–55)

Date	Sultan	Events
1826		Destruction of the Janissaries in Istanbul
1827		Sinking of the Ottoman fleet at Navarino Hüsrev Pasha's military reforms (*Asakir-i Mansure*)
1828–29		War with Russia
1829		Treaty of Adrianpole
1830–31		Revolt in Albania
1830		France occupies Algeria
1831		Mehmed Ali and Ibrahim attack Syria
1831		First Ottoman census
1832		Battle of Konya; Ottoman defeat.
1833		Russo-Ottoman Treaty of Hünkar Iskelesi
1834		Officer Training School and School for Military Science opened in Istanbul *Redif* reserve system established
1838		Anglo-Ottoman Trade Convention
1839	**Abdülmecid I** (1839–61)	Battle of Nizib; Ottoman defeat
1839–76		Gülhane Edict inaugurates the Tanzimat
1840		Treaty of London
1841		Straits Convention
1846		New conscription Law
1853–56		Crimean War
1856		Treaty of Paris
1861	**Abdülaziz I** (1861–76)	Hatt-ı Humayun
1863		Ottoman Bank established.
1864		Forced migration of Caucasus Muslims from Russia
1867		Sultan visits England and France
1869		New Conscription Law
	Murad V (1876)	
1876	**Abdülhamid II** (1876–1909)	First Ottoman Constitution
1877–78		War with Russia
1878		Congress and Treaty of Berlin
1881		Ottoman Public Debt Administration
1882		British Occupation of Egypt

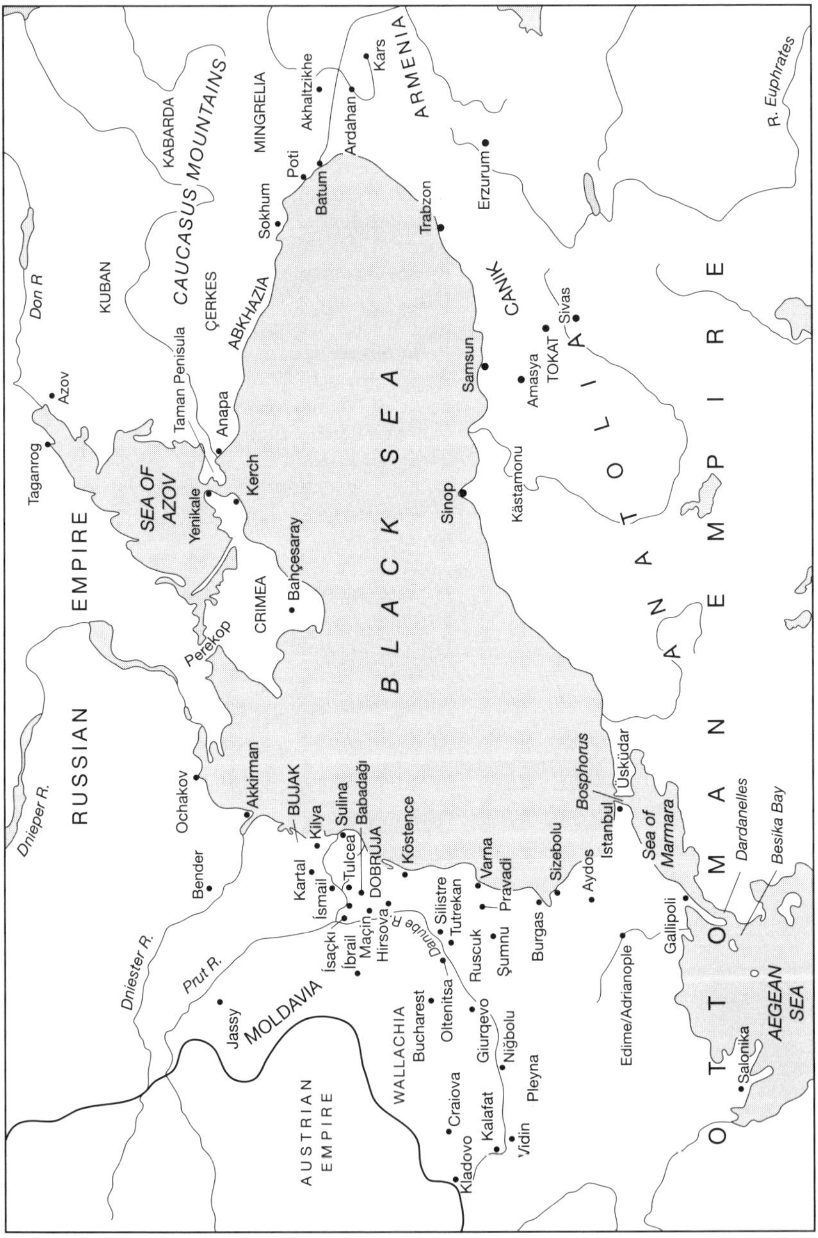

PLATE 1 *The Ottoman northern arc, Vidin to Kars, 1750–1830.*

PLATE 2 *The southern tier of the empire, 1800–50.*

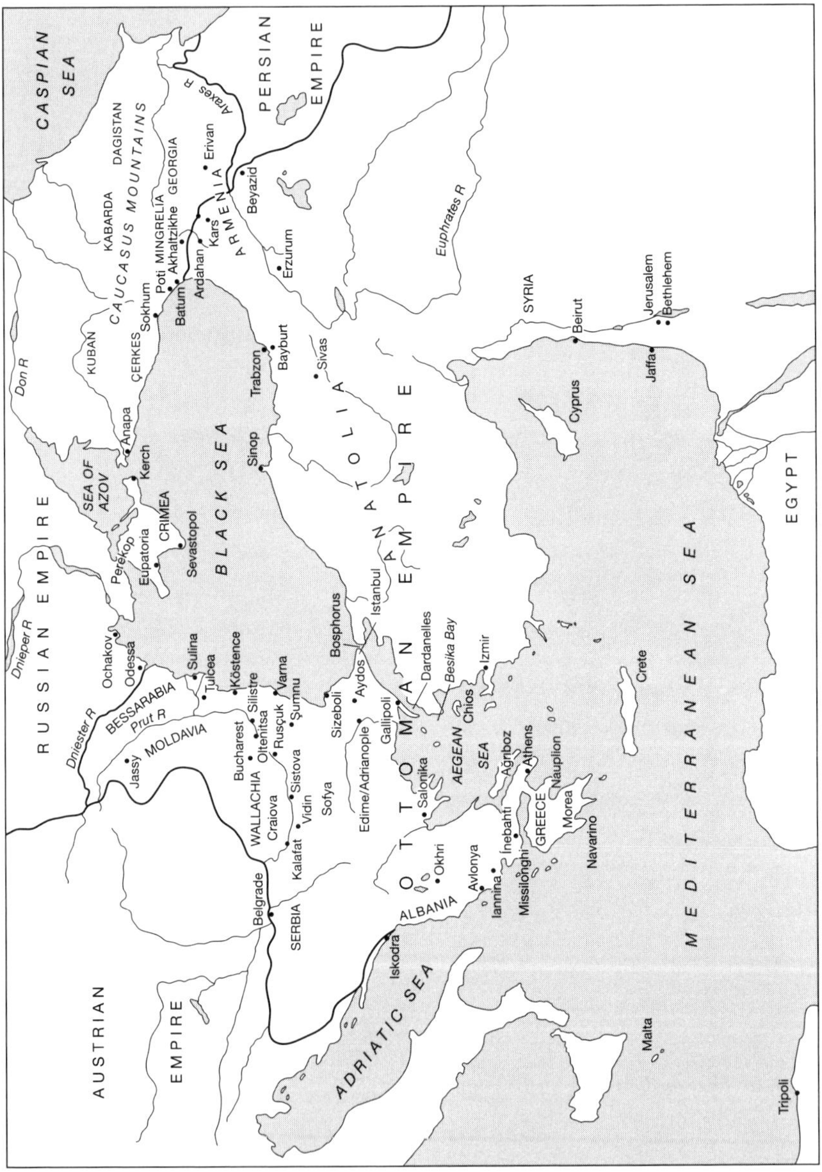

PLATE 3 *Greece, the Danube and the Caucasus, 1820–56.*

Introduction

It is without doubt the height of folly to undertake a history of Ottoman warfare which spans more than a century of conflict, bridges early modern and modern-style military systems, and describes an evolution of an army from the Janissaries to the Regular Army (*Nizamiye*) of the Crimean War, especially as for most of the wars to be discussed, campaign histories from the Ottoman point of view are few and far between.[1] To do so is also to fly in the face of a formidable body of historical studies which, with notable exceptions, have consistently overlooked the Ottomans as agents in their own development, generally attributed all military reform to French or Prussian assistance, and buried Ottoman military history under the carpet of European international relations.

Writing the story of Ottoman warfare after the treaty of Karlowitz in 1699 is a daunting task, as the nineteenth-century views on the nature of European warfare have strongly influenced the way we consider both the nature of violence and conflict in early modern societies and the Ottoman place within that system. Not only does a historian of the Ottoman Empire work against the stereotypes of decline embodied in the 'Eastern Question' historiography; he or she is faced with assumptions about state systems on the *raison d'état* that emerged in Europe after 1648, contrasted with Ottoman strategy driven by continuous holy war and impermanent peace with infidel (that is to say Christian) nations.[2] While both assumptions about early modern society are gradually unravelling in the twenty-first century, it remains difficult to engage meaningfully in questions of military imperatives when the debate is framed by the European narrative.

This is an oversimplified critique of a very large literature about an extremely important period in international politics. In keeping with the aims of this work, I will make a conscious effort to tell the story from the vantage point of Istanbul rather than London, or Paris, or Vienna, or St Petersburg. To the extent possible, my intent is to provide a more detailed version of the Ottoman 'Western Question' of the period. The core of this book is about the evolution of a military system from a Janissary to a conscript army, especially as events unfolded during the reigns of Selim III (1789–1807) and Mahmud II (1808–39).

There remain significant barriers, however, to telling the story properly from the Ottoman point of view. Not the least of these problems is that Ottomanists find themselves caught in the twilight zone between the empirical and the mythical. Much of the English and French material of the period (on which present-day western historians continue to rely) was written when fiction and fact had merged in the fantastic tales of eighteenth-century entertainments concerning the Turk and his world. The life story of Ottoman *ayan* Ali Pasha of Iannina, popularised by Byron, is the most obvious example of the genre and the conundrum.[3] Recent discussions have led to a reconsideration of nineteenth-century imperialism, and of representations of non-western, non-Christian societies as part of the discourse of empires. Gradually, a more balanced picture is emerging of the eastern Mediterranean in an important moment of transformation of the global order.

The opacity of the Ottoman record in surviving archives, in addition to its daunting volume, also proves a formidable barrier to such a project. For the period in question, archival records are chaotic and disorderly, perhaps reflecting the political confusion of the age. Multiple and ill-defined modern cataloguing projects have scattered related materials across confusing collections, and complicated the task of analysis. Nonetheless, we are gradually recovering the Ottomans through their archival history, and rethinking the role that nineteenth- and twentieth-century nationalist rhetoric has played in skewing our observations about the nature of pre-modern societies such as those of the Middle East.

Scholars working on the Ottomans in the Balkans, Anatolia, Egypt, and Syria have independently reached something of a consensus about the importance of the 1760–1830 period as a watershed for the Ottomans.[4] It was a moment when reform became necessary not just as a palace experiment, but as a matter of life and death for the dynasty and empire alike. Survival in the Ottoman dynastic context meant engaging in a series of multilateral negotiations, not only with foreign powers, but also with a host of new military, intellectual and religious elite groups, who emerged precisely because of shrinking borders, mobile military forces, and refugee populations. Arguably, the greatest challenge came not from external enemies, but from internal challengers who benefited from the central collapse and came to control the dialogue. Men such as Ali Pasha, as above, but also Alemdar Mustafa Pasha of Rusçuk, and Osman Pasvanoğlu, to mention only the most famous, engaged in an elaborate *pas-de-deux* with the dynasty over the defence of the frontier, and supplying men and arms for increasingly difficult campaigns on the Danube. As the eighteenth century evolved, an alteration of the imperial premise emerged, in response to the intensification of appeal of co-religionists in Russia and Austria to their 'countrymen' still under Ottoman rule. Mahmud II (1808–39) understood the need to reiterate Sunni Muslim orthodoxy not just as part of campaign rhetoric, but also as a means of responding to those resisting his reforms.

The challenge to the dynasty from within the Muslim world has only begun to be assessed, particularly as regards reform impulses such as the Wahhabis in Arabia, and the Naqshbandi in Anatolia and Central Asia. To what extent the reformation of the Ottoman Empire in the nineteenth century consti tuted the iteration of a Muslim 'colonising' or 'civilising' mission comparable to a similar process in Russia, Britain or France, remains a major question of this study.[5]

The passage from empire to nation in the modern world has been perceived as the natural trajectory of pre-modern empires. In that discussion, religious rhetoric has normally been couched as one aspect of proto-nationalism, or as one of the psychological tools of empire. Religious concerns were generally dismissed as irrelevant, or peripheral, to the concerns of *realpolitik*. By contrast, it has always been assumed that religion was the sole *modus vivendi* of the Ottomans, or of collaborationist groups, such as the Greek Orthodox community and patriarch of Istanbul, the Phanariots. Nationalist movements in parts of Europe, it has recently been argued, have evolved identities based exclusively on religion, which is assumed to be a distinctive, and resilient (read virulent) form of statehood.[6] The debate about the role of religion in the formation of the modern republic continues in Serbia and Greece, former territories of the Ottomans. More publicly, in Istanbul, where 'secularism' has been opposed to 'Islamism' in the political arena, the underlying assumption remains that a 'modern' state must be secular. My point is that the wholesale adoption of the western model to the history of the Ottomans continues to pervert discussion of vigorous internal debates over the conundrum of multi-national empires facing the modern age. It has prevented a more inclusive narrative of Ottoman history from emerging.

An intensification of Ottoman 'Islamism' was without a doubt one of the consequences of the revival of religious fervour stimulated by foreign Catholic–Orthodox rivalries, by missionaries, by the immense and complex web of sufi organisations, as well as by the heterodox exile communities thronging into remaining Ottoman lands from the Danube and Black Sea frontiers.[7] Reformer sultans such as Mahmud II were forced to recreate an 'orthodoxy' in the face of both Muslim and non-Muslim challenges to legitimacy.

There is another way of expressing the problem. In the Romanov and Ottoman contexts, military reform of the nineteenth century was invariably accompanied by religious symbolism, which military historians relegate to the realm of iconography (or regimental symbolism), *except* in the Ottoman (Muslim) case. Thus, for example, 'fanatic' as a descriptor belongs uniquely to the Ottoman armies of the period, although there are myriad contemporary European records of observers equally appalled at Russian troop ferocity, carried out in the name of the Virgin Mary, or Mother Russia, or Catherine the Great. What did it really mean for Mahmud II to ask of each of the surviving Janissaries after the massacre of the Istanbul troops in 1826, 'Are you a Muslim?' or to call his new army 'The Victorious Army of Muhammad?' Mahmud II's elimination of the Janissary corps in 1826 struck down not just an obsolete military force, but an entire social stratum, nominally Muslim, but sprinkled with fellow travellers (*yoldaş*) of all religious (and ethnic) affiliations, who benefited from tax relief, entitlements and protection. So too he destroyed the foundation of a society that had tolerated (often contemptuously) a multiplicity of ethnicities and religions, more by neglect than actual application of principles. This, I wish to argue, is social transformation indeed, and leads directly into consideration of the fundamental contradictions of late imperial 'Ottomanism'. Such contradictions were hardly unique to this empire.

So why take on the book at all? One reason is to fill a void. I was initially encouraged that Hamish Scott, the editor of the *Modern Wars in Perspective* series at Longman, felt the need to include a work devoted to the Ottoman military. My own dissatisfaction with the traditional historiography has also meant the reopening of a number of questions concerning Ottoman longevity and survival, part of ongoing discussions among Ottoman and world historians. There are three main aims in this particular narrative: one is to set the Ottomans among their neighbours, in a bid to contrast developments in the empire with those in Romanov and Habsburg domains. The second is to study the manning, maintenance, and sieges and battles in the fortresses along the Ottoman northern defensive line, which bristled from Belgrade to Ochakov on the Danube River and Black Sea, to Kars and Erzurum in the Caucasus. A third aim is to describe the transformation engendered by imitating the 'infidel' in both military and political spheres, which had an impact on social cohesion, and imperial logic.

Comparative empires

Ottoman pre-nineteenth century history is usually divided into two periods: a golden or classical Ottoman age, when the Janissary-style military system was at its strongest (1400–1600), then a long, slow, two hundred year descent into stasis and decay (1600–1800). The latter period is envisioned as a process of decentralisation, when powerful local notables emerged who challenged the Ottoman dynasty itself. The brief recovery of Ottoman absolutism in the nineteenth century is largely attributed to clever statesmen modelling themselves on the Europeans, with little reference to the internal dynamics which prohibited or facilitated the transformation of the military system. Recent work, partly enabled by increasingly accessible and better catalogued archives, has questioned the paucity of documentary evidence which underpins the historical narrative after 1800. The internal dynamics of change, strategies of negotiation between centre and periphery, and political legitimation of the late Ottoman state have emerged as questions for discussion concerning the Ottoman evolution after 1600. These large themes form a backdrop for this study on military strategy and imperatives, which is grounded in the discussions of comparative empires.

In our case, three land-based empires stood poised at the edge of the modern world, facing similar problems of finance and control over internal and external violence. For all three, the imperial army was the central organisation responsible for the survival of the dynasty. They were certainly universal empires at some point in their long histories, whether one defines that in political, economic or cultural terms. The ways in which they differed from the overseas empires of western Europe, the French, British and Dutch, has been a much belaboured subject in recent historiographical debates, but the ways in which they were similar to one another, or shared some similar historical trajectories, has been less studied. Taking up the story of Habsburg, Romanov and Ottoman rivalries in 1700, one contemplates huge expanses of territory, bewildering religious and ethnic variations and intricate real or constructed lineages and genealogies as justification for conquest. I do not intend to tell the story of all three in any completeness. Rather, the more developed historiographies of the Habsburg and Romanov cases will help to illuminate the less well-known Ottoman example.

The northern defensive line

The Ottoman defensive line is where most of the major campaigns with external foes of the empire took place after 1700. Because this is essentially a military history, a number of assumptions drive the discussion. The geopolitics of the region from Belgrade to Kars are of particular interest to me, as are the degrees of absolutism and legitimacy expressed in three different contexts, and the relationships between a dynasty and its internal challengers. Warfare was very often the catalyst of change, and very often the centre of periods of political and economic crisis, which could severely challenge dynastic legitimacy. Ottoman survival strategies most engage me in this context. To the extent that material is currently available to me, I have focused on describing select campaigns.[8]

At the beginning of the period under study, the three dynasties resembled one another more than they did at the end. In 1700, the Habsburgs had just triumphed over the Ottomans, acquired Hungary and a relatively stabilised border with their hereditary enemy. The victory was the greatest moment in the history of the Christian–Muslim struggle and the counter-reformation, and considerably enlarged the territories of the Habsburgs, bringing with it a destabilisation of the delicate balance of the federative, hierarchical *Reich*. The Ottomans, contained on their western front, had to come to terms with the erosion of both hegemony and legitimacy should they attempt to imitate the successful military systems of Europe. The Romanovs had stuck a claw in one end of the Black Sea, and were poised to burst on the European and Central Asian stages, with an arguably revolutionary modernised army and administration. In each case, there were western and eastern frontiers to subdue, requiring dexterity and ingenuity in styles of warfare. Ancient rights and privileged estates of various stripes continued to challenge the ruling houses for power. In each case, survival depended on the ability to mobilise and sustain long, debilitating and destructive, if episodic, campaigns. The conjunction of success and failure on the battlefields of the Danube, the Crimea and the Caucasus is a particularly apt way to compare and contrast the choices and consequences of power. Especially acute in the Romanov and Ottoman case was the question of vast, undefined frontiers, dominated by nomadic groups, or confederations, with their own proud traditions and styles of governance. These similar problems encouraged me to attempt to include the Ottomans in a normative discussion of limits and choices of early modern agrarian states, as exemplified in their military systems. There are limits to the comparative analysis, however, which are particularly apparent after 1750, when Austria began a long process of disengagement and accommodation with her neighbours to the north and east.

Military reform

This study argues that the struggle for survival was central to the changing dynamics of Ottoman society, especially in a period when war was generally proving too costly in human and economic terms, both for individuals as well as for the dynasty. Violence was endemic in the early modern period, when it was integral to daily life, accelerated when great campaigns were mounted, and never very far from one's doorstep, either in battle, or accompanying disease and famine. The problem was hardly uniquely Ottoman, but one which afflicted the Habsburg and Romanov dynasties as well.

The problem of moving from mercenary to native, conscripted armies was a universal pre-modern preoccupation, which proved more difficult to achieve in the Ottoman context than elsewhere. Over the period 1700–1830, an Ottoman military system had emerged, which was drawn from manpower overwhelmingly Muslim and peripheral (nomadic, mountainous and tribal) to an Ottoman urbanity. When the Ottoman sultans first entertained reform, they emulated Napoleon and Peter the Great. By the mid-nineteenth century, Tsars Nicholas I and Alexander II, Napoleon III of France, and Franz Joseph and Metternich of Austria were the contemporaries, and models for the reforms of Abdülmecid I (1839–61). The Ottomans had just begun the social transformation to a loyal instrument (*Nizamiye*) that would allow them to monopolise violence within Ottoman realms, when revolutionary nationalism and constitutionalism burst upon the absolutists of Europe in 1848. Trickling into the Ottoman Empire, and finding fertile ground in disaffected and abused non-Muslim communities, new ethno-religious identities competed with the 'Ottomanism' of the Tanzimat era, 1839–76, to result in the authoritarianism of Sultan Abdülhamid II (1876–1908).

It is the period 1760–1830, a pivotal moment in world history, which is most lacking in detail about Ottoman military developments, and which is the core of this story. It was a do-or-die moment for the Ottoman house, hardly the moment to introduce liberalism, or constitutionalism.

Reform, commonly *ıslah* in Turkish, was as much about improvement, or eliminating defects in the military system, as it was about Ottomanism. The parameters and imperatives of the late Ottoman state arose from the need to acquire and mobilise the means of war for survival and defence. The struggle by the late eighteenth century required a considerable alteration of the imperial premise, in order to convince the large non-Muslim population at home of the legitimacy of Ottoman power. The battle for the loyalties (suzerainty) of the Principalities (Wallachia and Moldavia) is particularly illuminating in that respect.

The Ottoman system militated against the development of a wealthy competitive or elite group in a number of ways which will be discussed. Nonetheless, provincial nobles play an enormous role in the success or failure of the war effort after the 1750s. While confiscation of estates, and more rarely, of charitable endowments, was within the power of the sultan, and frequently exercised, war profiteering generated powerful local elites who both challenged and cooperated with the late eighteenth- and early nineteenth-century Ottoman state. Local committees of notables and Muslim judges were responsible for raising troops, supplying grain, baking and shipping biscuit, maintaining the bivouacs, especially along the army routes to the battlefront, and supplying horses and wagons, in the myriad of contractual relationships which characterised Ottoman centre and periphery. Loyalty to the system persisted even in the midst of widespread corruption and abuse of taxation privileges. The toll on local society was large, and accounts of deserted lands, increasing banditry and decimated populations are legion, yet many of the territories of the empire were not directly within the war zones. The balance of loss and gain for local populations must be considered.

The outline of the narrative

In short, the perspective of this study will be partially integrative and comparative, in an effort to draw the empire into the historiography of Europe and the Mediterranean. Certain fundamentals about the Ottoman side will also have to be established, including mobilisation, logistics, financing, ideological claims, and the impact of military choices and demands on the diverse populations and social groups under Ottoman hegemony.

Each narration of the campaigns of the period 1700–1870 will include a brief survey of the origin, causes, progress and conclusion of the wars, highlighting significant battles, and discussing the treaties and their geopolitical implications. More importantly, those descriptions will be followed by an assessment of Ottoman military mobilisation, financing, supply, leadership and strategy, emphasising continuity and change, and the creation of alternative fighting forces to the obsolete and obstreperous Janissaries.

The transformation which occurred at the centre, especially from 1792 to 1830, has been much described but little understood. Massive documentation is slowly being brought to bear on many questions which have remained unanswered except in the most facile way. What we do know is that by the end of the period covered by this book, both Ottoman bureaucratic and military classes were trained in western systems and languages, and represented the new Ottoman elites with aspirations concerning the nation and constitutionalism. Part of the investigation here is to determine who made up those elites, and why they became attached to the reform agenda.

This work is primarily a study of the warfare on the Danube, Crimean and Caucasus battlefronts between the years 1700 and 1870. The Ottomans confronted armies of different calibres on the Iranian and southern frontiers in the same period, but it is the Danube frontier fortress line to which they devoted the most attention. In fact, the neglect of the other 'frontiers' (if one is willing to accept Greater Syria as a 'frontier' in military terms) cost the Ottomans dearly – for example, once Safavid Persian hegemony, always tenuous, dissolved in the 1720s. Or, when Mehmed Ali and his new Egyptian army reconnected the Egyptian–Syrian geographical entity in the 1830s, reconfiguring a centuries-old *topos* which predated Ottoman hegemony. While it is the northern tier that is the subject of most of this narrative, the southern, or Arab, tier becomes important to the story after the 1760s, but especially with Napoleon Bonaparte's invasion of Egypt in 1798.

This is also a study of land warfare, logistics and their integration into and impact on local societies. The Ottoman navy will not feature significantly in this story. The importance of the Danube fleet to the movement of men and goods from Istanbul to the battlefront, and the possible impact that had on local populations, is of interest to our story, but not central. Arguably, the Ottomans shifted revenues to land-based defensive systems, especially after the exhausting but successful conquest of Crete (1669), which cemented supply systems for the eastern Mediterranean and meant the beginning of the end for the Venetians. The navy does return in the nineteenth century, and has already attracted considerable interest among historians.[9]

Habsburg, Ottoman and Romanov diplomacy and politics in the seventeenth century open the story, followed by a description of the traditional military context circa 1700 in the second chapter. At the centre of the discussion are the challenge to Ottoman legitimacy in the Karlowitz Treaty of 1699 and the alteration to the sources of military manpower. Thereafter, the following chapters will include a brief review of the major diplomatic and military concerns of a particular period, with select campaigns, and select questions about Ottoman strategies and organisation as the focus of discussion. Under-utilised Ottoman texts and an attempt to describe warfare from the ground up, whenever feasible, will enhance the better-known political and diplomatic narratives.

For example, in discussing the 1737–39 Austro-Russian and Ottoman War, it seems appropriate to dwell on the Ottoman recuperative strengths and the nature of siege style warfare in the Balkans, in what proved to be the last great Ottoman war, and its illusionary triumph in the return of Belgrade. While the whole period is marked by monumental difficulties, notable successes, usually involving fortresses, and an ability to reestablish supply systems in desuetude suggest the resilience of an institutional apparatus, and, in fact, convinced contemporaries of the rightness of imperial aims. The 1768–92 Russo-Ottoman wars, the Küçük Kaynarca, Sistova and Jassy treaties and the significant loss of the Crimea, by contrast, will be followed by a discussion of the new methods of mobilisation, the complete dominance of provincial forces on the battlefield, and the bankruptcy of outmoded supply systems.

The 1768–74 Russo-Ottoman War remains the pivotal moment for the Ottomans, redolent in the literature as the inauguration of the 'Eastern Question', but also truly the greatest test to the traditional military systems, which had to be reorganised following the long, peaceful hiatus after 1740. That 'Eastern Question' generally would make of the Ottomans pawns on the international chessboard, disregarded except as a nuisance by the great powers. Three interrelated questions are embodied in the 'Eastern Question' as it concerned Britain, but also Russia and France: (1) access to India and trade; (2) access to the Dardanelles and the Black Sea for economic and strategic reasons; and (3) the problem of central and eastern European territories and the Ottoman Empire, more particularly the increasing threat of Russia.

The perception of the need to control Egypt, in order to have access to the trade routes to India, drove British policy discussions concerning the Ottomans during the Napoleonic period. Francophobe, and later, Russophobe British governments, often unwillingly, found British interests tied to the preservation of the Ottomans, colonial occupation and an informal empire that remained operative in the Middle East until the 1960s. The question of access to the Black Sea through the Aegean and Bosphorus straits, and all it implied for international trade and the balance of power after 1800, was a second underlying theme of the debates. The conflict between Mahmud II and Mehmed Ali (d. 1849) of Egypt was pivotal to the debates, especially during the the Syrian Crisis 1839–41, but beginning with the Greek Rebellion of 1821.

The third question concerned the struggle among Austria, Russia, and Prussia (and retreating Ottomans) over Poland, Ukraine and the Danubian territories. The general concern of European chancelleries was about the collapsing balance of power and prestige, expressed as concern over the fate of the Ottoman Empire but in reality driven by much posturing of empires and nation-states. Enlightenment views of liberation often conflicted with nineteenth-century religious and racial justifications for empire and domination.

Eastern questions and new Ottomans

The chronological boundaries of the Eastern Question in most surveys extend from the 1774 Treaty of Küçük Kaynarca, which ended the Russo-Ottoman War of 1768–74, to the collapse of the Ottoman, Habsburg and Romanov empires in World War I. It is possible, however, to break up the period 1760–1860 into two phases of European involvement in the East: the first, 1760–90, covers the period when investigative missions, diplomatic and military, characterised British and French involvement in eastern affairs. The French, long-standing friends of the Ottomans, pursued with vigour the development of trade networks in the east, both with the Russians and the emerging Greek entrepreneurial class. Secret initiatives and military missions gave the French a considerable tactical edge in the eastern Mediterranean, which facilitated the invasion of Alexandria in 1798 by Bonaparte and his army. Absorbed by intense rivalries with the French elsewhere, in North America and India, British policy makers failed to give the Mediterranean, the routes to India, and Russian aggrandisement serious consideration until the fall of Ochakov in 1788, the last Ottoman fortress on the northern coast of the Black Sea, served to forcibly remind Europe of the changed status of the region.

A second phase of the 'Eastern Question' after the 1790s involved the increasing importance of Britain as an ally to the Ottomans, in the long struggle with Napoleonic France, which was part of an emerging international consensus that the best solution to Russian encroachment was a live rather than a dead Ottoman Empire. This latter policy, which vacillated according to prime ministerial politics in London, committed Britain to an undeclared colonial domination of the Middle East crowned by the occupation of Egypt in 1882. The further consequence was the almost total eclipse of French political influence on subsequent events in the East except in greater Syria, especially Lebanon. The unresolved rivalries between France and Russia over the protection of Catholic and Orthodox communities resident in Ottoman territories, which centered around the holy sites in Jerusalem, were the *casus belli* of the Crimean War (1853–56). Ottoman, French, British, Italian and Russian armies became embroiled in one of the more tragic examples of pre-modern imperial warfare, fought, like the later American Civil War, in the full light of eye-witness journalism, photographs, and public opinion.

Caught up in global politics, the Ottomans struggled to survive and adapt to the new challenges of European financial and technical superiority. It is here that the book makes a contribution, by both injecting the Ottomans into the 'Eastern Question' and challenging two attitudes of enlightenment thought that have influenced the historiography concerning Ottoman military organisation and reform. One is the disregard of 'Muslim' protestations about 'infidel innovations', as nothing more than religious obscurantism. Most studies of the reform period to date have simply dismissed any protest cloaked in religious terms as non-progressive, backward and ignorant. The second inheritance is the discourse on 'despotism', particularly aimed at the kings of France in the eighteenth century, which settled on the Ottomans as the clearest example of the worst excesses of absolutist government, with China and Russia close behind.[10] Traces of both remain part of the narrative of empires into the present. My intention is not to nullify such generalisations, but rather to present more evidence from within the Ottoman context, as a way of balancing the picture.

If one measure of a modern state lies in its ability to centralise and monopolise the control of violence, the Ottomans failed to make the transition in the eighteenth century. They also failed to design a fiscal system as well-organised as their European foes around the business of war, another of the characteristics of European states of the same era. The consequences of the neglect of financing war made itself felt in 1768. Some long-term fiscal adjustment, and new taxation, however, can be traced over the period under discussion, at least until the disastrous 1787–92 war with Russia during the reign of Selim III. Bankruptcy faced Selim III, inhibiting the expansion of his 'new order' and accompanying economic reforms. Austria, meanwhile, reluctant partner of Russia throughout, withdrew from eastern adventures after the 1790 rapprochement of Prussia and Austria, and the signing of a Habsburg–Ottoman armistice. By 1791, the second 'diplomatic revolution' of the eighteenth century, and limited territorial gains from a costly war, ended centuries of Ottoman–Habsburg confrontation until the late nineteenth century.

I will argue for the continuity of indigenous military reform in both mobilisation and artillery systems from Selim III to Mahmud II, to counteract the historiographical reliance on works such as the *Memoirs* of Baron de Tott, the well-known technical adviser to Mustafa III (1757–74), who claims sole responsibility for artillery reform in Istanbul. (A colleague of mine once referred to Tott's memoirs as the French equivalent of Twain's *The Connecticut Yankee at King Arthur's Court.*[11]) Fictional or not, the cultural assumptions and full-blown Eurocentrism of Tott and others of his contemporaries, taken at face value, continue to influence the way this history is written. I do not evade a discussion of cultural difference, but rigid dichotomies such as civilised/barbaric, or occidental/oriental I have tried to avoid.

Thus, in examining the Napoleonic era, starting with the invasion of Egypt in 1798 and ending with the 1812 Bucharest Treaty, it seems particularly appropriate to emphasise the gradual redefinition (or perhaps more accurately, the reassertion) of the Muslim identity of the army. It had been re-fashioned on European models, particularly apparent in Mahmud II's 'Victorious Soldiers of Muhammad' which replaced the Janissaries after 1826. The success and failure of the military reforms of Mahmud II, and the construction of the new absolutism around the 'new' Ottomans, modelled on their near neighbours, deserves serious reconsideration from the Ottoman perspective, and without the assumption that the key to success lay in mimicking western Europe. One of the great turning points in Ottoman history is in 1828, when Mahmud II declared war on Russia and marched 40,000 new model army troops to the mouth of the Danube, two years after eliminating the Janissaries and one year after France and Britain sank his fleet at Navarino.

There was also continuity in the period in the growing peripheralisation of the Ottoman economy, represented in the commercial privileges system (capitulations) which extended to European powers and culminated in the 1838 commercial treaty with Great Britain, which forced free trade on Ottoman territories. The 1838–41 international crisis, when Mehmed Ali and his son Ibrahim threatened the very doorstep of Istanbul, represented the moment that indebtedness to Europe became a *fait accompli.* It is also known as the beginning of the era of the Tanzimat, which opened with the public promulgation of Ottoman intentions to reorganise society for the benefit of all subjects. The Crimean War worsened the debt problem, although foreign management of debt payments was only institutionalised in the 1881 Ottoman Public Debt Administration under French and British supervision. After the withdrawal of Mehmed Ali from Syria in 1841, the Ottomans and the Tanzimat bureaucrats clearly saw themselves in the role of 'civilising' monarchs in their peripheral and reoccupied territories such as Albania, Kurdistan, Iraq and Lebanon.

In the final chapter, I take a look at the origins of the Crimean War, and the Russo-Ottoman confrontations at Vidin on the Danube and at Kars in the Caucasus. What makes the Crimean War interesting is the confrontation of British and Ottoman military styles, and the drawing of the Ottoman Turk further into the European sphere of influence, and the global marketplace. For the first time on Ottoman battlefields, Turks, Albanians, and myriads of Caucasus tribesmen, fought side by side with, and were under the command of, dozens of foreigners, among them British, Polish, French, Hungarian, and Italian soldiers and officers. In truth, the military reforms of 1843 and 1846 mark the end of a particular period of military reform in the empire. In 1853–55, the 'Turkish' soldier, for better or worse, joined the modern historical record.

The impact of war and reform on the development of late Ottoman absolutism is of particular interest to me. This too will be comparative, especially with the Russian context, for though the balance of power had shifted to the northern empire by the end of this study, the process of transformation to a European-style absolutism looks somewhat similar. Ambivalence about foreign advisers, reappropriation of religious ideology as both a conservative and liberating force, increasing intolerance of multiethnic groups, and a more or less continuous problem with financing as suggested above, are some of the bases for comparison. The Austrian bid for empire, exemplified by the aggressive military build-up of Joseph II, could not be sustained when confronted at the turn of the century by the juggernaut of the Napoleonic 'citizen's armies'. Both Austrian and Russian contexts are instructive when examining the Ottoman evolution. So too is the Mehmed Ali era in Cairo, 1811–40, a phenomenal period of ruthless modernisation and militarism, culminating not just in altering Ottoman rule over Egypt, but also testing great power diplomacy at critical moments in world history. Arguably, Egypt was informally 'colonised' and drawn into world markets long before the British occupation of 1882, while the Ottomans found themselves colonised from within, by minority financiers and their international protectors, never more so than after the Crimean War. Austria, too, had a role to play in the final act of the three early modern empires, but it was not until the Balkan crisis of 1875–78 that Austria thrust east again, into Bosnia–Herzegovina. After 1848, and during the Crimean War, the Austrians played a tricky game of cards, which led to an increasing isolation of the Habsburgs from both eastern and western allies.

How to describe the new society that emerged by the end of Mahmud II's reign has long eluded historians of the Ottomans. By 1856, it is possible to think in terms of an Ottoman–Muslim absolutism that eschewed the 'secularism' of the Janissary era as once argued by Şerif Mardin. Mardin's 'secularism' might better be described as a Muslim universalism, which managed benevolence, if not strictly speaking equality, among its subjects. Abdülhamid II's reign is recognised as the age of ideology and the Ottoman Caliphate, especially in the work of Selim Deringil.[12] Later historiography has been so dominated by inter-related paradigms of imperialism and nationalism that the essential question of the re-Islamisation of Ottoman ideology in this earlier period has been overlooked. The reign of Abdülhamid II (1876–1908) is generally characterised as an era of extreme despotism, tremendous political upheaval, countryside violence and destruction, and a severe identity crisis, brought on by the 'turning Turk' of the military class. I argue that the crisis had its antecedents in the previous century of reform and the refashioning of the Ottoman ideological premises, which is a transition not yet very clearly delineated within Ottoman historiography.

Notes

1 The campaigns of the Ottomans in Hungary are the notable exception. See for example Géza Dávid and Pál Fodor, eds, *Ottomans, Hungarians, and Habsburgs in Central Europe: Military Confines in the Era of Ottoman Conquest* (Leiden: Brill, 2000), Caroline Finkel, *The Administration of Warfare: Ottoman Campaigns in Hungary, 1593–1606* (Vienna: VWGÖ, 1988), Caroline Finkel, 'The provisioning of the Ottoman army during the campaigns of 1593–1606', in *Habsburgisch-osmanische Beziehungen, CIEPO Colloque* (Vienna, 1983), 107–23, Rhoads Murphey, 'The functioning of the Ottoman army under Murad IV 1632–39/1032–49): key to the understanding of the relationship between center and periphery in seventeenth century Turkey' (PhD Thesis, Chicago, 1979). See also the Guide to Further Reading.

2 Bernard Lewis, *What Went Wrong? Western Impact and Middle Eastern Response* (New York: Oxford University Press, 2004). One critic suggested: 'It cannot be long before a US scientist discovers an Islamic gene, which will explain why they [Muslims] are so different from the rest of civilized humanity.' Alain Gresh, 'Malevolent fantasy of Islam', *Le Monde Diplomatique* August 2005, 16.

3 K.E. Fleming, *The Muslim Bonaparte: Diplomacy and Orientalism in Ali Pasha's Greece* (Princeton, NJ, PUP, 1999), 5–17. Another example is the anonymous life of George Stavrakoğlu, servant and master of Wallachian–Moldavian politics from the 1740s to his death in 1765, published in an estimated 15 editions between 1767 and 1869 (Guy Evans, 'George Stavrakoğlu and the rule of Ruşvet', in 'The Ottoman Empire in the eighteenth century', Kate Fleet, ed., *Oriente Moderno* 18 (79) n.s. 1 (1999): 147–60.

4 Ehud Toledano, 'The emergence of Ottoman–local elites (1700–1900): a framework for research', in Ilan Pappé and Moshe Ma'oz, eds, *Middle Eastern Politics and Ideas: A History from Within* (London, Tauris, 1997), 145–62.

5 See for a particularly interesting analysis of the question: Dominic Lieven, 'Dilemmas of Empire 1850–1918. Power Territory, Identity', *Journal of Contemporary History* 34 (1999), 163–200, especially 180.

6 Adrian Hastings, *The Construction of Nationhood: Ethnicity, Religion and Nationalism* (Cambridge: Cambridge University Press, 1997).

7 Gregory L. Freeze, 'Institutionalizing piety: the Church and popular religion, 1750–1850', in Jane Burbank and David L. Ransel, eds, *Imperial Russia: New Histories for the Empire* (Bloomington, 1998), 210–49, argues that the Russian Orthodox Church, relegated to the 'spiritual' realm by Peter and Catherine's secularisation of the autocracy, found itself in the end co-opting popular piety instead of controlling it.

8 I first addressed a number of these questions in 'Locating the Ottomans among early modern empires', *Journal of Early Modern History* 3 (1999), 1–32.

9 A good example of that is the collection of articles on the grand admiral: Elizabeth Zachariadou, ed. *The Kapudan Pasha, His Office and His Domain* (Rethymnon, Crete: Institute for Mediterranean Studies, 2002). It includes two articles by Rossitsa Gradeva, 'Shipping along the lower course of the Danube (end of the 17th century)', 301–24, and Svetlana Ivanova, 'Sketches from the life of a Kapudan Pasha on the Danube', 325–45, which effectively describe the Danube fleet and its functioning in the late seventeenth century. See also the Guide to Further Reading for the later period.

10 Thomas Kaiser, 'The evil Empire? The debate on Turkish despotism in eighteenth-century French political culture', *Journal of Modern History* 72 (2000), 6–34.

11 Michael Hickok, in a conversation.

12 Şerif Mardin, 'The just and the unjust', *Daedalus* 120 (1991), 113–29. Selim Deringil, *The Well-Protected Domains: Ideology and the Legitimation of Power in the Ottoman Empire, 1876–1909* (London: I.B. Tauris), 1998.

CHAPTER ONE

The political map of Central and Eastern Europe circa 1700

By the Treaty of Karlowitz of 1699, which ended years of intermittent warfare with the Holy League (Austria, Poland, Russia, Venice and the Papacy), the Ottoman Empire ceded all of Ottoman Hungary (except Temeşvar), Transylvania, Slavonia and parts of Croatia to Austria; Kamenice, Podolia and parts of the southern Ukraine to Poland; and Azov to Russia, which opened the door to the Caucasus for Peter the Great and his successors. The Venetians kept their gains in the Morea and parts of Dalmatia, although their victory was elusive, as the Ottomans would recapture the territories within two decades. By any standard, it was a crushing defeat for the Ottomans, and rightly considered over the centuries the major turning point for the empire. Significantly, the Ottomans reorganised and remobilised their army four times over the course of the campaigns from 1683–99, but were unable to establish long-lasting supremacy on the battlefield. While Eugene of Savoy's military command is generally recognised as the key to Austrian success, it is equally notable that Habsburg setbacks prompted much re-thinking of strategy and reform in the later decades of the eighteenth century. Russia, added to the League in 1686, emerged as the most significant player in these same decades, although the fruits of many of Peter the Great's reforms would not be immediately apparent.

Before discussing the Ottoman military organisation of the eighteenth and early nineteenth centuries, we must first examine the geopolitical theatre where much of the warfare unfolded. In what follows, with very broad strokes, I have tried to evoke an impression of place, and what it must have been like to live in that space and time. References to the

Habsburgs and Romanovs are intended to evoke the familiar among readers, in order to highlight both similar military problems and different cultural solutions in three imperial contexts on the northern arc of the Ottoman Empire. The last part of the chapter briefly introduces the sultan and his Janissaries in Istanbul circa 1700.

The geopolitics of the Danube

From the perspective of the third millenium, it is difficult to perceive how someone living in the late seventeenth century might have envisioned the map of the Danube region and the Black Sea. The borders of the unknown were rapidly receding in Europe, but territories beyond Poland–Lithuania, east of Vienna and south of Budapest and Belgrade, evoked the exotic and the dangerous. The traveller on the Danube was among the adventurers of the age, facing ill-charted territories, fluctuating frontiers, and threats to health and well-being from displaced populations, bandits, disease-ridden marshes, and the seemingly unending warfare between the three greatest opponents of the age: Habsburg, Ottoman, and Venetian. Russia had just begun the empire's relentless march south and east as this narrative begins.

Of the many accounts of diplomats, merchants and travelling ladies and gentlemen which provide much of our information and impressions of this territory, none is more thorough and colourful than the various reminiscences and other writings of Luigi Ferdinando Marsigli, nobleman, scientist, diplomat, cartographer, and factotum for both The Republic of Venice and the Austrian Habsburgs in the period. The Venetians, defeated in the long struggle with the Ottomans over control of Crete (1645–69), attempted to re-establish both diplomatic and commercial ties with Istanbul throughout the 1670s. As befitted young noblemen of his age, especially one interested in a future career in diplomacy, Marsigli first travelled abroad as a member of the entourage of the Venetian *bailo* to Istanbul in 1679. (Opportunities for foreign travel and adventure were generally arranged as part of an official entourage such as envoys to friendly or recently hostile neighbours.) Young and unencumbered with the task of making his way through the niceties of Istanbul diplomatic circles, Marsigli decided to return home overland through Bosnia and Dalmatia, territories within the Ottoman Empire. Thus began a twenty-year career in much uncharted territory. Within a decade, his knowledge of that region became essential to military strategists in Vienna, and his adventures in the Balkans were played out against the long series of annual campaigns between the Ottomans and the Holy League, stretching from 1683 to the treaty of Karlowitz in 1699.

William III, ruler of the Dutch Republic (1672–1702), and of England, Ireland, Scotland and Wales (1689–1702), was engaged in a long struggle with France's Louis XIV (1643–1715), and sought the aid of Austria. He instructed the English Ambassador Sir William Hussey to attempt a mediation between Leopold I, ruler of the Habsburg Monarchy (1657–1705) and Emperor of the Holy Roman Empire (1658–1705) and Süleyman II, Sultan of the Ottoman Empire (1687–91) in Istanbul. Hussey would normally have travelled by sea, but in this case, and for the first time in English diplomatic annals, the embassy travelled down the Danube, embarking at Vienna. Leopold I's handpicked representative, Luigi de Marsigli, joined the embassy at the great fortress of Esztergom (Gran) on the Danube in April 1691.[1] Esztergom had been captured from the Ottomans in the triumphant campaign of 1683, as the Ottomans retreated from the unsuccessful siege of Vienna and failed to regroup effectively to defend the fortress. That had ended an Ottoman residence of almost 150 years west of the great bend of the Danube, as is commemorated in a well-known Turkish ballad.[2]

The embassy proceeded downriver, passing Buda, surrendered by the Ottomans to the Holy League in September 1686, which likely accounted for the apparent desolation, lack of culture and livestock, and even of river traffic. Arriving in Belgrade, they were greeted with proper politeness and ceremony by the Ottoman fortress commander. Belgrade had already switched hands, first captured by Austrian forces in 1688 and then recaptured by the Ottomans following a disastrous Habsburg 1690 campaign. As Secretary to the English Ambassador, Marsigli, was rather nervous about his reception, but commented that the Turks treated him 'with perfect courtesy'. The embassy was to be escorted from Belgrade to Rusçuk (in present-day Bulgaria) by 150 Egyptian Janissaries who were on their way back to Cairo. The Ottomans themselves expressed fear of the *hayduts* (bandits) who might interfere with the smooth passage of the English envoy.[3] Eight days later Hussey and his party disembarked in Rusçuk, a ship-building port, to travel through the Balkan passes to Edirne to complete what would henceforth become the standard overland diplomatic passage to Istanbul. The embassy caught up with Sultan Süleyman II and Grand Vizier Fazıl Mustafa Köprülü (1689–91) in Edirne, as they prepared for the spring campaign of 1691. While their reception was cordial, the embassy itself ended in failure, except that Marsigli, sole member of the four participants still alive in 1699, influenced the direction

PLATE 4 *Nineteenth-century lithograph of the Upper Danube* (The Danube: Its History, Scenery, and Topography. Splendidly Illustrated from sketches taken on the spot, by Abresch and drawn by W.H. Bartlett. *By William Beattie, M.D. George Virtue, London, 1844*).

of the final Karlowitz negotiations by his long intimacy with the courses of the Danube and its tributaries.[4] It was he and an Ottoman commissioner by the name of Ibrahim (otherwise unidentified) who paced out the newly described fixed border of the treaty which opened a new era in Ottoman–European relations.

This was the terrain over which Habsburg and Ottoman fought for much of 200 years. The Danube River is emblematic of the trials and tribulations which pre-modern armies faced in the last frontier of Europe.[5] First and foremost, the topography and environment imposed different kinds of military strategies. Building bridges, crossing flooded streams, and drowning during retreats are constants of the chronicles of Danubian warfare. Vast stretches of marshes, hilly terrain, and difficult mountain passes mark the battlegrounds of this region. In addition, intense rivalries based on ethnicity and religion contributed to the normal strategic and logistical burdens which warfare of early modern Europe represented. The borders and much disputed territories of such regions as Croatia, Serbia, Transylvania, Wallachia, Moldavia, Bosnia, Albania, Bulgaria were home to a plethora of identities which we came to know intimately in the waning years of the twentieth century, in part because of resonating histories from the centuries in between. A 'hedgehog of frontier fortresses', built to prevent the further incursion of the Ottomans, and an 'unruly band of border states' resulted in 'acute political instability; but this, in turn, reflected the fact that each of the great powers, the Ottomans included, had at times to operate at its extreme range of effectiveness of action when seeking to bring any of these marginal polities under control.'[6]

The Holy League was the last crusading alliance which took up arms against the Muslim intruder to Europe. Consider, for example, the exhortation of the delegates of Leopold I at the Russian court in 1684 to join the League: 'Fight for the Cross of Christ! Lead your privileged peoples to destroy the cruel enemy . . . Occupy Constantinople, where your patriarch is forced to be a guest . . . Regain your seat of your Church, where now idols reign.'[7] The revitalisation of the Catholic Church, often attributed to the counter-reformation, gained impetus as well from the defence of Europe against the Muslim occupier. The Catholic Habsburgs were necessary to Europe as a buffer state; after 1699, and the eclipse of the Ottoman threat to Europe, western Holy League-style Christian coalitions became increasingly irrelevant, but sectarian-based rivalries continued, particularly in the contested territories of the Balkans.

At the end of the seventeenth century, there had emerged a recognisable 'concert' of powers in Europe whose primary mode of rhetoric of inclusion was Christianity, and whose primary enemy remained the Ottoman 'Turk'. However, the entangled communities of Orthodox, Catholic and Protestant Christianity which criss-crossed Holy Roman Empire and Ottoman territories alike remained a significant obstacle to collective action, and required a carefully-styled universality to appeal to the greatest number. Distrust of co-religionists continued to be a psychological barrier to Habsburg eastern ambitions, and a crucial aspect of their failure to follow up the striking victories of 1687 to 1690.

Arguably, it is also this war which began to weaken the loyalty of non-Muslims to their Ottoman rulers, and Christian and Muslim calls-to-arms remain an integral aspect of international relations for the next 200 years. Just how lodged the conflation of the Turks as Muslims is in the popular mentality can be demonstrated in an interview with Slav-speaking villagers from Ohri, observing Byzantine ruins in 1903. When asked if they knew who the builders of the ruins were, they responded: 'The free men, our ancestors.' 'Were they Serbs, Bulgarians, Greeks or Turks?' asked the interviewer. The young boys replied: 'No, they were not Turks; they were Christians.'[8] It should be noted that the word 'Turk' for 'Ottoman' was a European usage. The Ottomans thought of 'Turks' as uncultured 'hicks' or 'rubes' until well into the nineteenth century.[9]

Members of the Holy League

The main pillar of the Holy League was Habsburg Austria, whose rulers were also the Holy Roman Emperors, defenders of Christendom, head of a commonwealth of electorates and a myriad of medium and smaller powers. Some have argued that the Holy Roman Empire after 1648 was 'the great fiction of Europe', while others emphasise its robust endurance.[10] Military strategy and decision-making were determined by constant concessions (or lack of them) to individual dynastic demands of territories such as Bohemia or Bavaria, demands which, unsatisfied, could force revolt and warfare such as erupted in the Thirty Years War (1618–48). On numerous occasions during the long period of conflict 1683–99, Leopold I stood almost alone against his Ottoman foes, especially before Brandenburg and Russia joined in 1686. The federative nature of Austrian imperialism has been well documented, accounting for the fluctuating success and failure of the Habsburg military endeavour over the 16 years. The last-minute arrival of King Jan Sobieski and some 20,000 Poles at the siege of Vienna in 1683 helped to push the relief forces to the spectacular defeat of the Ottomans, just as they were prepared to breach the walls of the city. The effort was extremely costly: there were up to 100,000 war-related deaths, and much of the lower Austrian territory had to be repopulated.[11]

Historians of the period point to the vacillation of Habsburg policy between concerns for both a western front (against Louis XIV's encroachments) and an eastern front against the perpetual enemy, the Ottomans. External conflicts were complicated by a major threat from Hungary under Imre Thököly (d. 1705), who succeeded in being declared the 'King of Central Hungary' by the Ottomans just as the Habsburgs were negotiating a renewal of the 1664 Ottoman–Habsburg Treaty of Vasvat. By 1684, Austria, Poland, and Venice had united in the Holy League, intent on a three-pronged attack on Ottoman territories: Austria on the Hungarian and Croatian frontier; Venice in Dalmatia and Greece, and Poland in the Ukraine and the Black Sea. The addition of Russia to the Holy League in 1686, and the first major thrust of Russian troops into the Crimea in 1687, might ultimately have meant even further incursion into the vast Ottoman territories, but by 1697 the original three partners were exhausted, over-extended and desirous of peace. (Austria, for example, was also involved in the Nine Years War on its western front.) The Ottomans, who had regrouped and recaptured Temeşvar in 1695, were beaten decisively at Zenta in 1697 by Eugene of Savoy. Only Peter the Great, fresh from the conquest of Azov in 1696, was desirous of continuing, and withdrew from the peace negotiations at Karlowitz, begun in 1698. The remaining belligerents signed the treaty on the basis of *uti possidetis*. The treaty permanently altered the relationship between Habsburg and Ottoman.

The Ottomans ceded their territories in Hungary, Croatia and Poland as described at the beginning of this chapter, although unresolved border disputes remained. Similarly, the Venetians retained the Morea and parts of Dalmatia, although they would only hold them for two decades. It was a tremendous victory for the Habsburgs and their allies. The fixed border measured out by Commissioners Marsigli and Ibrahim, with minor, later fluctuations around Belgrade and Temeşvar, remained the essential Austrian–Ottoman line well into the nineteenth century. Still, even as the notion of permanent borders seeped into Ottoman–Europe diplomacy, this border remained porous. By 1800, the Habsburg frontiers extended 1,000 miles from the Adriatic coast to the northern end of Moldavia. The Habsburgs manned the extensive military frontier with Croatian and Serbian *Grenzer* (dragoons) until 1870, settling out immigrants from conquered territories as a simultaneous process of pacification and defence. Almost nothing is known about the impact of fleeing populations and retreating armies on the Ottoman side of the border. Danube fortress towns such as Vidin, downriver from Belgrade, as we shall see, became the front line of defence.

There are a number of striking aspects to the Treaty of Karlowitz. First of all, intercession by western European mediators was instrumental in its conclusion, establishing a pattern of treaty negotiations with the Ottomans which continued for at least the next century, with the exception of 1774 as we shall see. Secondly, the language suggested a treaty among equals, the Holy Roman Emperor and the Sultan of the Ottoman Empire styled as '. . . Most Serene and Most Powerful Prince and Lord Leopold, and the Most Serene and Most Powerful Prince of the Ottomans and of Asia and Greece and his glorious predecessors', a powerful evocation of the change in the balance of power.[12]

Thirdly, the diplomatic representation from Istanbul was headed by 'Lord Mehmet Effendi, Supreme Chancellor of the Ottoman Empire, and Lord Alexander Maurocordato, Privy Councillor and Secretary', the latter in his capacity as Grand Dragoman of the Porte, the second time that a Greek had negotiated with western powers on behalf of the Turks, and certainly not the last.[13] Throughout the eighteenth and well into the nineteenth century, as much as one-third of the diplomatic corps of the Ottomans was drawn from the Greek Orthodox families of Istanbul, obviating the need for the development of a more representative corps of professional diplomats trained in the European languages to serve the Ottoman court abroad. The Venetian *baili* dubbed the *tercüman* (translator, interpreter) one of the three evils of Istanbul, fire and plague being the other two.[14] The Ottoman reliance on the Orthodox Greek community has been described in many ways, often as part of a western assumption that Muslim cultural superiority militated against the learning of non-Muslim languages, thus requiring intermediaries from the non-Muslim communities in diplomatic practice.[15] While Muslim learning certainly preferred Arabic, Persian and Turkish (Ottoman) to Latin or Greek, the reliance on established communities also naturally grew out of the eastern Mediterranean rivalries, missionary and trade relationships, which formed part of Ottoman trading and hence diplomatic culture of the sixteenth and seventeenth centuries: 'An Orthodox Hellenic style of culture had emerged from the shadows of the first Ottoman centuries, enriched by a strong infusion of skills from the Latin (mainly Italian, largely Venetian) west.'[16]

This style of negotiation was particularly characteristic of what Hurewitz has called the second phase of 'The Europeanization of Ottoman Diplomacy', the 'unilateralism of a contracting empire', when 'dragomans or interpreters, who by definition should have been agents, [became] in fact principals.'[17] In addition to the use of tercümans and diplomatic staff, the Ottomans sent more special embassies and envoys to European courts in the eighteenth century than ever before: 34 Ottoman reports and travelogues from European embassies have been identified.[18] The unintended consequence of the reliance on an ethnic exclusivity for delicate negotiations was a marginalisation of the Ottoman ability to participate in the French-speaking culture of European diplomacy, as was abundantly clear by the end of the eighteenth century. In 1793, Selim III (1789–1807) inaugurated permanent representation in the major capitals of Europe, which continued, in fits and starts, until the end of the period covered by this book.

Rami Mehmed Efendi was chief of the Ottoman delegation at Karlowitz. Rami Efendi was a bureaucrat, the *Reisülküttab*, Chief of the Scribes, an office which was to evolve into the Ministry of Foreign Affairs by the nineteenth century. Heretofore, negotiations between belligerents had been conducted on the battlefield by the sultan and grand viziers. The Ottomans, decisively defeated for the first time in two centuries, were no longer able to dictate terms unilaterally, and began a long period of adjustment to European-style diplomacy. The *reis* and his staff became adroit at the niceties of diplomatic manoeuvring, as observed by diplomatic representatives of friend and foe alike. Posturing and feinting, offence and appeasement were all part of the theatre of international relations, which was often manipulated, sometimes successfully, by the Ottomans.

At Karlowitz, Rami Mehmed Efendi reputedly insisted stubbornly on the few concessions to be made by the victors, arguing that the clause concerning the evacuation-demolition of any forts in Ottoman territories be included exclusively for Ottoman benefit. In the event, the Habsburgs conceded to this small gesture which served to salvage Ottoman honour.[19] Far more interesting, however, is the justification for peace to be found in the preface to the Ottoman Turkish version of the treaty. The thrust of the language was this: The Austrians, Poles, Russians (called Moskov, or Moskova, from Muscovy, by the Ottomans) and the Venetians had so strongly united together and attacked the Muslim frontiers from all sides, from land and from sea, that it was impossible to divide and conquer them. Suspending hostilities and seeking peace was to be interpreted as the equivalent of jihad, and therefore good for the supreme state. This, it was argued, was why the men of state, the ulema and the sultan had agreed to it.[20] Jurists were reinterpreting the concept of holy war to permit a legal state of peace, basing it on the rationale of the good of the Muslim community – *maslaha* – a term much evoked in later treaties.[21] In other words, peace was another way to continue war. The justification was legitimated by the story of the prophet Muhammad, which became a cliché in Ottoman writings on war and peace. In the Treaty of Hudaybiya in 628 AD between Muhammad and the Meccans, the prophet was forced to concede a truce of ten years, in order to enable the new Muslim community to perform the pilgrimage in the city of Mecca. In the event, Muhammad and his community made a triumphal entry into Mecca the following year. In other words, such concessions were only a temporary stop on the way to the ultimate Muslim victory.

This obvious fiction, recognised as such or not, was in keeping with the Ottoman imperial notion of the permanently expanding frontier of the abode of Islam (*Dar-al-Islam*), and the inadmissibility of peace with infidels. It sufficed for the treaty of Karlowitz, amid considerable protest and rebellion, but became harder and harder to maintain as defeat piled upon defeat in the later eighteenth century. It was based on a vision of an empire surrounded by a wide buffer zone, where clientage and border raids, sometimes called razzia (from the Arabic *ghazi*, warrior, often translated as warrior for the faith, an interpretation hotly contested among Ottomanists),[22] protected the interior of the empire. In such a conception of empire, the key to military success lay in the ability to coerce or cajole the border communities to escalate frontier skirmishes into sustained campaigns. The problem of control over such autonomous populations was ever present, and a very significant component of the narrative of late Ottoman history. Legitimacy was argued by adherence to Islamic law, the shar'ia, and justified by the seal of approval of the chief religious officer (*Şeyhülislam*), although in actuality sultanic ordinances (*kanun*) had long coloured the nature of Ottoman administration. In some sense, in the period under discussion, the Ottoman loss of control over its ulema represented as great a challenge as the battlefield losses. The problem of imperial aims and religious justification was hardly unique to the Ottoman context. The Russian Orthodox church, by way of comparison, became an extremely important arm of the civilising mission of the Romanovs in the very same era.

The rhetoric of justification for the Karlowitz treaty took one form; what the treaty actually meant on the ground, another. By the realities of defeat, the Danube assumed the function of the final frontier, especially after Temeşvar was ceded to the Austrians in the Belgrade treaty of 1739. After 1700, preservation of the fortress line from Belgrade to Azov became the primary strategy of all future Ottoman campaigns and treaty negotiations. Neutrality, or at least disengagement, was also a part of the tools of the new diplomacy following the Karlowitz treaty, especially in the middle years of the eighteenth century.

The Russians signed their own treaty with the Ottomans in 1700, withdrawing claims to the Crimea, having learned the hard lessons of the logistics of campaigning on the northern shores of the Black Sea and in the Caucasus, lessons which continued until 1783, when they were rewarded with success and the addition of the Peninsula to the Russian empire. 'The fundamental fact of Russian history was colonisation in a great plain', according to the Russian historian V.O. Klyuchevsky, which he divided into three periods: first, of Kievan Rus and the river Dnieper; second, of the upper Volga and of a consolidation around Moscow by the late fifteenth century. The third period of colonisation, by his account, began in the seventeenth century, with the Romanov entrance onto the European stage, the third side of the triangle which included the Poles and the Ottomans.[23] Here, again, topography and environment drove the style of warfare as much as the wavering loyalties of Tatar and Cossack. The territory between the Prut and Dnieper to the Don and Kuban and beyond to the Volga Basin was contested for over a century, with significant dislocation and resettlement of populations on both sides.

The Principalities

The western end of the Ottoman frontier zone of influence was dominated by periods of war and peace with Poland and fluctuating loyalties of the tributary territories of Transylvania, Wallachia and Moldavia, the latter two coming under Ottoman suzerainty by the mid-fifteenth century. As Belgrade and Azov represented the east–west axis of defence, Hotin on the Polish–Ukraine border assumed that role as the northernmost pivot of Ottoman imperial interests, so that one of the focal points of Ottoman–Russian campaigning in the eighteenth century became the Prut River valley. With 1699, the Poles had thrust the Ottomans back on that defence line; by the end of the century, the estuary of the Danube itself was under siege.

While the 1700 treaty recognised Peter the Great's enterprise in capturing Azov and establishing a naval base at Taganrog, his gains were lost at the spectacular Ottoman victory at Prut in 1711, played out against Peter's Great Northern War with Sweden (1700–21), the flight of Charles XII into Ottoman territory and increasing Russian interference and control of Poland. When a final agreement was reached in 1713, Azov was repinned onto the Ottoman defence map, but the Russians had achieved the right to a permanent representative in Istanbul for the first time. In the mid-eighteenth century, the Ottomans attempted to remain neutral in the face of the occupation and partition of Poland which began with the succession struggles in the 1730s. In 1768, however, Russian–Polish–Cossack skirmishes on the Ottoman frontier served as the excuse for the Ottomans to return to war against Catherine the Great.

Much is often made of the Russian support for the Orthodox co-religionists in Ottoman territories in this period, as reflected in the behaviour of the leaders of the Principalities: Dimitrius Cantemir in Moldavia and Constantine Brancoveanu in Wallachia were both accused of collaborating with Peter. In fact, their assumed treachery resulted in the establishment of the Phanariot regime in the Principalities, an Ottoman attempt at centralisation quite contrary to almost two centuries of rule of the two semi-autonomous territories. Effective Russian manipulation of the rhetoric of defence of co-religionists as means of intervention in Ottoman affairs was a product of the treaties at the end of the eighteenth century. In the language of the Prut peace terms, there is little hint of religious protectionism: 'The Sublime Porte for its part and the Russians for theirs shall promise that the inhabitants, subjects, or other persons who may be under their protection shall nowhere be molested or caused anxiety.'[24]

Susceptible to conflicting religious ideologies as were the Principalities of Wallachia and Moldavia, so too were the Christian populations in other, more fully integral Ottoman territories of the Balkans, such as Croats, Greeks, Serbs, Macedonians, and a host of other ethnicities. A similar cultural style and Ottoman adaptation of local boundaries and governance sustained the empire until later in the seventeenth century. As Ottoman administrative systems began to break down, and local rivalries heated up, religious rhetoric challenged the Ottoman imperial worldview. The degree of their influence on the flow of events remains a much-debated subject, but contemporary writers expressed the Turk–Christian confrontation as the pre-modern equivalent of the 'clash of civilizations', however much we wish to deny it. Ethnic nationalist fervour, sometimes contesting, sometimes subsuming religious affiliation by the nineteenth century, has anachronistically been applied to the period under discussion.[25]

Ottomans and Tatars

Since the late fifteenth century, the Tatar–Ottoman relationship was a fruitful one, with the vast northern Black Sea steppe frontier serving as the Ottoman–Moscow border. Tatar horsemen were understood to be the legitimate descendants of the Golden Horde, who had settled in the Crimea and dominated Moscow for over two centuries. Tatar raids into Russian territories, for both plunder and slaves, were legendary and constant. Their renowned horsemanship, and durability in harsh conditions, are the subject of pervasive legends. The relationship between the Ottoman dynasty and the Tatars was equally driven by the recognition of the potential value of the Tatar cavalry to Ottoman warfare. They were particularly adept at harassing the enemy territories, armies and supply lines. The Ottomans could count on the Tatar Khan, who was generally appointed with Ottoman consent, to serve as the vanguard for any new campaign not only in the northern steppe regions, but also in Poland and the Principalities, as was the case in the Holy League wars. In the period under discussion, the relationship between Tatar and Russian, as well as Tatar and Ottoman, altered significantly and permanently.

The Ottomans preferred clientism to colonisation in their dealings with the Tatars. In periods of crisis in Ottoman dynastic succession, acute in the first half of the seventeenth century, the Khanate in Bahçesaray (Crimea) was often proposed as the potential heir of last resort to the line. As this study opens, Tatar–Ottoman relations had become more complex, with the sultans intervening in appointments, supporting exiled Tatars in Istanbul, and paying considerable sums to the current Khan to incite him to the northern raids which preceded major Ottoman campaigns on the Danube. In spite of Tott's expressed admiration for the Tatar nobles in the late eighteenth century, they had, in fact, become a liability and obstacle to Ottoman military reform, expressly so in the 1768–74 Russo-Ottoman War, although not all Ottomans would have agreed with that sentiment.[26] All of the treaties between the Romanovs and the Ottomans took up the matter of control of the Tatars, and just as often they were the excuse for breaking the peace between the two empires. Ottoman campaigns on the northern shores of the Black Sea were rare after the sixteenth century, as the Tatars served the Ottomans effectively as a military buffer zone until 1700. Tatars were instructed to wreak havoc on the borderlands to preface the arrival of the imperial army. We shall see how the army organisation fared when they were forced to alter their strategy concerning the Tatars after 1774, as the reality of the loss of the Crimea became apparent.

Eastern border strategies

The eastern end of Ottoman territories, stretching from Baghdad into Central Asia, was another zone of nomadism, religious confrontation and warfare for the Ottomans, this one on very different terms. Since the early 1500s, the Ottomans had been in competition in the east with the Persian Safavids, a dynasty of Sufi mystic origins which adopted Shiism as the state religion and challenged the very basis of Ottoman rule. The sixteenth-century struggle between Ottoman Sunni and Safavid Shiite was an eastern echo of the reformation underway in western Europe. The Religious Peace in Augsburg and the Treaty of Amasya, by coincidence both negotiated in 1555, legislated the right of individual princes (Christian or Muslim) to determine religious preference within their own territories. This had followed on the Ottoman defeat of the Mamluks in Cairo, which gave them access to Mecca and Medina, the most sacred sites of the Muslim world. The Safavids challenged the world ambitions of the Ottomans, who on numerous occasions responded to requests for assistance from other enemies of the Safavids, such as the Sunni Shaybanids in the north, numerous rulers on the shores of the Indian Ocean, and even from the rival Sunni Mughal house in Delhi. This geopolitical context of the Ottoman/Safavid territories for the early modern period has been very little studied from the Muslim point of view; it is generally discussed in the larger context of the international spice trade rivalries, especially after the arrival of the Dutch and English in the early 1600s. Even the papacy was not above trying to ally itself with Asian enemies of the Ottomans, a strategy with a long history beginning with the Mongols.

The Ottomans were clearly aware of the threat from both land and sea by 1600. They had made significant incursions into Eurasia to prevent the further advances of Muscovy under Ivan IV, and failed. They countered the Portuguese off the coast of East Africa and in the Indian Ocean, temporarily, with some success. They established the district (*sancak*) of Akkirman in 1538 to close the circle of defence on the Black Sea. Confronting the Ottoman and Tatar threat made for some curious bedfellows: the Pope and the Muscovites; the Persians and the Muscovites; subsequently the Persians and the English. As this study gets underway, the Russians, British and Persians, for whom the primary issue was trade, began the long struggle for economic supremacy in Iran and Central Asia, the latterly named 'Great Game' of the nineteenth century.

By 1700, the border between the Ottomans and the Persians was defined by the Treaty of Zuhab, 17 May 1639, following the Ottoman recapture of Baghad in 1638 by Murat IV (1623–40). The treaty is notable for its language, for the recognition of the shah of Persia as '. . . the ornament of the Persian Throne, the splendour of the kingdom of Djem, and whose magnificence is equal to that of Darius, the great prince and illustrious Lord, the Precious Pearl of the Sea of Royalty, the sun of the sky of Sovereignty, the noble Eagle of the high region of the Dignity of Shah, the Most Illustrious and Majestic Prince', and for the reiteration of the role of the Ottomans: '. . . the most glorious Padishah who is the Defender of the faith, whose Majesty is as great as that of Solomon, who is the substitute of God in the world, and who has justified the maxim that An equitable Sultan is the shadow of God on earth; . . . the supporter of Islamism and Musulmans, the exterminator of heresies and of the polytheists, the Sovereign of the two Orients and the two Occidents, the servant of the two Holy Cities, the Treasure of Mankind and the apple of the age, who is protected by the Supreme Being whose divine assistance men implore, and favoured by the most High and propitious God.'[27] Much has been made of the Ottoman reluctance to fight with brother Muslims, prohibited as it was by Islamic law. While the threat to the Shiite 'heretics' is implied in the text of this treaty, it reads far more like a statement of divine right along the lines of James I of England, and reflects the rhetoric of dynastic politics of the early modern age.

This border agreement survived into the nineteenth century, but unrest along the Sunni–Shiite frontier zone was continuous, exacerbated by the rise of Russia and the resultant migration and emigration of populations. Immigrant nomadism was accelerated by the shifting balance of power and increasing intolerance of religious and ethnic difference that characterised both Russian and Ottoman imperial centres. Significantly, the east was a border of few fortresses, where formidable and forbidding mountain ranges and a climate of extremes served just as well to prevent conquest. It was also a mix of ethnic and religious communities: besides the Sunni/Shiite, Kurd, Armenian, Georgian, Çerkes, Laz, Abkhazia, Tatar, Kalmyk and Kazakh to name a few.

The geopolitical realities of such vast territories, nominally Ottoman, drove not only the language of diplomacy, but also the style of and limitations on warfare, the creation of real or imaginary frontier zones, and greater or lesser central control. Distance, seasonal change, and unpredictable but regularly expected disasters: famines, floods, and plague, made sustained campaigning unrealistic. The harsh and stark setting,

Safavid scorch-and-burn strategies, and increasing Janissary resistance to eastern campaigns kept Ottoman major incursions in the east to a minimum in the seventeenth century. (It is worth remembering that Baghdad was 1,334 miles from Istanbul, Belgrade 587).[28] The spectacular campaigns of Murad IV to Baghdad and to Erivan a mere three years earlier (1635) were unusual, attributable to a sustained lull on the western frontier as Europe was distracted by the Thirty Years War, and to considerable robustness in the treasury as a result of Murad's fiscal reforms. Such campaigns, mobilising an estimated (and usually exaggerated) 100,000 troops, were difficult to organise and costly to sustain.[29]

The tenuous Safavid hold on Persian hegemony fell apart after 1719, especially so by 1722 when Isfahan was conquered by the Afghan leader Mahmud Ghazlay. The Ottomans became further embroiled on the eastern frontier, with some predictable results, such as a further loss of control over revenues, with an increasing degree of nomadisation of eastern and northern Anatolia. A well-documented Janissary revolt, led by Patrona Halil in Istanbul in 1730, was directly related to the campaign mobilisation effort for that particular frontier. Uncharacteristically, the Ottoman thrust into the Caucasus in this period was sustained, driven by Russian encroachments in Georgia, but just as much determined by the Ottoman need to recoup increasingly unruly territories. Sporadic and inconclusive confrontations continued until 1746, when the 1639 borders were essentially re-established. Distance, heterodoxy and nomadism remained powerful obstacles to rationalised warfare in the east, and total war unjustified in the popular mind against co-religionists, even those officially viewed as Muslim heretics. This book ends with the Ottoman sultans of the Tanzimat period still attempting to consolidate control over eastern tribal confederations.

The Habsburgs and the Romanovs chose, by contrast with the Ottomans, to establish military frontiers as a means of defence of increasingly closed or contentious borders. The settlement and employment of distinct populations was part of the strategy. The Russians also had a series of military colonies, as defence lines against Tatar and Cossack, which were enlarged precisely in the first period of our survey, and were double in number by the mid-seventeenth century.[30]

Imperial aims: Ottoman, Habsburg, Romanov

The Habsburgs ruled over a complex federation of well-entrenched nobilities, economically self-sufficient for the most part, but tied to an imperial centre through a common culture, counter-Reformation Catholicism. In the period under discussion, especially the 1680s–1740s, the Austrian model of empire had its moment of fruition, even as more disparate populations were added to its territories, though never in the bewildering variety of either the Russian or the Ottoman case. In an attempt at centralisation and reform, Maria Theresa and Joseph II continued the basic contradictory policies of the counter-reformation age, even as they laid the foundation for the Austro-Hungarian monarchy of the nineteenth century. Roman Catholicism remained the bulwark of their universal message. The great struggle for the Habsburgs after 1700 was the challenge from Prussia and later from Napoleon Bonaparte, which led to the dimunition and total reconfiguration of Habsburg territories.

In Russia, vast expanses of territory and multiethnic populations in the hundreds were added to the Russian dominions in the period of study, and the construction of a universal empire was intricately bound up with local military and administrative service and supply of men and arms. Peter the Great is credited with creating the modern Russian army, but the process of reform had begun under his predecessors. Peter's major contribution lay in the 1722 Table of Ranks, which brought coherence to the Russian administration and created a 'new service nobility'.[31] Catherine the Great's Charter of 1785 continued the process by establishing a corporate structure and legal privileges, but reasserted Orthodoxy as the religion of empire, and the rights of the nobility over the serfs. Both compacts required the kind of negotiations familiar to Ottoman and Habsburg rulers, which included concessions to the aristocracy and the populace at large. Russification and conversion were integral parts of the equation of Romanov universalism, strengthened in nineteenth-century Orthodox missionary movements in central Asia, and often opposed vigorously.

The Ottoman rise and eclipse starts and ends earlier than either Habsburg or Romanov. Ottoman history of the seventeenth and eighteenth centuries is generally viewed as a time of stasis and decline, a paradigm that has resisted modification until recently, when more research has focused on internal dynamics of change.[32] At the beginning of our period, the Ottomans had just suffered a humiliating defeat, accepted treaty conditions which challenged their political legitimacy, and faced the imminent collapse of the centre in 1703, in one of the most severe Istanbul revolts of the period.

The Ottoman universal message continued to be Muslim, an Islam generally tolerant (in theory more than pratice) of multi-ethnic polities: to become Ottoman was to assume a cultural, rather than necessarily an ethnic, identity. Ottoman patronage of the shar'ia, and the persistence of a belief in the impartiality of Islamic law among large segments of the population, however imaginary, constituted a powerful unifying ideology. Growing intolerance towards non-Muslims is one of the principal trends of the era under discussion, imposed by shrinking borders, the dislocation of Muslim populations, and the increasing volume of rival Christian rhetoric justifying imperial territorial claims discussed above. Ottoman dynastic, household imperatives resisted the inclusion of the new sources of power emerging in the countryside. The new challenge of provincial elites arguably emerged precisely in response to the exigencies of the needs of the battlefront.

In this they differed sharply from their Habsburg and Romanov rivals, who strove to include newcomers, albeit in very circumscribed terms. Maria Theresa's and Joseph II's long internal struggle with the Kingdom of Hungary and the international conflict with Prussia are well known. Habsburg policies were governed by Maria Theresa's profound commitment to Catholicism, although they were later tempered by Joseph II's more moderate views. Discussion of Romanov colonialism and negotiation with the 'others' in the east is just emerging in post-Soviet Central Asian studies.[33] True assimilation in the Russian context required conversion to Orthodoxy. By contrast, the Ottoman re-appropriation of exclusively Muslim symbols and Turkish ethnicity was belatedly undertaken in the nineteenth century, when they had already lost the loyalty of a majority of the non-Muslim subject peoples.

The Ottomans resisted the inclusion of emerging provincial powers into the royal household for a number of reasons, most of which require a definition of the dynastic household and its unifying icon, the Janissaries. In 1700, the opening moment of this study, the sultanic household faced possibly the greatest challenge to its survival to date, following upon multiple defeats on the battlefield and the signing of a deeply humiliating peace treaty with the Holy League. More importantly, the failures accompanied and followed more than half a century of Ottoman reform masterminded by the vigorous reforms of the Köprülü family of grand viziers.

The Istanbul revolts that occur throughout the war period, especially after Grand Vizier Kara Mustafa's spectacular failure at Vienna in 1683, are indicative of a significant power struggle between the sultan and and his senior officers: the grand vizier, the commander of the Janissaries (*Agha*), and the *Şeyhülislam*. Mustafa II (1695–1703), under tutelage of his mentor, the *Şeyhülislam* Feyzullah, had envisioned himself as the new Süleyman the Magnificent, and had led his army onto the battlefield at Zenta in 1697. Although successful in small field battles and sieges, the army under the sultan was annihilated by Eugene of Savoy in open field confrontations, which finally forced the resolution of the conflict. An attempt was made to keep the reality of the treaty of Karlowitz from the public for a considerable period. As part of that strategy, Mustafa II decreed the move of the seat of power from Istanbul to Edirne, where he took up residence in 1701, ostensibly to facilitate the signing and implementation of the conditions of the treaties.

Revolts in Istanbul circa 1700

In 1703, a very serious revolt broke out in Istanbul among the military, who complained of overdue pay, and of the sultan's absence in Edirne. This was not the first nor the last such Janissary revolt of the period, but there were some new elements who were added to the ranks of the rebels. Observers noted an interesting mix of civilians, a new coalition of soldier, artisan and members of the religious class, disdainful of any attempts by the sultan to mask the loss of political legitimacy. One of the real causes for grievance was the perceived lack of balance in power. *Şeyhülislam* Feyzullah, who continued to be patronised by Mustafa II, was the focus of much of the anger, his corruption and nepotism excessive even for the time, and his influence over the sultan considered too great. Furthermore, he overstepped the boundaries of his position as head of the religious arm of the household, establishing corporate relationships traditionally the domain and prerogative of the vizierial and pasha households. He antagonised the potential candidates for the post of grand vizier, the head of the bureaucracy which had begun its consolidation under the famous Köprülü family in the mid-seventeenth century.

The demands of the rebel leaders in Istanbul were articulated by members of the religious class, the ulema, who, like the *Şeyhülislam*, were servants of the sultan, but rarely participants in the unruly behaviour of the military arm. They operated as the arbiters of justice in interpreting the shar'ia, which was enforced in the local courts scattered throughout the Ottoman realms. The courts, and the kadis, or local judges, were the single, most consistent representation of Ottoman rule in the provinces. In this instance the religious officials first attempted mediation, and then provided a justification for the rebellion, once they saw that the rebels intended to march on the sultan in Edirne. Couched in the question-

and-answer style rhetoric of the religious ruling called a *fetva*, four questions were posed and answered: the first concerned Mustafa II's neglect of his 'trust' in looking after his subjects, 'allowing injustice and inequity to reign' while he went hunting, wasting the public treasury. The second legitimated the right of a Muslim community to stand up to an unjust ruler. The third condemned those who sided with an unjust ruler. The fourth charged Mustafa II '. . . with having compromised his mandate by accepting the peace treaties' and conceding so much territory to the Christian powers.[34]

This manifesto, as with many others of the period before and after, is couched in the religious rhetoric of Islam, arguing that the sultan's chief duty lay in maintaining the balance and stability of the Muslim community, while defending and extending the borders of the territories of Islam. Its consistent use in such manifestos has blinded historians to the reality of dynastic politics, which often operated at odds with such justificatory rhetoric. The sultan was interested in personal and familial survival. For that, he depended upon the Janissaries, whose loyalty had to be reacquired at each accession. They represented '. . . the necessary ceremonial action and illusions upon which sultanic rule depend[ed].'[35]

As a result of the 1703 revolt, Mustafa II's brother, Ahmed III (1703–30), became sultan. Military confrontation outside of Edirne was avoided, as the imperial loyalists, mostly troops recruited from the Balkan countryside, deserted Mustafa and joined the ranks from Istanbul. The bloodshed which followed, once Feyzullah was brutally eliminated, did not centre on Mustafa II or his household; he spent the remainder of his life in seclusion in the palace. But the violence in Istanbul became more generalised, resulting from three related problems: the lack of discipline and control over the disorder and destruction; the dissolution of rebel unity, amidst rivalries concerning the balance of power; and finally competition for the coronation accession gifts, the traditional reward for the Janissary pledge of allegiance to a new sultan.[36] By this time, the ceremony of submission by which the Janissaries swore allegiance to the new sultan was a theatrical gesture masking the real power of the corps to control events in the imperial capital. As one historian has noted, the results of this particular rebellion indicate 'no significant interest among the military . . . to exercise leadership of the government . . . a lack of the need to wield sovereignty.'[37] The consequence for military affairs was in the reassertion of the power of the grand vizier over the commander of the Janissaries, and a willingness by the rank-and-file Janissaries to accept the return to the normal balance of power.

Janissary loyalties and ethos

The key to Janissary acquiescence lies, once their immediate demands of payment were satisfied, in the Ottoman notion and operation of households (the preferred term for Ottoman-style elite courts). As *kapu kulu* ('household-slave', the term used for the imperial troops), the classical imperial Janissary served the sultan, and operated within a corps structure for whom the virtues of 'unswerving gratitude, loyal servility to the sultan and unflinching heroism'[38] represented the ideal. For the earlier period, the relationship of sultan to his troops has often been described as that of father to his sons. This also gave them the right to challenge their father, so to speak, when his authority was too lax, too oppressive, or otherwise interpreted as inappropriate. Servitude was characterised by periodic insubordination, more of a constant than contemporary histories of the earlier period allow us to believe. The revolt I have just described was by far the most sustained to date in the capital itself. While it would not be the last, it restored the traditional balance, which more or less guaranteed the continued mediocrity of military leadership, subordinated to the emerging service elite of the bureaucracy, and the continued appointment of the commander of the Janissaries by the grand vizier. By the mid-eighteenth century, the grand vizier himself was not, with rare exceptions, drawn from those with military prowess, or experience in provincial administration.[39] Rather, the grand viziers were increasingly palace appointments as bureaucrats, made and unmade by palace corporate politics.

By 1700, the provinces of the empire were ruled by court-appointed governors (*valis, pashas*) who had established significant networks of power in extended households which mirrored the royal household in Istanbul. As such, they organised small, Janissary-style mercenary armies in their own entourages, called the *kapu halkı* ('household folk'). The significant power struggle even before 1650, but intensified thereafter, lay in the competition over the control of tax revenues among Ottoman officials such as the provincial governor, generally the career route to grand vizierial appointment; the Janissaries, posted to the countryside and in major fortresses, as well as the *ayan*, who emerged as the result of this rivalry, and are a prevalent part of the landscape in the period under discussion. The imperial Janissaries of Istanbul functioned after 1700 as the imperial guard of the sultan. The provincial households became the source of armed manpower, and as such could not be ruled as much as co-opted by the imperial circle. As with European monarchs, such as the

Habsburgs, the sultan had to contend with independent armies from a very early date.

In fact, the whole latter half of the seventeenth century represents the culmination of a long process of evolution in Ottoman dynastic politics, from rule based on charisma, and legitimated by ancestral authority, to a more impersonal 'institutional monarchy . . . Decentralizing the decision-making process, abandoning direct taxation based on regular land registration . . . and promoting the spread of "tax farms" . . . were connected with these changes in the style of sultanic rule.'[40] Fruitful comparisons can be made between the Ottoman and European contexts, such as the French, whether the processes are described as 'decentralisation' in the Ottoman or 'centralisation' in the European case. The question remains of whether or not the Ottoman house ever succeeded in monopolising the control of violence, which is one of the more interesting arguments about early modern European state formation, and the rise of modern armies.[41]

Of equal importance to sultanic legitimacy was the traditional notion of the household on campaign, perhaps a legacy of their Central Asian origins in the distant (and highly mythologised) past. The crescent formation order of the military forces which the Ottomans employed in major open field confrontations, even as new technical and strategic developments dictated otherwise, was driven by the idea that the military apparatus of the state was just that: an expression of the order and proximity to the privileges granted by the sultan, or his substitute, the grand vizier. That is to say, that the deployment of Ottoman assembled forces on the battlefield was representative of 'the political hierarchy of the Ottoman state'.[42] Such representation of founding myths remained evocative well into the eighteenth century.

Admittance to the Janissaries meant acceptance into a circle of privilege, and many of the rebellions of the period under discussion were driven by the desire to see the circle of privilege enlarged, and admittance to a carefully maintained, if largely fictional, distinction between the *askeri*, or tax-exempt military, and the *reaya*, or taxable peasants, preserved. The major dilemma for the structure and functionality of the Ottoman state was that expanding the number of Janissaries meant a considerable drain on sources of revenues, so the tension between those who wanted in and those who wanted to keep them out is one of the leitmotifs of later Ottoman household dynamics. The sultans adapted many different strategies for mobilising and financing campaigns in order to solve the problem of their imperial guard, strategies which will be discussed in subsequent chapters. Suffice it to say that sustained challenges to the Ottoman ruling household of the seventeenth and eighteenth century were driven as much by evolution in the source and composition of the Ottoman armies as by the factional politics and the 'rule of the women' in the palace.

To go to war was to take the government on parade, and, it could be argued, stimulate the economy. The presence of the entire imperial household in frontier zones did guarantee a temporary restoration, voluntary or coerced, of the tributary relationships which characterised Ottoman administration. Theatrical ceremonies conveyed sultanic benevolence and power and reinforced the tenuous kinship bonds of independent and entrepreneurial troops. It was an empire-wide endeavour, and could be sustained only as long as local populations were not disadvantaged. As Stoye adroitly notes: 'campaigning on a large scale justified enlarging the army to a maximum, and within this expanded force it was easier to contrive a balance of power which subdued the more refractory elements',[43] a very considerable argument for the regularity with which the Ottomans mounted large campaigns until the period with which this book begins. The eighteenth century is remarkable for the restraint from warfare, especially the years 1740–68. The usual explanation is that the Habsburgs and Romanovs were preoccupied elsewhere, but it is also important to point out that the 'limitless' Ottoman horizons were no more. We now turn to an examination of what Sugar once called 'A near perfect military society'.[44]

Notes

1 John Stoye, *Marsigli's Europe, 1680–1730* (New Haven: Yale University Press, 1994), chapter 3, *passim*. Stoye has assembled an incredible amount of material on the life and works of Marsigli, using the bulk of the unpublished diaries and studies found mostly in Bologna and Vienna.

2 Herewith the song in Turkish, followed by a translation: 'Estergon kalesi, bre dilber aman, su başı durak; Kemirir gönlümü, bre dilber aman, bir sinsi firak; Gönül yâr peşinde, bre dilber aman, yâr ondan ırak; Akma Tuna akma, bre Şahin aman, ben bir dertliyim; Yâr peşinde aman da gezer koşar yandım kara bahtlıyım.' Esztergom fortress, aye beauty aye, a stop at the fountainhead; A sneaking grief at separation, aye beauty aye, gnaws at my heart; Heart in pursuit of the beloved, aye beauty aye, beloved far away; Don't flow Danube, don't, aye falcon aye, I am the troubled one; I wander, run, exhausted, my dark fate.

3 Stoye, 106–07. *Hayduts* evokes mercenaries, marauders or freedom fighters depending on one's point of view.

4 Hussey died of plague; Mustafa Köprülü died on the battlefield at Slankeman; Süleyman II died before the campaign got underway.

5 William H. McNeill expressed it in the title of his book: *Europe's Steppe Frontier 1500–1800* (Chicago: Chicago University Press, 1966).

6 William H. McNeill, *Venice: the Hinge of Europe, 1081–1797* (Chicago: Chicago University Press, 1974), 182–83.

7 Staatsarchiv Detmold L 114A (von Blomberg-Iggenhausen, Nr. 737). Hans-Heinrich Nolte, 'The European system', unpublished paper.

8 Basil G. Gounaris, 'Social cleavages and national "awakening" in Ottoman Macedonia', *East European Quarterly* 19 (1996), 421.

9 The Ottomans never considered themselves Turks – it was applied early by Europe to the empire, in part because of the long association from the crusades, when Turk replaced Saracen, or Arab, as the generic term for the Muslim infidel. European diplomacy therefore referred always to Turkey, the Turks, etc., a term only recognised and utilised in the nineteenth century by the dynasty as part of its identification of its subjects. Turkish nationalism arose as an ideology of resistance only in the twentieth century.

10 Raymond Birn, *Crisis, Absolutism, Revolution: Europe 1648–1789*, 2nd edn (Fort Worth, TX: Broadview Press, 1992), 114.

11 Michael Hochedlinger, *Austria's Wars of Emergence 1683–97.* (London: Longman, 2003), 155–60. See also Thomas H. Barker, *Double Eagle and Crescent: Vienna's Second Turkish Siege and its Historical Setting* (Albany: SUNY, 1967) for the blow by blow account.

12 An attempt had been made in 1606, at the negotiations for the peace of Zsitva-Torok, which concluded the 1593–1606 Ottoman Habsburg War, to require the Ottomans to call the Habsburg rulers the 'Roman Emperors', but the Ottomans had ignored the clause. The semiotics of Ottoman translations of Latin or Italian treaties, the usual common language between Muslim and Christian, remains virtually untouched.

13 The first was Panagioti Nicoussias with the Venetians over Crete (1669). William H. McNeill, *Venice*, 213–14. This text of the treaty is from Fred L. Israel, ed., *Major Peace Treaties of Modern History*, vol. 2 (New York: Chelsea House, 1967), 869.

14 'In Pera sono tre malanni: peste, fuoco, dragomanni.' Mary L. Shay, *The Ottoman Empire from 1720 to 1734 as Revealed in the Dispatches of the Venetian Baili* (Urbana, IL: University of Illinois Press, 1944), 38.

15 Bernard Lewis, *The Muslim Discovery of Europe* (New York: Norton, 1982), 78–88, but especially 105: 'Dealing with infidel foreigners was a dirty and dangerous business and best left to other infidels . . . For such dealings as were felt to be necessary with the infidel world, most Muslim rulers were content to rely on infidels calling on them – a natural tribute from the lower to the higher order – and to make use, even at home, of intermediaries to shield them from too close contact.' Lewis's sentiment is clear.

16 McNeill, *Venice*, 215, goes on to note the development of two 'heads' of Ottoman society and civilisation: one Muslim, one Christian.

17 J.C. Hurewitz, 'The Europeanization of Ottoman diplomacy: the conversion from unilateralism to reciprocity in the nineteenth century', *Belleten* 25 (1961), 460–61.

18 Faik Reşat Unat, *Osmanlı Sefirleri ve Sefaretnameleri* (Ankara: Türk Tarih Kürümü, 1968).

19 This is the theory advanced by Rifaat Abou-El-Haj, 'Ottoman attitudes toward peace making: the Karlowitz case', *Der Islam* 51 (1974), 131–37; see also Virginia H. Aksan, *An Ottoman Statesman in War and Peace: Ahmed Resmi Efendi, 1700–83* (Leiden: Brill, 1995), Chapter 1 *passim* on the development of the office of *Reisülküttab*.

20 Abou-El-Haj, 'Ottoman attitudes', 136.

21 Thomas Naff, 'The Ottoman Empire and the European states system', in Hedley Bull and Adam Watson, eds, *The Expansion of International Society* (Oxford: Oxford University Press, 1984), 152.

22 See Cemal Kafadar, *Between Two Worlds* (Berkeley, UC Press, 1995), 62–90, for the best summary of the debate.

23 Paul Dukes, *The Making of Russian Absolutism, 1613–1801*, 2nd edn (London: Longman, 1990), Chapter 1.

24 J.C. Hurewitz, *The Middle East and North Africa in World Politics: A Documentary Record*, 2 vols (New Haven: Yale University Press, 1975), vols 2, 36.

25 George F. Jewsbury, *Russian Annexation of Bessarabia: 1774–1828, A Study of Imperial Expansion* (Boulder, CO: Westview Press, 1976), 8.

26 François de Tott, *Memoirs of Baron de Tott*, 2 vols (London, 1785; reprinted New York, Arno Press, 1973). As a French consul in the Crimea in 1769, Tott observed and was impressed with Khan Kırım-Giray (vol. 1, pt. 2, 190 ff.). At the Siege of Vienna, Tatars may have captured as many as 70,000 souls used as forced labour in building the siege works (Barker, *Double Eagle*, 283–84).

27 Hurewitz, *The Middle East*, vols 1, 26.

28 Rhoads Murphey, *Ottoman Warfare 1500–1700* (London: UCL Press, 1999), xiv.

29 Murphey, *Ottoman Warfare*, 5–6.

30 From 8 to 17 garrison towns; Carol Belkin Stevens, *Soldiers on the Steppe: Army Reform and Social Change in Early Modern Russia* (Dekalb, IL), 20.

31 Dukes, 90: of 262 ranks, 126, or 48 per cent, were military.

32 Rhoads Murphey, 'Ottoman resurgence in the seventeenth century Mediterranean', *Mediterranean Historical Review*, 8 (1993), 186–200.

33 Daniel R. Brower and Edward J. Lazzerini, eds, *Russia's Orient: Imperial Borderlands and People, 1700–1917* (Bloomington, IN: Indiana University Press, 1997), a recent example. Michael Khodarkovsky, *Russia's Steppe Frontier: The Making of a Colonial Empire, 1500–1800* (Bloomington, IN: Indiana University Press, 2004), is a lively glimpse of the imperial/nomadic struggle.

34 Rifaat Abou-El-Haj, *The 1703 Rebellion and the Structure of Ottoman Politics* (Leiden: Brill, 1984), 71.

35 Stephen Turk Christensen, 'The heathen order of battle', in S.T. Christensen, ed., *Violence and the Absolutist State: Studies in European and Ottoman History* (Copenhagen, 1990), 95.

36 Abou-El-Haj, *The 1703 Rebellion*, 79.

37 Ibid., 85.

38 Christensen, 'The heathen order', 96.

39 One exception is Muhsinzade Mehmed Pasha, who assumed command of the army on the Danube in 1771 when an old man.

40 Rhoads Murphey, 'Continuity and discontinuity in Ottoman administrative practice during the late seventeenth century', *Poetics Today* 14 (1993), 425.

41 As described in the work of Charles Tilly, *Coercion, Capital and European States, AD 900–1990* (Oxford: Blackwell, 1990), especially his notion that localising revenue extraction creates new loyalties to centralising monarchies, and Ariel Salzmann, 'Towards a comparative history of the Ottoman state, 1450–1850', *Archiv orientální* 66 (1998) 351–66. (Proceedings of the XIIth Congress of CIEPO, 1996, 351–66.)

42 Christensen, 'The heathen order', 95.

43 Stoye, *The Siege of Vienna* (New York: Holt Rinehart and Winston, 1964), 30.

44 Peter Sugar, 'A near perfect military society: the Ottoman Empire', in L.L. Farrar, ed., *War: A Historical, Political and Social Study* (Santa Barbara, CA: UC Press, 1978), 95–104.

CHAPTER TWO

The Ottoman system

The Ottoman Empire has long served as the particularly egregious example of the dangers of a pre-modern society where militaristic values and demands predominated, and civilians were subordinated to the military class, a situation which 'discourages independent and original thinking, which in turn precludes innovation'.[1] By this definition, the economy was also driven predominantly by the demands of the military. When a 'high percentage of its [society's] surplus is committed to nonproductive enterprises, usually related to the military, economic adjustment is consequently difficult and progress slow'.[2] Individual soldiers in the Ottoman context became the 'bandits of God', the instrument of a state whose chief *raison d'être* was 'to extend the realm of Islam'. The failure of the Janissaries was equally the failure of the state: 'The Ottoman Empire, a military society, declined when its army failed and disappeared when its army disintegrated'.[3] By the parameters outlined above, the empire should have collapsed after 1699. This chapter will explore the reasons why the Ottomans persisted, by arguing that the state of military development in all three imperial centres was at parity in 1700.

Most military histories emphasise the imperatives of the battlefield, the evolving use of gunpowder and the necessity for organising large-scale warfare which typifies European states for the seventeenth and eighteenth century. In the period 1660–1700, the Ottomans had been deeply engaged with the Christian powers of Europe, establishing a twenty-year peace with the Habsburgs after the Ottoman defeat at St Gotthard on the Raab River in 1664; finally subduing the Venetians in the siege of Kandia on Crete in 1669, after 24 years of expensive and inconclusive confrontations in the Mediterranean,[4] and extending the furthest into Poland, occupying Podolia for a brief moment, in a contest over what became Ukraine, with Russian, Polish–Lithuanian Commonwealth and Cossack forces (1671–99). Major coalitions of European powers, as we have seen, checked Ottoman power by 1699.

The 1703 crisis in the capital described in the previous chapter was the capstone of a fifty-year period of continuous upheaval in the countryside, driven in large measure by the pressing need for men and money. The ability to restore order and reorganise the Ottoman forces in the 1680s was largely the work of a remarkable family of bureaucrats, the Köprülüs, Ottoman Richelieus and Mazarins, who served off and on as grand viziers and commanders of the army in the field into the first decade of the eighteenth century. Experience on the battlefields of the Danube tended to refresh Ottoman knowledge of evolving battlefield conditions.

Major European confrontations of the mid-eighteenth century, however, in which the Ottomans did not participate, significantly altered both military organisation and technology, especially as regards small arms and artillery, changing the balance of power by the time the Ottomans confronted the Russians on the Danube in 1768. It is generally conceded among military historians that the Ottomans could hold their own on the battlefield until at least 1700. What holding their own meant in that context will be explored in this chapter. As Chandler notes: '. . . the Turk remained a doughty opponent, if only on account of the size of his forces and the fury of his initial attacks, earning the grudging admiration of his Austrian and Russian foes.'[5]

Rethinking imperial military history

Military historiography after Clausewitz has generally been dismissive of pre-Napoleonic warfare, exhibiting 'a tendency to underrate the determination and ability of aristocratic societies',[6] and making even shorter shrift of the Ottomans, concluding that they never faced significant military opposition until the armies of Montecuccoli, Prince Eugene of Savoy, or the Russian Field Marshals Rumiantsev and Suvarov. Yet for all three empires, it was the age when weather, crop failure, disease, and distances from sources of supply were just as important in determining success on the battlefield as weak central command, or incompetent leadership. Such factors were great levellers of military competence. As whole societies came to be mobilised to sustain dynastic goals, resistance to collaboration with imperial aims emerged. Re-emphasis on warfare as one of the catalysts of modern state development has produced a number of works on the evolution of armies of large empires, with universalist pretensions, pre-industrial and multi-ethnic societies, whose legacy can be traced into the modern period. In that context, Ottoman historiography has taken some interesting turns, among them a reassessment of Sugar's 'near-perfect military society'.[7]

One of the products of this new approach is an appreciation of the relationship between ruler and ruled in absolutist contexts, as well an understanding of the societal costs of mobilising the countryside, and the impact that the constants of war (shortages, hoarding, starvation, disease, and nomadism) exerted on rural populations. The most successful early modern proto-state was the one which monopolised the control of internal and external violence, including the organisation of capital.[8] Dynasties could not continuously over-extend their extortion of the very provincial manpower and resources which sustained their bid for hegemony, any more than they could overburden those who chose to support the enterprise. Negotiation over rights to collect taxes, and other such privileges, were an integral part of going to war, and understandably the source of wealth of many of the prominent families of Europe of the seventeenth and eighteenth century. The delicate balance between bellicosity and the ability of the state to finance it has recently been graphically described: 'The simplest, most effective check on the king's power to tax arbitrarily was the peasants' refusal to pay. In the late 1630s and 1640s, this refusal to pay reached staggering proportions and it served as a major impediment to the extension of the king's power.'[9] This is no less true for the Ottoman Empire, and part of this study will be devoted to the ways in which the empire attempted to finance warfare in the later period.

Scholarly interest has recently reawakened concerning the use of myth and symbols to promote dynastic power, albeit in the late Ottoman context.[10] In order to understand the internal dynamics of the Ottomans, and by contrast, the other two very complex and extensive enterprises which confronted one another across the Danube, we must also consider the use of psychology and ideology as part of the rhetoric of imperial identity. The justification for excessive exaction was generally accompanied by an articulation of an imperial worldview emanating from the dynastic household. Subscription or submission to a potent ideology which merged dynastic and religious ideals was part of an array of strategies available to potential competitors in early modern societies. Such forces certainly influenced the Ottoman context, and recognising them allows us to see more clearly beyond the Janissary and holy war script which has satisfied the community of historians for so long.

The nature and composition of the Janissaries

That competent bureaucrats could revive the economy is only one dimension of the story of Ottoman perseverance. The Ottoman military system consisted of a standing army, the Janissary corps, and the artillerymen; a fief-based cavalry (*sipahi*) and various auxilliary forces, most often drawn from the proximate terrain of the battlefield, and from among both Muslim and non-Muslim populations. In order to understand how the dynasty survived the crisis of legitimacy described in the previous chapter, and managed to assemble within a decade the reputedly huge force which outnumbered and overawed Peter the Great on the Prut in 1711, we must examine the nature and composition of the Ottoman army of the period.

The Janissaries were the salaried imperial infantry corps, founded in the mid-fourteenth century, which was the heart of a military system which was the 'terror of Europe' until well into the eighteenth century. Accurate statistics for the corps after 1650 are hard to come by, reflecting the problem of military statistics in general.[11] The number of active or effective members of any corps of the early modern period is hard to extract from either the official rolls (generally acknowledged as inaccurate reflections of reality), or contemporary observers, whose desire to inflate the numbers of the enemy, dead or alive, is understandable. The problem is particularly acute in the Ottoman context, because of the tendency to view every member of society except the peasant as a member of the Janissary corps. Take the lament of Thomas Thornton, author of a history of the Ottomans, writing in 1807:

> *Sir James Porter considers the army to be composed of the body of the people, and the janizaries to amount to two to three hundred thousand men, independently of those who get themselves enrolled to enjoy the privileges. Pey[s]sonnel supposes they may consist of many millions. Baron de Tott calculates them to be four hundred thousand: and finally, Mr. Eton . . . determines them to be an hundred and thirteen thousand four hundred. But the number of effective janizaries is best determined by the amount of their pay. Two thousand four hundred purses are issued every six months from the treasury; a sum which allows thirty piastres a man for an army calculated at forty thousand.*[12]

Thornton's own estimates prove the most durable. Figures that exist for the earlier period indicate that the palace Janissary infantry corps grew from 11,000 to 12,000 in 1527, rising to ca. 20,000 in 1567[13] and to 37,000 by the end of the Long War (1593–1606).[14] Thereafter, the *mustered* fighting force of the imperial infantry may never have exceeded 40,000, the *effective* force probably being considerably smaller, especially in the later period when desertion became endemic. For the 1683 campaign with the Holy League, Marsigli reckoned 30,000 Janissaries.[15] An Ottoman source on the Vienna siege lists about half that number as actually in combat.[16] (The vanguard of the army which arrived in Ösijek in June of 1683 was composed of 3,000 Janissaries, 500 *Cebecis* (Armorers), 20,000 cavalry and 8,000 Tatars).[17]

Similarly, for the 1768–74 Russo-Ottoman war, the estimates range from 60,000 to 400,000 for the total assembled force in that case.[18] A figure of 30,000–40,000 mustered Janissaries is consistent with Thornton's evidence as well as that compiled by Rhoads Murphey, who notes, '[i]t cannot be said of any period of its [Ottoman] history that military institutions dominated civil society.'[19] The sheer cost of maintaining the highly privileged Janissary corps constrained its enlargement and forced the creation of alternative, more temporary forces, a practice widely evident in the western European context. In fact, one of the most interesting contrasts among the Ottomans, Habsburgs and Romanovs is that just as the latter two were centralising the administration of warfare, the Ottomans decentralised it, though the social costs only became evident in the 1790s.

The Janissary system was originally organised by the conscription and forced conversion of young non-Muslim men, known as the *devşirme*.[20] New recruits were added to the *acemioğlan* corps, essentially the reserve for Janissary corps, and used to top up Janissary numbers as necessary. Manumission was automatic upon ceremonial admission into the Janissaries. Until 1600, the Ottoman army and palace bureaucracy were drawn almost entirely from this source, from lowest foot soldier to grand vizier. By 1700, however, the *devşirme* was finished as a means of recruitment. One of the last of the *devşirme*-style round-ups was in 1666, when 300–20 youths were reputedly gathered from the European territories. Official records indicate that, in 1687, only 130 new Janissaries were added to the corps in the traditional manner, promoted from within the *acemioğlans*.[21] These numbers are insignificant for the campaigns that preoccupied the Ottomans in that period. In other words, the military slave system had long since been abandoned as the chief source of Janissary recruits. Pressure from within the ranks of the regiments to admit outsiders, mostly Muslim-born, sometimes children of the corps itself, had in fact gradually replaced what were no doubt much resisted, and less than productive, round-ups.[22] The central dilemma of the Ottoman state for the next century was how to contain the size of the imperial infantry corps, whose members, initially deployed beyond Istanbul to the provinces to quell rebellion, merged with and contested the administrations of both local and Ottoman officials. Janissary elites as part of powerful local households dot the landscape of the late eighteenth century. Yet the dynasty had still to mobilise effective fighting forces for the far-distant, seasonal and siege warfare characteristic of the age, and maintain a series of fortresses along the north frontier. As the tightly-controlled recruitment system fell into desuetude, other means of recruitment and retention had to be employed.

The dilemma was not entirely the issue of payment, discipline and training, always offered as causes of discontent by Janissary rebels themselves. First and foremost to be a Janissary was to enter tax-free status: all those in the military were exempt from taxation. Discontent was equally a reflection of changing power structures, the rise of competing households, and the challenge to the privileged position of the Janissary. The Janissary barracks, organised as *ocaks* (literally, hearths); the symbols of the kitchen, the great soup cauldrons, and ceremonial meals shared with the sultan or his substitute, the grand vizier, on the eve of battle, continued to symbolise the patrimony of the Ottoman dynastic household. The demand for accession gifts from any new sultan signalled the voluntary submission to the new sovereign by his unruly children, the Janissaries. Yet, by 1700, there were reproductions of such patrimonial households, with their own armies, in all the major fortress cities and towns of the northern empire. The southern frontier, i.e. the largely Arab and Kurdish territories of the empire, as we shall see, operated somewhat differently. The assumption of political power by increasingly autonomous imperial troops is a constant refrain of latter-day Ottoman history.

Belonging to the Janissaries became an entitlement and an honorific. Extra privileges accorded the corps included annual allotments for uniforms; incentive bonuses before and after battles; 'healing money' for meritorious battle wounds; and money for retirees and widows. Such entitlements were regularly distributed to members of the corps in its heyday, and continued to be part of the demands in the eighteenth century. The Janissaries retained the status of a privileged caste, a Praetorian guard, long after their usefulness on the battlefield had declined.

Individual regiments developed an *esprit de corps* that resembled an extended family. To encourage participation in campaigns, provincial governors promised registration in the corps. Artisans of all kinds found it

PLATE 5 *Members of various pre-reform Ottoman forces* (*Chevket Pacha* (*Mahmond*), L'organisation et ces uniformes de l'amée Ottomanne (Depuis sa création jusqu'à nos jours), *Vol. 1, 1900*).

convenient to become attached to the corps, as noted by Raymond and others, for physical and economic protection.[23] Tott, in the Crimea as French consul, was made an honorary member of the corps in Perekop in 1770, a common practice of the period.[24] By the time Thornton was writing in the early decades of the nineteenth century, and just prior to the official elimination of the Janissaries in 1826, who they were and what they had become was difficult to determine.

The surviving Janissary rolls are a reminder of a well-documented problem of controlling registration in the famed corps: for example in 1688, the rolls were reduced by 20,000 names;[25] again in 1771, 30,000 names were stripped from inaccurate, often corrupted registers, to mention only two examples.[26] In reality, reforming grand viziers had always to contend with inflated muster rolls, often benefitting themselves from pocketing salaries of non-existent soldiers. Sultan and grand vizier alike faced the increasing debt burden not just of army wages, but also of the aforementioned accession gift, what had become the obligatory contractual agreement between the Janissaries and any new sultan who took the throne. The scale of that one-time payout simply ballooned. Uzunçarşılı gives several examples: in 1687, 4,557 *kese* were distributed: roughly one-third came from the sultan's inner treasury, and two-thirds from the annual Egyptian tribute to Istanbul. At Ahmed II's (1691–95) accession, because of the war, instead of a two *akçe* increase to salary, and a 3,000 *akçe* gift, 6 *kuruş* (720 *akçe*) was distributed to each of 17,000 Janissaries, with a promise of the remainder when they returned home. Sarı Mehmed Pasha, Chief Financial Officer six times between 1703 and 1716, was said to have melted the palace silver to make up the accession payment for Ahmed III (1703–30). Mahmud I (1730–54) distributed 29,530 *kese akçe*.[27] Obviously, this was a ruinous policy, which fell by the wayside as the economic crisis deepened at the end of the century, but the payment for valour, special duties, captives, wounds, and numbers killed in battle continued until the mid-nineteenth century. The Janissaries embodied a pre-modern social welfare system which extended throughout Ottoman society. Small wonder, then, that there was a clamour to join the ranks as volunteers, and a vigorous resistance to change.

Seeing beyond the Janissaries has been one of the largest challenges to traditional Ottoman historiography, especially in the eighteenth-century period, when they seemed to be everywhere. For most of the principal campaigns, however, the palace corps represented only one-fifth of the assembled forces, which were overwhelmingly composed of cavalry rather than infantry as our period opens. For the campaigns of the second half of the eighteenth century, the Janissary corps had to be rebulit by adding raw recruits on the march.

Janissary command and discipline

The Janissary Agha was in charge of an organisation which was a law unto itself, but as was previously noted, he was appointed by, and subject to, the authority of the office of the grand vizier, who was considered the sultan's absolute deputy until the dissolution of the Janissaries in 1826. Within the organisation itself, promotion was based almost entirely on seniority and favouritism, as such influences were pervasive throughout the Ottoman system. It proved fairly easy to buy one's son into the corps, and hence, promotion likely followed much the same pattern. As we will see in later chapters, the development of a general staff, and the inculcation of military merit and discipline in the officer corps, were the least successful of the reforms even after 1826, because of the entrenched resistance to the training and educating of non-palace, i.e. untrustworthy (and unknown), men.

Military regulations were included in the *kanunnames* of the sultans, quite distinct from the shar'ia. Military justice was overseen by the two military judges (*kaziaskers*) of Anatolia and Rumelia, with the latter presumed to be the senior of the two. For the period under discussion, we have one example of the military regulations, called *Eyyubî Efendi Kanûnnâmesi*, after the presumed author, about whom we know next to nothing. From internal evidence, it would appear to have been compiled in the reign of Mehmed IV (1648–87). A very comprehensive description of Janissary rules and financing, it also includes a description of the internal discipline of the Janissaries. Miscreants were subjected to shaming and the bastinado, applied to the feet. The severity of the crime determined the number of strokes. Very severe crimes (not specified) called for clapping the offender in chains. If execution was warranted, this was carried out in a humiliating fashion. The Janissary was choked by an executioner, and unceremoniously dumped into the sea, with the death formally announced by firing a cannon.[28] There is no other evidence, in this manual at least, about other forms of military discipline: attendance at roll call; restriction to barracks, even training and drilling. Archery practice was known, as the sultans often participated in such displays, but all other forms of military discipline seem to have fallen by the wayside by 1700. We will take up the attempts to reinstate the former strict and effective discipline with the discussion of the reforms of Selim III and Mahmud II.

Provincial forces: sipahis

The oldest element of the Ottoman forces was the military organisation based on the *timar*, the land grant (usufruct) to a *sipahi* who received his estates in return for military service, but who retained the right to the fief for his lifetime only. The *sipahi's* position as a feudal landlord was 'precarious and revocable'.[29] The main obligation of the ordinary *sipahi* was to report to campaign in person. The size of the *timar* determined how many squires, armed retainers and tents he was required to supply. *Sipahis* were also obliged to furnish their own horses, arms and sustenance, and soldiers from each *sancak* often collectively organised their own supplies. Governors of the provinces were responsible for mobilising the *sipahis* for campaigns, and army headquarters maintained registers, updated as each cavalryman arrived on the battlefront. Those who did not report were in danger of losing their entitlement, initially temporarily; as time went by, and more and more *timariots* failed to appear, the authorities responded by striking the defaulters from the lists. The demise of the *timar* system was bound up with the changing nature of warfare, as well as the parsimony of the dynasty, slow to respond to the changing economic condition of the provincial forces.

The story of the dissolution of the *timar* system starts far earlier than 1700, in fact, as part of Ottoman realisation of the need for more massed firepower against the Austrians. The evolution to favour infantry and massed formal battles in western European arenas was not as evident in Danubian warfare. The Hungarian campaigns of 1684–99, for example, involved a series of sieges, with occasional massed open-field battles. We will see the same pattern unfold in the 1736–39 war in the contest over Belgrade. It is the case, however, that Ottoman officials were aware of the advantage of trained infantry as early as the Long War (1593–1606) between the Habsburgs and the Ottomans. Arming the peasantry, first as regiments of soldiers during that struggle, and then as provincial militias to protect themselves against rebel bands, complicated the scene. The relationship between displaced *timariots*, or other demobilised combatants, and the very serious extended rebellions (the Celali revolts) that were endemic between 1591 and 1611 and episodic thereafter, has long been a source of discussion among Ottoman historians. Ottoman fear of an armed countryside is palpable in the documents that survive which were issued in an attempt to control the distribution and use of handguns.[30] Sorting out how these potential soldiers were reabsorbed into the system, or forced into banditry, is one of the tasks of the military historian looking at the evolution of the Ottoman army after 1650. It has proved no less difficult for European military historians in like contexts.[31]

The more serious problem for Ottoman stability was that the *timar* land could no longer support the cavalryman's obligations to report to campaign. By the end of the seventeenth century, the revenues of entitlement from the *timar* were not sufficient. Many *timariots* simply could not afford the cost of campaigning. Those who did turn up for a campaign were ill-equipped to face the mercenary firepower of the Habsburgs, were increasingly reluctant to fight, and the first to want to leave to return to their holdings. The question of who owned the *timar* lands became decidedly muddy after the early 1600s, and the right to distribute them more diffuse. The tendency was to convert the fiefs into tax farms, and large numbers fell out of the jurisdiction of the *sipahis*, part of a general trend of the seventeenth century, by which revenue collection devolved to emerging gentry households, consistent with similar trends in other parts of the early modern world.[32] Beneficiaries of the evolving system were the imperial household, administrative officials and Janissaries, the latter coincidentally, who, as noted above, were sent into the provinces to quell the Celali revolts; by the mid-seventeenth century, reassigned *timars* regularly went to members of the Janissary corps, sometimes in lieu of pay.

Sipahis were initially drawn from the earliest and central Ottoman territories of Anatolia and parts of Rumelia. Tributary and frontier territories added at a later date were governed more loosely, or submitted annual tax revenues, called *salyaneli*. Here again, as with the Janissaries, we must be cautious about figures: during the long Süleymanic age (1520–66), *sipahis* numbered around 80,000; by the early 1600s, their number had dropped to half that amount. A realistic count of *timariots* would hover around 50,000 for smaller campaigns, and 80,000 for sultan-led campaigns, which were far fewer for most of the seventeenth century and non-existent in the eighteenth.[33] It would appear to tally with Marsigli's figure for the field army mobilised from the provincial *timariots* for 1683, roughly 65,000.[34] How many of them showed up is of course another story. A hundred years later, Ahmed Resmi noted a few stragglers in 1769, old men calling themselves '*timarlıs*', encumbered with too much baggage and too many retainers. No official Ottoman counts of soldiers for the 1768–74 war includes *timariots*. Instead, Janissaries were continuously demanding that they be given *timars*, as supplementary income.[35] Most of the remaining *timars* had long been converted into taxes to replace military service, or otherwise absorbed into tax revenue farms, and no longer represented soldiers required to report to the battlefront.

Frontier defence systems

What happened, then, to the potential military manpower represented in the *sipahis* in the century following the treaty of Karlowitz in 1699? This remains one of the most vexing problems in Ottoman military history. The broad band of frontier territory on all fronts required numerous levels of contracts for particular services for a given campaign, sometimes on a permanent tributary basis, for protection and/or tax relief, sometimes on a contingency basis. The latter might include siegework diggers (sappers), crews to repair roads and build bridges, or transport crews, i.e., wagons and drovers, as auxiliaries. These needs were most often met on the local level, by recruitment or coercion, often negotiable as part of the tax burden on villages and towns. The Ottoman sources use the term *serhad kulu*, 'frontier troops', sometimes for this kind of mobilisation, which might include cavalry frontier guards and permanent garrison troops, as well as campaign-contingent auxiliary corps. We will return to them shortly.

Another category of provincially-mustered troops which became a common feature of the battlefront in the eighteenth century were the local regiments (sometimes called *yerli kulu*, or *yerliye*) raised by governors and other officials of the provinces. The categories often overlap, and the distinctions are more obvious in the seventeenth than the eighteenth century, when central control over the provinces became more and more tenuous. These latter soldiers, recruited locally, became the private armies of ambitious local lords.[36]

The evolution from *sipahis* to private armies was accelerated by the Ottoman use of local irregular bands, mercenaries variously known as *levend*, *sarıca* and *sekban*, as early as the sixteenth century.[37] *Sarıca* and *sekban* refer to armed infantry musketeers, similar to the militias of Europe. They were drawn from among the *levend*, a term which originally connoted armed, vagrant, and landless peasants, and then evolved into its military usage. All functioned as 'independent soldiery companies'.[38] The mobilisation and demobilisation of these companies have been linked, as with the *sipahis*, to major revolts in the seventeenth century. Such roving bands, when organised into fighting forces in the eighteenth century, were called household *levend*, or state *levend*, the distinction being whether or not they were part of the provincial governors' forces (*kapu halkı*), paid from provincial revenues, or directly by the state (*miri levend*).[39] By the 1720s, the provincial governor was expected to arrive on the battlefield with 200 of his private entourage (*kapu halkı*) and 1,000 to

2,000 recruits, infantry and cavalry, the latter increasingly paid out of the inner treasury of the sultan. Specific battlefield statistics are still in want of study, but between 1683 and 1769, the number of mustered state militiamen, paid directly from the centre, rose from roughly 10,000 to 100,000, a remarkable change in the style of recruitment of manpower for the Ottoman army.[40]

This came about largely because of the need for the Ottomans to organise provincial security in cooperation with the state-appointed religious official, the *kadi*, to curb the abuses of the military administrative class and to control countryside violence.[41] That imperative had sent Janissaries into the provinces in the attempt to curb local revolts. I have alluded to the struggle to establish local notable households which was endemic in the eighteenth century, a struggle if not generated, certainly exacerbated by Ottoman attempts to maintain order. It was common practice for the Ottomans to eliminate the militia-turned-bandit by arming the countryside for its own protection, and then enlisting the resulting bands for the next campaign, what has been described as the Ottoman effectiveness in 'embodying within itself the potential forces of contention'.[42]

In military terms, this process resembles to a degree what John Lynn, in studying the policies of Louis XIII and XIV, has called the evolution from aggregate contract ('off the shelf') to 'state commission' armies.[43] State commission armies, according to Lynn, represented a stage between rapidly raised, unreliable contingency services and trained, standing troops, also a transition to conscription rather than contract. In the Ottoman context, I would argue, the loss of control over the distribution of the Janissary and *timariot* entitlements forced the evolution of a system which resembled the Bourbon case, especially after the crises around the sultan in 1700 determined that devolution was to continue. Over the next hundred years, the increased use of such troops was instrumental in the Ottoman ability to make the transition to a (modern) conscript army. Of course, in the Ottoman context the reliability and loyalty of such troops was an acute problem, especially as determined by the financial stability of the state, and its willingness to redistribute provincial revenues for the benefit of local officials. Two factors militated against making possible a full transition to a 'state commissioned' army: justifiable Ottoman paranoia of all forms of banditry and rebels, combined with an elite disdain for the upstart. Both impressions are reinforced by examining surviving Ottoman archival documentation, and by the official and semi-official chronicles that label all insubordination as rebellious.

Nationalist historiography of the Balkans and Republican Turkey once prevented a more realistic picture of the role and manipulation of the Ottoman rural flotsam and jetsam from emerging. Consider, for example, the word *haydut*, represented in Bulgarian historiography as a nationalist movement, '. . . the longest-lasting resistance to Ottoman rule in the Bulgarian lands . . . a symbol of the national spirit.'[44] Ottoman attempts to curb their excesses were continuous for two hundred years, but just as assuredly, those bands represented potential cannon fodder, and Habsburg, Romanov and Ottoman made use of them at various times. Ambitious local leaders also did so, and it is this tango of ambivalence that so characterises Ottoman eighteenth-century history. From 1700 to 1826, there is scant evidence of Ottoman attempts at settlement of borderlands for the purpose of defence, though, as we will see, the remaining lower Danube fortresses were regularly reinforced and refortified. The town of Vidin was reorganised shortly after 1699, for example, as it had become the pivotal fortress on the western end of the Belgrade–Azov line.

Various degrees of incorporation of rebellious ethnic forces into their regimental structures distinguish the Habsburg and Romanovs from the Ottomans in this later period. The Cossacks, frontier colonists par excellence and a free population that escaped serfdom by settling and defending the fringes of the Russian empire, could serve as one illustration. The process of settlement of the southern regions of Muscovy was accelerated after the Russo-Ottoman War of 1677–81, when new infantry regiments successfully defended Ukraine against the Ottomans. For example, the Russian foot soldiers in Belgorod region numbered under 8,000 in 1674; after 1681, the regiments included 17,000 men. Russian Field Marshal Rumiantsev is credited with reorganising the Cossacks into his regimental structure in the 1768–74 Russo-Ottoman War, part of an extensive Cossack settlement system implemented by the Romanovs and often vigorously resisted.[45] A plethora of historical obligations, based on hereditary service and land entitlements, conflicted with the new military demands in a territory that was sparsely populated and agriculturally poor. It also forced considerable negotiation between the state and the southern tier which was only partially solved by relocation of Russian and German civilian populations into the province of New Russia in the late mid-eighteenth century.[46]

The Habsburg military border was similar in its initial functional use as a buffer zone in ill-defined border regions, and as a long-term mechanism for drawing unruly and mixed ethnic populations into the imperial orbit, and settling them on abundantly available land. Military service was tied to land allotments in Croatian territories as early as 1535. Settlers were given basic autonomy, freedom to practise their own religion, and tax relief in exchange for permanent military service. Two regions developed: one centred on Karlstadt fortress between the Adriatic Sea and the Sava River; the other was Varadin between the Sava and Drava Rivers. The command of these border troops was regulated by the Hofskriegsrat first in Vienna, and then decentralised in the Inner Austrian administration in Graz. The troops numbered 5,913 in 1573. In the 1683–99 war, they served in an auxiliary capacity, raided Bosnia and extended their (and Habsburg) control over parts of southern Croatia. A similar system was established along the Tisza, Maros and Danube Rivers, the Slavonian Military Corridor in the Habsburg acquisitions after 1699, mostly made up of Serbian Orthodox refugees, who were also added to older districts, and represented a considerable light infantry force. Eighteenth-century reforms gradually incorporated Grenzer units into the imperial army, and the military frontier became a cheap source of trained troops for the state. Some 88,000 Grenzers served in the Seven Years War. Before its dissolution, the military border system extended 1,000 miles, 20 to 60 miles deep, from the Adriatic Sea to the Carpathian Mountains.[47]

Survival strategies

To suggest that Russian or Austrian frontier strategies were more successful than the Ottoman version is to lose sight of the human cost. The life of the Grenzer was miserable; they and their families 'lived continually on the brink of starvation'.[48] Attempts to incorporate the Orthodox Grenzers into Habsburg forces faced stiff resistance to the Catholic proselytising and revocation of special rights which accompanied such incorporation, and provoked rebellions in 1695, 1714, 1719, and 1728. In 1735, a serious insurrection in Varadin was directly related to attempts at conversion. The loss of frontier freedoms represented by more closely-regulated regimental structures was the root cause of a Slavonia-wide revolt in 1765.[49]

Tsarist colonisation of the Crimea and Caucasus was predicated on the same disdain and contempt for non-Russians as attributed to Ottoman elites, coupled with an intolerance for Muslims that meant the voluntary and involuntary exile of hundreds of thousands of Tatars and other Caucasian Muslim groups into Ottoman territory after 1783, when the Crimean Peninsula was annexed by Catherine II.[50] Apostasy to Islam of baptised Tatars (*Kriashens*) in the Volga region is an example of a strategy of resistance that occurred throughout the nineteenth century.[51]

Subversive strategies abounded, and the lines drawn by ethnicity, style of life (nomad, mountaineer, bandit), and religious persuasion of such groups in the period under discussion are ambiguous and multi-layered. This suggests that struggle was as much over centralisation and loss of autonomy as over religion. Military imperatives can be a great leveller, especially in the service of consolidation of territory. A standing army required a census: conscription and forced settlement often went hand in hand, although military colonies were freer than mobilised regiments, especially in Russia. Nonetheless, the ideology of unity and exhortation of Romanov and Habsburg, however fictive, was couched in religious garb: inclusion ('citizenship') implied Orthodoxy in the Romanov context, as it implied Catholicism in the Habsburg realms, and conversion was a matter of eighteenth-century dynastic strategy. Muslimness was assumed if one served in the Ottoman forces, a badge of exclusivity, if one was in the Ottoman circle, but conversion was simply not part of Ottoman Balkan military strategy in the eighteenth century as it so clearly was in the other two dynasties.

The demographics of exile described above, however, meant a tremendous influx of Muslim peoples into highly mixed territories in Anatolia and the Balkans, reflected in the increasing public clamour for a reiteration of Ottoman Muslimness by 1800. Shrinking boundaries, and mounting distrust of the loyalties of Christians forced an increasing reliance on the peoples from the peripheries of empire: on the exiles as well as on the long-standing use of Albanians and Kurdish tribal groups as the source of manpower for the militias of later campaigns. Particularly problematic in the east was the unreliability of religious sensibilities of the eastern frontiersmen, Muslim co-religionists for the most part, but historically heterodox and despised by the Ottoman elite. The volatility of this particular mix is exemplified in the problems of command and insubordination which most characterise the military campaigns of the late eighteenth century.

The northern frontier

The Ottomans preferred zones of influence, and clientage, to defensive corridors for their territories north of the Danube and the Black Sea. The Tatar Khanate, centred at Bahçesaray in the Crimean Peninsula, and the territories of Wallachia and Moldavia were required to contribute men and supplies to the major campaigns, the latter two in return for semi-autonomous, non-Muslim status. No more than a token presence of

Janissaries was legislated as part of the equation in the case of Moldavia and Wallachia. The Muslim Tatars, potential heirs to the Ottoman house, by contrast, were generally incited to raid enemy territories at the beginning of the campaign season, and remained useful at harassing supply trains and small columns of Russian soldiers even in the 1768–74 War. The relationship was ambivalent at best; the Khanate was in fact maintained by large influxes of money from Ottoman coffers, a system which guaranteed Ottoman continual interference in Bahçesaray politics. Once the Russians began the invasion of the Black Sea littoral and penetrated Moldavia on a regular basis, Tatar usefulness on the Danube was considerably lessened. After 1774, when Tatar independence was secured by the Russians as part of the Küçük Kaynarca treaty, the issue of Tatar survival was subsumed in the greater public debate and pressure to recover a large chunk of lost Muslim territory. This accounts for the sustained and debilitating attempts to recover the northern shores of the Black Sea that drove Ottoman foreign policy until 1792.[52]

Moldavia and Wallachia were essential for providing grain and livestock. In both, hereditary princes had been allowed to rule their Christian populations by contributing both men and supplies to the imperial campaigns. In practice, this meant they maintained their own private armies, and were susceptible to competing loyalties, the situation faced by the Habsburgs once they acquired a majority of the territory in Hungary after 1699. It is precisely in the period following 1700, especially after 1711, when Moldavians under Dimitrius Cantemir joined the Russian army at Prut, that the Ottomans attempted to alter the relationship in Moldavia and Wallachia by appointing ruling households from within the Istanbul Phanariot community, and by curbing, and sometimes eliminating, the local armies. Thereafter they could no longer depend on the manpower from those territories. The attempt at imposing central control demonstrates Ottoman awareness of how essential the Principalities were as a bufferzone, and as the breadbasket of the army and Istanbul, but the new regime forced them into other regions for mobilising campaign irregulars.[53] As to the numbers involved in 1683, Marsigli estimated the tributary troops in the field army at 28,000: 12,000 Tatars, 8,000 from Transylvania, and 4,000 each from Wallachia and Moldavia. He added another 10,000 Tatars who were paid to help in the defence of Buda.[54] The one notable Ottoman experiment in ethnic regiments, the Zaporozhian Cossack regiment of Silistre, was established only as part of Mahmud II's reforms of the 1820s, from Cossacks who had relocated into Dobruja.[55]

The southern frontier: Egypt and the Arab Provinces

Practically speaking, Egypt and the Arab/Kurdish provinces of the empire, predominantly Muslim, can be conceptualised as the southern Ottoman frontier, especially by 1800. The entire eighteenth century, especially after 1750, was largely characterised by massive campaigns on the northern frontier. What that led to in areas around Baghdad, Damascus, and the Levant coastal cities, was the emergence, and in some cases the re-emergence, of nomadic or semi-nomadised tribal groups and mountain men, such as the Türkmen, various Kurdish populations, the Bedouins, and the Druze. This was particularly troublesome for the empire, which depended on the success of the pilgrimage from Damascus to Mecca for both status and finances, and the stability of Egypt for its annual tax revenue, and for the 3,000 Janissaries normally supplied for large campaigns. The physical and financial neglect of the Janissary garrisons in cities of the southern tier seems to have been fairly profound. What arose in the void as a consequence will be discussed later, but some trends are worth noting. Istanbul relied on the dwindling numbers of Janissaries to make their own way. To do so turned them into entrepreneurs, bandits and even gun runners. City defence mechanisms involved creating local armies, such as the *yerliye* of Jerusalem, or the *mamluks*, of Cairo fame, who competed for local resources and attacked the Janissary garrisons on a regular basis. The centrally-appointed Ottoman governor generally walked into a maelstrom of competing militias, and could find himself besieged in his own fortress. Another consequence was the gradual sorting out of the locals (Arabs, Bedouins) from the 'Turks', troops sent from Istanbul or Anatolia. Hence, local rulers such as Cezzar Pasha, defender of Acre against Napoleon, created and financed their own mercenary armies, which very often incorporated foreign (including European) renegades, precisely because of the unreliability of local networks and loyalties. Hence the Arab provinces, and especially Cairo, were undergoing their own introduction to a different military culture at the same time as Istanbul, which would have different outcomes.[56]

Auxiliary forces: the religious factor

Ottoman use of Christian and Muslim auxiliaries as local militias, in garrisons and in the fleets on the Danube, was one which served them well on the European battlefronts. It was an integral part of the strategy of conquest of earlier centuries. Such auxiliary forces could occasionally acquire advantageous tax status, but more often were simply levies of the local population. These militia could be cavalry and infantry units, or local peasants assigned to many of the less glamorous tasks of warfare, such as building defences or digging ditches. The plethora of names, the range of definition and the longevity of the associations in the popular mind and in Ottoman documentation are phenomena familiar to those who write the military history of the early modern age, when shovels were far more numerous than cannons in fortress inventories. As an example, Murphey's close study of campaigns and sieges in the east in the seventeenth century are especially evocative: for the siege of Baghdad in 1638, 24,000 *beldar* (trench diggers) were levied from one in twenty Anatolian households, and marched to the front.[57] Karaferye *kaza* (present-day Greece) was ordered to supply 100 *beldars* for the 1768–74 war in February 1769, suggesting the continued practice of levying non-Muslim villages for auxiliary units.[58]

Although it is difficult to sort out the mobilisation and financing of the Ottoman borders after 1650, we do have information on their organisation of the Hungarian and Bulgarian fortress systems of the previous hundred years from which to draw some examples: *martolos*, generally assumed to be Slav, Christian cavalry, served in campaigns, as crews of river fleets, and even occasionally as paid garrison soldiers. They formed the backbone of the defence of the borders in the sixteenth century, along with the *akıncı*, who were irregular raiding parties. Such local troops were tax exempt, and the *martolos* sometimes received a salary.[59] Another example of a local group was the *voynuk* corps, largely in interior Bulgarian territories, especially around the Filibe–Sofia–Belgrade axis. Again, service meant tax exemption – these were border forces called into the service of the sultan on campaign.[60] Other troops, massed in Hungary more than Bulgaria, especially after the conquest in the sixteenth century, included the *Ulufeciyan-i süvari*, *beşlüs*, *farisan*, various terms for salaried cavalry regiments, most likely, but not always, Muslim. *Azabs*, in the Balkans and Hungary, were generally a lower class of infantry, distinguished from the *müstahfizes*, local fortress guards, who in the sixteenth century were probably non-Muslim, and *not* Janissaries. The *müstahfiz* corps was commanded by the *dizdar*, the local fortress commander. *Yerli* (meaning 'local') came into use in the seventeenth century for various local services, generally hired on a contingency basis, for building bridges, etc.

Yerli, like *müstahfiz*, implied non-Janissary until the mid-seventeenth century. Their tasks were various. *Yamak* (recruit), a term of opprobrium by the nineteenth century, came into greater use in the post-1700 period. Initially it appears to have meant civilians mobilised for specific tasks in the times of war. Equally, the term seems to have replaced the *gönüllüs* ('volunteers') of earlier centuries who wanted to join the Janissary regiments, and who were used for a variety of tasks. In reality, *yamaks* became the ill-paid, ill-trained Muslim garrison guards of Danubian, but more especially of the Black Sea and Bosphorus, fortresses who were the source of much military unrest by the end of the eighteenth century, and very often blamed for resisting attempts at reform, most especially in the downfall of Selim III in 1807–08.[61]

The Ottomans, at least initially, made a clear distinction between major fortresses, with resident regiments of Janissaries supplemented by Istanbul Janissaries at the inauguration of campaigns, and lesser lines of defence, largely the domain of local troops. Estimates for the sixteenth and seventeenth centuries indicate that one tenth of the manpower in the Hungarian fortresses represented salaried Janissaries. The rest were local.[62] Marsigli puts the garrison Janissaries at 21,426 at the time of the 1683 campaign.[63]

An official record for the period 1771–79 is illustrative of the continued commitment to manning the strategic fortresses: Belgrade averaged 5,798 resident guards between 1771 and 1779; Vidin 9,238 for the same period; Ochakov averaged 7,795, Bender 6,735.[64] By that time, deployment of military manpower was divided along religious lines – one must presume the *non*-Janissary salaried soldiers in these fortresses, the *yamaks* mentioned above, to be norminally Muslim, as they also were for the Black Sea fortresses. Still, the use of Christian auxiliaries continued, especially in areas where large Christian populations remained under Ottoman jurisdiction. They served, or were coerced into the necessary tasks of campaigning, be it manning the Danube fleet, digging siegeworks, or driving wagons. Ottoman documentation is very clear about their right to payment, or tax relief. In most surviving documentation, military occupations are not identified by religion, but by entitlement, or as a village collective with historic tax obligations and exemptions. Some relocation occurred: Gradeva argues that physical separation of Christian from Muslim populations occurred in Vidin, by creating separate districts, once the city returned to border status after 1699, as part of *serhad* strategy. Local troops were reassigned to outlying minor fortresses, perhaps another indication of separating soldiers of different religions.[65]

Realities of Danube warfare

Paper trails and life on the ground are two different realities, however, and daily life is almost impossible to recover. Murphey argues that the military historian's concentration on major confrontations such as St Gotthard in 1664 neglects the importance of frontier warfare, what he calls 'a pattern of uncontested or minimally contested offensives', common to both sides, in which the local border forces play a very considerable role. His example is the spring 1664 Ottoman offensive to recover and repair fortresses in territories damaged by Croatian and Hungarian winter raids, after the main Ottoman army had been demobilised. The unopposed sieges of Pecs and Ösijek by the *Ban* of Croatia, Nicholas Zriny, in January and February of 1664, although abandoned, caused considerable damage to two fortresses pivotal to the Ottoman defence system, and required the rebuilding of the outer walls at Pecs and the bridge at Ösijek, and large crews of mobilised local manpower over several months, a process not quite completed when Grand Vizier Fazıl Ahmed Paşa arrived with the army in mid-May. Zriny had also besieged Kanizsa on the Tisza River in April, which forced the army to repair the recovered fortress, and neutralise seven lesser fortresses of the Croatian border, by a series of swift attacks on them throughout July 1664. With the exception of a two-day siege at Pölöske, the other garrisons surrendered or fled. Conclusion of these encounters was generally accompanied by voluntary surrender, and the promised safe passage being integral to the negotiations. After the investment of Uyvar by the Ottomans in 1663–64, the conditions of surrender included supplying the retreating soldiers with grain and wagons for transport, as well as a letter to Habsburg Emperor Leopold confirming the bravery and determination of the garrison's resistance.[66] In this case, the Davids and Goliaths shared a code that operated when the balance of power was lopsided. This was probably as typical as the stories of violations of such a code which dot the narratives.

The human cost and dislocation of Danubian warfare can be glimpsed in the rare autobiographical accounts such as that of Osman Agha, a cavalryman captive of the Habsburgs for a number of years, whose tale best illustrates the nature and uncertainty of border warfare. As a young man, Osman participated in border raids around Temeşvar. When he matured and inherited some property, he equipped himself properly with horse and weapon, and was assigned as the head of a detachment of light cavalry with ten men under him. At that time, after the 1683 campaign, the Habsburgs were besieging the fortress line along the Tisza and Maros

Rivers, had attacked the fortresses of Çanad and Arad north of Temeşvar, then burned the perimeters and withdrawn. 'All the Serbian people who had run away became bandits or mercenaries attached to the Austrians . . . Almost every day, they would attack Temeşvar residents, killing or enslaving travellers, farmers and the like, taking them to their forts to be imprisoned, later ransomed. We, too, attacked their forts and took prisoners'. This was the situation when Osman's detachment was asked to deliver the Janissary salaries to Arad fortress. Their task was to escort the salary money as far as Lipova, a lesser fortress to the east of Arad on the Maros River, a ten-hour ride from Temeşvar. Osman was among 80 soldiers assigned for the task which they accomplished without incident.

It was cherry season in Lipova, so the detachment dallied an extra day to enjoy the abundance. The defence of the fortress, Osman declared, was reduced to '. . . less than three hundred men, more than a hundred having been killed in previous battles. In short, including us, there were fewer than 500 soldiers in the garrison . . . The Austrians had sixteen infantry and cavalry regiments.' It proved impossible for the visiting detachment to escape as over 2,000 Serbian bandits and other irregulars blocked the Maros bridge crossing. Osman and the others retreated to the fortress, after burning the bridge. The siege of the fortress began the second day before dawn. '. . . [T]hey began to pound the walls with cannons and mortars, and demolished a large enough section to get through by afternoon. A few of the company inside responded by filling the gap with pillows and mattresses.' As the Austrians closed, most of the defenders concentrated their fire on the breach, fighting with light cannons and hand guns, and leaving the other gates unattended. These were forced by the Serbian and Hungarian bandits and irregulars. The defenders found themselves besieged in the inner fortress, the outer perimeter and the town of some 2,000 houses having been burned. For three days and nights, the Austrians pounded the walls and picked off defenders from atop minarets and roofs. Osman's detachment was assigned the most dangerous part of the walls. Finally, the fortress raised the white flag. The defenders asked to be allowed to depart in peace for Temeşvar. This was refused, the Austrians reserving the right to take prisoners. Fighting was resumed for another two days, after which the fortress surrendered unconditionally.

'According to the agreement, men and women in the fort began to go out unarmed. We were watching the process from the battlements. Austrian soldiers had lined up from their army to the fort in two rows. The distance between the Austrian army and the fort was a quarter of an hour . . . The soldiers standing on the road forcibly pulled at the men, women and servants and stripped them. Those who resisted were killed and robbed of everything . . . Their carcasses were spread all over the road. However much the Austrian officers wanted to stop this, it was not possible. Even as we watched, generals on horseback shot and killed some of those who stripped and killed Muslims.' Osman and his companions stripped themselves so as to avoid a similar fate.

About sixty of the garrison fighting force were kept as captives for potential ransom, while the rest of the population of Lipova was reduced to slavery. Only women, the sick and the most poor were allowed to depart for Temeşvar. Osman and a companion were sent to officers in the General Prince Louis Guillaume de Bade regiment as captives. Osman's master, a Lieutenant Fischer, had as a servant a 'Saxon' girl from Erdel province, with whom 'he fulfilled some of his bodily needs'. She served as translator. Before long, the Lieutenant proposed that Osman go home to Temeşvar and arrange his ransom. Osman secured a guarantor in return for raising a second ransom, and actually returned to Temeşvar for that purpose. Five of them crossed an abandoned landscape to their homes. In the promised seven days, they returned to Austrian territory, with the ransom, only to find their captors had been reassigned to the area around Ösijek. Again, they marched across boundless wasteland, 'passing towns left abandoned and derelict by the heels of the Austrian soldiers, using the wells dug in their encampments, and encountering Serbian bandits, who questioned them, but did not harm them.' Then, they had the misfortune to run into some Hungarian bandits.[67] In the end, Osman was not released by his Austrian master, even after the Karlowitz treaty. He finally escaped after eleven years of slavery, seven of them spent in Vienna, and returned to a much changed Temeşvar, which most of his family had left, where he served the local governor as a translator for a number of years, and appears to have been involved in the commissions established to solve border disputes arising from the new treaty line. By the time Temeşvar itself surrendered to the Austrians in 1717, Osman had lost everything, including his wife. Afterwards, he served the Vidin commander as translator, then emigrated to Istanbul in 1724.[68]

Logistics of warfare on the Danube: the Belgrade–Azov fortress line

Osman Agha's story must have been typical in many ways for the period, although he appears to have come from a background of some wealth and education. His adventure could be duplicated for the entire fortress system of the Belgrade–Azov line, the one which will preoccupy us throughout the period under study: Belgrade, Vidin, after 1699, Rusçuk, Silistre, İbrail, Tulcea, İsmail, Kilya, Hotin, Bender, Akkirman, Kilburun, Ochakov, Yenikale and Azov. Edirne served as the mobilisation and supply base for Danubian campaigns; Sinop served as the base and port for transport to the northern Black Sea garrisons; Salonika served that purpose for campaigns into Greece. The Black Sea and the Danube were absolutely crucial for the movement of supplies for the army, and blockades and inaccessibility of fortresses are determinants of failure in the campaigns of the eighteenth century.

The logistics of maintaining such a long frontier, as well as supplying the larger campaigns in the field, was an undertaking of enormous proportions for any pre-modern dynasty, but particularly so in the geographical context we have described. To the issues of terrain, distances, and infrastructure must be added the question of unpredictable seasonal variations, dry and wet, which plagued all military reliance on grain. For the Ottomans to organise and go on campaign was a cumbersome and lengthy process. The call to arms was sent out on the main campaign routes to Anatolia and the Balkans in December of the year before the campaign, often signalling to opponents the intention before the declaration of war.

Troops in Istanbul mobilised and left the city in early spring, while those raised from the countryside were ordered to join the main army on the march. The tremendous distances, and an average daily march of no more than ten miles, meant that reaching the actual battlefields in Hungary or in eastern Anatolia might not occur until mid-June.[69] From Edirne to the Drava crossing in the spring of 1663, the army took 85 days. This included a sixteen-day stop in Sofia to pasture the animals on new fodder and wait for troops to arrive, and another eleven days in Belgrade to gather supplies. By the time the army reached Esztergom to make the final Danube river crossing to Uyvar, 119 days had passed.[70] Major confrontations were often confined to July through October, when field conditions and the lack of fodder generally forced the suspension of hostilities.

A particular genius at logistical systems is generally acknowledged as part of the Ottoman success even well into the period under discussion. While the story of changes and dissolution of the system belong to later chapters, a few observations seem in order here. Overabundance of both men and supplies was one of the characteristics of the army organisation, and one that might have had an impact on the outcome of particular battles, such as Zenta in 1697.[71] Excess of manpower, in particular, is one of the complaints of Ottoman observers in the eighteenth century, a phenomenon which tended to overwhelm even well thought-out supply systems. Commanders and provincial pashas came with large entourages, in imitation of the grand vizierial staff, which included the bureau chief of most of the central offices of the bureaucracy. Campsites were also plagued with followers of all stripes, reckoned in the thousands, who set up stalls and coffee shops. By contrast, though camp followers were a persistent problem in European armies of the eighteenth century, frugality and mobility had become an essential part of their military strategy.

The Ottoman soldier, even at the worst of times, was often better fed than his opponents. Sharing a meal with one's soldiers remained an important part of Ottoman military ceremonial, and keeping the Janissaries well fed and supplied was a constant of Ottoman campaign strategy. Bread and/or biscuit were imperatives, with rice and mutton, and barley for individual horses and pack animals forming part of the rations. For the 1683 campaign, this probably meant as much as 32,000 lbs of meat and 60,000 loaves of bread per day to maintain the troops.[72] In 1768, the initial requisition of biscuit (*peksimed*) was for 22,400,000 kilograms from Istanbul, Gallipoli and İsakcı, the main supply centre at the mouth of the Danube. These central depots were supplied by biscuit produced by villages all over the empire.[73]

Camels, mules, water buffalo and oxen drew both supply wagons and artillery, the largest of the cannon requiring up to twenty oxen. The successful Baghdad campaign of 1638–39 required 77,444 camels carrying 542,113 *kile* (almost 14,000,000 kilograms) of grain, primarily barley for fodder, in territories notorious for their harshness and scarcity.[74] In one report for 1769, the Chief Drover reported that 962,353 camels and 52,578 mules were leased for the campaign season, the camels assigned to the imperial army and the grand vizier, while the mules were primarily for the transport of supplies. Of these camels, 48,306 were distributed to the entourage and officials of the grand vizier alone.[75]

Supporting the numbers described above forced early on the creation of a system of warehouses and well-stocked way-stations (*menzils*), a responsibility of the towns and villages of the area, which were theoretically paid at fixed prices for the goods they brought to the army, or whose taxes reflected their obligation to military supply. This is in keeping with the general principle of communal obligation, or guarantee, which operated in all transactions between the state and subjects. Villagers were obliged

PLATE 6 *Examples of Ottoman transport circa 1700 (Marsigli)* (L'etat militaire de l'empire Ottoman: ses progrès et sa décadence/par Mr. le comte de Marsigli. *A La Have: Chez Pierre Gosse & Jean Neaulme, Pierre de Hondt, Adrien Moetjens: A. Amsterdam: Chez Herm Uytwerf, Franc. Changuion, 1732).*

to stand surety for local militias, should they desert, for way-stations, for grain, for wagons, for drovers, among a myriad other expectations around pre-modern communities and warfare. There were just as many opportunities for the entrepreneur as there was the potential for ruin of village economies. The way-stations' primary purpose was to stable spare horses to maintain speedy communications with Istanbul. For the post-1700 period, military supply and its impact on the local economies is a vastly under-studied field, rife with historians' assumptions about extractions, devastation and oppression. One late account for the way-stations on the road from Kırkkilise in Thrace to the mouth of the Danube at İsmail, broken into ten stages gauged at 97 hours march, records both the central state and village contributions (*imdadiye*, campaign tax). The total *imdadiye* was 49,500 *kuruş*, and the additional central treasury contribution was 34,500 *kuruş*: not exactly matching funds, but not insignificant. Villages would have been required to stock the way-stations with other supplies as well as horses. By 1700, pack animals were supplied on the basis of rent and wages of drovers. By 1800, cash substitution and market prices had become the order of the day for much of what had previously been exchanged in kind, and as tax relief. This discussion of the impact of imperial campaigns on local populations will be continued in subsequent chapters.[76]

At least until 1800, representatives from all the guilds of Istanbul were required to accompany the army on campaign. In an age when all expected to be tapped to support the campaigns in some fashion, the community of merchants and tradesmen were obligated to do their share, both by producing a collective levy, and by accompanying the troops, with a tent and goods to sell. In theory, they stood to make a profit on the battlefront. For the proposed Persian campaign of 1730, for example, one or two representatives of 28 trades with 85 tents were required to report for duty with the cash contribution of each guild, called the *ordu akçesi*. Included were leather workers, saddlemakers, butchers, bakers, grocers who supplied dried and fresh nuts and vegetables, mattress and blanket makers, soup makers, and blacksmiths of various sorts, who produced horseshoes, utensils and cauldrons, etc. In 1730, however, the trades collective, while willing to produce the required levy, officially protested against the misuse of the funds by the government when the campaign was delayed, and demanded the return of their *ordu akçesi* in the Istanbul court – one of the underlying explanations for their participation in the Patrona Halil revolt of 1730, which is almost invariably described as a reactionary revolt of Muslim fanatics against the westernisation of the court of Ahmed III (1703–30).[77] The targets of the mob were indeed the fancy, western-style pleasure palaces of the Ottoman court, but the real issue was elite excess and inequity. Later chapters will assess the call for reform represented by the post-1700 revolts in the capital of the empire, and the success or failure of the Ottoman bureaucracy in responding to the crises.

Campaign financing

Imperial household budgets and the general financing to sustain the dynasty have received considerable scholarly attention in the last decade. Before turning to the campaigns of the eighteenth century, I would like to point to a number of the significant trends which recent studies have elucidated. First and foremost, it must be stated that all Ottoman reform efforts were directed at resisting the growth of rival power bases, which were considered a threat, and thus the fiction of *miri* or state land was maintained, in order to mask the real growth of private property in the eighteenth century. Tax devolution to provincial tax farmers shifted from annual to lifetime revenue farms (*malikane*), bid for at periodic auctions.[78] This had the dual effect of further reducing the actual return to the imperial treasury and creating a gentry class of brokers in the countryside, who were powerful enough to negotiate their share in the war effort, be it for men or supplies. While full information is scarce and may never prove to be completely recoverable, the central treasury budgets for the 1500–1700 period appear not to have risen or fallen dramatically, causing Murphey to argue for a degree of continuity unaffected by warfare, although he does acknowledge that the sultan's Inner Treasury (Privy Purse) tended to be tapped to cover campaign expenses.[79] The budgets extant for the 1768–74 war period, again fragmentary, indicate significant minting of coins, and the financing of the provincial non-Janissary militias almost exclusively from Mustafa III's (1757–74) private purse funds.[80] I would argue that these trends, accelerated by the demands of increased army size and more sophisticated arms of the eighteenth-century battlefield, led directly to Selim III's moment of bankruptcy, which prevented the full implementation of his planned reforms after 1792. Ottoman military reforms of the later period are invariably accompanied by treasury reform and an acute shortage of cash.

The imperial Janissary troops in the fortresses were always paid, although often late, by the normal tri-monthly increments of the system which was underwritten by the non-Muslim poll tax (*cizye*). Local garrison forces were supported in a variety of ways, but often by defining a group of tax farms as an *ocaklık*, revenues assigned to the wages and maintenance of the garrison forces (*ocak*, or regiment). The Bosnian militia, with a proud history of service with the empire, were organised in this fashion; so too were fortresses like Vidin, particularly after great influxes of Janissaries following Vidin's reversion to a border fortress in our period.[81] By the end of the eighteenth century, the Janissaries of the border fortresses were supported by a combination of the poll tax and tax farm revenues (*mukataa*) from Muslim and non-Muslim alike, both gathered from far-flung districts. Considerable [re]organisation and ad hoc arrangements were needed in order to get the job done. The impact of declining revenues increased the encroachment on provincial tax bases which supported the local militia arrangements.

Shortage of money meant that the tax contributions which village collectives had often paid in kind, be it for horses, wagons, grains or other supplies, were converted into cash substitutes, a cycle which Balkan historians have noted increased village indebtedness and enriched local power brokers. A system which involved an Ottoman official, the *mubaayacı*, or state purchaser, who was instrumental to this process, has received little study. These individuals determined the village contributions by distributing them over a tax base, saw to the payment and shipping of war supplies, most particularly grain, and attempted to maintain the balance between state needs and village realities.[82] This function was devolved even further to provincial officials in the eighteenth century, though occasionally centrally-appointed overseers were also sent out as inspectors, which was a potential source of aggrandisement and abuse. One alternative for the villager was simply to flee, and join the ubiquitous bands of tax rebels, who then turned up on the battlefront as recruits.

Another way to finance war-related enterprises was to give local officials, governors and their representatives (district tax and law enforcement officials: *mütesellims*, or *mutasarrıfs*) the right to levy provincial taxes. These included a bewildering variety of extraordinary taxes, which were reorganised by 1700 into a levy of village households, called *avarizhane*. The local judge, combined with a local official (elected, state or self-appointed – all three were possible), determined the equitable distribution of such taxes across the tax base. Such extraordinary taxes were assessed generally on a six-month basis in the provinces and might include paying for visits of pashas on the way to the battlefront, support of local officials in fulfilling their duties, and a host of other minor exactions.

Emerging out of the difficulties of financing the campaigns of 1683 was another category of taxation to assist provincial officials called to war. This was the *imdadiye*, or *imdad-i seferiye*, the extraordinary campaign tax previously mentioned. The *imdadiye* was a right of taxation granted in response to their pleas that they could no longer finance their entourages out of their revenues. The campaign tax was imposed for specific campaigns, and attached to particular commanders, as a means of raising troops and supplies. It, like all such extraordinary taxes, became ordinary over time. Thus, as our study opens, the right of extraction from local revenue sources, most stipulated for the use of the appointed governor, or equivalent, was in the process of devolving into the hands of a new class of local magnates.

That the sultan and his court might be more than willing to tap this rising wealth in times of need should be familiar to military historians. The sultan viewed the campaign tax as a loan: records indicate an increasing reliance on the governor's personal contribution to war efforts over the eighteenth century, and attempts by the sultan to recover the 'loan'. Confiscation of estates was also practised, for those either dismissed from office, or who failed on the battlefield. To counter the exactive power of the state, more and more of taxable territory was alienated by converting it to charitable uses. Individuals of means set aside their estates by creating endowments (*vakfs*). Such tax-exempt, large-scale charities offered the corollary benefit of social welfare to the Muslim community at large, in hospitals and schools, but impoverished the central treasury. The royal household itself, especially the women of the harem, had, in earlier centuries, supported war efforts with contributions from their own endowments.[83]

Protection from state oppression came also in the guise of entitlements, and goes a long way to explain the continuation of the Janissary system. Documentary entitlements, be it the right to pay and rations of the Janissaries (*esame*), or licences (patents) of privilege (*berat*) for almost any category of service and trade, could imply tax exemption, or status, and were the badges of membership in Ottoman pre-1800 society. Each group took care of its own: each Janissary regiment had its own treasury, and court cases are full of inheritances recorded as reverting to individual regimental treasurers. At the village level, appeals for tax relief took the form of judicial affidavits (from provincial courts, *hüccets*) which appealed to the sultan's justice, and often received attention. Increasingly, the study of village organisations reveals similar types of collective protection which developed in response to increasing tax coercion. Regions in the war zones and border fortresses were often exempt from extraordinary impositions as they already supported local garrisons, in the *ocaklık* system described above.

Ottoman logistics depended on the will of the population to participate in the supply system as it has been described. Archival documents reveal a plethora of obligations concerning all manner of military purchase and supply, which cemented the contractual relationship of sultan and subject. In other words, there were some ways of negotiating respite from the long arm of grievous exactions, at least for those who could afford it, or had access to capital or a network of influence. Little could otherwise prevent the strong-arm tactics imposed on peasants and townsmen by rapacious overlords.

These were the general conditions of the Ottoman military and its financial base at the opening of our more detailed study of the eighteenth-century campaigns. Western sources universally accord Grand Vizier Baltacı Mehmed Pasha's victory over Peter the Great at Prut in late July 1711 as owing to the overwhelming forces he had at his command, which outnumbered those of the Czar by four to one. Mehmed Pasha was severely criticised, indeed ultimately dismissed, by the sultan, for having let the Russians escape stiffer penalties. As we turn to the Prut campaign in order to discuss the Habsburg, Romanov and Ottoman confrontations on the Danube and Black Sea in the first half of the eighteenth century, we may discern other reasons for Mehmed Pasha's reticence.

Notes

1 Peter F. Sugar, 'A near-perfect military society', in *War: A Historical, Political and Social Study*, L.L. Farrar, ed. (Santa Barbara: UC Press, 1978), 104.

2 Ibid., 104.

3 Ibid., 96–7.

4 Halil İnalcık and Donald Quataert, eds, *An Economic and Social History of the Ottoman Empire* (Cambridge: CUP, 1994), 424–31.

5 David Chandler, *The Art of Warfare in the Age of Marlborough* (London: Batsford, 1976), 74.

6 Jeremy Black, ed. *European Warfare 1660–1815* (London: UCL Press, 1994), 70.

7 John A. Lynn, *Giant of the Grand Siècle: The French Army 1610–1715* (Cambridge: CUP, 1997); John Keep, *Soldiers of the Tsar: Army and Society in Russia 1462–1874* (Oxford: Clarendon, 1985); Rhoads Murphey, *Ottoman Warfare 1500–1700* (London: UCL Press, 1999); Karen Barkey and Mark van Hagen, eds, *After Empire: Multiethnic Societies and Nation-building* (Boulder, CO: Westview Press, 1997).

8 Charles Tilly, generally, in *Coercion, Capital and European States, AD 990–1990* (Oxford: Blackwell, 1990), but also Brian M. Downing, *The Military Evolution and Political Change: Origins of Democracy and Autocracy in Early Modern Europe* (Princeton, NJ: PUP, 1992).

9 Collins, James B., *Fiscal Limits of Absolutism* (Berkeley: UC Press, 1988), 214.

10 Selim Deringil, *The Well-Protected Domains: Ideology and the Legitimation of Power in the Ottoman Empire, 1876–1909* (London: I.B. Tauris, 1998).

11 Janissary is a word that covers a wide variety of soldiers: imperial Janissaries consisted of infantry and four regiments of cavalry, also referred to as *Sipahis*, not be confused with the *sipahis*, holders of *timars*, discussed later in the chapter. Some statistics combine them, others separate the two. Furthermore, a full range of auxiliary forces, such as sappers, miners, and artillerymen, were also members of the corps. Official Ottoman records kept them strictly separate, but foreign observers were less in tune to the nuances. Hence, we have the wide variance in numbers. The auxiliaries will be taken up in due course. Janissaries of all sorts served in all the garrisons and will also be discussed later.

12 Thomas Thornton, *The Present State of Turkey* (London, 1807), 173–74. English Ambassador Sir James Porter's *Observations on the Religion, Law, Government and Manners of the Turks* was published in London in 1768. Peyssonnel and Tott were both French consuls in Ottoman territory, and authors of studies of the Ottomans; Tott's *Memoirs of Baron de Tott*, published in English and French in 1785–86, was a European best-seller. Eton's famous diatribe against the Turks, *A Survey of the Turkish Empire* (London, 1798), long served as a major source of information for nineteenth-century European historians. An intensive exploration of the Ottoman archives of the kind I undertook for the 1768–74 war needs to be undertaken for all the subsequent confrontations of the pre-Tanzimat period, to be more secure in our understanding of the size of this force.

13 Gulda Káldy-Nagy, 'The first centuries of the Ottoman military organization', *Acta Orientalia Academiae Scientiarum Hungaricae* 31 (1977), 167.

14 Halil İnalcık, 'Istanbul', *EI2*, CD edition.

15 Luigi Marsigli, *Stato militare dell'Imperio Ottomano. État militaire de l'Empire ottoman* (The Hague, 1732), 183. Marsigli's work, full of lists of Ottoman forces, remains curiously difficult to use for specific estimates, because of the confusion of terminology, but the work is full of interesting details and pictures of camp life and Ottoman ordnance.

16 Murphey, *Ottoman Warfare*, 48, fn. 21.

17 John Stoye, *The Siege of Vienna* (New York: Holt, Rinehart and Winston, 1964), 20–22.

18 See my *An Ottoman Statesman in War and Peace* (Leiden, 1995), 134–36, for a discussion of the problem of statistics for the 1768–74 war.

19 Murphey, *Ottoman Warfare*, 49.

20 Vasiliki Papoulia, 'The impact of *devshirme* on Greek society', in Gunther E. Rothenberg, Béla K. Király and Peter F. Sugar, eds, *East Central European Society and War in the Pre-Revolutionary Eighteenth Century* (Boulder, CO: Social Science Monographs, 1982), 549–62. 'Certainly, it cannot be coincidental that the evolution of the Ottoman society is characterized by a deep antithesis that led to the creation of fighting gangs in mountainous regions whose members eventually became the liberation armies of the Christian population', 557, represents the general Balkan view of the *devşirme* system.

21 Murphey, *Ottoman Warfare*, 45–46.

22 The debate about the nature and impact of the *devşirme* occurs within the larger debate about Mediterranean slavery in general, and Middle Eastern military slavery in particular. Enslavement was generally legally justifiable in the context of newly conquered territories. See Irina E. Petrosjan, 'The *Mabda-i kanun-i Yeniçeri ocaği tarihi* on the system of *devşirme* in György, Kara, ed., *Between the Danube and the Caucasus* (Budapest: Akadémiai Kiadó, 1987), 217–27 for a 1606 view of the system.

23 André Raymond, *Le Caire des Janissaires: l'apogée de la ville ottomane sous 'Abd al-Rahman Katkhud* (Paris: CNRS, 1995); also, 'Soldiers in trade: the case of Ottoman Cairo', *British Journal of Middle Eastern Studies* 18 (1991), 16–37; Antonios Anastasopoulos, 'Imperial institutions and local communities: Ottoman Karaferye, 1758–74', PhD, Cambridge University, 1999; Evgenii Radushev, 'Les dépenses locales dans l'empire ottoman au XVIIIe siècle selon des données de registres de cadi de Ruse, Vidin et Sofia', *Études balkaniques* 1980/3, 74–94.

24 Tott, *Memoirs*, vol. 1, pt. 2, 70.

25 Rhoads Murphey, 'Continuity and discontinuity in Ottoman administrative theory and practice during the late seventeenth century', *Poetics Today* 14 (1993), 425, from the historian Rashid's history.

26 Sadullah Enverî, *Tarih* (1780), Istanbul University MS TY 5994, folios 328–29. Enverî spent the entire 1768–74 Russo-Ottoman War on the battlefront.

27 İ. Hakkı Uzunçarşılı, *Osmanlı Devleti Teşkilatından Kapukulu Ocakları*. 2 vols. Ankara: Türk Tarih Kurumu, 1944, vol. 1, 342–43. On Sarı Mehmed, see Virginia Aksan, 'Ottoman political writing, 1768–1808', *International Journal of Middle East Studies* 25 (1993), 53–69, and W.L. Wright (ed.), *Ottoman Statecraft*, an English translation of Sarı Mehmed's *Counsel for Viziers and Commanders (Nasaih ül-Vüzera ve ül-Ümera)* (Princeton, NJ: Princeton University Press, 1935). Distinguished Seyyids (Muslims with honourable lineages) in the thousands were also given a gift of 1,000 *akçes*

each. See Halil Sahilliağlu, 'The income and the expenditure of the Ottoman treasury between 1683 and 1740', in *Studies on Ottoman Economic and Social History* (Istanbul: Organisation of the Islamic Conference, IRCICA, 1999, 76–77).

28 Abdülkadir Özcan (ed.), *Eyyubî Efendi Kanûnnâmesi* (Istanbul: Eren, 1994), 47–48, also reproduced in Uzunçarşılı, *Kapukulu Ocakları*, vol. 1, 622–23.

29 J. Deny, 'Tīmār', *Encyclopedia of Islam*, vol. 8, 772 (767–76). I am using the term *timar* here to apply to all categories of such fief-style land-holding, which had different names depending on the size of the grant. The English term *timariot* and *sipahi* are commonly encountered in early European texts about the Janissaries.

30 Karen Barkey, *Bandits and Bureaucrats: The Ottoman Route to State Centralization* (Ithaca: Cornell University Press, 1994); Halil İnalcık, 'Military and fiscal transformation in the Ottoman Empire, 1600–1700', *Archivum Ottomanicum* 6 (1980), 283–337; R.C. Jennings, 'Firearms, bandits and gun-control: some evidence on Ottoman policy towards firearms in the possession of the *Reaya* from judicial records of Kayseri, 1600–27', *Archivum Ottomanicum* 6 (1980), 339–58.

31 McNeill's observation about drill and discipline alludes to the problem: 'Drill, dull and repetitious though it may seem, readily welded a miscellaneous collection of men, recruited often from the dregs of civil society, into a coherent community . . .' William H. McNeill, *Pursuit of Power: Technology, Armed Force and Society Since A.D. 1000* (Chicago: University of Chicago Press, 1982), 131; also his drill book, *Keeping Together in Time: Dance and Drill in Ottoman History* (Cambridge, MA: Harvard University Press, 1996).

32 Linda T. Darling, *Revenue-Raising and Legitimacy: Tax Collection and Finance Administration in the Ottoman Empire, 1560–1660* (Leiden: Brill, 1996), 120–22. Yusuf Oğuzoğlu, 'Osmanlı şehirlerindeki askerlerin ekonomik durumuna ilişkin bazı bilgiler', *Birinci Asker Tarih Semineri Bildirileri*, vol. 2 (Ankara: Genelkurmay Basımevi, 1983), 169–76.

33 Halil İnalcık, *The Ottoman Empire* (London: Orion, 1973), 47–48; Murphey, *Ottoman Warfare*, 41–42.

34 Marsigli, *L'État militaire*, 183.

35 Aksan, *An Ottoman Statesman*, 190–92.

36 The Ottoman term for these provincial figures is *ayan*, sometimes derebey.

37 İnalcık, 'Military and fiscal transformation', 292; Mustafa Cezar, *Osmanlı Tarihinde Levendler* (Istanbul: Güzel Sanatlar Akademisi, 1965), 351–56.

38 İnalcık, 'Military and fiscal transformation', 295.

39 Cezar, *Levendler*, 214–16.

40 See Virginia H. Aksan, 'Whatever happened to the janissaries?' *War in History* 5 (1998), 23–36, and 'Ottoman military recruitment strategies in the late eighteenth century', in Erik J. Zürcher (ed.), *Arming the State: Military Conscription in the Middle East and Central Asia 1775–1925* (London: Tauris/Palgrave Macmillan, 1999), 21–39.

41 İnalcık, 'Military and fiscal transformation', 301, 307–08.

42 Karen Barkey, *Bandits and Bureaucrats: The Ottoman Route to State Centralization* (Ithaca: Cornell University Press, 1994), 241.

43 John A. Lynn, *The Giant of the Grand Siècle: The French Army 1610–1715*. Cambridge: Cambridge University Press, 1997, 5–9.

44 Bistra Cvetkova, 'The Bulgarian haiduk movement in the 15th–18th centuries', in Rothenberg, Kiraly and Sugar (eds), *East Central European Society and War in the Pre-Revolutionary Eighteenth Century* (New York: Columbia University Press, 1982), 301–37.

45 Zenon Kohut, *Russian Centralism and Ukrainian Autonomy* (Cambridge, MA: Harvard University Press, 1988), 110–11, 115–20.

46 Carol Belkin Stevens, *Soldiers on the Steppe: Army Reform and Social Change in Early Modern Russia* (DeKalb: Northern Illinois University Press, 1995), 82.

47 Gunther E. Rothenberg, 'The Habsburg military border system: some reconsiderations', in *War and Society in East Central Europe, vol. 1: Special Topics and Generalizations on the Eighteenth and Nineteenth Centuries*, Béla K. Király and Gunther E. Rothenberg, eds. (New York: Columbia University Press, 1979), 364–69. This discussion is a recap of his two books on the subject, indispensable for any consideration of the Grenzer: the figure of 88,000 (an estimate of one-quarter of the army), is from Rothenberg, *The Military Border in Croatia, 1740–1881: A Study of an Imperial Institution* (Chicago: University of Chicago Press, 1966), 40–41.

48 Rothenberg, 'Habsburg', 373.

49 Ibid., 366–69.

50 Austin Lee Jersild, 'From savagery to citizenship: Caucasian mountaineers and Muslims in the Russian Empire', in Daniel R. Brower and Edward J. Lazzerini, eds., *Russia's Orient: Imperial Borderlands and People, 1700–1917* (Bloomington, IN: Indiana University Press, 1997), 102. Some 300,000 Tatars were exiled from the Crimea in the eighteenth century; perhaps as many as 500,000 Circassians from the Russian-held Caucasus in the 1860s (103).

51 Agnes Kefeli, 'Constructing an Islamic identity: the case of Elyshevo village in the nineteenth century', *Russia's Orient*, 271–91.

52 This story is succinctly told in Michael Khodarkovsky's *Russia's Steppe Frontier: The Making of a Colonial Empire 1500–1800* (Bloomington, IN: Indiana University Press, 2004).

53 The Phanariot regime is described in detail in Illie Ceauşescu, 'Military aspects of the national struggle of the Romanian Principalities in the eighteenth century', in *East Central European Society and War in the Pre-Revolutionary Eighteenth Century*, 371–85. Hospodar Nicolai Mavrogheni is noted as an exception, having fielded an army in service to the Ottomans 1786–90 (372).

54 Marsigli, *État militaire*, 135, 184.

55 Avigdor Levy, 'Formalization of Cossack service under Ottoman rule', in *East Central European Society and War in the Pre-Revolutionary Eighteenth Century*, 491–505.

56 Amnon Cohen, 'The Army in Palestine in the eighteenth century – sources of its weakness and strength', *British School of Oriental and African Studies Bulletin* 34 (1971), 36–55.

57 Murphey, *Ottoman Warfare*, 116. By the time the siege was finished, they had dug five miles of trench, at 50 yards a day.

58 Antonios Anastasopoulos, 'Imperial institutions and local communities: Ottoman Karaferye 1758–74', PhD (Cambridge, 1999), 146.

59 Evgenii Radushev, 'Ottoman border periphery (*serhad*) in the Nikopol Vilayet, first half of the sixteenth century', *Études balkaniques* (1995), 141–60 (150–51). '. . . [O]nly Christians served in the ranks of *martolos* along the entire Ottoman *serhad* from Southern Hungary to Giurgiu.'

60 Radushev, 'Ottoman border periphery', 154.

61 Thornton, *The Present State of Turkey* (1807) 171–72, claims they were unpaid, and '. . . though enrolled, are not embodied into *odas* [equivalent of company], are dispersed throughout the empire, living as burghers, mixed with the people, and following different trades and professions, or idle vagabonds, or at best but laboring peasants.'

62 Most of the information here on Hungary can be found in Klara Hegyi, 'The Ottoman military force in Hungary', in *Hungarian–Ottoman Military and Diplomatic Relations in the Age of Süleyman the Great*, Géza Dávid and Pál Fodor, eds. (Budapest: Loránd Eötvös University, 1994), 131–48; esp. 136–38. Mark Stein also reviews the categories of occupants of Hungarian fortresses in his 'Seventeenth century Ottoman fortresses and garrisons on the Habsburg frontier', PhD, Chicago, 2001, a study of Kanizsa and Uyvar.

63 Marsigli, *État militaire*, 181. This is in addition to his 30,000 on campaign.

64 BA, D.BŞM 4274, an account of salaries, ration and clothing allowance of the resident guards in the *serhad* fortresses, beginning date 1185 (1771), ending date various, but latest is 1193 (1779).

65 Rossitsa Gradeva, 'War and peace along the Danube: Vidin at the end of the seventeenth century', in Kate Fleet (ed.), The Ottomans and the Sea, *Oriente Moderno* n.s. XX (LXXXI) no. 1 (2001): 149–75.

66 Murphey, *Ottoman Warfare*, 126–28, from an Ottoman chronicle by Mühürdar.

67 Osman of Temeşvar, *Kendi kalemiyle Temeşvarlı Osman Ağa*, ed. Harun Tolasa (Konya, 1986), 44–55. A German translation was published (Cambridge, 1980): *Autobiographie der Dolmetschers Osman Ağa aus Temeswar*. Frédéric Hitzel, translator of the French edition of Osman's story, uses it to discuss the problem of war captives in general on the Muslim–Christian line: 'Osmân Ağa, captif ottoman dans l'Empire des Habsbourg à la fin du XVIIe siècle', *Turcica* 33 (2001), 191–212. The French translation by Hitzel is *Prisonnier des infidèles. Un soldat ottoman dans l'Empire des Habsbourgs* (Arles: Sindbad, 1998).

68 Osman of Temeşvar, *Kendi kalemiyle*, 21.

69 Perjes, Geza 'Army provisioning, logistics and strategy in the second half of the seventeenth century' (*Acta Historica* 16, 1970, 1–51), estimates 20 kilometers per day; Murphey, *Ottoman Warfare*, 65, contrasts rearguard wagon trains who managed 2 miles per hour to crack vanguard troops who managed 3 miles per hour, with 4–6 hours/day an average. Generally, European armies counted on 5 miles per day.

70 Murphey, *Ottoman Warfare*, 66–70.

71 Murphey, *Ottoman Warfare*, '. . . the encumbrance resulting from oversupply in both men and equipment seems to have played a critical role in the Ottoman defeat', 73.

72 Stoye, *The Siege of Vienna* (New York: Holt, Rinehart and Winston, 1964), 20–22.

73 Virginia H. Aksan, 'Feeding the Ottoman troops on the Danube 1768–74', *War and Society* 13 (1995), 1–14, 6.

74 Murphey, *Ottoman Warfare*, 77. 1 *kile* is equal to 25.659 kilograms.

75 BA, D.BŞM 3949, 4 and summary, 16–17. The number of camels seems fantastic, and may represent a multiplicity of service (lease or rent) of individual camels. Still, the statistic suggests the scale of the enterprise. In fact, the statistic is based on number of days service; the figure was probably more like 5,000–10,000 camels: they were reckoning 237 days for the campaign, at 20 akçe a day, but length of service was various. Discussion in Murphey's book is useful for those interested in calculations about camel loads and the difficulties of transport: *Ottoman Warfare*, 75–78.

76 BA, D.BŞM 5665, dated 1787, the beginning of the second Russo-Turkish War. 11 *kuruş* was the equivalent of 1 pound sterling at the time (İnalcık and Quataert, *Economic and Social History*, 968).

77 M. Münir Aktepe, 'Ahmed III devrinde şark seferine iştirâk edecek ordu esnafı hakkında vesikalar', *Tarih Dergisi* 7 (1954), 17–30, reproduces three documents illustrating the system. Uzunçarşılı reproduces a document from the 1768–74 war, listing 77 different crafts with 173 tents, representing the contributions of the three cities of Istanbul, Edirne and Bursa *kapukulu ocakları* (vol. 1, 371–72).

78 See the works of Mehmet Genç, and Linda Darling for further information.

79 Murphey, *Ottoman Warfare*, 50.

80 Tott observed in this regard: '. . . in case of War, the Grand Seignior is obliged to supply the deficiency by the Treasures he has in reserve. Those of Sultan Mustapha were now nearly exhausted, and though he had expended more than twenty-five millions Sterling, he found but little effected to console him for so great a Sacrifice' (*Memoirs*, vol. 2, pt. 3, 236).

81 Michael Hickok, *Ottoman Military Administration in Eighteenth Century Bosnia* (Leiden: Brill, 1997).

82 The *Mubayaacı* resemble most the *munitionnaires* of the French administration. Ottoman fixed prices for goods rarely reflected market and inflation realities. The wage for *miri levendat* militias remained at 2.5 *kuruş* per month for 50 years. Likewise the Janissary salaries rarely inched upward. See the Bibliography and Guide to Further Reading in this volume for important studies on the Ottoman economy by Pamuk, Cezar, Tabakoğlu and Salzmann.

83 Leslie Peirce, *The Imperial Harem* (New York: Oxford University Press, 1993), 200–20.

CHAPTER THREE

From Prut to Belgrade: illusions of success and failure

The political setting 1700–50

The first half of the eighteenth century saw the continuation of the closing of the eastern frontiers of Europe, in a series of campaigns in which Ottoman, Habsburg and Romanov armies played out victory and defeat along the Danube. Between 1700 and 1739, the Ottoman army participated in two swift and successful campaigns: on the Prut in 1711 and against the Venetians in the Morea in 1715, but suffered costly defeats by the Austrians in 1717. In a series of confrontations which reiterated the magnitude of the problems in the military organisation, the Ottomans faced Russian and Austrian armies again in 1736–39, on contested frontiers around Belgrade, Ochakov and Azov. That is the subject of this chapter.

The triumph of the Treaty of Belgrade in 1739, which returned the pivotal fortress for the entire northern defensive line to the Ottomans, resulted after three years of Habsburg hesitancy, incompetence, splintered command, and insufficient men and supplies. The Russian thrust into the Crimea in early 1737 under Commanders Münnich and Lacy was equally wasteful, but psychologically important as Münnich also penetrated deep into Wallachian territory before being forced to withdraw as peace was concluded.

Ottoman mobilisation and supply systems, slowly organised, proved resilient on the Danubian front, less so in Ukraine and the Crimea. The victory accorded them at Belgrade prompted complacency among Ottoman bureaucrats at the court concerning the need for reforms, which became increasingly essential as the century wore on. Istanbul, no less than Vienna, suffered from the factionalism and psychological blows of victories followed by retreats which characterised 1737 to 1739. St Petersburg watched in astonishment as most of the territories captured by Eugene of Savoy were ceded to the Ottomans by the Treaty of Belgrade. Perhaps the greatest enemy of all was disease, which crippled the campaigns of all three armies in 1738.

The Europeanisation of Ottoman diplomacy continued, as each new diplomatic or hostile confrontation was accompanied by mediation by other western powers. France, Great Britain, the Dutch Republic, and latterly Prussia offered to intervene in the eastern European quarrels. Ottoman embassies to Europe became more commonplace, and concerted efforts at mid-century to stay out of European conflicts, especially during the vizierate of Mehmed Ragıb (1756–63), known as Koca Ragib, are also indicative of a new diplomatic worldview.[1]

The Ottoman case

Religious rivalries, Catholic, Orthodox and Muslim, played a role in the movement of populations in the Balkans, most especially on the Bulgarian–Wallachian border after 1715. The impact of retreating Muslim populations behind newly-defined Ottoman boundaries, however, is less well understood.[2] The increasing use of Albanian troops in the Morea and on the Black Sea fortress line, whose undisciplined bravado and ferocity generally went unchecked, did little to foster sympathy for the empire, and a great deal to alienate the populations of the periphery. Traditional formations, such as the Bosnian Muslim militia troops, remained especially loyal to the Ottomans, and proved very effective during the 1736–39 War.

The consolidation of local dynasties continued in the eighteenth century, and one of the more vigorous historiographical debates among Ottomanists concerns the genesis and significance of such families in the Balkans as well as the Arab provinces. The significant figures of the latter half of the eighteenth and into the nineteenth century, such as Canikli Ali in Trabzon, Mustafa Alemdar Pasha in Rusçuk, Osman Pasvanoğlu in Vidin, or Ali Pasha of Iannina, emerge from the context of conflicting loyalties, and skillful negotiation of demands for men and supplies from Istanbul.

Significantly, the grand viziers of the 1736–39 war period were some of the last of the centrally-trained military commanders with empire-wide experience of governance. One such, Hekimoğlu Ali Pasha (1732–35;

1742–73; 1755), was three times Grand Vizier, and successively defended Bosnia against the Habsburgs in July of 1737. Others include Muhsinzade Abdullah Pasha, Bender Commander against Münnich and the Russians, briefly Grand Vizier in late 1737, and İvaz Mehmed Pasha, Commander of Vidin, Grand Vizier in 1739–40, who proved both a competent commander against the Habsburgs at Grocka, and a wily and tenacious negotiator with Austrian representative Neipperg over the surrender of Belgrade.

Warfare, however, particularly burdensome in the Balkans in this first half of the eighteenth century, accelerated both the redistribution of tax revenues to local power sources, and the decentralisation of governance. By the end of the period, local elites were chiefly reponsible for manning and supplying the army, and in some cases, for defending the borders almost single-handedly. Local families profited from the needs of the state, and their challenges to Ottoman sovereignty itself by 1800 are well documented, but new allegiances to the dynasty were also created, a process of Ottomanisation which allowed for the nineteenth century bid for absolutism under Mahmud II (1807–39) and Abdülhamid II (1876–1908). Many of the revolts of the period represent the struggle not just against Ottoman oppression, but also the rivalry between older elites, such as the centrally-appointed and locally-despised Ottoman governors mentioned previously, and the newer gentry who profited from proving their reliability as suppliers of men and food. Balkan and Arab provincial areas shared such trends, with nuances of difference driven by the ethnicity and religion of their respective populations, and the degree to which local Janissaries played a part.

The centuries-old gradual nomadisation of eastern Anatolia, for example, was exacerbated in this period by the dissolving Safavid–Ottoman borders and the new alliances of tribal with urban elites. In the first half of the eighteenth century the Ottomans engaged in a long struggle with Nadir Shah and the Afsharids in the 1720s and 30s, a struggle finally settled only in the treaty of 1747. War against fellow Muslims, however heretical, was never popular, so major campaigns could seldom be mounted to the east. The obdurance of the Janissaries in Istanbul, who expected little profit from warfare in the east, proved an effective deterrent. This led to the increasing reliance of the grand viziers on local military leaders, such as the Jalili family in Mosul, who were responsible for preventing the Persian occupation of Baghdad in 1743.[3] The breakup of Safavid hegemony preoccupied not just the Ottomans, but also the Russians in the Caucasus.

Austria's position and ambitions

The Austrian surrender of Belgrade to the Ottomans in 1739 is much debated among historians. In spite of a better-organised and financed standing army, the Habsburg failure in high command, the demands of European theatres of war, and a considerable underestimation of Ottoman recuperative powers, led to the loss of face in the treaty of Belgrade, which became an impetus for the reforms of Maria Theresa, especially after Frederick the Great invaded Silesia the following year. A consensus appears to assign mediocrity to much of the Austrian military leadership of the eighteenth century, the notable exception being the triumphs of Eugene of Savoy at Varadin in 1716 and Belgrade in 1717.[4] Without generals of the stature of Eugene, who died in April, 1736, Habsburg military command floundered. Significant reforms would not be undertaken in Vienna until after 1748, once the disastrous confrontations with Frederick the Great and the Prussians had revealed the shortcomings of an organisation relying on federative arrangements with the provincial estates. Prussian and Russian recruitment systems outclassed the Austrian, and under-funding was a chronic problem. Hochedlinger's study emphasises the defensive nature of Austrian strategy, which varied according to whether they faced an eastern or western enemy. The eastern front was characterised by ferocity which was exacerbated by the so-called 'locust strategy' of regional armies.[5]

The Habsburgs were engaged on both western and eastern fronts in the first half of the century, a circumstance which drove much of their strategic planning of the period. By 1739, the need to incorporate and stabilise newly-acquired territories such as Italy and the Banat (Temeşvar), generated a realignment of Austrian diplomacy by mid-century. The 'diplomatic revolution' of 1756, when traditional rivals France and Austria joined forces to the astonishment of Europe, was contingent upon a gradual withdrawal of the Habsburgs from much of the west of their empire. The Franco-Austrian alliance itself is certainly emblematic of the changes in the Habsburg geopolitical position in western Europe, which was enabled by an earlier reconciliation of Bourbon–Habsburg interests in the 1752 Aranjuez pact.

Justified concern over the rise of Prussia (Hohenzollern) was certainly influential in forcing a reversal of alliances, as were the world-wide British–French conflicts which were pending. Closer to home, the Prussian threat imposed uneasy alliances with Russia, the new northern giant with eyes on expansion south and west. The Hungarian bid for independence in the Ferenc II Rákóczi war which opened the century was an additional incentive to consolidate imperial holdings, especially once the Banat was surrendered to Austria by the Ottomans in 1718.[6]

Austrian fear of Russian expansion and of 'Orthodox internationalism' meant that adhering to the Russo-Austrian defensive alliance of 1726 would be tempered by the desire to keep the Russians from the permanent occupation of the Danubian Principalities, even if that required supporting the survival of the Ottomans.[7] Much hesitation and ambivalence in Austrian strategic thinking and diplomacy is evident in the disengagement from the Balkans in 1716, 1739 and again in 1791. When Frederick II struck Saxony in 1756, marking the beginning of the Seven Years War, the Prussians inaugurated the first great age of global imperialism, which France, Britain and their allies played out in India, America and the Middle East as well as Europe.[8]

Russian expansion and the pre-modern impasse

The expansion of the Russian empire which begins in this period is truly remarkable. Great distances, untamed, wide and depopulated territories, as well as fluctuating loyalties of Orthodox populations, added to the complexities of establishing a military presence on the Danube. By the end of the century, Russian garrisons lined the frontier from the Baltic to Black Sea with ports on both.[9] The Great Northern War, which ended in 1721, established the Russians on the Baltic. Azov, the strategic post on the north and easterly end of the Black Sea, like Belgrade at the western end, passed back and forth between belligerents, but was the one substantial gain for the Russians in the 1739 Belgrade Treaty. Odessa was the new Russian port at the western end of the Black Sea at the close of the eighteenth century, a project made possible by the unilateral annexation of the Crimea in 1783. Security, however tenuous, on the Far Asian border was also achieved by an agreement to consolidate the South Siberian frontier with China in 1717, by basic Chinese 'indifference' to the Russians in the period, and by the voluntary submission of some of the Kazakh steppe nomads to Russian hegemony (in 1730).[10] Thus, much like the Habsburg version, the first half of the Romanov eighteenth century passed in the consolidation of borders. Both the Habsburgs and the Romanovs participated in the breakup of Poland, and one of the reasons for the Ottoman declaration of war against Russia in 1768 was most certainly the fear of the Russian eventual occupation of Poland and western Ukraine. By the end of the century, it was generally conceded that no one country could or should control the contested territories, an impasse dramatically demonstrated on many bloody battlefields, and mandating the continuation of balance of power diplomacy as the 'Eastern Question'.

It is within this context that we need to examine Ottoman successes and failure in the first half of the century, that is, before Habsburg and Romanov participation in the European wars at mid-century (1740–48, War of Austrian Succession and 1756–63, the Seven Years War) which by all accounts inaugurated a different age of warfare and widened the technological and organisational gap between Europe and the Ottomans. Frederick the Great, in particular, learned a valuable lesson: before 1759, he paid little attention to rising Russian might; afterwards, he cautioned his brother Prince Henry about Catherine II's potential 'to make Europe tremble'.[11] The undeniable, single-handed victories by Catherine II's armies over the Ottomans from 1768, however, brooked no interference, and the issue of equilibrium was a moot point. European territorial and diplomatic considerations thereafter stretched to Istanbul.

Centralisation of the control of violence

Peter the Great's military reforms began in 1699 with the elimination of the traditional infantry forces, the *streltsy*. They were gradually replaced by a conscript army, based on service for life. The new conscripts were drawn primarily from Russian serfs. '[Peter] initiated the change from a semi-standing, semi-regular army to a standing, regular army by disbanding much of the Moscow garrison, converting 24,000 southern frontier infantry into taxpaying farmers, and ordering the raising of a new army of three divisions of nine regiments each. Attaining peace with Turkey on one day, he used his new army to attack Swedish possessions on the next.'[12] This system, in spite of its cruelty, generally served the Russian war aims well. The Romanovs, like the Ottomans, had a ready supply of recruits, the former conscripted, the latter voluntary. Both sides appeared immune to huge battlefield losses, something which 'amazed and horrified contemporaries'.[13] Peter reorganised the military and civil bureaucracy in the 1722 Table of Ranks, though the extent and effectiveness of his military and civil revolution is still debated. Russian military command in the Seven Years War has been described as 'uninspired and notably cautious',[14] much as in the Habsburg case. The middle ranks of command in that war, however, included future brilliant commanders of the Danubian frontier in 1769 and 1770, such as P.A. Rumiantsev.

Other areas of military organisation did not receive the same kind of attention. The lack of a consolidated supply system failed the Russians repeatedly on campaign, demonstrably so in 1711 and at Ochakov in 1738, and even in the Seven Years War, when famished Russian soldiers, forced to haul wagons like draught animals, with nothing but oak leaves to eat, were dying daily by the hundreds.[15] The settlement of Ukraine in the second half of the eighteenth century, the extended fortress line, and access to warehouses in newly acquired territories did much to alleviate the chronic Russian shortages, but could not prevent the patterns of pillage and scorched earth tactics which were prevalent parts of pre-modern warfare on the Danube, less so on the Habsburg than on the Russo-Ottoman frontiers.

Peter the Great instituted a central military command (by 1718 the Military College), one of the most influential of his military reforms. In the late seventeenth century, there had been eighteen military chancelleries in Russia.[16] Harnessing warfare with an appropriate bureaucracy was a feature of eighteenth-century European politics, and the lack of it one of the chief defects of the Austrian structure in mid-century. In the Russian context, centralisation of command was part of Peter the Great's redefinition of the bureaucratic and military obligations of the aristocracy.

The consolidation of Austrian arms

Prince Eugene of Savoy was responsible for the reorganisation of the Austrian military as well as for its brilliant successes on the battlefield. As president of the Imperial War Council from 1703 until his death in 1736, Eugene had an enormous influence on the course of affairs in Austria. He introduced regimental and disciplinary reforms comparable to those elsewhere in Europe, although from this historical distance his legacy is viewed as rather more conservative, resisting significant structural changes. Revenues increased by the acquisition of new territories in the Treaties of Utrecht–Rastatt (1713–14) also allowed for maintaining a larger standing army, one which stood at over 300,000 by 1790.[17] With such renewed forces, Austria was able to defeat the Turks and make further gains in Serbia and the Banat by the mid-eighteenth century, but suffered reverses because of indecisive command and conflicting aims of the participants in what was still essentially, if waning, coalition warfare of the Holy Roman Empire. Consolidation was impeded with the acquisition of further border territories, and populations accustomed to the clientage policies of Ottoman-style sovereignty. Persuasion, especially of the Hungarian nobility, required much alteration to the Habsburg imperial premise, no more evident than in the creation and elaboration of the military corridor (*Militärgrenze*).[18]

Ottoman administration of warfare of the period is best illustrated by focusing on contemporary campaigns: Prut against the Russians in 1711; the Morea (Peloponnesus) in 1715, and select confrontations of the 1736–39 Russo-Austrian–Ottoman War followed by a discussion of the Treaty of Belgrade.

Origins of the Prut campaign

Frederick the Great once asked Ottoman diplomat Ahmed Resmi (d. 1783) why the Ottomans had treated the Russians so leniently on the Prut River in 1711. Resmi replied cautiously, as one might expect, stressing the sultan's generosity and satisfaction with the restoration of the Black Sea fortress line by Peter's surrender of Azov and the surrounding fortresses. In a further comment, Resmi noted that Frederick had good reason to question Ottoman strategy, having been extremely pressed himself by the Russians in recent times.[19]

The tales of leniency, conspiracy, betrayal and bribery which accompany almost every version of the Prut campaign of 1711 make it worth re-examining from some contemporary sources, as a way of entering the world of Ottoman, Russian and Austrian warfare in the first half of the eighteenth century. The story involves two audacious and impetuous military adventurers, Charles XII (1697–1718) of Sweden and Peter the Great of Russia, whose long struggles over the Baltic Sea and territories from Livonia to the Ukraine involved Poles, Cossacks, Tatars, Moldavians, and ultimately the Ottomans. The European context of their headlong confrontations, culminating in the famous battle of Poltava on 9 July 1709, when Peter's victorious troops outnumbered those of Charles XII by almost four to one, is well known, and need not be reiterated here.

The Swedish monarch took refuge in Ottoman territory after Poltava, with a few of his Polish allies, where he was subsidised by the Ottoman government and installed in the border fortress of Bender. His natural, if unreliable, allies were the Crimean Tatars, who had lost the Russian subsidy they felt their due by the treaty of Karlowitz and were anxious to regain their influence and status as the frontier zone between Ottoman and Russian.

These two forces have often been credited with forcing Ahmed III (1703–30) to war with Peter, a common assumption made about Tatar influence throughout the eighteenth century. While Tatar politics and refugees in Ottoman territories considerably influenced the Ottoman–Russian confrontations in the latter half of the century, in this instance it was less so. Even though the Tatar contingent of 70,000 horsemen was integral to the Ottoman victory, their demands were ignored in the treaty which followed. There are two other important aspects of Ottoman involvement in this campaign: one was the restoration and maintenance of the fortress line from Belgrade to Azov; the other was the repositioning of the relationship with the Principalities of Wallachia and Moldavia. When Moldavian Prince Dimitri Cantemir, a presumed loyal ally to several sultans, deserted to the Romanovs, it must have been particularly galling.[20]

Cantemir's own history of the period records Ahmed III's concerns. More disturbing to the sultan than Charles XII or the Tatars was the news that Peter was building a great fleet at Taganrog, especially as the Russians, 'who had been before entirely ignorant of naval affairs, had now gained such knowledge therein, that they ventured to send their ambassadors to Constantinople in ships of war.' The sultan was of the opinion that this czar aspired to be Alexander the Great and had to be 'chastised'; otherwise 'he will cause trouble when we are engaged elsewhere.' Ahmed III was also persuaded that Constantin Brancoveanu (d. 1714) of Wallachia was plotting against him with the Russians. Cantemir himself was sent in late November 1710 to replace Nicholas Mavrocordato in Moldavia and seize Brancoveanu in Wallachia. Among other things, Cantemir was promised Wallachia in perpetuity and assured that he would not have to pay the annual tribute (*pişkeş*) while he was master of Moldavia. His grievance, when he arrived on the scene and learned that, on the contrary, he must not only pay the *pışkeş* but also raise troops, build bridges and give aid to Charles XII, is palpable on the page: '[P]erceiving how little faith was to be expected from the infidels, [he] throws off his attachment to the Turkish interest . . . [and] sends a trusty messenger to the Czar, with an offer of himself and his Principality.'[21]

Cantemir is often characterised as a nationalist seeking Romanian independence, whose transfer of loyalties had 'wide popular backing and attracted large numbers of volunteers into his army . . . The breadth of support for the prince's bid for independence justifies the assertion: "The war fought against the Turks was, as always, a popular one." '[22] Whatever the extent of the uprising, Peter certainly recognised Cantemir's support in the charter of 13 April 1711: 'We, the great sovereign, Our Tsarist Majesty, having invoked the aid of the Lord, and trusting in the justice of our arms . . . have ordered our armies to invade the Turkish lands under our personal leadership, in the hope that God will give us victory over this perjurious and hereditary enemy not only of ourselves but of all Christendom . . . Dimitri Kantemir, being a pious Christian and zealous in the cause of Jesus Christ, has seen fit to offer to collaborate with us toward the liberation of the glorious Wallachian people, which is under his rule, as well as of other Christian nations now suffering under the barbarous yoke.'[23] Peter's call for the liberation of the orthodox Christians, in separate manifestos to Serbia, Slavonia, Macedonia, Bosnia and Herzegovina, illicited little support. It was probably drafted by Raguzinsky, a Serb with experience of Istanbul, who like Cantemir encouraged Peter's southern aspirations.[24] In spite of the initial promises of men and supplies from Cantemir, Brancoveanu, and Raguzinsky, a general rebellion during the brief campaign of July never materialised, at least in part because the countryside of Moldavia was in the midst of a famine and many of the residents had fled to the safety of the hills. Brancoveanu defected early, and turned the war supplies he had promised the Russians over to the Ottomans.

An interesting manuscript on this campaign by a Janissary named Hasan Kurdi, one of the last generation of the traditional system,[25] gives us his version of the Romanov initiatives in the Principalities. The manuscript begins with an imaginary speech by Peter, a narrative technique the author uses throughout in a very dramatic way. Peter begins: 'All the lands of the Ottomans are inhabited by Orthodox Christians, who speak a single language. The whole population is so ready, they will arise before our arrival, and attack the cities. The Ottomans have no power; they have no artillery or treasury. They may be able to raise 50,000 soldiers, but those soldiers have suffered so much because they haven't been paid, they are weary, and disobedient.' Therefore, the author comments, Peter had sent word to the Christians (*Rum taifesi*) that they should rise up and enslave the Turks. On another occasion, he has Peter saying: 'All the peasants of *Rum* are my soldiers: when I set out, they will attack every province. The Ottomans are slow to act; before they arrive, I will cross the Danube.'[26] Later, Hasan imagines the Czar's observing the arrival and encampment of the Ottoman army on the Prut, which Hasan describes as 'an endless spread of white, ornamented tents', and has him say in amazement: 'They didn't tell me it was like this. The Ottomans do not even have the strength of 50,000, so what is this army?'[27] Part of that army included Hasan Kurdi's estimated 50,000 Serbs and Croats, soldiers he calls *miri levendat* (state-funded militias), under the command of Hersek (Hercegovina) Governor Durmuş Pasha.[28]

The miscalculation and misinformation implicit in this characterisation of Peter proved costly. The czar, in fact, was neither well-endowed with friends at the moment, nor well-prepared for an all-out campaign against the Ottomans. Most of Europe was absorbed by the War of the Spanish Succession, lasting until 1714, when the Treaty of Utrecht temporarily settled the Habsburg–Bourbon rivalry, at least concerning Spain. Furthermore, the Poltava victory had alerted western Europe to the dangers of an expanding Russia. Efforts were made to bring the Great Northern War to an end by mediation. English and Dutch representatives also served as negotiators in the long effort to get the 1699–1720 Russian–Ottoman northern border conflicts settled by mediation.

After the Swedish defeat at Poltava, Augustus II was restored as King of Poland under Peter's protection. That did not guarantee the Diet's agreement to participate in a new Russo-Ottoman War, and in fact the Polish nobility refused to break off relations with the Ottomans. While Augustus remained steadfast to the Czar, his subjects did not. The Zaporozhian Cossacks, already riven by Hetman Mazeppa's support of Charles XII at Poltava, could be found on both sides in 1711. Peter used Cossack regiments as mercenaries, under their own leaders, to his advantage in the course of the Great Northern War.

Great distances and imperial over-extension characterised the campaign on the Prut River. The Romanov new-style army had proven its mettle at Poltava, but had been in the field since 1700, and in 1711 was scattered over a wide stretch of territory between the Baltic and the Black Seas. In 1695–96, the traditional *streltsy* infantry had been instrumental in taking Azov from the Turks, but a revolt in their ranks in Astrakhan in 1705–06 was excuse enough for their final elimination.[29] In 1699 Peter ordered the construction of new infantry regiments as an alternative force; by 1705 the standing army, raised primarily by levy, totalled 168,000, 130,000 of them foot soldiers.[30] The new levy system required one man in fifty households; the new 'immortals', liberated but in military service for life, were raised primarily from among household serfs and slaves. In 1705 the number was one in twenty; by 1711 when the new war with the Ottomans threatened, one man in three from amongst household servants was raised in Moscow. By the end of Peter's reign in 1725, 53 levies had raised close to 300,000 recruits. To acquire such recruits was one thing; to keep them necessitated various measures such as chaining them together in carts for transport to their units, or the branding of individual recruits, introduced in 1712.[31]

That Peter had built a disciplined, effective force in so short a time was a cause of considerable admiration as well as criticism: in 1710, Austrian diplomat Otto Player commented that '. . . they have reached an amazing degree of proficiency thanks to the incessant application and efforts of the tsar and to severe punishments and marks of favour and distinction, as well as to the experience of foreign officers of all ranks drawn from many nations. It is surprising what perfection the soldiers have attained in military movements, how orderly and obedient they are to their superiors' commands, and how bravely they conduct themselves in battle.' Still, '. . . little thought is given in Russia to the preservation of soldiers. Bad organisation and inadequate supervision of essential military stores are almost the only, though they are the most serious, defects.' Scarcity of men was hardly the problem. The issue was supply: '. . . peasants are taken away from the soil and often die of hunger in the field for lack of food.'[32] Faulty supply and intelligence systems were the real cause of the failure on the Prut. The organisation of provisioning systems was the least successful of the military reforms of the Romanovs. Supply breakdowns were the primary cause in the failure of military manoeuvres again in the 1736–39 Russian Crimean campaigns, and during the Seven Years War.

This was in direct contrast to Ottoman attitudes to the individual soldier. Paramount to successful Ottoman warfare was the perceived necessity to have adequate provisions in well-defined, well-placed warehouses, and the judicious application of wages and bonuses throughout the campaigns. This implied an adequate transportation and communications system. Neither the supplies nor the roads were an easy problem to solve in this particular period, or in this particular arena. Nor was it much different from the rest of Europe. By Chandler's account, the wars of Louis XIV are reckoned as embodying an era of warfare when 'wars became matters of elaborate manoeuvre', and armies '. . . were too preoccupied with problems of supply to achieve a truly total victory.'[33] Supply trains and grain dominate much of the military thinking of the first half of the century. Central and eastern European warfare was equally subject to the unpredictability of the elements which contributed to a successful campaign: good weather, fair crops, and a population large enough to sustain the considerable impact of foraging troops and their hungry mounts. An outbreak of plague, or, as in this case, a plague of locusts which left the Moldavian countryside bare, produced more casualties than actual confrontations. The three days of fierce hand-to-hand combat on 20–22 July 1711 were preceded by months of march, desertion, exhaustion and hunger, the latter a Russian rather than an Ottoman problem. In fact, once a truce was declared, the Russians purchased their much-needed food from the Ottoman stores.[34]

On the march – Prut 1711

The preparations for the campaign on the Ottoman side began in late November 1710 with the declaration of war, signalled by arresting Tolstoy, the Russian representative, as a guarantee against the behaviour of his government. Simultaneously, orders were sent to Tatar Khan Devlet Giray to penetrate and raid Russian territory in the direction of Moscow immediately.[35] Charles XII was sent 500 purses and 36 horses.[36] The call to arms went out to both Anatolian and Rumelian governors and provincial commanders, with orders to assemble by March of the following year. At the same time, orders concerning the assembling of supplies and munitions were issued. Hasan Kurdi noted that the gathering and parade of the troops in Istanbul occurred between the 12 and 16 March; salaries were distributed to the Janissaries on 1 April; by 20 April, the Grand Vizier Baltacı Mehmed Pasha and Janissary Agha Yusuf Pasha had arrived in Edirne. Here, the horses were allowed to graze on the spring grass for more than twenty days, as the Istanbul contingent awaited the arrival of troops from Anatolia. Those from the Rumelian provinces travelled via Sofia to rendezvous at İsakcı on the southern shores of the Danube. Soldiers from the Arab provinces, and the 3,000-strong Egyptian contingent, demanded for every major campaign, came via Gallipolli. The governors of Trabzon and Erzurum in eastern Anatolia were assigned to defend the Crimea, under the command of Commander-in-Chief İsmail Pasha, stationed at Ochakov. His task was to protect the passage into the Crimea in case of a Russian attack, or to sail to the aid of the troops on the Danube, if necessary.[37]

The march from Edirne to İsakçı took 23 days. By 15 June, the first of the troops was crossing the bridge on the Danube. The Grand Vizier crossed to Ismail on 28 June.[38] By early July the Ottomans confronted the Russians on the Prut, outnumbering them, by one estimate, six to one (100,000 cavalry and 140,000 infantry to Peter's 7,000 cavalry and 33,000 infantry).[39] The Ottomans succeeded on this occasion in crossing the Danube before the Russians could prevent it.

Peter had left Moscow with the main army on 17 March. Even earlier, in mid-January, Field Marshal Sheremetev was ordered to march to the Russian border at the Dniester, and prevent the Ottoman crossing of the Danube, along with regimental guards sent from Moscow. By 6 June

Sheremetev was at the Dniester; by 16 June, harassed by Tatars, low on supplies due to the famine in Moldavia, and discouraged by his calculations that the Ottomans would arrive at the Danube before him, he marched to Jassy, where Cantemir and his estimated 5,000 troops and supplies awaited. By 25 June, the main army with Peter was assembled at Shorofi (Soroki) on the Dniester, where they received Sheremetev's news that he had not prevented the Danubian crossing of the enemy. Angered, Peter pressed on to Jassy during the first week of July, where he found himself stranded without further promised supplies and men from Moldavia, and a camp already exhausted of its own meagre supplies. On the march, water was scarce. Peter elected to send a contingent with Cantemir to attack and capture Ibrail, an important Ottoman depot on the Danube, in order to feed his army. Meanwhile, another force under Sheremetev would proceed along the right bank of the Prut to prevent the Ottoman crossing. This is how the forces stood on the evening of 19 July.[40]

In the Ottoman camp, Janissary Agha Yusuf Pasha and his vanguard (*serdengeçti*) were busy building the pontoon bridges over the Prut at Falçi. In order to prevent the Russian advance guard from interfering, troops crossed the river by barge; just as many, mostly Tatars, swam across and engaged the enemy during the night of 19 July, with Çerkes Mehmed Pasha and Diyarbakır Governor Ali Pasha appointed to the task. Fierce fighting ensued, and the Ottoman soldiers were rewarded with 'tongue and head' money, at first 100 *altın* a head, but by afternoon, only 30 *kuruş*. Russian soldiers were captured; the rebels from the Principalities were killed.[41] By morning, the three bridges had been built, and the rest of the Ottoman army had crossed to the right bank of the Prut.

Peter, fearing for the advance guard, moved the main army forward, but was unable to relieve them and entrenched himself at the rear. As darkness fell, fighting ceased. The advance guard joined Peter's troops. They decided to retreat to a safer position and set up a defensive position at Novostanilişti, with their backs to the Prut. Here, Peter found himself completely surrounded, unable to get supplies or water from the Prut because of continual artillery barrages from the Ottoman emplacements. Fighting continued for over three hours on 21 July, but the Russian position and fierce resistance prevented a decisive conclusion to the battle. Considerable casualties had occurred on both sides, which were badly demoralised. The Janissaries, who had reputedly been reduced by 7,000 men that day, had lost the will to continue. Hasan Kurdi noted the hesitation when the troops were ordered to regroup and besiege the Russian entrenchments, and attributed it to fatigue and lack of morale. 'A huge cry went up, like the day of judgment. "We have no recourse [or guide], so be it," and they stood at the ready, their hands on the bridle. Some stood in prayer, some in kneeling position; some recited the glory of God, while others sat and prayed. Still others humbly raised their hands in supplication, while many were heedless, some of them shouting vociferously.' On 22 July, the grand vizier attacked the Russian rearguard at dawn. The Ottomans rained cannon fire on the Russian camp. The Russians were starving and thirsty; the horses dying. Hasan Kurdi imagines the Russian soldiers dropping their weapons, saying: 'We have no power to fight the Ottomans. It is better to save ourselves and submit than all be destroyed, as we have no strength to resist this army, and these soldiers and so much artillery.'[42] Peter 'retired to his tent overcome with grief . . . and forbade anyone to enter.'[43]

Sheremetev offered peace in Peter's name, and it was accepted after some delay, by Baltacı Mehmed. The Ottomans first pressed for the surrender of the traitorous Cantemir (and Raguzinsky), but Peter was adamant about protecting them. The treaty was opposed by the Tatar Khan and Stanislas Lesczynski, the deposed king of Poland, Charles XII's representative on the battlefield. The King of Sweden himself was swimming the Prut at the time in a last-ditch effort to reach the negotiations.[44] Baltacı Mehmed had just received the news that Peter's General Ronne, with 12,000 cavalry, had besieged the fortress at İbrail, the supply depot, which, combined with his uncertainty about the stamina of the Janissaries, made him hesitant to renew fighting. Ronne, in fact, occupied İbrail on 24 July, and surrendered it as part of the treaty settlement three days later.

Peter's humiliation was slighter than expected, but the Ottomans achieved their aims: Azov would be returned intact; the newly constructed forts were to be demolished. Russian troops were to leave Poland immediately. Peter was denied access to the Black Sea, and denied a representative in Constantinople. Charles XII was free to return to Sweden.[45] Moldavian and Wallachian rebel leaders had been eliminated. It is not insignificant that the Ottomans henceforth appointed the governors of the Principalities from within trusted Greek families in Istanbul.

The treaty was signed on 23 July 1711, but required two more years of bellicose threats, counterthreats and negotiations before the final version, known as the Treaty of Adrianople, was ratified in June 1713. The Romanov southern aspirations had to wait to another day, in 1774, when the battlefields were the same, but the balance of power had shifted to the victorious troops of Catherine II.

Hasan Kurdi's narrative demonstrates less interest in higher politics than in the difficulties the army now faced in returning to Istanbul. For reasons that are not entirely clear, this took until the end of the year. First, on the battlefield, the wounded and valorous were rewarded: the Janissary Agha was given 50 *kese* to distribute to the wounded, and many of the veteran soldiers were rewarded with stipends in cash or as land grants. The grand vizier and his officers prepared a full report to the sultan, and bridges had to be built in order to move from the Prut to the Danube itself. At the beginning of September, the main army was still on the northern shores of the Danube, where wages were distributed. A constant of Hasan Kurdi's narrative is bad weather: high winds and tremendous amounts of rain; swollen rivers and difficult passages. The passage to Edirne lasted until early November, where the army was again paid (6 November 1711). Here, Baltacı Mehmed was dismissed, and Janissary Agha Yusuf Pasha was temporarily appointed grand vizier. Not until early December did the Janissaries enter Istanbul.[46] By that time, Ahmed III was already contemplating another campaign to force the acceptance of the treaty arrangements, as Peter resisted abiding by them. In fact, as noted, the sultan declared war two more times before the final settlement in 1713.

The recovery of the Morea

Once peace with Russia was assured, the Ottomans turned their attention to the Morea, which they wished to recapture from the Venetians. The army, revitalised by the Prut campaign, required only an excuse for the call to arms. In December 1714, Ahmed III declared war on the Venetians by arguing that Venetian support of Montenegrin rebels and interference with Ottoman merchant vessels in the Mediterranean was in breach of the Karlowitz Treaty. Scholarly consensus is that the Ottomans were keenly aware of both the weakness of the Venetian hold over the Peloponnesus, and of the general disarray in the western allies following the War of the Spanish Succession. The centuries old Venetian–Ottoman rivalry over the Morea had involved repeated capture and surrender of strategic fortresses, and was likely viewed as restoring the Mediterranean frontier.[47] Habsburg–Ottoman relations prior to the Morea invasion were cautiously cordial: the Austrians were preoccupied in western Europe; the Ottomans maintained a healthy respect for the victors of the Karlowitz Treaty. Even as they invaded the Morea, an Ottoman delegation was sent from Istanbul to Vienna to reassure the Habsburgs of their desire to maintain peace.[48]

Venice stood alone against the Ottoman invasion in the spring of 1715. The record of Venetian occupation of the territory ceded to them by Karlowitz in 1699 is mixed. Catholic and Orthodox animosities coloured much of the politics, but military policies, relying on local taxes and conscript labour to construct new or revitalise old garrisons, also alienated the largely Greek population. New fortifications at Corinth and Nauplion were insufficient to prevent the swift reconquest by the Ottoman army in a few months. Thereafter, Venice withdrew from international conflict, and became a Habsburg client and ally until Napoleon abolished the Republic in 1797.[49]

The Ottoman army which invaded the Morea left Salonika on 22 May and arrived at Thebes on 9 June, where the Janissary advance guard paraded for the Grand Vizier Damad Ali Pasha (also unown as Silahdar Ali Pasha, 1715–16). The troops marched an average three to five hours, covering nine to fifteen miles per day. On 13 June, a council of war was held in the grand vizier's tent. Kara Mustafa Pasha was sent to İnebahtı with 15,000 Janissaries; Yusuf Pasha was sent ahead to Corinth with the Agha of the Janissaries and the remainder of the army; the grand vizier was to follow. The targets of the main army were the fortresses at Corinth and Nauplion, the latter heavily fortified by the Venetians. By one contemporary estimate, the total number of troops in camp on 9 June was 14,994 cavalry and 59,200 infantry, numbers, as always, to be used with considerable caution.[50] Between 16 June and 13 September when the grand vizier's return journey to Istanbul began, Corinth, Nauplion, Koron, Modon, and İnebahtı had been captured or capitulated, with little resistance. Venetian control of the Peloponnesus was at an end: capture of some of the islands ended more than 300 years of Venetian rule in the Aegean.

By 22 September 1715, the grand vizier was within a day's march of Nauplion, when the sultan's official congratulations and gifts reached him. Parades and ceremonies to celebrate the victory continued for a week, during which time the normal three months pay was distributed to the Janissaries. On 10 October, the standard of the Prophet, the official signal that the Ottomans were on campaign, was formally replaced in its ceremonial casket. On 17 October, six months more pay was distributed near Larissa. Then, by slow stages, the army returned to Istanbul, arriving in Edirne on 5 November. The grand vizier's entrance into Istanbul occurred on 2 December 1715.[51]

There are a number of points to be made about this brief campaign, and the earlier one on the Prut. First, Venetian lack of preparation and inability to finance the war meant that it was basically a walkover for the Ottomans. Venetian fortification included much war materiel, but little manpower. The nine-day siege of Nauplion (12–20 July) involved probably no more than 1,000–1,500 Venetian and Greek soldiers, but the fortress yielded very large quantities of munitions.[52]

Second, success was determined by the superiority, even a plethora, of numbers of Ottoman troops and the continued willingness to tolerate large losses and considerable desertion. For example, on 2 August, the grand vizier ordered the passage to Corinth closed to prevent desertion.[53] Brue estimated that 8,000 were killed and another 6,000 wounded during the nine-day siege of Nauplion, forcing the grand vizier to move camp because of the stench.[54] At the siege of Modon on the western side of the Peninsula in mid-August, the number of Janissaries had been reduced to 11,000.[55]

Yet a third factor contributing to the success was the constant communication and apparent cooperation between the Ottoman naval and land forces. The Grand Admiral (*Kapudan Pasha*) not only captured a number of the smaller islands; he also ferried the largest of the siege guns (the number at Nauplion estimated at seventeen, the largest of them weighing 390 *okka*)[56] from Ağrıboz to Nauplion and then to Modon. In both cases, while the damage inflicted by the guns was negligible, their arrival tipped the balance in favour of the Ottomans. The role of the artillery on the battlefields of the Balkans will be taken up at the end of this chapter.

A final factor has to do with command. Brue dismissed the grand vizier on this campaign, saying he could not control the Janissaries or the quarrels among his officers. Damad Ali Pasha apparently admitted early on in the campaign that he was a novice, and left the conduct of affairs to his officers.[57] In that, Brue misapprehended the culture of the Ottoman camp. He records councils of war in the grand vizier's tent on a number of occasions, after which orders were carried out, and officers and regiments dispatched to various duties. Intelligence received from spies and captured Venetians, as well as representatives of local populations seeking Ottoman protection, was analysed and incorporated into the command. On at least one occasion, in late June, the grand vizier deprived himself of transportation in order to send food and fodder to the advance guard.[58] There are an equal number of entries in the Brue journal of the imposition of discipline on officers who disobeyed commands, abused privileges, etc. Nonetheless, the ill-defined command hierarchy, and rivalries among the grand vizier, the Grand Admiral and the Agha of the Janissaries, could lead to battlefield disarray. The accident of command, based more on proximity and loyalty to the sultan than on competence, and the rivalry between various branches, such as the artillery and the infantry, are all evidence of the cracks in the Ottoman military edifice that continued to widen in the eighteenth century.

The real question of control concerned the Janissaries who, in the course of the siege at Nauplion, attacked against orders and essentially mutinied, blaming the grand vizier for a lack of leadership. Control also proved difficult in the matter of enslaving populations, and in enforcing capitulation conditions, if the fortress surrendered. In the case of Nauplion, which did inflict large casualties on the Ottoman troops, that meant considerable abuse of the local population, although even Brue notes the frequent grand vizierial intervention in the matter of slaves.[59] Incentives remained cash bonuses for valour and wounds as well as booty, human and otherwise, for what had become essentially a voluntary army, though technically bolstered by fictive registers of a professional remnant, with a totally inadequate pay scale.

Austria enters the war

In the spring of 1716, the Habsburgs reaffirmed their defensive alliance with Venice, and declared war on the Ottomans for the continued violation of the stipulations of the Karlowitz treaty. The massive Ottoman failures at Varadin in August of 1716, the capitulation of the Banat in 1717, and the loss of the fortress of Belgrade after the capture of the city in that same year, shocked Ottoman bureaucrats, after the relatively easy successes on the Prut and in the Morea. At Varadin, Grand Vizier Damad Ali Pasha, the veteran of the Morea, who had assembled the same kind of force as before, faced Eugene with evenly matched forces, numbering in the area of 60,000 on each side.[60] ‘The German cavalry proved their superiority to the Asiatic in regular charges, and the victory of the Christians seemed secure, when the Janissaries on the Turkish left broke the Austrian infantry, routed the wing opposed to them, and pressed hard upon the centre. Eugene immediately brought up a reserve of horse, with which he charged the Janissaries, and retrieved the fortunes of the day.’[61] The grand vizier, credited with charging into the final clash, lay among the casualties, as well as great quantities of booty. The battle lasted five hours. Within 20 days, Eugene turned his attention to Temeşvar, which capitulated by the end of the year. Serbian irregulars were instrumental in subduing the countryside.

The new grand vizier Halil Pasha (1716–17), previously Commander of Belgrade, was appointed upon the death of his predecessor on the field and charged with reassembling the army for the spring campaign. Eugene again took the initiative, and besieged Belgrade on 15 June 1717 before the large Ottoman relief army arrived from Niş. Finding his army encircled, Eugene saved the day by confronting the arriving troops and scattering a far superior army. Eugene ordered an attack on the Ottoman forces at two in the morning of 16 August: 'The Turkish outposts were negligent; the discipline of their whole army was lax; they had slept in careless confidence; they woke to panic confusion; and when once the Christian columns were within their works, the greater part of them fled without even attempting resistance.'[62] The Belgrade garrison, deprived of relief, surrendered the next day. Casualties on the Ottoman side numbered as many as 15,000 to 20,000, with less than 6,000 Austrian losses.[63] By September 1717, both sides were anxious for peace. The 1718 Treaty of Passarowitz completed the Austrian acquisition of the remainder of Hungary, begun with Karlowitz in 1699. Most importantly, the city of Belgrade was transferred into Austrian hands. Temeşvar (Banat) and Western Wallachia (Oltenia) were also ceded to Austria.[64] The Peloponnesus remained Ottoman territory, and ended the 200-year-old Venetian–Ottoman rivalry. Eugene of Savoy's great triumph at Belgrade proved temporary, however, when later Ottoman–Habsburg rivalries forced another series of lengthy and costly confrontations on the Orsova stretch of the Danube.

Renewal of Ottoman–Habsburg conflict on the Danube, 1736–39: stalemate

Within 20 years of 1718, much of the territory ceded to the Habsburgs had been returned to Ottoman rule. By contrast with the 1716–18 campaigns, in 1737, 1738 and 1739 the Austrians were indecisive and defensive, while the Ottomans showed unusual sustaining power, the single most important objective for them the return of Belgrade. As before, this coverage of the 1736–39 campaigns will emphasise the Ottoman side, as both the Austrian and Russian contexts have been well covered.[65]

Uneasy alliances

Russia declared war on Turkey in April of 1736. Austria, allied with Russia since 1726, at first attempted to mediate between the belligerents, reluctant to mobilise once again on the eastern frontier. The ostensible reasons for the Russian declaration were two-fold: Tatar raids in the Ukraine and the Ottoman–Persian struggles in the Caucasus, to which the Tatar Khan and his army were ordered as auxiliaries. To reach the frontier, the Tatars were forced to cross Dagıstan, then occupied by Russian troops and claimed by St Petersburg.[66] The Russians struck first. By April, Russian Commander Burkhard Christoph Münnich was within a few days' march of Perekop, the mouth of the Crimea, with 54,000 troops, including 5,000 Don Cossacks, 4,000 Ukrainian Cossacks and 3,000 Zaparozhian Cossacks.[67] They had grain enough for two months only, and too few horses, expecting to find both in the Crimea. Perekop, the key to the Peninsula, fell in May 1736. The Ottoman commander of the garrison and his remaining 2,544 men surrendered and were made prisoners.[68] Azov, on the mouth of the Don, was also invested under the command of General Peter Lacy, and the Ottoman garrison surrendered after a magazine blew up in the fortress. On 1 July 1736, 3,463 men, 2,333 women and children were allowed to leave the fortress unharmed.[69] In spite of those early successes, neither army was able to follow up with complete control over the Crimea. Tatar strategy was to harrass supply trains, make small raids, and burn all fields and villages on the Russian route. After Perekop, there were few significant engagements, although Bahçesaray was sacked and burned before the Russians withdrew in frustration. Both the Ottoman and Tatar troops had retreated to Kaffa, where supply ships promised both relief and the possibility of defence. The campaign of 1736 ended in Russian withdrawal into the Ukraine, after a tremendous loss of life, an estimated 30,000 deaths from hunger, famine and disease. Only 2,000 died in battle.[70]

Surprised by the preemptive strike in the spring of 1736, the Ottomans had hastened to assemble a force to protect the Crimea and Ochakov. In April of 1736, Grand Vizier Silahdar Seyyid Mehmed Pasha (1736–37) was ordered on campaign, and the Governor of Trabzon Yahya Pasha was put in charge of Ochakov, key to Crimean defence. In the normal course of affairs, four to six months were required to fully mobilise the Ottoman troops from long distances. Levies of soldiers from Bosnia and Albania were ordered in the summer of 1736, one in twenty from the Bosnia militia, estimated at the time at over 20,000 strong.[71] By the following summer, the garrison at Ochakov included 7,000 Albanian and Bosnian recruits.[72] Consequently, the ill-prepared Ottomans relied heavily on the Tatars in 1736, as they always had in the Crimea. Until that time, the Russians had not seriously attempted to conquer the peninsula. The

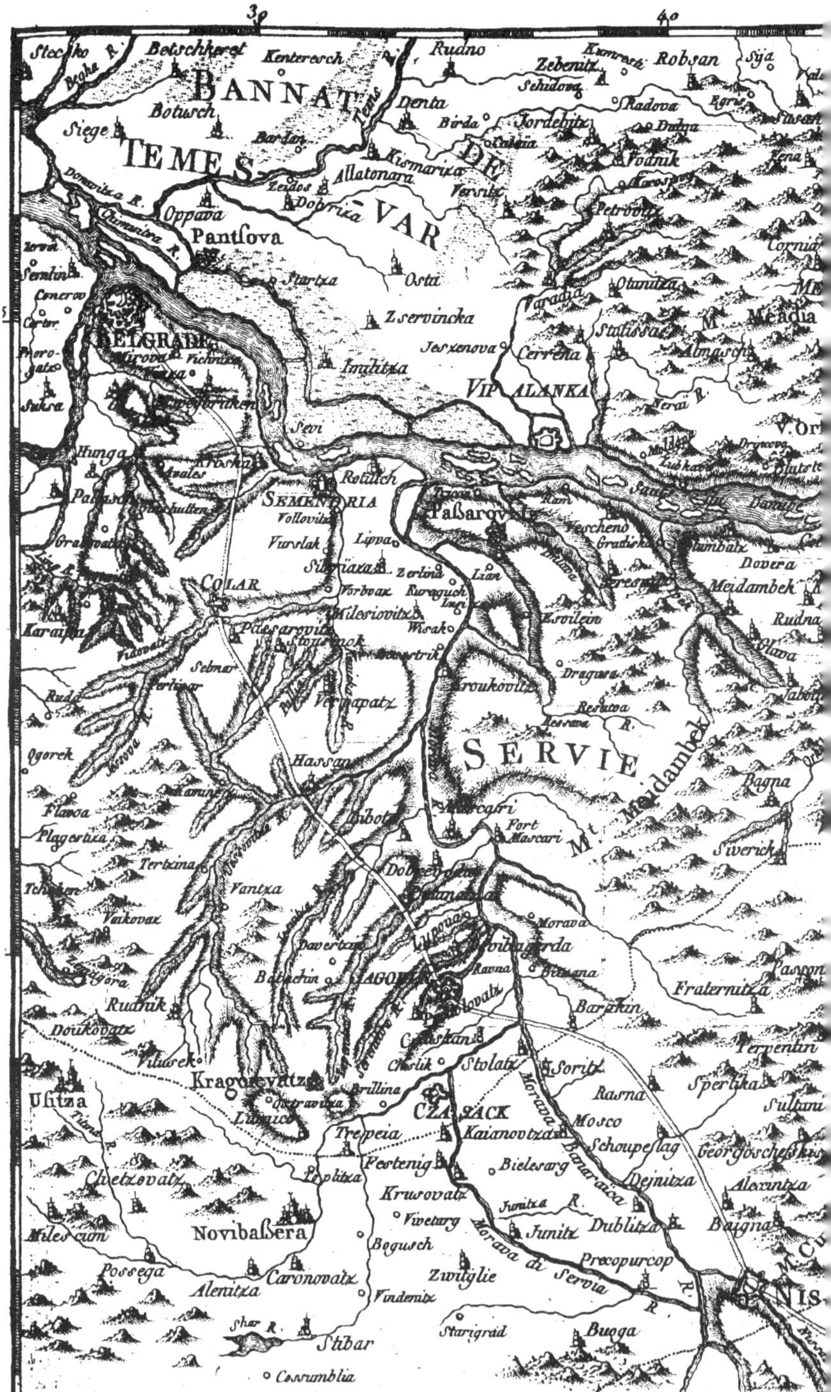

PLATE 7 *Map of Lower Danube at time of Habsburg-Ottoman campaigns, 1*
*(*Keralio, Louise Felix, Histoire de la guerre des russes et des imperiaux contre le
en 1736, 1737, 1738 en 1739, *v vols., Paris, 1780).*

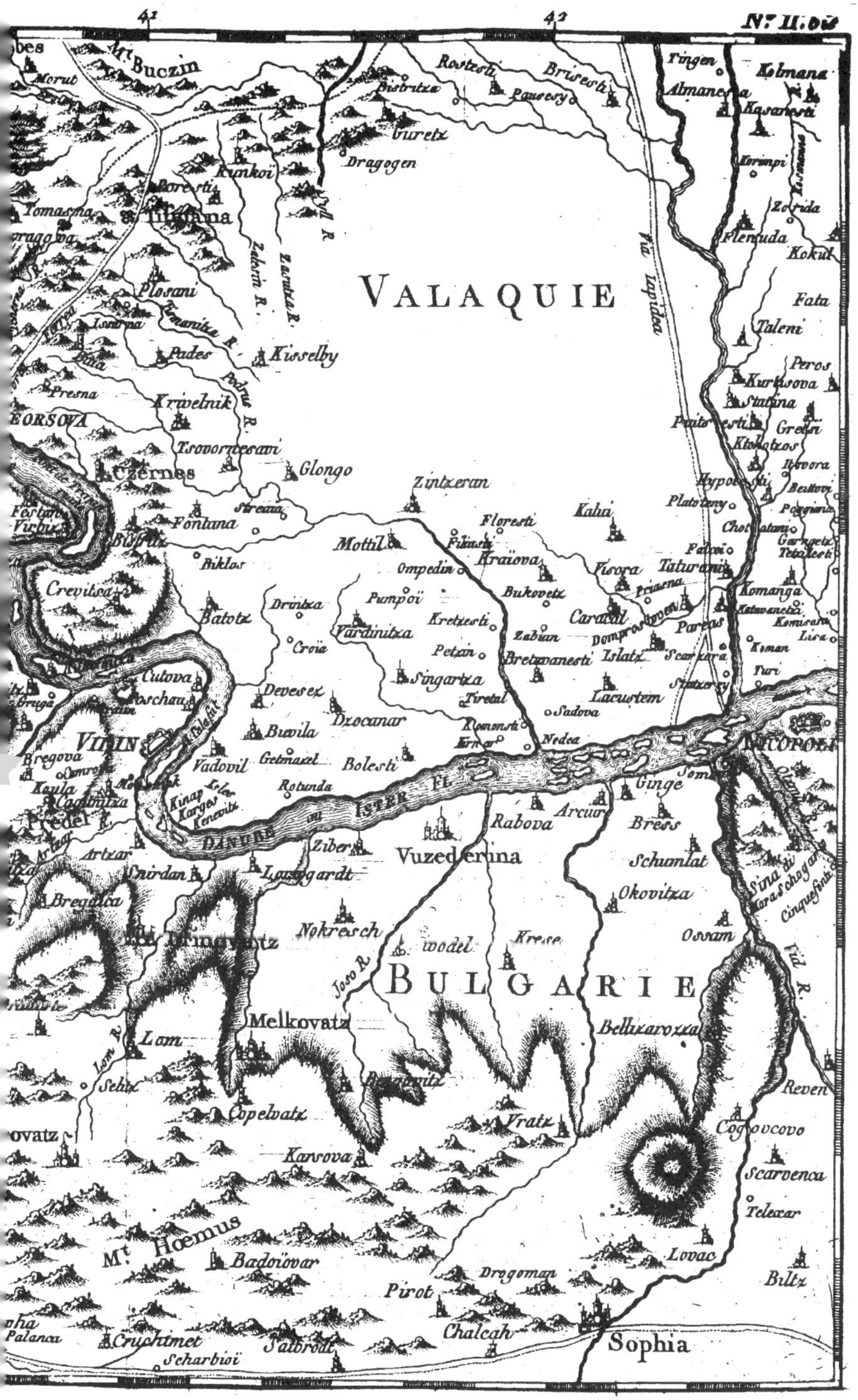
VALAQUIE
BULGARIE
Mt. Buczin
Mt. Hoemus
ISTER FL.
DANUBE
VIDIN
NICOPOLI
Sophia
Vuzederina
Melkovatz
Rabova
Czernes
Glongo
Kisselby
Krivelnik
Zintzeran
Kraïova
Caracal
Florești
Ompedin
Pumpoï
Vardinitza
Singartza
Bretwanesti
Lacustem
Sadova
Nedea
Islatz
Bolesti
Rotunda
Vadovil
Batotz
Biklos
Fontana
Crevitza
Drintza
Croïa
Devesez
Bivila
Dzocanar
Ziber
Lom
Lam
Copelvatz
Karsova
Vratz
Badoïovar
Pirot
Dragoman
Chalcah
Scharbioï
Cruchimet
Lovac
Bilts
Telexar
Scarvenca
Cognovcovo
Reven
Bellizarozza
Okovitza
Schumlat
Bress
Ginge
Ossam
Kokul
Fata
Taleni
Peros
Kurtisova
Kalia
Visora
Platoteny
Rostesti
Brisesti
Bistritza
Guretz
Dragogen
Tingen
Kolmana
Almanesta
Kasanesti
Flenduda
Plosani
Pades
Presna
Tsovornesavi
Dragogen
Vil. Laptilea
Wodel
Krese
Nokresch
Joso R.
Vid R.
Lom R.

Ottomans were also preoccupied in the Caucasus, where a treaty with Nadir Shah and the Persians was not finally in place until October of 1737.[73] Peace with the Persians provided the opportunity to redirect soldiers to the Danubian battlefields.

Silahdar Mehmed Pasha is universally condemned as obstinate and inept, perhaps an additional reason for the apparent inactivity of the Ottoman command in the summer of 1736. His vizierate was actually run by Osman Halisa Efendi, Deputy to numerous grand viziers, whose arrogance and lack of preparation for the war in 1736 and 1737 were the cause of his dismissal and execution in 1737. Silahdar Mehmed escaped execution but was replaced by Muhsinzade Abdullah Pasha in August of 1737.

Throughout late 1736 until the spring of 1737 at Ottoman headquarters at Babadağı, Austria's representative Leopold von Talman and Ottoman officials had futiley attempted mediation and construction of a peace congress, which would finally be held at Nemirov, near the Ottoman and Russian borders, beginning in August of 1737. All the while, preparations for war continued in Vienna, and in January 1737, the Austrian Council agreed to support their ally by opening up a Russian campaign on the Danubian frontier. The Habsburgs assumed that the Russian army, with Münnich in charge once again, would concentrate on Wallachia and Moldavia, which caused much apprehension in Vienna. Austrian forces, meanwhile, would attack the Bosnian and Serbian frontiers, east and west of Belgrade, especially Vidin, the primary fortress protecting Ottoman Wallachia after 1718, and Niş on the Morava River, another important Ottoman base, close to the Habsburg border.

General European opinion was that the 'fatal moment' of the Ottomans had arrived. Religious fervour against the infidel had not entirely died, making it easy to stir up popular opinion in Habsburg lands. On paper, Austria assembled an army of 122,540 combatants, divided into three divisions: one for Wallachia and Moldavia, one for Bosnia, and the third to attack Vidin and Niş.[74] Field Marshal Friedrich Heinrich von Seckendorf was in command of the main army attacking Vidin; Joseph Friedrich Hildburghausen was posted to Bosnia; George Olivier Wallis to the left bank of the Danube. Division in command, exacerbated by Emperor Charles VI's accommodation of the Catholic opposition to the Protestant Seckendorf, is often cited as a reason for the failures of 1737; more likely indecision on the ground and poor communications about enemy whereabouts contributed as much to the outcome.

Ochakov and the Crimea, 1737

Austria declared war on the Ottomans on 14 July 1737, having been in the field since May, and concentrated immediately on reducing Vidin and Niş in order to penetrate into Ottoman territory, counting on the proximity of the Russian forces in Wallachia. As it turned out, however, Russian objectives in 1737 did not match those of Austria. Münnich spent the spring months descending the Bug River in the Ukraine in preparation for an attack on Ochakov, which he reached in early July. Ottoman spring mobilisation concentrated on Ochakov, to defend against the Russian siege. This time, Münnich came fully supplied, but at the cost of delays, meaning that it took over three months to reach Ochakov. By the time they got there, they found Ochakov well defended and stocked, with over 20,000 men and sufficient artillery and supplies for a sustained siege.[75] Turkish sources claim that Commander Yahya Pasha had made an appeal to the grand vizierial camp on the Danube for both men and supplies, having only 6,000 in the fortress, an appeal which was ignored by Grand Vizier Silahdar Mehmed Pasha. Instead, 3,000 troops were sent by Bender Field Commander Muhsinzade Abdullah Pasha.[76] Even at 20,000, however, the Ottomans were outnumbered by a ratio of three to one.

The first storm of the fortress by the Russian forces was repelled with heavy losses, though the fortress Commander, Yahya Pasha, failed to follow up the advantage gained over the Russians.[77] The Russians were without their large artillery, had to cross a considerable ditch which had been dug around the fortress, and lacked the proper equipment for scaling the walls. As a result of Russian mortar fire on the houses within the fortress, fire broke out, and on the second day of the siege, the powder magazine within the city blew up, killing an estimated 6,000 defenders. Thereafter the fortress capitulated, and in the ensuing slaughter, in spite of the white flag, all but 3,000 of the garrison were killed. '. . . [Within] twenty-four hours the stench of decaying corpses was such that the Russians had to withdraw fifteen miles from the fortress.'[78] Russian casualties were estimated at 4,000, but disease and hunger took an additional toll, making their losses more the equal to those of the Ottomans.[79] At the end of the 1737 campaign session, during which Münnich's forces proved unable to move on to Bender, the Russians left a garrison of 5,000 to defend Ochakov, and returned to winter quarters in Poltava.[80] Lacy, with 40,000 troops, had much the same fortune in the Crimea in 1737, withdrawing after deep penetration of the peninsula to winter headquarters.

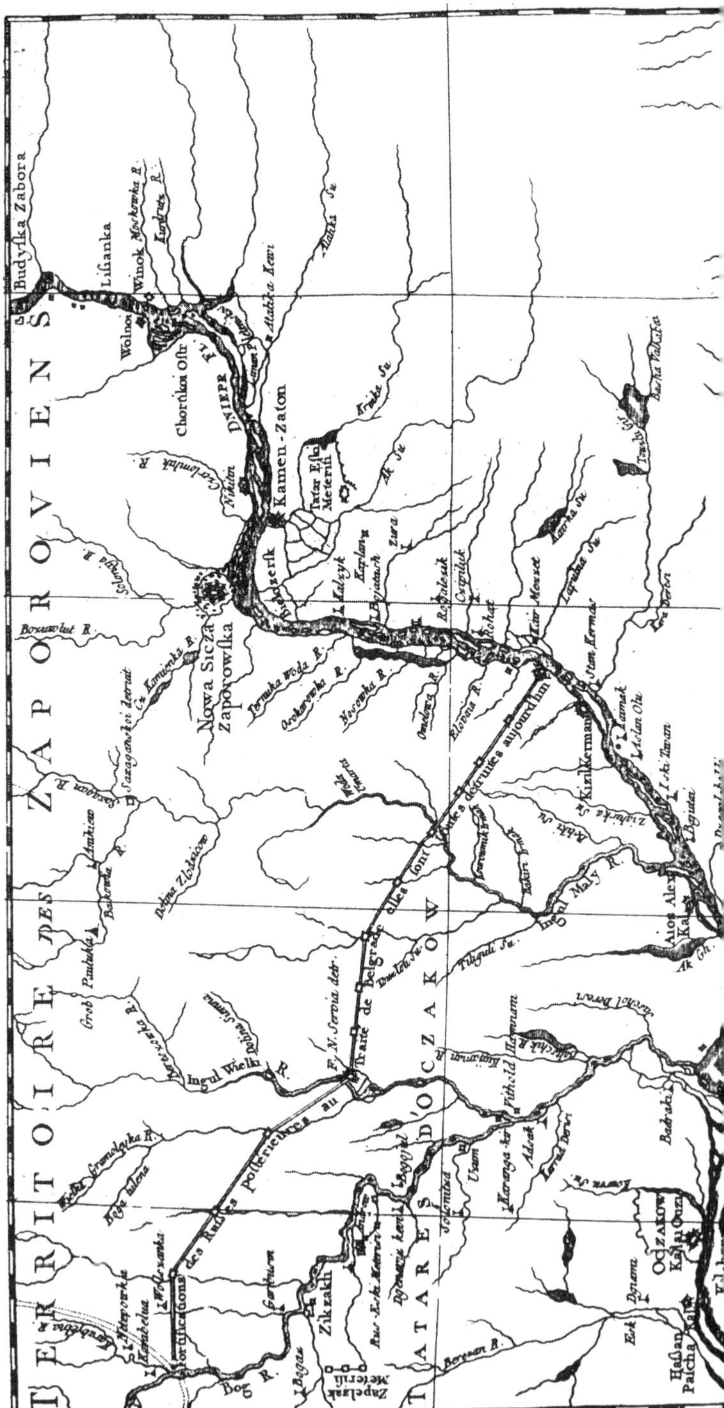
TERRITOIRE DES ZAPOROVIENS
TATARES D'OCZAKOW
Nowa Sicza Zaporowska
Kamen-Zaton
DNIEPR
Ingul Wielki R.
Bog R.
Oczakow

PLATE 8 *Map of Crimea, northern Black Sea Coast, 1737–39 (Keralio, Louise Felix,* Histoire de la guerre des russes et des imperiaux contre les turcs en 1736, 1737, 1738 en 1739, *v vols., Paris, 1780).*

In late October, the Turks attempted to retake Ochakov, besieging it with an estimated 20,000 Ottoman soldiers and an additional 20,000 Tatars, an unusual step for so late in the campaign season.[81] Huge losses, torrential rains and desertions on all sides forced the Ottomans to abandon the siege by 10 November. Lack of initiative on the part of the Danubian fleet, which failed to attack the Russian ships defending Ochakov on the seaward side, was also one of the reasons for the failure.[82] Grand Vizier Muhsinzade Abdullah Pasha was replaced by Yeğen Mehmed (1737–39), notable as belligerent and determined to continue the conflict into 1738.

The Bosnian army under Governor Ali Pasha

Yet another reason for the Ottoman failure in the Crimea was the necessity of maintaining a second front around Belgrade. The combination of ineptitude, lack of supplies and poor communications appears to have afflicted both sides of the conflict over Vidin and Niş. At first, Seckendorf's army was successful, and easily captured Niş, as Ottoman forces melted away from the various frontier garrisons and *palankas* (border outpost). Seckendorf's complacency prevented him from following up with a quick strike at Vidin, the original aim of Austrian strategy, where the Ottoman army had begun to assemble. By August 1737 the Nemirov Peace Congress was underway, the Ottomans had fortified Vidin and Hildburghausen had been bested by an Ottoman relief force at Banjaluka under Bosnian Governor Hekimoğlu Ali Pasha.

The Austrian attack aimed at Izvornik and Banjaluka began in July 1737. Ali Pasha assembled a counterattack force from the local militia population which had been reduced in numbers owing to the 1736 levies for the Ukrainian/Crimean battlefront.[83] Following orders from Istanbul, Ali Pasha instructed local commanders to appoint officers, and assemble troops in the area of Travnik, the provincial capital, to replenish the provincial troops who had already left. These troops were crucial, as mentioned above, in defending besieged Banjaluka on the Verbas River in late July 1737.

The Croatian army under Hildburghausen moved from Gradiska south to encircle Banjaluka, just a few days after a company of the local militias had arrived to reinforce the Banjaluka garrison. The garrison decided on a preemptive strike, attacked the 8,000-strong Austrian vanguard and pushed them back across the Verbas. Austrian losses on that occasion were large, but even worse when the main relief force under

Ali Pasha, probably no more than 20,000, dispersed a similar-sized force of Austrians before Banjaluka on 4 August, breaking the back of the Hilburghausen offensive in Bosnia, and forcing a retreat to the opposite side of the Sava River.[84]

A third front for the Ottomans developed at Vidin at the end of the year. Once the threat to Bender from Ochakov failed to materialise, the grand vizier ordered troops to assemble at Vidin. In early August, a corps of 9,000 Austrians was sent to Vidin under Khevenhüller. Approached about capitulation, the Vidin garrison under İvaz Mehmed refused to surrender. On 14 August, when the Austrians tested the defences of the fortress, İvaz Mehmed proved resilient and resisted the attack. This setback further disheartened Seckendorf, who held a council of war on 29 August and decided to abandon the attempt on Vidin, especially after hearing of the Ottoman victories in Bosnia. Thereafter he ordered his own troops to the defence of Bosnia, and then back over the Sava, while Wallis withdrew into western Transylvania, leaving Khevenhüller with 5,000 men, and General Doxat at Niş exposed to the Ottoman advance at Vidin, under the command of Ahmed Köprülü. Doxat reputedly had close to 4,000 effective troops, and close to another 3,000 too ill to fight. The fortress was short of water.

Khevenhüller then withdrew to Radhujevac in Habsburg territory, still able to defend the Timok halfway up to Niş. He did not have long to wait. On 29 September, the Ottoman forces, 15,000 strong, crossed over the Timok and forced Khevenhüller back from his strategic position. İvaz Mehmed had 50,000 men assembled to retake Niş. Stranded by his commander-in-chief, General Doxat surrendered Niş to the Ottoman forces without a struggle on 16 October, and was given safe conduct to Belgrade. Late in the season, the 15,000-strong garrison at Vidin, commanded by the newly promoted *Serasker* İvaz Mehmed, pressed their advantage upstream with army and navy as far as the treacherous rapids known as the Iron Gates. Raiders invaded Habsburg territory around Orsova, as the island fortresses were assaulted on 11 November. The blockade lasted a week, but was lifted as the season was too late to prolong it, and the Ottomans withdrew to Vidin.[85]

For refusing to stand and fight, Doxat was executed in March of 1738. Seckendorf was placed under house arrest throughout the remainder of the war.[86] A less than glorious season of warfare for the three imperial giants came to a close. The Russians occupied Ochakov, but the Ottomans had managed to recapture Niş and repulse the attack on Bosnian territory. Belgrade and Temeşvar remained in Austrian hands.

While the Austrians are castigated in hindsight for weaknesses in command, other problems point to endemic difficulties regarding mobilisation and supply. By all accounts, the Austrians had to rebuild the army on the Hungarian front from scratch, as the imperial army had been deployed on three different fronts for too long a period. The Russians suffered acutely by underestimating distance and supply. Until the spring of 1737, the Ottomans were deceived by Austrian protestations that they did not intend war, accounting in part for the unusually late mobilisation on the upper Danube. All sides were crippled by disease and insanitary conditions. The period 1737–39 are notable plague years, and disease accounted for more deaths than the fighting. By the time of the Belgrade treaty in 1739, for example, the Russians had been forced to withdraw from Moldovia and the Crimea because of the impossibility of maintaining healthy and well-supplied garrisons in the area. At Ochakov alone, an estimated 60,000 Russians were lost before it was razed and abandoned in late 1738.[87] Ottoman loss and desertion figures are practically impossible to estimate with any certainty, although archival documents will occasionally reveal the continuing problem of getting men to the battlefield and keeping them there.

Diplomatic negotiations at Nemirov were suspended when the Ottoman legation left in mid-October; relations between the partners in Vienna and St Petersburg were fractious and fraught with mistrust, as both sides persisted in secret diplomacy. For the remainder of the war, Ottoman diplomacy aimed first at an alliance with France, and when that proved unrealisable, at French mediation.[88] On both the diplomatic and military fronts, the Ottomans exhibited a tenacity which exasperated diplomats and discouraged both opponents.

Ottoman victory at Orsova, 1738

The pattern of victory and retreat continued in the few confrontations of 1738, but one thing is certain. Ottoman exertions had one aim in mind: the capturing and fortifying of the string of garrisons that marked the Danube path to their main objective, Belgrade. The campaign season opened in February, when Ali Pasha sent 15,000 men into Serbia, capturing strategic border garrisons in the Morava valley, which re-established Ottoman control in Serbia.[89] As the campaign season opened in Austria, the Austrian forces on the Danube were badly depleted; most garrisons were undermanned, and the imperial treasury stretched. This was especially true of the garrisons around Orsova. The Austrian intent to avoid open confrontation with the Ottoman main army resulted from poor command as much as from insufficiencies in men and supplies, notably transport.[90] Lothar Königsegg, President of the Ministry of War, not noted as a military genius, was appointed actual Commander-in-Chief, although nominal command was left to Charles VI's young son-in-law, Franz Stephan of Lorraine. A significant dearth of competent commanders plagued the Habsburgs after the death of Eugene of Savoy, who had discouraged the emergence of potential rivals.

Most of the 1738 campaign season was spent in investing and surrendering the strategic areas around Orsova (Old Orsova), specifically Adakale (New Orsova), an island just downriver from the Iron Gates.[91] Old Orsova on the northern shore fell to *Serasker* İvaz Mehmed in early May. By June the Ottomans had captured Mehadia, allowing them deep access into the Banat. On 15 August 1738, the Austrians surrendered Adakale, in spite of their valiant resistance and recapture of Mehadia. The surrender of the island followed two unsuccessful efforts the same summer, and was achieved through the tenacity of *Serasker* İvaz Mehmed, and with the support of the main artillery corps, as well as through the initial failure of the Austrian field command to appreciate the Ottoman single-mindedness concerning the control of the lower Danube around Vidin. When the two armies, evenly matched at 15,000 men, did confront one another, on 4 July at Cornea, 25 miles north of Orsova, the Austrians beat back the advancing Ottoman forces, and regained Mehadia on 9 July. Thereafter logistics kept the Austrians from advancing further, even though the victory was followed by a quick dissolution of the Ottoman will to fight, a general flight toward Fethülislam, and the abandonment of the Ottoman camp, notably 1,500 wagons, 3,000 heads of cattle, 50 cannons and 14 mortars.[92] Franz Stephan was hailed as the second Prince Eugene in the streets of Vienna.[93] *Serasker* İvaz Mehmed was removed from command and confined to Vidin.

The Ottoman main army under Grand Vizier Yeğen Mehmed had arrived in Vidin by early July, but was hampered by having to await the arrival of transport and supplies. Austrian commanders inexplicably disregarded intelligence concerning Ottoman troop strength and preparations for what would prove to be the main offensive of the war. This omission might be explained by the eighteenth-century preference for open-field confrontation rather than the defile strategy demanded by the mountains and valleys of the Orsova area, and the generally-accepted notion that open-field battles were to be avoided unless victory was certain. Most compelling are the reports on logistics, scarcity of food and transport which particularly afflicted the Austrian side. Königsegg found himself in the middle ground of suspect loyalties and banditry so characteristic of Ottoman borderlands of the period. Even as the Ottomans closed in on Orsova, the Austrian Commander ordered retreat of the relief force.

An Ottoman assault on Adakale was repelled by cannon fire again on 13 July, but a general rout was prevented by the forceful command of İvaz Mehmed, who reorganised the scattered bands and was reinstated as *Serasker* by Grand Vizier Yeğen Mehmed. By 18 July, the siege of Adakale was once again underway, this time with the entire Ottoman arsenal, perhaps as many as eleven batteries of 120 cannons and 14 mortars.[94] More significantly, the Ottoman sappers undermined the defences, and the dry summer lowered the Danube level sufficiently to allow the besiegers easier access. Capitulation occurred on 15 August, just in time for İvaz Mehmed, who was facing massive problems on his side.[95] On this occasion, it was Istanbul's turn for jubilation, despite its preoccupation with shortages and rebellion. The two belligerents spent the remainder of the year rehearsing the final confrontation beneath Belgrade the following summer. By the end of the year, French mediation was accepted as a *fait accompli* by both sides. Hertz rightfully argues that the fall of Adakale was pivotal to the events of the following year, pointing to its significance in all subsequent peace negotiations.[96] The 1738 campaign left the Ottomans in a position of strength.

The 1738 season was just as disastrous for the Russians as the Austrians. Münnich once again failed to cross the Dniester in his march on Bender. Having overloaded his army in preparation for the desolation of the proposed march, he did not reach the banks of the Dniester until mid-July, where he faced an entrenched Ottoman army of 60,000. Disease forced the Russian garrison to abandon Ochakov by the end of the year. General Lacy made the same trek into Crimea as in the previous two years, retreating as before with very little accomplished.[97]

With a new resolve, Vienna and St Petersburg decided on closer coordination in the 1739 campaign. The empress pledged 20,000 Russian troops to aim at Hotin and Transylvania to divert the Ottoman forces from the Danube, but international pressure on all sides was also urging an end to the hostilities. In Istanbul as well, the peace party prevailed by mid-March in the appointment of İvaz Mehmed as Grand Vizier (1739–40). The Ottomans inclined to end the conflict with diplomacy. By early May, Sultan Mahmud I (1730–54) gave the French Ambassador, the Marquis de Villeneuve, the power to negotiate the treaty on behalf of the Ottomans.[98]

For the 1739 campaign, Wallis was Commander-in-Chief of the Austrian forces, while nominal supreme commander Franz-Stephan remained far from the disease-ridden battlefront. Wallis arrived at Belgrade by mid-July. İvaz Mehmed was stationed south-east at Semendre, Habsburg headquarters from the previous year, with an advance guard at Grocka, a day's march to Belgrade, on the southern Danube shore. Wallis marched to Grocka with his cavalry on the night of 21–22 July. The same night the Ottomans moved their entire army to defend the narrow pass and road at Grocka. The most significant confrontation of the year occurred there, with the Austrians gaining the upper hand after an exhausting day-long contest and the arrival of cavalry reinforcements under Count Reinhard Wilhelm Neipperg. The carnage left over 3,000 Austrian dead, with another 2,500 wounded. The following year a traveller would note that '. . . one cannot go ten steps without stepping on human corpses piled on top of one another, all only half decomposed, many still in uniforms.'[99] Wallis retreated in the middle of the night of 22 July, fearful of the Ottoman strength. Remarkably, the Ottomans did not harass the withdrawal, which would have been more typical. İvaz Mehmed's restraint must have been dictated by the certain knowledge that they had won the day.[100] By August Belgrade had been surrounded, and Commander-in-Chief Wallis was in slow retreat. Included in the Ottoman ranks was Hekimoğlu Ali Pasha, the Governor of Bosnia who had been so instrumental in securing Serbia earlier in the war,[101] with 30,000 men.

Early in August, negotiations got under way between the Habsburgs, represented by Chief Negotiator Neipperg, and Villeneuve on behalf of the Ottomans. In the meantime, Schmettau, veteran of the 1737 campaign, had been sent by the War Ministry in Vienna to assess the real situation, and to invigorate the dispirited and inactive Wallis. Schmettau was appointed Commander of the Belgrade garrison, and discovered a fit garrison of almost 12,000 men. He achieved a small victory over the Ottomans with a pre-emptory assault on the Borsha redoubt across the river on 29 August, which encouraged Wallis to redirect the efforts of the main army against Belgrade. News from Hotin was encouraging: Münnich was about to cross the Dniester. In other words, the position of the two allies called for a halt to negotiations and an attempt to change the status quo on the battlefield.

Isolated (a virtual hostage) in the Ottoman camp, however, Neipperg remained ignorant of the changing environment, and operated on instructions from the emperor himself, dated from 18 August. On 29 August, as Schmettau attacked Borsha, the plenipotentiaries drafted a peace treaty, which was followed by another ten days of arguing over the peace with Russia, resulting in treaties which all three sides officially signed at Belgrade on 18 September. On 6 September, as part of the treaty stipulations, the demolition began of the outer, new fortifications erected by the Habsburgs at Belgrade.[102] On 8 September, the news arrived in Vienna. Münnich, who had routed the Ottoman army at Hotin, received the news in astonishment at his new headquarters in Jassy, Moldavia on 24 September. All sides were forced to accept the triumph of the Ottomans and Villeneuve by December of 1739.[103]

By the conditions of the treaty, the Ottomans acquired Belgrade, Old Orsova and Adakale. Azov remained under Russian control, although the fortress was razed and the territory became a no-man's-land. Russian merchants would be allowed the right of free trade in the empire. Villeneuve's reward was the renewal of the French capitulations in 1740, guaranteeing substantial French trading privileges, with the Ottomans. Small wonder that the Ottomans rightfully viewed the Belgrade treaty as a vindication of their half-century of efforts on the Danube and the Black Sea. How illusionary the victory actually was would be borne out on the battlefields of 1768–74.

We should not underestimate what exchanging the territories in the Danube area did to local populations, Muslim and Christian alike. One western observer of the exchange of power in Belgrade noted: 'Earlier one thought himself to be in Germany because of the architecture, administration, commerce and manufacture, clothing, citizens, spirit, Christians and Jews – all were German; and now: All European culture has disappeared, the city and its inhabitants have been devoured as if from earth and the city seems possessed by a foreign, Asiatic being.'[104] Bias aside, the observation points to the human costs of year after year of imperial campaigns in this region, especially in the case of the Serbians (or Rascians, as they are often called in the texts), whose loyalties were suspect, who fought on either side, or as bandits as it suited individual groups.[105] Belgrade became the western end of a middle ground frontier for Janissary and Serbian *haydut* alike. As McNeill so acutely notes:

By 1740 or thereabouts, therefore, one may properly say that only very narrow interstitial areas remained unincorporated into one or the other of the great agricultural empires that had long bordered upon Danubian and Pontic Europe. An almost uninhabited stretch of territory still separated the Russian Ukraine from the Crimean settlements and Turkish border forts at the mouths of the Dniester, Dnieper, and other strategic points.[106]

Ottoman military strengths and weaknesses

I have spent considerable time on the narrative of these campaigns to demonstrate Ottoman military strengths as well as their weaknesses. Baltacı Mehmed's army at Prut had recovered from the revolts endemic in Istanbul after 1700. The Janissary Corps appears to have reorganised itself without difficulty. Still, increasingly, military service drew more heavily on voluntary, temporary mobilisation, even to fill the Janissary ranks, and the bulk of the troops on the Prut were raw and undisciplined, willing but untrustworthy, as Hasan Kurdi so vividly demonstrates. The problem of discipline had not been adequately addressed, possibly the main reason for keeping the troops in Edirne months after the campaign was finished. The technical difficulties of troop movement and supply seem to have been overcome, as evidenced by the crucial bridge-building at the time of the Prut encounter. Brue's description of the 1715 campaign in the Morea, where the Ottomans faced little resistance from the Venetians, illustrates much the same thing: the continued unruliness of the Janissary troops, whose impetuous fighting style remained dangerous to friend and foe alike. He was witness to an army that succeeded when the odds were overwhelmingly in its favour, and when sufficiency of supply and ready cash were used as incentives for encouragement of individual valour. Again, the cooperation between army and navy, and the adroit psychological use of artillery, militarily effective or not, suggests a still formidable organisation.

The success on the Danube I have just described as part of the 1736–39 War is grudgingly accorded the Ottomans only after the weaknesses of their enemies have been catalogued. In particular, emphasis on Villeneuve as engineer of the coup during negotiations over Belgrade has masked a considerable body of evidence that the Ottoman negotiating team was clear-sighted and tough. Not only was Hekimoğlu Ali Pasha involved, but also *Mektupçu* Koca Ragıb Pasha, later grand vizier under two sultans, already a veteran of crafting peace with Persia, and *Reisülküttab*, Mustafa Tavukçubaşı Efendi, Chief Scribe, functioning as foreign secretary. Both men were instrumental in keeping the Ottomans out of European wars for the next two decades, Koca Ragıb especially in the long flirtation between Frederick the Great and the Ottomans over the possibility of a

Prusso-Ottoman alliance. Koca Ragıb's account of the siege of Belgrade and the resulting negotiations became one of the most widely circulated manuscripts of the eighteenth century.[107]

The resources of the empire were still robust enough to allow for the massive organisation and transport of supplies, food and materials such as pontoons, but not in a way which compensated for the essential vacuum in military command. In the course of any battle, the role of the Janissary commander and the provincial commanders seems crucial, and indeed valour and success were regularly recognised in the battlefield parades, official gifts from the sultan, and the consultations between grand vizier and commanding officers that punctuate narratives of campaigns such as this. As with any other early modern army, the disjuncture between field command and dynastic or factional politics remained acute. The consequences of neglecting certain aspects of military organisation, such as the creation of an officer class, or the organisation of trained and disciplined infantry and artillery troops, would not become apparent until the latter half of the century. In this regard, the Ottoman enemies were quicker off the mark. It is one of the curiosities of Ottoman history that their innovations in firepower and logistics of the sixteenth and early seventeenth centuries served them well enough, in spite of Karlowitz, to engender complacency that long outlived the reality of the lack of societal investment in military technology and training.

Some few Ottoman intellectuals were aware of the growing gap, notably İbrahim Sarı Mehmed Pasha, who was the chief financial officer (*defterdar*) of the empire six different times between 1703 and 1716. He was particularly concerned with the well-being of the enlisted men, but equally insistent that good leadership and knowledge of the strategy of the enemy were just as important. His contemporary İbrahim Müteferrika (d. 1745) went even further, citing the model of Peter the Great who had rebuilt his army by imitating those of victorious nations.[108] This included the instruments of warfare, in particular the victorious arms.

These are questions that would become particularly pressing with the outbreak of war in 1768. Before then, with relative peace on both eastern and western fronts, and considerable pressure to maintain it exerted among the bureaucrats of Istanbul, the empire experienced an almost unprecedented quarter-century of calm before plunging once again into an interminable struggle with the Romanovs over control of the northern Black Sea coast. By that time, both Habsburgs and Romanovs had engaged in the first era of global conflict instigated by Anglo-France rivalries and the rise of Prussia.

Notes

1 Aksan, *An Ottoman Statesman in War and Peace: Ahmed Resumi Efendi 1700–83* (Leiden: Brill, 1995) – Koca Ragıb was a member of the negotiating team at Nemirov in 1736 and then at Belgrade in 1739.

2 See Virginia H. Aksan, 'Whose territory and whose peasants? Ottoman boundaries on the Danube in the 1760s', in F.F. Anscombe, ed., *The Ottoman Balkans* (Princeton: Markus Wiener, 2006), 61–86.

3 Robert W. Olson, *The Siege of Mosul and Ottoman–Persian Relations 1718–43* (Bloomington, IN: Indiana University Press, 1975), and Dina Rizk Khoury, *State and Provincial Society in the Ottoman Empire, 1540–1834* (Cambridge: CUP, 1997).

4 Cassels, *The Struggle for the Ottoman Empire, 1717–40* (London: John Murray, 1966); Karl A. Roider, *Austria's Eastern Question 1700–90* (Princeton, NJ: PUP, 1982).

5 Civilian populations referred to Tatar auxiliary forces sent in to terrorise enemy countryside as '*Renner und Brenner*' ('runners and burners'). Michael Hochedlinger, *Austria's Wars of Emergence: War, State and Society in the Habsburg Monarchy, 1685–1797* (London: Longman, 2003), 139–44; also H.M. Scott, *The Emergence of the Eastern Powers, 1756–75* (Cambridge: CUP, 2001), 14. See Charles W. Ingrao, 'Habsburg strategy and geopolitics during the eighteenth century', in Gunther E. Rothenberg and B.K. Kiraly, eds, *East Central European Society and War in the Pre-Revolutionary Eighteenth Century*, 49–66 for an overview.

6 Ingrao, 'Habsburg strategy', 52; see also a recent treatment of these questions by Ivan Parvev, *Habsburgs and Ottomans Between Vienna and Belgrade, 1683–1739* (Boulder, CO: East European Monographs, possibly the worst edited book I have ever encountered).

7 From Ingrao, 'Habsburg strategy', 62, but based on Karl A. Roider's *The Reluctant Ally: Austria's Policy in the Austro-Turkish War, 1737–39* (Baton Rouge, LA: Louisiana State University, 1972), who argue as well that finance was the key to Austrian defeats as well as lack of leadership; see also A.V. Florovsky, 'Russo-Austrian conflicts in the early eighteenth century', *Slavonic and East European Review* 47 (1969), 94–144, who uses the term 'Orthodox internationalism' (95) but cautions that it represented a cultural, rather than an imperial, ideal.

8 See C.A. Bayly, 'The first age of global imperialism, c.1760–1830', *Journal of Imperial and Commonwealth History* 26 (1998), 28–47, which focuses not just on violence and the size of armies, but also ideology and the economic domination of the British in India.

9 H.M. Scott, 'Russia as a European power', in R. Bartlett and Janet M. Hartley, eds, *Russia in the Age of the Enlightenment: Essays for Isabel de Madariaga* (Basingstoke: Macmillan, 1990), 8–9. See also his *The Emergence of the Eastern Powers, 1756–75* (Cambridge: Cambridge University Press, 2001).

10 Emmanuel Sarkisyanz, 'Russian imperialism reconsidered', in T. Hunczak, ed., *Russian Imperialism from Ivan the Great to the Revolution* (New Brunswick, NJ, 1974), 50; also Lanham, MD: University Press of America, 2000. Scott makes use of the of term 'indifference', in 'Russia as a European power', 9.

11 Scott, 'Russia as a European power', 15, quoting Christopher Duffy, *Russia's Military Way to the West* (London: Routledge and Kegan Paul, 1981), 74 and 124.

12 Richard Hellie, 'The Petrine Army: continuity, change and impact', *Canadian–American Slavic Studies* 8 (1974), 237–53.

13 Scott, 'Russia as a European power', 14, used statistics from Kunersdorf in 1759 when one-third of the Russian troops died; casualties at Ochakov in July 1737 were almost totally on the Ottoman side, 22,000 compared to 4,000 of Münnich's Russian troops: David Chandler, *The Art of Warfare in the age of Marlborough* (London: Batsford, 1976), 310. These were exceptional battles: one in four or one in five casualties seems to have been a more realistic expectation of western European military strategists.

14 Scott, 'Russia as a European power', 14.

15 John Keep, 'Feeding the troops: Russian Army supply policies during the Seven Years War?' *Canadian Slavonic Papers* 29 (1987), 31, describing a Russian drive into East Prussia in the spring of 1757. The disaster forced a consolidated effort to reform the supply system (33). Keep argues that the Russians did not consider the physical well-being of their soldiers a priority of military management until after the Crimean War, when railroads helped to solve the issue of transport.

16 Hellie, 'The Petrine Army', 244.

17 Jeremy Black, *European Warfare 1660–1815*, 149. He notes however that the number was half of that in 1740. Bérenger, however, argues the inability of the Austrians to raise more than 110,000 to 120,000 regular army troops for most of the century. (Jean Bérenger, *A History of the Habsburg Empire, 1700–1918* (London: Longman, 1990), 43. He is contradicted by Hochedlinger for the second half of the century, *Austria's Army*, 297ff. In 1740, real strength stood at 107,892 foot and horse soldiers; in 1748, again, effective strength stood at 120,000–130,000. By 1760–61, 180,000 regular army effectives is Hochedlinger's estimate. Under Joseph II, the regular peacetime army stood at 220,000 men, but at peak, when Austria was at war

on three fronts (Turkey, Prussia and Poland), an inclusive number of 497,000 was reached (inclusive meaning border units, the *Militärgrenzer*, artillery and special corps). For Hochedlinger, this represents the militarisation of Austrian society, resembling, although not fully parallel, to the Prussian model so favoured by late eighteenth-century military theorists.

18 '[D]ynastic/state control over the concept of honor was one attempted over much of Europe, not least because it helped with one of the major problems of the age: persuading the nobility to govern in the interest of the state . . . the switch to honor in service rather than birth.' Black, *European Warfare, 1660–1815*, 131.

19 See my discussion of Ahmed Resmi's embassy to Berlin in *An Ottoman Statesman in War and Peace*, chapter 2, esp. 89–90, for the interview which occurred in late 1763.

20 Cantemir was familiar with the Ottoman court, having been educated in Istanbul, where he and his rival Brancoveanu had resided in the 1680s and 1690s as guarantors of the good faith of their families in Moldavia and Wallachia. Demetrius Cantemir, *The History of the Growth and Decay of the Othman Empire* (London, 1756), as indicated in the biographical material at the end of the narrative which stops with the Treaty of Adrianople in 1713 (454 ff).

21 Ibid., 450–52. This indignation has an ingenuous and self-justificatory tone.

22 Florin Constantiniu, 'Tradition and innovation in the eighteenth-century military structures of the Romanian lands', *East European Society and War in the Pre-revolutionary Eighteenth Century*, 389.

23 George Vernadsky et al., eds, *A Source Book for Russian History from Early Times to 1917* (New Haven, CT: YUP, 1972), 335.

24 B.H. Sumner, *Peter the Great and the Ottoman Empire* (Hamden, CT: Archon Books, 1965), 46.

25 Hasan Kurdi calls himself the son of a Janissary, who was added to the corps in 1657 (probably as a child, a typical practice), who was on campaign in 1671 and 1672, but thereafter served as a clerk in Janissary offices on Crete, Midilli, Ağrıboz and İnebahtı from 1682 to 1705, when he returned to Istanbul.

26 Russian National Library, St Petersburg Fund 933, TNS A Collection 155, folios 24–47, listed as *Anonymous History of Prut*, but author identifies himself as Hasan Kurdi, folio 47a. (Hereafter Hasan Kurdi.) This passage is from folio 27. The Ottomans applied Rum (or Byzantines) to Greeks, or other Orthodox Balkan Christians.

27 Hasan Kurdi, folio 30–31.

28 Hasan Kurdi, folio 26b. He watches them arrive by river to the fortress of Isakcı in late June 1711.

29 John Keep, *Soldiers of the Tsar: Army and Society in Russia, 1462–1874* (Oxford, 1985), 101.

30 David Chandler, *The Art of Warfare*, 66. Keep has figures for 1711 indicating much the same preponderence of infantry, but divided into field and garrison troops: 33 cavalry regiments (43,824), 42 field infantry (62,454), and 43 garrison infantry regiments (64,769). This excludes the native (or oriental) frontier mercenaries.

31 Keep, *Soldiers of the Tsar*, 103–7.

32 Vernadsky et al., *Source Book*, 332–34.

33 Chandler, *Art of Warfare*, 20–23.

34 Voltaire, *Russia Under Peter the Great*, M.F.O. Jenkins, trans. (Rutherford, NJ: Fairleigh Dickinson University Press, 1983), 161. The original work was published in two parts 1750–63. In this account, Voltaire's venom is reserved not for the Turks, but for Charles XII. There are dozens of readily available accounts of the actual battle and its aftermath. The review of the sources from the Ottoman point of view is in Akdes Nimet Kurat, *Prut Seferi ve Barışı 1123* (1711) (Ankara, 1951). I have deliberately chosen to stay with the contemporary accounts.

35 Sumner, *Peter the Great*, 39.

36 Cantemir, *History*, 451.

37 Necati Salim, *Prut [1711]: Türk Ordusunun Eski Seferlerinden bir İmha Muharebesi* (Istanbul, 1931), 16.

38 Hasan Kurdi, folio 25b–26. Salim has the main body of the army crossing on 1 July (27). I have generally checked the manuscript version against Salim's account, which is one of a series of campaign histories published by Askeri Matbaa (Military Press) in the 1930s. Salim's sources were the Ottoman historian Mehmed Raşid, several Russian memoirs and Von Hammer. The navy, both Mediterranean and Black Sea fleets, was deployed at the same time (Salim, *Prut*, 15).

39 Chandler, *The Art of Warfare*, 305; the number probably includes the Tatar contingent. Salim, 27, lists 120,000 standing, 130,000 provincial and 70,000 Crimean forces. Hasan Kurdi estimated 97,000 Janissaries (includes infantry, cavalry and artillery, divided into 7 columns with 25 cannons apiece), and 40,000 provincial troops from the Balkans crossing the bridges on the Prut on 19 July (folio 33). 70,000 troops (mostly Tatars) from the Crimea (arriving from Bender that same day) went further up the left bank

of the Prut, where some crossed and closed the circle on the czar's camp on eastern and northern flanks. Peter later exaggerated the ratio further, claiming Turkish forces had 270,000 men against his 22,000. Voltaire says that if that is so, Russian losses numbered 16,246, as Peter noted he had 31,554 infanty and 6,692 cavalry on 20 July (Voltaire, *Russia*, 158).

40 Salim, *Prut*, 28–30. Voltaire notes that Charles XII was only 25 leagues away from the czar at Bender, where he stayed throughout the campaign.

41 Hasan Kurdi, folio 31b–32a. An extraordinary amount of money, if the account is to be believed. '*Dil*' very often has the sense of 'informant', so could be referring to captives as sources of information rather than body parts.

42 Hasan Kurdi, folio 36.

43 Voltaire, *Russia*, 158. Hasan Kurdi's account parallels that of Robert Sutton, then British Ambassador in Istanbul, who noted Ahmed III's tremendous relief; see Akdes Nimet Kurat, ed., *The Dispatches of Robert Sutton* (London: Royal Historical Society, 1953), 60.

44 Salim, *Prut*, 40–41, footnote 2; his source is Von Hammer.

45 Charles XII left owing various creditors and the sultan three million thales which was finally repaid by Sweden with a shipload of muskets and explosives in 1738. See Caroline Finkel, *Osman's Dream* (London: John Murray, 2005; Basic Books, 2006), 336.

46 Hasan Kurdi, folio 42–43.

47 N.A. Bées, 'Mora' *EI2* 7 (1993), 236–41, is an informative summary of the long rivalry. See also Fariba Zarinebaf, John Bennet and Jack L. Davis, *A Historical and Economic Geography of Ottoman Greece: The Southwestern Morea in the eighteenth Century* (Athens: American School of Classical Studies at Athens, 2005), which includes an appendix describing the fortress town of Navarino, an important port for the Ottoman navy expeditions against Crete in the seventeenth century and the site of the destruction of the Ottoman fleet in 1827. Zarinebaf's translation of the *kanunname* on the Morea, compiled by Ottoman officials in 1716, is included.

48 Karl A. Roider, *Austria's Eastern Question*, 41–42, a good survey of the choices facing Austria in the period under discussion.

49 William H. McNeill, *Venice*, 220–23, argues that the Venetians were in fact financially over-extended in the territories acquired in 1699, and after 1715 'ceased to dream of resurrecting their imperial greatness', with considerable relief (221).

50 Benjamin Brue, *Journal de la campagne que le Grand Vesir Ali Pacha a faite en 1715* (Paris, 1870), 67–68. Brue was an interpreter in French service at the time. Brue estimated the passage at 1 league per hour, the total of leagues from Salonika to Corinth at 195.5 (1 league equalling 3 miles or 4,445 meters, i.e., 9–15 hours a day). The estimated total number of men, including those who were either advance guards, or joined up after 9 June, was 110,364 (22,844 cavalry and 87,520 infantry): 67. Such figures should be only considered a guide.

51 Brue, *Journal*, 62–64. The Janissary Corps was paid three times a year: the final payment was actually two quarters, representing six months. Late pay, sometimes no pay, was a common feature of pre-modern army life, but the Ottomans seem always to have found a way to pay out something. Whether the individual amounts distributed sufficed is another story entirely.

52 Brue, *Journal*, 34. A total of 141 cannons head the list.

53 Brue, *Journal*, 37.

54 Brue, *Journal*, 31–33.

55 Brue, *Journal*, 42. By that time, of course, contingents of Janissaries had been ordered to capture the fortresses of Corinth and Nauplion, both larger and more significant than Modon. Many of the casualties were a result of the Janissary-style headlong rush against well-fortified walls, against the instructions of the grand vizier.

56 Brue, *Journal*, 27.

57 Brue, *Journal*, 31.

58 Brue, *Journal*, 9–10.

59 Brue, *Journal*, 30–34. The fortress of Nauplion was defended by '700 disciplined troops, 5–600 slaves and some Greek militias', (34), according to the submission of Commander Alexandre Bon. That throws into considerable doubt Brue's estimate that 25,000 inhabitants were enslaved by the victorious Ottomans. I have suggested caution about all his figures and descriptions.

60 Chandler, *The Art of Warfare*, 305. Bérenger, *History*, 30, lists the number of Ottomans assembled at Belgrade at 120,000 men. Hochedlinger, *Austria's Wars*, puts the Austrian paper strength at 86,000 men and 88 guns at Varadin.

61 Creasy, *History of the Ottoman Turks* (London: R. Bentley and Son, 1877), 342. He has obviously consulted Von Hammer, who himself used Mehmed Raşid, the historian of the period.

62 Creasy, *History of the Ottoman Turks*, 345.

63 Chandler, *The Art of Warfare*, 305. Bérenger, *History*, 30–31.

64 Hochedlinger, *Austria's Wars*, 196, for the most recent historiography.

65 Especially good in English are Cassels and Roider's two works. See Roider, *Reluctant*, for the best retelling of the diplomatic manoeuvres of the Austrian court throughout the war, although he has been criticised for overlooking the Serbian context. In French, Keralio's retelling of the campaigns is particularly interesting: *Histoire de la guerre des russes et des impériaux contre les turcs en 1736, 1737, 1738 and 1739, and de la paix de Belgrade qui la termina*. 2 vols. (Paris, 1780). He emphasises the sectarian differences which divided the Austrian command, and is scathing about Münnich's self-importance and the wastefulness of the Crimean campaigns of the period. One of the Keralio maps is reproduced here (Plates 7 and 8). Equally interesting as a contemporary text is *Memoirs of the Life of the Late Right Honorouable John Lindesay, Earl of Craufurd*, by Richard Rolt, a eulogy to Crawford, fatally wounded at Grocka (London, 1753). The latter gives the reader a sense of the culture of foreign, high-born command which was characteristic of both the Russian and Austrian officer corps of the age.

66 Roider, *Reluctant*, 44.

67 Keralio, *Histoire de la guerre des russes et des impériaux* vol. 1, 27.

68 Keralio, *Histoire de la guerre des russes et des impériaux* vol. 1, 39.

69 Keralio, *Histoire de la guerre des russes et des impériaux* vol. 1, 67.

70 Roider, *Reluctant*, 70.

71 Hickok, *Ottoman Military administration in Eighteenth-Century Bosnia*. (Leiden: Brill), 26–27. Hickock utilised all the Ottoman sources on this campaign.

72 Keralio, *Histoire de la guerre des russes et des impériaux* vol. 1, 107.

73 This treaty was signed in spite of the objections of the Russians, who had signed a similar treaty in 1735, on the condition that the Persians not deal with the Ottomans without prior Russian approval.

74 Keralio, *Histoire*, vol. 1, 96.

75 Cassels, *The Struggle for the Ottoman Empire*, 132. See also Virginia H. Aksan, 'Manning a Black Sea garrison in the eighteenth century; Ochakov and concepts of mutiny and rebellion in the Ottoman context', *International Journal of Turkish Studies* 8 (2002), 63–72.

76 These from the levies described above. Uzunçarşılı, *Osmanlı Tarihi* 4:1, 264. Cevat Erbakan, *1736–39 Osmanlı-Rus ve Avusturya Savaşları* (Istanbul, 1938), 27–30, says 4,000. Muhsinzade Abdullah had a distinguished career as a soldier and statesman, and is the father of Muhsinzade Mehmed Pasha, the most experienced of the generals of the 1768–74 War, grand vizier on

the Danube 1771–74. Erbakan's work is untrustworthy; more reliable is *Türk Silahlı Kuvvetleri Tarihi Osmanlı Devri* vol. 2, pt. 4: *Ikinci Viyana Kuşatmasından Nizâm-ı Cedidin Teşkiline Kadar Olan Devre* (Ankara, 1982), 238–47.

77 Keralio, *Histoire*, vol. 1, 111.

78 Cassels, *The Struggle for the Ottoman Empire*, 132.

79 Edward S. Creasy, *History of the Ottoman Turks*, still a generally lively and passionate account, 365–68. He and Cassels blame the Cossacks for the slaughter of the Ottoman side. General Münnich had a well-known reputation for severity and brutality.

80 The costs were high for the one prize; apart from manpower, Keralio notes 15,000 pairs of oxen for the artillery alone perished in that season (vol. 1, 129).

81 Probably an overestimation; Roider, *Reluctant*, 123–24.

82 Keralio, *Histoire*, vol. 1, 144–52.

83 Bosnia's standard commitment to Ottoman European campaigns generally numbered 2,000 infanty and cavalry. Both Hickok's account of this Bosnian example and the archival records indicate the increasing percentage of the local militias in the main field army (Hickok, *Ottoman Military Administration*, 26–27).

84 Hickok, *Ottoman Military Administration*, 30–35; Uzunçarşılı, *Osmanlı Tarihi* 4:1, 274–77. I have been very conservative about numbers, as both troop and casualty numbers vary greatly for this confrontation.

85 This is well summarised by A.Z. Hertz, 'Ottoman conquest of Ada Kale 1738', *Archivum Ottomanicum* 6 (1980), 155–59. [151–209] Hertz is also the author of 'Ada Kale: the key to the Danube (1688–90)', *Archivum Ottomanicum* 3 (1971), 170–84.

86 Roider, *Reluctant*, 112–14.

87 Cassels, *The Struggle for the Ottoman Empire*, 152. At the Belgrade siege the same year, an estimated 80–100 Austrian troops died of plague, malaria and dysentery each day (153).

88 Roider, *Reluctant*, 132.

89 Uzice and Rudnik (Hickok, *Ottoman Military Administration*, 35). A.Z. Hertz, 'Ottoman conquest', 159, has made very significant use of contemporary German and Ottoman descriptions of this campaign.

90 Starting a campaign that early meant that an army had to remain stationary until '. . . the Habsburg high command deemed the spring growth of fresh green grass [sufficient] to nourish the thousands of horses necessary . . .' (Hertz, 'Ottoman conquest', 163).

91 Adakale (New Orsova) had become Austrian territory in 1717, and their fortress, St Elisabeth, had taken over a decade to build (Hertz, 'Ottoman conquest', 152–53).

92 Hertz, 'Ottoman conquest', 169; also Hochedlinger, *Austria's Wars*, 215.

93 Roider, *Reluctant*, 137.

94 Hertz, 'Ottoman conquest', 170–77.

95 More than half the army had disappeared from death and disease, and the Ottomans had resorted to leaving empty tents erected to disguise their reduced strength. Hertz, 'Ottoman conquest', 187.

96 Hertz, 'Ada Kale', 164, and 'Ottoman conquest', 194–96.

97 Roider, *Reluctant*, 138–40.

98 Roider, *Reluctant*, 153–54.

99 Roider, *Reluctant*, 160–61. The memoirs of the Earl of Crawford include a list of the dead and wounded (Rolt, 204). Hochedlinger, *Austria's Wars*, 215, lists 2,000 dead (including five generals) and 3,000 wounded.

100 Crawford was nonplussed as to why Wallis retreated, and equally surprised at the Ottoman restraint (Rolt, *Memoirs*, 201–2), but complimentary about İvaz Mehmed's command.

101 Cassels, *The Struggle for the Ottoman Empire*, 168–69.

102 This was a demand insisted upon by the Ottoman side, that the troops on site be involved in and witness the destruction, forcing the hand of Neipperg by accelerating the imposition of the stipulations. I believe Ottoman negotiating strengths are downplayed in the narratives of these events in favour of the bribery school, which claims that nothing was accomplished without the requisite '*bakshish*' (Cassels, 193). Hochedlinger calls the Habsburg diplomacy 'amateurish and uncoordinated' (*Austria's Wars*, 215).

103 The Belgrade treaty is another text which needs to be analysed in its original and translated versions: it took more than a year to negotiate the final language of the treaties, largely because of the haste and ambiguities of the first version (Roider, *Reluctant*, 176). For those keen to read the standard contemporary account, see Marc Antoine Langier in bibliography.

104 Roider, *Reluctant*, 175–76. Bruce McGowan has a good article on 'population and migration' in Halil Inalcik and Donald Quataert, eds, *An Economic and Social History of the Ottoman Empire* (Cambridge: Cambridge University Press, 1994), 646–57.

105 Rolt, *Memoirs*, where Crawford describes Grocka, notes that 300 Rascian [Serbs] cavalry were both an integral part of the valiant stand at the ambush (181), and responsible for it ('. . . the treachery of the Rascians, whom the

Turks permitted to pass . . .' 186). He expresses a pretty universal sentiment about the reliability of local militias in this era.

106 McNeill, *Europe's Steppe Frontier*, 178.

107 Aksan, *An Ottoman Statesman*, 8–9. See Kemal Beydilli's work on the Prussian connection: *Büyük Friedrich ve Osmanlılar* (Istanbul: I.Ü. Edebiyat Fakültesi, 1985), and *1790 Osmanlı-Prusya İttifâkı* (Istanbul, 1984). See also Hickok, *Ottoman Military Administration*, 8–9.

108 Both are examined in Virginia H. Aksan, 'Ottoman political writing, 1768–1808', *International Journal of Middle East Studies* 25 (1993), 55–56 (53–69). Sarı Mehmed's work has been translated: *Ottoman Statecraft*, ed. and trans. by W.L. Wright (Princeton, NJ: PUP, 1935). Müteferrika's *Usul ül-Hikem fi Nizam ül-Ümem* was published in French translation in 1769: *Traité de Tactique*, trans. K.I.S. Revicsky (Vienna, 1769).

CHAPTER FOUR

The Austrian–Russian–Ottoman Danube waltz, 1768–92

Overview

This chapter deals with two pivotal wars in Ottoman history, known to the Russians as Catherine II's First and Second Turkish Wars, rightly celebrated (at least the first one) in St Petersburg and Moscow as the triumph of her southern strategy to acquire access to the Black Sea. For the Ottomans, it is the moment of greatest collapse and humiliation after Karlowitz, and inaugurates the military reform period we describe generally as the New Order of the two sultans Selim III (1789–1807) and Mahmud II (1808–39). Among historians of Europe, this period is generally regarded as stimulating an international relations debate around the 'Eastern Question', on the future of the Ottoman Empire.

Bearing in mind that anything written about the Ottoman Empire after 1760 peeks out from behind the Eastern Question curtain, this chapter briefly looks first at the political and social settings in the Balkans in 1768, then examines mobilisation, logistics and financing in the 1768–74 and 1787–92 campaigns. Of particular emphasis will be the complete collapse of the Ottoman supply systems. Descriptions of major confrontations allow for an examination of the strengths and weaknesses on the Austrian, Russian and Ottoman side. The chapter concludes with some reflections on the impact of this period on the imperial order in all three contexts.

1768–92 and the Ottoman Empire

The period from 1740 to 1768 was one of the longest periods of peace for the Ottoman Empire in its entire history. With the exception of the Iranian frontier, which erupted in warfare and rebellion after the break-up of the Safavid order in the 1720s, the empire rested on the triumph of Belgrade in 1739. Led by an astute grand vizier, Koca Ragıb Pasha, Vizier 1757–63, who convinced several sultans of the virtue of peace and multilateral negotiations, the empire achieved two decades of tranquillity and economic recovery. The Ottomans benefited from the expanding world economy of the period as well.[1]

Part of the reason for prosperity was that the Ottomans sat out the international upheavals of the series of continental and worldwide campaigns that permanently altered the balance of power of Britain and France in North America and India, and established the importance of Russia and Prussia in eastern Europe. The Seven Years War (1756–63) is also acknowledged as pivotal to the military history of Russia, producing generals like Rumiantsev and Suvorov who were to provide such decisive leadership in the two Ottoman–Russian wars at the end of the century. Austrian struggles with Ottoman and Prussian armies alike shook Habsburg Vienna from its lethargy, leading to a vigorous reorganisation of the military.[2]

In the 1760–1830 period, Ottoman country-wide gentry, or 'Ottoman-local elites', figured enormously in the success or failure of the Ottoman war effort. War profiteering was a constant stimulus for temporary, if reluctant, loyalties, and generated powerful regionalisms, manipulated by colourful and controversial figures, who both challenged and cooperated with the late-eighteenth, early-nineteenth century Ottoman state. The conflict between the Janissary loyalists and the new upstarts is one of the constant motifs of the chronicles, most widely studied by scholars of Ottoman Egypt and Syria.[3]

The growth of a provincial elite was facilitated by the decentralisation of tax collection, especially the widespread use of tax farming (*iltizam*) and the concentration of both mobilisation and supply in the hands of many of the same tax officials, frequently referred to as *mutasarrıfs*.[4] By 1768, local 'committees of notables' and Muslim judges were responsible for campaign preparations: mobilising troops, requisitioning grain and biscuit (*peksimed*), supplying draught animals and wagons, and provisioning the bivouacs (*menzil*) along the route of the army's passage. In the midst of the widespread corruption and abuse of taxation privileges which accompanied warfare, honesty occasionally prevailed, in the reports of concerned bureaucrats which dot the account books of the war periods, allowing us glimpses into the 'realities' of Danubian logistics. Still, one of the most obvious consequences of shrinking territories was that the tax burden was redistributed to smaller, often dislocated populations. This too is reflected in the local court records, where distribution of the extraordinary taxes across village collectives is evident (*avariz*, and *tevzi*).[5] While we are not yet in a position to talk about the burden of taxation for the whole period, there is clearly emerging evidence of severe pressures on local resources after 1760.[6]

When the Ottomans took up arms again in 1768, it was to inaugurate almost a century of confrontation on the Danube line, the 'waltz' in the title of this chapter. The pattern of events indeed took on a certain rhythmic regularity. Austrian or Russian armies invaded a Danube Principality, which precipitated a hasty declaration of war from the Ottomans. Russian army headquarters were established in Jassy, from which the commanders could deploy troops down the Prut to pivotal garrisons on the mouth of the Danube. In the years after 1774, the Russians maintained troops in Moldavia (Jassy) or Wallachia (Bucharest) almost at will. The Ottomans did not maintain a regular field army, and were slow to mobilise, although considerable and fairly constant attention was paid to the garrisons in the line from Belgrade to İsmail. The Ottomans considered the investment in that line absolutely essential. After 1812, when for example the Rusçuk fortress and its town were completely destroyed by the Russians, and again in 1828, Mahmud II initiated reconstruction projects of the fortress system that lasted until the loss of Bulgaria in 1877. Such attention accounts in part for the initial success of the Ottomans in the Danubian battles of the Crimean War, as well as for the stiff resistance of the Ottoman forces under Commander Osman Pasha at Plevne in the 1877–78 War.

The two wars of the period under discussion were characterised by a long and bitter series of sieges, by numerous bloody attacks and counter-attacks, mining, and sapper activity. The Ottoman garrisons were eventually worn down by endless artillery barrages, and often a massacre of the occupants of the fortress towns followed. For example, Ochakov, the last Ottoman garrison in the Crimea, was invested and captured in 1788, and İbrail, a major garrison (and source of supplies) protecting Bulgaria, in 1791: these were very significant victories for the Russian army. Russian observers were impressed by Ottoman last-ditch efforts, which often resulted in a terrible toll, with troops dying to the last man.

PLATE 9 *The Danube basin (a map of the Russian dominions in Europe)* (McMaster University Library, Rare Maps Accession #9464. A Map of the Russian Dominions in Europe, with Adjacent Countries extending Westward beyond the Vistula. London. Published by J. Cary, No. 181 Strand May 1st 1814).

BENDER
Tiraspol
Odessa
Oczakow
Kinbourn
Ovidiopol
Akkerman
Ismail
Kilia
Tuldscha
Kostendsche
NOTE
Rußian Territories are colored Green
Turkish D° Blue
Austrian D° Yellow
Dutchy of Warsaw Red
Prußian Territories Orange
Swedish D° Purple

Şumnu, in present-day Bulgaria, became Ottoman headquarters in the late eighteenth century, and figures largely in the wars of the first half of the nineteenth century. Increasingly, the Russians focused on subduing the Trans-Caucasus, a secondary diversion for the Ottomans which assumed primary importance by the Crimean War. Kars and Erzurum in eastern Turkey were first attacked and occupied by Russian armies in 1828. Peace was often hastily constructed on the battlefield – between commanders – in response to Istanbul pressure, but just as often because of grand vizierial consultations *in situ*. During the confrontations of 1828–29, the Russians came perilously close to Istanbul, a fact which led to panic and quick capitulation by the city and dynastic household.

Once actual conditions of peace were agreed upon, carrying out of the terms often lasted decades, especially in the 1760–1830 period. The loss of the Crimea in 1783 was bitterly resented in Istanbul. Arguments over borders in the eastern Black Sea region were long and acrimonious after 1807. Similarly, Mahmud II stiffly resisted external interference with populations he considered his subjects, notably the Serbians and then the Greeks. It is easy to argue that the Ottomans and Russians were in a constant state of hostility after 1768, with the possible exception of the uneasy (and brief) alliance against Napoleon at the turn of the century.[7]

By 1768, the Ottomans had largely moved from a professional to a 'volunteer', militia-based army, paid for by a combination of local, sometimes extraordinary taxes, combined with direct support from the sultan's treasury. Even the Janissary regiments, who still manned the border fortresses, were newly reconstituted for the campaigns that began in 1768. The hiatus had meant the almost total collapse of any system of discipline and recruitment, and the Ottomans were completely unprepared for a major campaign when they made the initial declaration of war against Russia in 1768. Their tenuous economic recovery was halted by the two decades of warfare on which they embarked, impeding any efforts at military and fiscal reform until the peace of Jassy in 1792. Between 1760 and 1800, prices trebled, deficit budgets became the norm, and the state occasionally resorted to forced loans from its nobles and confiscation of their estates in order to continue to finance warfare.[8]

The period after 1750, it has been argued, was one of the most significant of the phases of the development of the modern European army.[9] The Ottomans missed a generation of developments by remaining outside the battlefields of the Seven Years War. Thus, the technology gap significantly hindered their ability to counter the massed firepower of the mobilised rapid field artillery characteristic of the post-1756 period. Ottoman sultans like Mustafa III (1757–74) were well aware of the problem and commissioned foreign experts such as Baron de Tott to rectify the deficiency, but systemic economic and organisational problems proved insurmountable. It is possible to argue, however, that erratic but fairly consistent attention to the artillery as an essential auxiliary corps characterised the period 1760 onward. Reconfiguring the armed forces generally became a primary preoccupation of generations of Ottoman statesmen only after the utter collapse of the Danube defence system after 1787.

1768–92 and the Russians

A glance at Russia for the period under discussion reveals similar problems but a very different outcome. The most striking aspect of the period 1760–1830 in the Romanov Empire is the massive amount of territory it acquired under Catherine II (1762–96), including not only Ukraine and chunks of Poland but also great parts of the northern Black Sea coast, the Caucasus and Central Asia. An expanding rather than a contracting empire, Russia under Peter the Great and Catherine added multi-ethnic, multi-religious populations in great numbers, challenging and ultimately reinforcing Russian orthodoxy as the state ideology of the nineteenth century. Similarly, Peter the Great's early eighteenth-century radical transformation of the Russian military into a professionally run and empire-wide mobilisation project, ultimately forced the continuation of serfdom as the bedrock of an imperial system until the middle of the nineteenth century. The construction of universal political sovereignty, Russian-style, required the cooperation of the local military and administrative service and supply of men and arms. Peter rationalised military and civilian service in the Table of Ranks (1722). Catherine the Great's Charter of the Nobility of 1785 formally liberated the nobles from servitude to the state, and guaranteed their property rights, and by extension, their serfs.[10] Both rulers inaugurated the Russian social transformation, anticipating the Ottomans in the process of modernisation by a half-century.

To insist on Catherine's enlightenment agenda is to blind us to the realities of imperial politics, however, and the horrors of the dreadful battlefields of the northern Black Sea coast. Her difficulties are best demonstrated by her policies concerning the Crimea, where an administrative reform agenda conflicted with the realities of accommodating new settlers in Muslim territories. The unilateral incorporation of the Crimean Peninsula into the empire in 1783 offered an opportunity to test policies that later operated in the expanding frontiers of Muslim Central Asia. Efforts to draw Tatar nobles and Muslim religious officials into state service, however, were accompanied by colonisation measures which forced the significant exile of Tatar populations into Ottoman territories, from land already much depopulated by decades of warfare, into lands which would be contested by Tatars, Circassians, Kurds and Christians (Armenians and Georgians) until the twentieth century. By the nineteenth century, the Crimea was predominantly Russian/Orthodox territory.

Expansion into the Ukraine and the Crimea considerably enabled the Russian war effort by stabilising a notoriously unruly frontier and gradually bringing the Cossacks under state control. Late eighteenth-century warfare on the Ottoman frontier was costly and wasteful, however, no more so than in the confrontations in the Principalities and the Crimea. Financing became an acute problem for the Romanovs in the 1780s, when the brilliant, and savage, conquests of Suvorov climaxed two decades of victories in the south for the Russians. Europeans remained impressed with the courage and tenacity of the Russian infantryman, proved on the battlefields of the Seven Years War.[11] The Russo-Ottoman wars of 1768–74 and 1787–92 (the latter including a reluctant Austria from 1788–91), however, proved very destructive to Catherine's legislative and economic reforms, and generated opposition from noble and peasant alike.[12] The 1773–74 Pugachev Rebellion in particular threatened to bring down the dynasty altogether. The Russian army, composed of mobilised and 'freed' serfs, increased threefold in the decades under discussion, with a similar increase in expenditure.[13]

It is thus possible to consider the Russian imperial drive entering a different phase after 1790, when the three contenders for control of the Danube and the Black Sea paused to catch a breath before being drawn inexorably into the international context of the Napoleonic period. The Ottomans struggled to regroup under Selim III and Mahmud II; the Romanovs, under Catherine's successors Paul and Alexander, slowed the process of reformation, restored the predominance of the military, and reinstated the role of the Church in imperial affairs.[14] Adherence to Orthodoxy, and Russification, became integral parts of later Romanov subjecthood.[15]

1768–92 and the Austrians

The Habsburgs, by contrast, after acquiring prodigious amounts of territory between 1714 and 1739, were exhausted by several decades of expansion and campaigns on multiple fronts. During the 1737–39 War, as we have seen, Austria fielded 140,000 regular troops as well as 45,500 irregulars from the military border settlements, but had proved unable to dominate the struggles with the Ottomans on the upper Danube, resulting in the loss of Belgrade in 1739. That humiliation was followed immediately by Frederick II's attack on Silesia in December 1740, which continued in the War of Austrian Succession until 1748. Both it and the Seven Years War which followed proved a trial by fire for Maria Theresa (1740–80) and her son Joseph II (1780–90). Extensive Prussian-style reform of agrarian estates and military conscription led to an increasing militarisation of Austrian society. By one reckoning, 50 per cent of state expenditure went to the peacetime Austrian army in the 1780s.[16] Austrian strategy in the 1760s and 1770s was to explore diplomatic alternatives in preference to open conflict with the Ottomans as well as Prussia and Russia. 'Vienna had come to value the Turks as peaceful neighbors, and the Carpathian Mountains constituted a natural border between Austria and Poland.'[17] This may have prompted the July 1771 treaty with the Ottomans, by which the Ottomans promised a subsidy to Vienna and the cession of Little Wallachia, which the Austrians had surrendered in 1740, in return for guaranteeing Polish territorial integrity. It was abrogated by Austria in 1772, never seriously implemented, and in any case, rendered void by the first Polish partition.

An uneasy rapprochement with Prussia, equally fearful of the substantial Russian gains in the 1769–70 campaigns against the Ottomans, allowed for the first partition of Poland in 1772 by which Austria added Galicia to its territories. An opportunistic seizure of Bukovina from the defeated and vulnerable Ottoman Empire was territory gained at little expense. The theft was formally acknowledged by the Ottomans in 1775. By 1781, Joseph II had signed a secret alliance with Catherine the Great, which would eventually draw Austria into the second Russo-Ottoman War of 1787–92. The alliance followed discussion between the two monarchs concerning the division of the Ottoman territories, especially Moldavia, Wallachia and Bessarabia. Known as the 'Greek Project', in reality it was the testing of waters about the division of spoils should the Ottomans begin war anew, an early stage of the Eastern Question. Catherine aimed at liberating the three territories above, an independent state to be called Dacia, while Joseph II was to have a free hand in the rest of the Balkans. As events unfolded, Austria aimed at a border approximating that of Treaty of Passarowitz in 1718; Catherine's interests lay primarily in the Crimea, and Ochakov, key to the northern Black Sea littoral. In spite of an ambitious conscription programme, and no less than six Austrian armies spread along the eastern frontier, the Ottomans proved a resilient foe in the rough and disease-ridden terrain of the Austro-Ottoman border. French loss of influence in Europe, and the deterioration of relations with Austria, left Joseph II vulnerable by 1790. He died in February of that same year, and was succeeded by his brother Leopold II (1790–92). Prussia chose that moment (January 1790) to conclude a defensive alliance with Selim III. Austria found itself on the brink of war with Poland and Prussia, and forced into an early withdrawal from the war with the Ottomans by a September 1790 armistice, in spite of notable territorial successes on the battlefield in the year. The upshot was the July 1791 Austro-Prussian treaty, and the August Sistova Treaty with the Ottomans. By special convention the Austrians acquired Old Orsova at last, a very costly enterprise for a postage-stamp, but a vital piece of cross-Danube territory, and gateway to the Banat.[18]

The extremely taxing and costly campaigns against Napoleonic armies in 1796–97, and the peace of Campoformido of 17 October 1797, capped an age of Habsburg military engagement and defeats which would lead to a complete reordering of Austrian Habsburg territories by 1803.

1768–74: Russian miracle, Ottoman debacle, and Austrian test

There are two persistent ideas about the 1768–74 war, both of which concern Poland. The first is that the outcome of the war could only be determined after Austria, Prussia and Russia decided on the first division of Polish territory in 1772. The second is that French diplomats were responsible for bribing the Ottomans to precipitate the hostilities in late 1768, as part of their effort to secure their candidate in the succession question in Poland. The first idea owes its predominance to the historiography of high politics, arguing the shifting balance of power in eastern Europe consequent upon the challenge of Frederick of Prussia, which unsettled all of Europe. The second view makes France the earliest and longest 'friend' of the Ottomans, and justifies Ambassador Villeneuve's diplomatic triumph over the Habsburgs in the Treaty of Belgrade, with which I closed the previous chapter.

France was an ally of Sweden and the Ottoman Empire, and latterly the Austrians, after 1756, while Britain proved a distant, often frosty friend of Russia. For the 1760s–1780s period, protection of trading networks remained paramount for both powers, while the Colonial Wars played out the future of Franco-British rivalries on the world stage. For the three territorial empires of eastern Europe, the disposition of Poland was a more acute and pressing problem, arguably one that inaugurated the 'Eastern European Question'. Frederick the Great's expansionist aims served as the catalyst after 1740. Poland, as a political entity, disappeared off the map of Europe until the twentieth century, following the third and final partition in 1795.

Neither view of the 1768–74 war in itself is particularly wrong, merely incomplete: they ignore Ottoman strategic imperatives. Ottoman diplomacy, conducted until the end of the century from Istanbul, depended on information circulating in the hothouse of the foreign community, using the infamous dragomans, or interpreters, as their linguistic go-betweens. Another source of intelligence was the Tatar exile community, whose size and influence expanded enormously as a result of the Russian conquest and subsequent annexation of the Crimea. Sifting through the conflicting reports from the Crimean Khan and his rivals, as well as from the governors of Moldavia and Wallachia, was the task of the newly emerging foreign affairs bureaucracy of the eighteenth-century Ottoman court, which had became the responsibility of the *Reisülküttab* and his staff.[19] The *Reis* and the Grand Vizier were fully aware of the problem of Poland, but more concerned about the territorial aggrandisement of Russia, as numerous political memoranda of the period acknowledge. Furthermore, the lengthy negotiations which Frederick the Great carried on for a decade with Mustafa III and Grand Vizier Koca Ragıb, culminating in Ahmed Resmi's official embassy to Berlin in 1763, demonstrate a level of awareness generally not accorded the Ottoman side. Resmi's route took him across Poland, and he described the current disorder there following the death of Augustus III.[20] French pressure to intervene aside, the Ottomans preferred neutrality in 1764, when a note was circulated to that effect to the diplomatic community in Istanbul.[21]

In fact, the gradual Russian incorporation of the various parts of Cossack territories, especially its encouragement of the settlement and active fortification of Zaporozhian borderlands by the Russians after the 1750s, alarmed the Ottomans. By 1764, the northern frontier of the Zaporozhian Cossacks had become an imperial province known as New Russia, which organised colonists on a military frontier. It is no small coincidence that the first Governor General of the new territory was P.A. Rumiantsev, whose outstanding performance as Commander-in-Chief of the Russian forces in 1768–74 was built on his detailed knowledge of the problems of mobilisation and logistics in the newly-acquired imperial territories. The last of the Zaporozhian territory was incorporated into the Russian empire in 1775, immediately after the victories of the 1768–74 War gave Catherine access to the Black Sea.

Poland, Russia and the Ottomans were equally engaged in attempting to pacify unruly borderlands, the first two on the right and left banks of the Dnieper and the Ottomans in the Principalities to the southwest, Moldavia and Wallachia. Poland and Russia feared the potential coalitions of Orthodox Cossacks and peasants (*haidamaks*) which periodically erupted in the territories described. Polish Catholics resisted Russian influence, particularly after the Russian installation of Stanislaw Poniatowksi as King of Poland in 1764.

In 1768, the Bar Confederation, an insurrection of Polish Catholic nobles, resisting the imposition of the Russian puppet-king, stimulated a counter Cossack/*Haidamak* revolt of significant size. A Russian army, already stationed in Poland, moved south to suppress both, and violated Ottoman territory at Balta. This was too close for Ottoman comfort, and a protest was lodged, which then became an Ottoman declaration of war. France's leading minister, the Duc de Choiseul, may have crowed that French influence and bribery forced the Ottomans to the battlefields, but the politics of the Danube proved equally influential.

The centre of the conflict: the lower Danube River

A closer look at the state of the Principalities will perhaps make this assertion clear. Nicholas Mavrocordato was the first Phanariot ruler (*voyvoda*, later *hospodar*) of the Principalities, appointed by the Ottomans first to Moldavia (in 1711) and then to Wallachia (to 1715). Thus began the semi-colonial rule of the territories, subservient to their masters in Istanbul, and initially demilitarised. The Ottomanisation of the influential Phanariot families, part of the creation of the empire-wide Ottoman-local elites I alluded to earlier, has prompted much debate about the subsequent development of Romania. Ottoman policy was inspired by strategic concerns, driven by repeated evidence of the disloyalties of the quasi-independent *voyvodas* and *boyars* of earlier centuries, and by the increasing vulnerability of the Ottoman–Russian frontier. During the Phanariot period, which lasted into the 1830s, the countryside was occupied four times by Austrian and Russian armies, and the populations subjected not just to the hardships of warfare, but increasingly odious tax and service burdens imposed by their new rulers. Stripped of their private armies, the governors were responsible nonetheless for buying, often at fixed prices, and delivering, the agricultural products demanded by Istanbul and Russia. The result was peasant flight, with one estimate for Wallachia and Moldavia suggesting a loss of half of the peasant families to territories south of the Danube in the 1740s. The alternative for those of means was to sell to the expanding European market upstream, a trend effectively argued by Stoianovich decades ago.[22]

Ottoman bureaucrats were not completely oblivious to the problems on the Danube. One convincing piece of evidence is an Ottoman investigative report filed in Istanbul in 1759 or 1760. It concerned the illegal occupation of Wallachian farmland by Ottoman soldiers and vagrants engaged in farming, who had crossed from the south of the Danube. This represented an encroachment on peasants' agricultural rights. The investigative committee surveyed and mapped the territory from Fethülislam to İbrail, and restored property, or compensated the rightful owners. The report is a rare indicator of the concern of the central authorities with the stabilisation of the Danubian territories.[23]

The fortress of Vidin, a pressure point of the northern border and scene of numerous bloody conflicts since 1718, was the source of the problem. Large numbers of Ottoman soldiers (described as Janissaries, *yamak*, and other *mevacibli*, or salaried), stood idle at Vidin, their numbers swollen by the exodus of Muslims from newly-acquired Habsburg territories, Hungary and Transylvania. They appear to have been supporting their fellows (*yoldaş*, comrades, also called *serseri*, or vagrants, in the documents) in Craiova, a town north of the Danube in present-day Romania. Demobilised soldiers, local militias and Janissaries were all involved in a network of agriculture and commerce. This was contrary to the agreements with the Principalities, dating from the 1500s and renewed after the Treaty of Passarowitz in 1718, which prohibited the settlement of Muslims on Christian lands north of the Danube. The soldiers were equally guilty of violating peasant rights, the defence of which was one of the principal tenets of the Ottoman dynasty. The commission confiscated some 1,300 small farms, which is one measure of the size of the problem. Returning the lands to the original owners, and resettling peasants who had fled, was also part of their charge.[24]

There is little doubt that such encroachments were now a constant problem of the region, exacerbated by a system of mobilisation that encouraged the creation of private armies, or warrior bands, upon campaign demobilisation. In Sofia, on the eve of the declaration of war in 1768, one such band was responsible for attacking a group of Muslim and Jewish merchants travelling to Samokov, killing two of them and wounding several others. The striking aspect of the names of the culprits is that they are mostly identified as *beşe*, making them fellow travellers of the Janissaries.[25]

Thus, the Ottomans were sensitive to border disturbances in the Balkans, and wary of the gradual creeping imperialism of Russia in Ukraine and Polish territories. The Bar Confederation, led by Potoski, appealed to Sultan Mustafa III, who found a large Russian army close to Moldavia alarming, and had continuous news from the Crimea of Russian fortress repair and building. Notes were exchanged, but Russian equivocation about the size of the force in Poland dissatisfied the Sultan, who was already more than convinced that it was time for war. Grand Vizier Muhsinzade Mehmed Pasha urged caution, arguing forcibly that the empire was simply unprepared for a major campaign, but to no avail. He was dismissed, and replaced by Hamza Pasha, a man more amenable to the wishes of the court, but extremely incompetent, and likely mad. Hamza Pasha was quickly replaced by Mehmed Emin Pasha, himself a product of the court system. It was he who commanded the troops as they left Istanbul. The Russian resident Obreskov, with ten of his entourage, was locked up in the Istanbul fortress of Yedikule, the traditional Ottoman signal that hostilities were about to begin. The official wording blamed the Russians for violating the treaties, territory and persons of the Ottoman sultanate, while interfering in the sovereignty of Poland.[26] The date was October 1768.

Declaration and preparations for war, 1768

Neither side was prepared for engagement on the Danube or in the Crimea, but the precipitate declaration of the Ottomans meant that the belligerents had six months to prepare before serious confrontations could begin the following spring, a fact which gave the Russians, with troops already stationed in Poland and Ukraine, something of an edge. Catherine II was preoccupied in Poland with civil conflict, but quickly organised her resources for war. Putting her legislative reform agenda on hold, Catherine II created an informal council to run the war and negotiations. The sultan's hasty declaration had the unintended effect of refocusing Russia's expansionist policies to the south, and realigning the relations between Prussia and Austria, which led ultimately to the first partition of Poland in 1772. Both Frederick II and Austrian Chancellor Kaunitz considered that the declaration of war had fundamentally altered Europe's political system.[27]

The sultan had previously ordered 6,000 troops to each of the two pivotal fortresses of Hotin and Ochakov. He also ordered Tatar Khan Kırım Giray to make raids into the new Russian territories in the Ukraine as early as January of 1769.[28] Further Ottoman mobilisation efforts can be followed in a unique source, the record of the actions of the central *Defterdar*, Sarım Ibrahim Pasha, at the beginning of the campaign.[29] The first thing to be considered was to ensure that the fortresses on the strategic line – Hotin, Bender, Ochakov, İbrail and Killi – were supplied with 100,000 *kile* each of barley, wheat (*hınta*) and flour (*dakik*). This would normally be secured by forced purchase at fixed prices (*mubayaa*). The manuscript includes a very extensive list of the supplies and animals considered essential to the beginning of a campaign. In the matter of hardtack (*peksimed*), for example, 400,000 *kantar* were requisitioned from bakeries in Istanbul, Salonika and Gelibolu, with a price tag of 2,708 *kese*, by far the largest expense on the list. Other examples include: 820 pairs of water buffalo, for 69 *kese*; 1,450 draft horses, 58 *kese*; 1,000 trench diggers (*beldar*) two months wages, 58 *kese*, and so forth. Fortress repairs, lumber for Danube bridges, army supplies, tents, etc., are all carefully recorded in Mustafa Kesbi's account.

The second matter was securing the sufficient tax revenue from the countryside, and minting coins to pay for the campaign. Insufficiency drove the minting of coins: 2,000 nisfiye (half sequin) and 2,000 zer-i mahbub (gold coin worth 25 *kuruş*) were produced. The next stage was mobilising the countryside, through the *levendat* militia system.[30] Deployment of the troops thus mobilised was envisioned in the following proportions: 6,000 infantry and cavalry under Albanian Kahraman Pasha, to guard Moldavia; 14,200 for Ochakov fortress; 17,560 for Hotin fortress; 3,100 for Bender fortress; 24,000 for Crimea; 2,000 for the Khan's entourage; 15,000 palace *sipahis* for the Bender Commander's entourage – 81,760 troops in total.

If we add some 60,000 Janissaries and related corps assembled in Babadağı, plus 40,000–50,000 Tatars raised by the Khan, we arrive at the anticipated need of 150,000–200,000 men for the Danubian and Black Sea frontier in this war. Kesbi has an all inclusive total of 254,900 cavalry and infantry on the same page, then later details the numbers of Janissaries, and close to a thousand officers and their entourages.[31]

A sample listing of the regional commanders ordered to report to Hantepesi mustering grounds (north of the Danube crossing at İsakcı) suggests that the soldiers for this war were drawn from the Balkans, Anatolia and the Caucasus: Arnavut (Albanian) Kahraman Pasha, Abaza (Abkhazian) Mehmed, Tarsuslu (South-east) Koca Agha, Canikli (on the Black Sea) Ali Pasha and sons; Kütahyalı (Aegean) Rıdvan Agha, Dagistanlı Ali Pasha; Çerkes Hasan Pasha, etc. (Caucasus). In the list as well are major family names, important to later events: Karaosmanoğulları, Çapanoğulları among others. Token troop numbers from Egypt (3,000 were automatically – on paper – sent from Cairo as part of the tributary arrangements), and the Arab provincial recruits, coming from long distances, arrived after the main body of troops was already in place at Hantepesi. There, a roll-call determined actual battlefield numbers for the distribution of salaries and supplies as above.[32] Artillery, and the large corps of *topcus* (gunners), *cebecis* (armourers) and *arabacıs* (waggoners) accompanied the army, although the biggest pieces were first transported by ship to Varna. Guns including 150 ten-pounders, 50 heavy guns, and an equal number of mortars are estimated to have been the original requisition of artillery for the 1769 campaign.[33]

Estimates in western sources about the size of the Ottoman armies are generally out of proportion to the realities, largely because the Ottomans still insisted on travelling as of old, with huge baggage trains and large numbers of camp followers. Late eighteenth-century Ottoman camps may be likened to disturbed beehives. It was also the case that the government accompanied the Grand Vizier, so the Chief Financial Officer and his staff, for example, were in command headquarters, while stand-ins held their offices in Istanbul. An acute shortage of trained officers characterises this war and the following ones until well into the mid-nineteenth century, so much so that state bureaucrats often were appointed officers in the heat of the moment to fill the breach.[34]

The fighting potential of these forces on the battlefield is also generally exaggerated. For this campaign, probably 70 per cent of both the central and provincial forces were raw recruits; insubordination was endemic; the encouragement of individual prowess and immediate cash rewards for valour persisted. While it appears from the documentation that manpower was over-abundant, the vast armies vanished into thin air on the road, and almost immediately upon a major confrontation with the enemy. One significant reason appears to have been the serendipitous application of discipline, severe when meted out but generally not part of the daily life of the Ottoman soldier. There was no equivalent to the ubiquitous gauntlet in use in European armies, reputedly most severely applied in the Russian case. Ottoman punishment took two forms: banishment to one of the castle-prisons, such as in Gelibolu – or immediate execution.[35] Ahmed Resmi's observations of Frederick the Great's military exercises while he was at Potsdam are instructive. He remarked on the miserable soldier (using the word *sultat*) forced to drill, . . . bewildered and confused . . . worse than slaves, starving and ill-dressed. He was horrified by the use of the gauntlet for discipline.[36] The *bastinado* is conspicuous by its absence in contemporary Ottoman military reports, even though it was the traditional disciplinary weapon of the Janissaries. The single most persistent problem for the Ottomans was the disdain of Janissary officers and officials for the provincial forces, a refusal to respect the demands of the uncultivated recruits and their upstart commanders. By gradually turning their peasants into soldiers, the Ottomans violated one of the basic principles of the state ideology, which advocated a strenuous demarcation of the line between peasant and soldier. From the 1790s, the reformers advising Selim III tacitly recognised the blurring of categories when advising the sultan about using raw Anatolian recruits for his *Nizâm-ı Cedid* troops.

By contrast, the Russians had introduced considerable European order and discipline into the post-Petrine army, as well as the beginnings of a military command structure. This was more evident after they began to emulate Prussia in the 1740s, but the system was further extended during this war under Field Marshal P.A. Rumiantsev, who was given sole command of the battlefield after his victory at Kartal in 1770. He was responsible for the strategy of installing military settlements in the newly-acquired territories of the south, and his instructions concerning recruitment and training, especially field formations and discipline, became operational over the rest of the century. Rumiantsev was known as 'the Transdanubian', as the first Russian officer to break the formidable Ottoman fortress line along the Danube in 1773–74.

Russian mobilisation was based on an elaborate conscription system, which raised 300,000 troops for the 1768–74 campaigns alone. This was accomplished with great difficulty, and targets were rarely reached. Desertions were a constant, although not as high as the 10 per cent of Peter's earlier army. Villages surrendered their ill and old for the levies when possible, although the strong and young were first choice of recruiters as well as of landlords. Mutilation and buying one's way out, or finding a substitute, offered a means of escape. To be conscripted was to be given a death sentence, as the term of service for life was only reduced to 25 years at the end of the century (1793). Between the 1680s and the 1790s, the

Russian standing army almost doubled in size, from 200,000 to perhaps as much 350,000, making it reputedly then the world's largest. The army was financed on one-fifth of the revenue of the French monarchy, by contrast, making it also one of the most parsimonious.[37] In 1769, the Russians fielded two armies: the first, 60,000 strong, commanded by A. Golitsyn in Poland and aimed at Wallachia and Moldavia, and the second, 40,000 troops, under Rumiantsev across the Crimea. More than half of the army in the Crimea was made up of irregulars, especially Cossack and Kalmuk cavalry.[38] For artillery, the first army in Poland had 120 twelve-pounders, and the second army 48 pieces.[39] The Russian command had the advantage of experienced soldiers, but the government still ran on a very small provincial bureaucracy, fewer than 5,000 officials for the entire empire, who served as officers during the wars.[40] Financing war was a constant problem, and deficits in Russia jumped from 2,000,000 roubles in 1769 to 9,300,000 in 1772. The government resorted to printing money and borrowing from foreign creditors.[41] On the Ottoman side, military expenditures doubled during 1768–74.[42]

While command gave the Russians considerable advantage over their Ottoman foes, supply remained the greatest imponderable, affected by dreadful terrain and uncertain weather, and consequently levelling the presumed advantages of either side. We have already seen how difficult the Crimean Peninsula proved to be for Russian armies in the 1736–39 campaigns. Wallachia and Moldavia were equally difficult to manoeuvre, criss-crossed with gorges and ravines. The Danube regularly flooded, forcing constant vigilance at strategic crossings. The Ottomans had to rebuild the bridge over the Danube (at İsakcı) in 1769, and again in the spring of 1770, when it was washed out by floods. Spring thaws brought oceans of mud. In 1773, Rumiantsev could not besiege Rusçuk and Silistre until mid-July because of excessive rain.[43] Drought often preceded or followed flooding, making crop availability uncertain. And everywhere there were insects. Ahmed Resmi commented about the main army's arrival in Bender in 1769 that it was as if '. . . the world was covered with flies.'[44] An unexpected blizzard at the same fortress in 1788 killed one quarter of the Russian troops.[45] Plague continued its devastation in such a setting: in August 1771, 400–500 Russians a day were dying in a particularly virulent outbreak caused largely by the war conditions of the southern frontier.[46]

Writing to Catherine II just prior to a major confrontation with the Ottomans, Rumiantsev's observations about the conditions in Wallachia and Moldavia in June 1770 are particularly apt: the land was wasted, the villages depopulated, the inhabitants, fleeing both the enemy and the plague, had abandoned their fields completely. There was no evidence of their ability to feed themselves, much less supply the troops. Since the previous September, the Russians had deprived the local populations (north of Hotin) of their horses, oxen and wagons, to transport the only food supply the troops had had over a long and wretched road.[47] Supply problems were alleviated as the Russians captured well-stocked depots along the Prut, in their 1770 march to the Danube, suggesting both a sufficiency on the Ottoman side and a continued ability of Wallachia to produce the needed grain supplies, if effectively managed.[48]

For the Ottomans, of course, the Russian occupation of fortresses like Hotin, İsmail, Kilya and Bender was disastrous, not just from a strategic but also from a logistical point of view. The Ottomans were justly famous for, and dependent on over-supply, but the 1740–68 hiatus in campaigning had meant a neglect of the roads and warehouse systems which guaranteed their ability to function in the Danube region. Furthermore, they maintained a Danube fleet, which in the seventeenth century numbered 52 vessels.[49] The 1769 campaign required the restoration of fortresses, supply depots, and means of transportation which disabled much of the field action of that year. In 1770, the Ottomans suffered tremendous losses, and abandonment of much of their supplies to victorious Russian armies. A further disaster in 1770 was the occupation and blockade of the Dardanelles by the unprecedented appearance of a Russian fleet, which defeated and set alight the entire Ottoman Mediterranean fleet at Çeşme, near present-day Izmir. The net effect was to strangulate the supply system of Istanbul. Combined with disorderly retreats, and the invidious habit of fleeing troops to plunder their own camps, the renowned Ottoman supply system fell apart.

The collapse of the Ottoman supply system

Behind these catastrophes lies another story remaining to be told about how the supply system actually worked from fortress to fortress. Such an investigation requires addressing the enormous surviving documentation of the Ottoman archives concerning the problem of acquisition and distribution of foodstuffs to the troops in the field. I have consulted some of those for İsakcı and Rusçuk, both fortresses, ports, and supply warehouses on the south shores of the Danube, which were absolutely pivotal to the well-being of the Ottoman army as it marched north into the territories between the Prut and the Dniester. The frontline supply depot was at Hantepesi, north of Jassy and south of Hotin. The passage between was punctuated with bivouacs, which were also generally supplied with pack animals and food levied from the countryside.

İsakcı Warehouse Superintendent Hatibzade Ahmed Agha reported in early 1769 on the structural repairs that had been ordered in late 1768 for the warehouse. He had been spending 2,500–3,000 *kuruş* a month over the last year on salaries for trench diggers, ironmongers and carpenters, and listed considerable expenses for daily supplies he was drawing from Moldavia.[50] Similarly, orders to purchase a supply of 4,500 wagons of straw, 40,000 *kantar* of hay and 9,000 wagons of firewood at fixed prices were sent to Chief Commissar Mehmed Tahir and the İsakcı judge in January 1769. The Warehouse Superintendent and the Commissar were also responsible for seeing to the purchase of grains (barley, wheat and flour) as well as biscuit (*peksimed*).[51] This was shipped from all over the Balkans. Multiple orders to individual state provisioners (*mubayaacı*) in late 1768 indicate the concern of Istanbul about provisioning the fortresses from the recently harvested crops. The orders contain specific cautions about sifting the grain for impurities.[52] Similar orders were sent to the bivouacs with explicit reference to the amounts and prices in the records for the previous campaign of 1736. By December 1769, projections and records of purchases for the 1770 campaigns were received from the new İsakcı Warehouse Superintendant, Ali Agha. This document is particularly revealing, because it notes the cancellations of numerous orders due to peasant resistance, and chaotic conditions up and down the shores of the Danube.[53] By February and March, former Deputy Grand Vizier Ahmed Resmi was instructed to buy supplies at market rather than fixed prices and ship them to the Burgos depot on the Black Sea because of the impossibility of obtaining supplies in Wallachia and Moldavia. He was given 330,333 *kuruş* to do so, an enormous sum to be handed over to an individual bureaucrat.[54] It suggests a measure of desperation that is borne out by multiple records for the same period. One records the fact that 60,000 *kantar* of biscuit was shipped from the naval storehouses of Istanbul because rebellions in the Principalities made it impossible to collect the grains usually supplied from those regions.[55]

It only got worse: in August of 1772, the Ochakov *Defterdar* wrote in desperation about his debts accumulated because of the matter of supplies and rations: daily barley and flour rations were arriving, but there was no back-up supply for the winter. The biscuit supplies were low because of the last siege. The fresh meat ration had been cut from all soldiers. There had been two meat deliveries at 14,000 *okka* each, but they had not sufficed. Cash substitutes for non-existent supplies had caused him to take loans and issue promissory notes to the clamouring soldiers. He was already 150,000 *kuruş* in debt, and needed 110,000 more to get through the winter. A subsequent set of documents record him as the 'late' *Defterdar* and includes a list of creditors, most of whom are identified as commanders of the various corps.[56]

Another cri-de-coeur from Rusçuk to the Grand Vizier, in December of 1772, reported that the commissary had managed to distribute only *nan-i aziz* ('white' bread for the Janissary corps and central government officials) for the two months since rations were sent from army headquarters. Foraging in the neighbourhood had produced enough barley for nine days only and the suffering had been dreadful. An additional 10,000 *kuruş* was forwarded to Rusçuk in response to this request, an insufficient gesture in a countryside which had no supplies to buy.[57] In late 1772, the Ottoman and Russian armies were idle, while diplomats engaged in the long, fruitless negotiations for peace that continued until the spring of 1773. By that time, Tott noted, Mustafa III had spent more than 25,000,000 (sterling) on the war.[58] Ahmed Resmi, a scathing critic of Ottoman logistic efforts in this period, notes how ill-prepared the chief administrators were, referring particularly to the 'demented' Commissar Tahir Agha, who failed to see that the warehouses on the passage to the front were adequately prepared, causing needless death and suffering of the soldiers.[59] As Rumiantsev noted in his despatches, the Grand Vizier could not move all his forces forward to Hotin, largely because his main supply depot was two hundred *versts* away.[60] After 1770, the Ottomans had to rely entirely on their own territory for supplies, a significant influence on the outcome of events.

Two major confrontations dominate the 1768–74 Russo-Ottoman War. The first, the siege and ultimate capture of Hotin in 1769, was characterised by the preponderance of cavalry. The second, at Kartal (Kagul) in 1770, was the major confrontation of the war, with the fully arrayed Ottoman army backed up against the Danube across from İsmail and the significantly smaller Russian army under Rumiantsev confronting them. As elsewhere in this book, I will tell the story out of contemporary chronicles, of which there are a significant number for this war.

Hotin lay at a junction of the Dniester and Prut rivers, in heavily forested territory, making it difficult for cavalry actions, and requiring significant logistical talent for coordinating the necessary fording and foraging. Under Golitsyn's command, the Russians made an early crossing of the Dniester in April, 1769, to face a heavily defended Ottoman fort, with possibly as many as 12,000 troops, and 30,000–40,000 cavalry under four commanders. Within two leagues of the fortresses, after a difficult forced march, Golitsyn elected to withdraw across the Dniester. His ostensible reason for retreating was the news of the impending approach of the main Ottoman army. In fact, the fortress had just been rocked by a rebellion of the Ottoman fortress guard, who had killed the garrison commander. Rumeli Mehmed Pasha, one of the more successful of Ottoman cavalry commanders in this war, was credited with chasing the Russians back over the Dniester. Golitsyn's indecisiveness would cost him his command later in the fall.[61]

Phase two of the battle for Hotin involved a siege. In early July Golitsyn's troops crossed over the Dniester again, and besieged the fortress. Battlefield chronicler Enverî takes up the story: in the face of the Russian assault, Rumeli Mehmed and his estimated 10,000 troops withdrew into the fortress. The Russian siege of Hotin lasted 27 days. By the twenty-third day of the siege, all supplies of barley, fodder and biscuit were exhausted. The Russian camp was equally desperate for food, and foragers were travelling two days' distance to find supplies. Moldovancı Ali Pasha, commanding a relief force, was known to be marching north from Hantepesi. Golitsyn, worried about the relief force and about supplies for his troops, crossed back over to the Dniester. Moldovancı Ali, credited with forcing the retreat of the Russians, was appointed Grand Vizier in place of Mehmed Emin Pasha, universally excoriated as a complete incompetent.[62]

The third phase of the siege and abandonment of Hotin involved the Ottoman army building a bridge across the Dniester, and sending advance guards across the river. The main army could not make the crossing until early September. Already entrenched for the winter near Kameniçe, the Russians gained the upper hand, routing the Ottoman troops, who managed to recross the bridge before it collapsed. Golitsyn estimated Ottoman losses at 3,000 during the day-long confrontation. By mid-September, Golitsyn had ordered his soldiers to cross the Dniester again, only to find Hotin fortress entirely deserted. Ahmed Resmi noted that the Grand Vizier simply walked out and left the doors open, largely because of the lack of supplies, but also because of desertions and the quarrelling among the commanders.

Golitsyn credited the Polish and Albanian troops under his command for their valour. He also had ten battalions of grenadiers who cut down the Ottoman cavalry. While the Ottomans had firearms and artillery with them, both proved ineffective. The problem was always the same:

Ottoman generals ordered skirmishes but failed to mount a real resistance to the Russian advance. To fight in this arena, the Ottomans were now relying almost entirely on unruly cavalry, called up as *levend*, or as part of the Khan's entourage. This would be typical of the rest of the war. By the end of the year, the Russian vanguard was established at Jassy, and had penetrated deeply in Wallachia, attacking the fortresses on the Danube. The main Russian army, under Rumiantsev's command, spent the winter to the north of the Dniester. The Grand Vizier and the main Ottoman force had withdrawn to the south of the Danube at Babadağı.

Rumiantsev argued that the Ottomans would reassemble the army in the spring, and that his own troops needed to have comfortable winter quarters meanwhile. Furthermore, he noted, the Ottomans would increase the number of their infantrymen, that having been their disadvantage in the Hotin confrontations, and that his own light-horse cavalry had suffered tremendous loses which it would be impossible to restore before the spring campaign. In fact, the surviving Ottoman documents, mentioned above, does in fact reveal a preoccupation with supplies, along with a list of 20,000 infantry and 11,500 cavalry troops who were ordered to be mobilised from Rumeli and Anatolia.[63]

The year of disasters: Çeşme and Kartal (1770)

The winter of 1769–70 was unusual for the Ottomans in that the Grand Vizier remained in campaign headquarters at Babadağı. Russian regiments harried the Danube fortresses, especially around Giurgevo, and in early February 1770, Ottoman troops from Craiova, Niğbolu, Rusçuk and Giurgevo, numbering perhaps 20,000, confronted the Russians at Bucharest and were routed. By mid-February, General Stoffeln had occupied Giurgevo. His report on the confrontation at that fortress reveals the extent to which Ottoman commanders could mount a defence, but were unable to resist a determined advance. A total of 10,000 Ottoman cavalry appeared at the first engagement, along with 5,000 infantry with artillery support. The cavalry dispersed almost immediately. The infantry abandoned three cannons and pulled back to another set of trenches, fortified with a moat and a high rampart from which artillery fire continued. Some 3,000 Ottomans reputedly were killed in the attempt to defend the town of Giurgevo, which they finally abandoned. Later despatches report that the Russian troops had left Giurgevo, unable to sustain the occupation of the area. Stoffeln ordered the firing of the town, all the warehouses and magazines and some hundred vessels, along with all the surrounding villages.

Contrast Stoffeln's account with that of Ottoman chronicler Vasıf Efendi's description of the attempt on Giurgevo. It is likely that Vasıf was part of the Ottoman contingent. Commander Sarım Ibrahim Pasha crossed the Danube to Giurgevo in later January. The confrontation began with equal force, when a fog enveloped the combatants and hid earth and sky. The Ottoman forces were paralysed by the fog, and a Russian battery of twenty cannons took advantage of their surprise. Ibrahim had only five artillery pieces with him, which he had not had the time to install properly. His troops fled. Three hundred of the enemy attempted the assault, but after three days, they saw the futility of the siege, fired the surroundings of the fortress, and retreated toward Bucharest.

Since the Russians had entered Wallachia, Vasıf continued, they had actively tried to seduce the subjects of the sultan. They, in turn, subscribing to the maxim that 'one should recognise as master the conqueror rather than the conquered', sided with the Russians and provided the enemy with supplies. Such was the case at Slatina on the River Olt. A small corps of the enemy was harboured in that village and threatened Craiova. The *ban* of the village was a Greek called Manolaki, who had been appointed by the *Hospodar* of Wallachia, himself a traitor who had fled to the Russians. This Manolaki, incensed by the treason of his Prince, requested aid from the pasha of Vidin. Gathering around him the Albanians scattered throughout Wallachia, and with some help from the Vidin pasha, he successfully defended Craiova. His fame reached the ears of the Grand Vizier, who was persuaded that any individual, no matter his religion, merited compensation for his loyalty, and commanded his appearance at Babadağı in order to be recognised.[64]

Vasıf's account of the events of fall 1769 repeatedly reports the merits, exploits and loyalty of individual commanders such as Manolaki and Moldavian Commander Abdi Pasha. The latter's bravery in repulsing the Russian conquest of Galatz, and his subsequent defence of İbrail, two absolutely vital fortresses for the Ottomans at the mouth of the Danube, are both documented by Vasıf. As Abdi Pasha and his troops approached Foksani, to which the Russians had withdrawn, he harangued his troops: 'Do not take prisoners and count their heads until the enemy has been routed; do not plunder until you are assured of victory. Think of the glory and merit you will acquire by fighting with the pure intention of serving the religion of the prophet.' All appeals were without avail. At the first volley of enemy fire, most fled. Nonetheless, a stiff resistance at İbrail, which was well fortified with perhaps 5,000 to 6,000 troops, forced a Russian retreat and their subsequent attacks on Bucharest and Giurgevo, as attested above. Rumiantsev's despatch is terse: İbrail's resistance was too strong for the fortress to be taken with bare hands. But the Grand Vizier, pressed for reinforcements in Babadağı, could only muster 200 men, who could not cross the Danube because of great blocks of ice that were hurled down the fast-running river.[65] Uncoordinated command, and the consistent insubordination of the troops, who preferred pillage and desertion to sustained attack, paralysed the Ottoman war effort.

The damage to the Danubian defence system as a result of the Russian successes of 1769 was incalculable, and also promised to cripple the Ottoman supply system for the second year in a row. Furthermore, even in May 1770, Russian spies were reporting that the Grand Vizier had only 20,000 troops, and that his forces were depleted by desertion as much as they were increased by new arrivals from Anatolia and the Black Sea coast. The Ottomans were also having difficulty building the bridge across the Danube, as the waters were very high in the spring of 1770. It was a year of famine as well, with crop failures all over the region.

A sustained revolt in Morea, stirred up and supported by Russian agents and arms, as a prelude to the arrival of the Russian fleet in the eastern Mediterranean later in the year, compounded the problems of mobilisation. The series of rebellions that broke out all over the peninsula preoccupied Morean Governor Muhsinzade Mehmed after March 1770, forcing him to divert to the Morean revolt Albanian irregulars (*levend*) headed for the Danubian battlefields. The revolts are significant not just because of the emerging Orthodox politics of resistance to the Ottomans, but also because of the ferocity of Christian attacks on the Muslims and the equally ferocious retaliation by the Albanian troops. The result was a considerable flight of the Morean (Orthodox and Muslim) population to the Aegean coast of Anatolia and elsewhere.[66]

The campaign season of 1770 thus compounded the disasters of the previous year, even though in April, Commander Abdi Pasha had some success in pushing the Russian advance guard back from the area around Craiova to Jassy, where Rumiantsev had established Russian headquarters. In August, the two armies faced one another at Kartal, with the Ottoman forces backed up against the Danube in the marshy delta of the Kagul River. By late July, Rumiantsev, greatly outnumbered, had already succeeded twice in routing the Ottoman combined forces along the Prut River, commanded by Abdi Pasha and Crimean Khan Kaplan Giray, and stood poised to lay claim to the entire chain of fortresses along the estuary of the Danube. Grand Vizier Halil Pasha had crossed the Danube in July to join the other forces. The Russians may have been outnumbered by as much as five to one (100,000 to 20,000).

Vasıf, who was in Abaza's entourage, described the army as resembling the waves of an ocean. Rumiantsev remained astonished by the Ottoman ability to dig trenches overnight, and the considerable resistance of what he described as the last of the formidable Janissaries in the inner ring of trenches. In his report to Catherine, Rumiantsev described the confrontation with the Janissaries as fierce, and accompanied by five hours of continuous firing from the heaviest cannons.[67] The Khan's cavalry fled first, sparking the desertion of the infantry and leaving Abdi Pasha's infantry and the Janissaries to bear the brunt of the confrontation. The entire baggage-train and 150 cannons and carriages were left behind and captured by the Russian forces. Some 3,000 Ottoman soldiers are said to have died at Kartal, but worse followed. Those fleeing had to cross the Danube by boat – and the Russians fired on the fleet assembled for that purpose, sinking many vessels. 'Fleeing troops were crushing and slashing each other, some climbing about the ships, others clutching at the ropes and planks. The greatest loss was there – as evidenced by the drowned bodies floating in the river.'[68] The estimates ranged from 20,000 to 40,000 dead, after the final confrontation before İsmail a few days later, when it too was captured.

The news of the worst military disaster ever suffered by the Ottomans arrived in Istanbul simultaneously with the news of the complete destruction of the Ottoman navy at Çeşme, burned after an attack by the Russian flotilla, which astonished all of Europe by its presence in the Mediterranean. The 24 Russian ships, commanded by Alexis Orlov and several naval officers of British origin after an extensive refitting in British naval dockyards, sailed through Gibraltar and attacked the Ottoman fortress towns in the Morea without opposition from the Ottoman fleet, which withdrew into the safety of Çeşme harbour. Caught there, the twenty-odd Ottoman ships were destroyed.[69] Even so, the Morean revolt was suppressed before it could materialise, but left a legacy of animosity between the Greek and Albanian populations which played itself out during the Greek Revolution later.

Peace negotiations and the Crimea

The campaign season of 1771 was desultory, with considerable skirmishing in and around the fortresses stretched along the Danube west from

İsmail to Belgrade. The Russians had occupied Bucharest and remained there for the duration of the war. While small Russian detachments could push to the south of the Danube, the territory could not be invested in any permanent way. By November 1771, when Muhsinzade Mehmed was reappointed as Grand Vizier and Commander-in-Chief, both sides needed a 'time-out'. Catherine II, from the first determined to negotiate without intermediaries, steadfastly opposed the mediation of interested parties such as the Austrians, the Prussians or the British. The Ottomans settled on Prussian and Austrian mediation, and negotiations got under way in July 1772, although the mediators were ejected by the Russians at the first meeting. By late August, negotiations had broken off over the question of the independence of the Tatars. Field Marshal Rumiantsev wrote the Grand Vizier directly to try to salvage the moment, and together they arranged a peace conference in Bucharest in November that extended the truce until March of 1773.

For historians more informed concerning the Polish Question than the Tatar Question, August 1772 resonates as the date of the first partition of Poland. Some argue that the outcome of the Russo-Ottoman peace negotiations was dependent on the settlement of that question, which temporarily resolved the differences among the Russian, Austrian and Prussian courts. For the Russians, it meant the Russian military presence in Poland could be reduced or redirected to the Ottoman frontier. Catherine spent most of the winter of 1771–72 negotiating an alliance with the Crimean Khan, that effectively broke the centuries-long Ottoman–Tatar connection. The alliance also fractured already divisive Tatar factions who took refuge in Ottoman territory and incited Istanbul to continuous belligerence over the question of Crimea and the strategic fortresses of the northern Black Sea coast.[70]

Plenipotentiaries Abdürrezzak for the Ottomans and Obreskov for the Russians started negotiations in November 1772, but were unable to break the impasse over Tatar independence. By February 1773, Obreskov made the following offer: Tatar independence under a Russian guarantee; continued Russian occupation of the fortresses of Kerch and Yenikale at the egress of the Sea of Azov, and free navigation of the Black Sea and Bosphorus Straits by Russian ships. This proved totally unacceptable to Istanbul, even though those conditions formed the basis of the final treaty of 1774.

Neither side wanted to go back to war, but Catherine II insisted that in doing so, Rumiantsev press his troops south of the Danube for the *coup de grâce* which would force Ottoman acquiescence to her treaty conditions. There was some urgency in her request, as 1773–75 were also the years of the Pugachev Rebellion, when a pretender to the throne, and his Cossack supporters, represented a very real threat to Catherine's rule and distracted reinforcements from the southern frontier.[71] The campaign year of 1773 was punctuated with significant confrontations up and down the banks of the Danube, as recorded in Rumiantsev's dispatches. In early June, for example, General Weissman, after crossing the Danube at Tulcea, reported attacking 12,000 of the enemy at Karasu. The Ottoman commanders were Arnavut Abdullah Pasha, Çerkes Hasan Pasha and Tatar Sultan Başmet Giray.[72] The Ottomans appear to have been able to move supplies by ship from Rusçuk to Silistre, as reported by Rumiantsev's informants: some 40 vessels had been seen arriving in early June 1773.[73] Silistre was the focus of attack by the Russians until early July.

The fighting moved along the Danube to Hirsova, where Suvorov (of later Ochakov fame) was put in command. Reinforcements from the Grand Vizier, deep in Bulgaria at Şumnu, were sent to Silistre, which was finally abandoned by the Russians. Ottoman troops proceeded to Hirsova under the command of Numan Pasha. Battlefield impressions, as well as diplomatic dispatches from Russian field commanders, warned that the Ottomans had ceased to think of peace, and that the Russians were facing choice and fierce troops. It is one of the maxims of Ottoman warfare that here, as in the 1736–39 War, the Ottomans were able to recuperate from tremendous losses with a greater resolve to resist, if only temporarily. This is especially the case along the Danubian garrison line, as we shall see again in 1787–92, 1806–12, 1828–29, and 1853–54. On this occasion, they no doubt used the extended truce to regroup. It is also important to remember that the commanders were increasingly local, defending home territory with their own troops.

Supplies were a constant worry for Rumiantsev, who competed with the Ottomans for victuals from the fertile plains of Wallachia. Bulgaria proved daunting for the Russian regiments. 'Steep mountains and deep pits define its conditions. We had taken entire camps, baggage-trains, and artillery from the enemy, but there was no possibility to defeat entirely his troops, who found shelter in forests and ravines where . . . our cavalry and infantry . . . could not move. On the contrary, nothing is so easy in that land as to make an ambush and to attack unexpectedly'.[74] By October, however, the Ottoman army was once more in disarray. Numan Pasha was replaced by Dağistanlı Ali Pasha, who was forced to defend Bazarcık, not far from the Grand Vizier at Şumnu, against Russian attack. Winter brought a hiatus. Rumiantsev reported:

The rains which have been pouring here uninterruptedly for three weeks, sometimes with snow and violent storms, have brought a cold and very severe weather, unusual for the local climate at this time of year. The water from overflowing rivers has flooded many villages and even washed way some of them together with inhabitants, as well as all the bridges. Our vessels located in many places in the Danube and other rivers . . . have been damaged by the stormy element and part of them loaded with victuals have been wrecked. For many days, communication across the rivers was interrupted, and couriers going to me or from me were waiting for a week to cross.[75]

Rumiantsev's assessment of the situation in late 1773 reveals both Ottoman strategic initiatives and the arenas that would engage the Russian forces in the second of the two Ottoman Russian Wars, when Catherine II's southern policy, under considerable attack in St Petersburg, would bear further fruit. By his account, the Ottomans were firmly entrenched with skilled warriors in the Rusçuk, Niğbolu, Vidin and Belgrade garrisons, and had proved their mettle in holding those fortifications. Extending the Russian forces south into the area around Varna simply left Wallachia vulnerable to Ottoman attack. The Ottomans had complete mastery of the Black Sea. As long as the Ottomans continued to hold Ochakov and Kilburun, they could supply the Crimea – even wood was delivered that way – and keep the Tatar hopes alive. In spite of the Russian success in the Crimea and Bucak, the Ottoman fleet guarded the Danube estuary, and hence maintained control of the Crimean Peninsula. Building a bridge across the Danube remained unrealistic because no one could control the flooding, which had been excessive throughout 1773. Crossing had to be by vessel.[76] The final capture of Ochakov lay in the future (1788), but Kilburun, across the bay at the mouth of the Dnieper River, was captured by the Russians before the treaty was signed at Küçük Kaynarca in 1774.

In spite of spirited regrouping in fall 1773, the Ottomans were unable to mobilise an army of sufficient size to resist the Russians in the spring of 1774. Grand Vizier Muhsinzade Mehmed commanded a completely demoralised and disintegrating army. The prospects for 1774 were bleak. Camp revolts were common, looting of supplies normal, and desertion endemic. In spite of the exaggerated statistics of Ottoman strength, neither side had more than 50,000 troops (the Russians far fewer than that) in the final set of confrontations around Şumnu in late June 1774. Completely surrounded, and hopelessly outnumbered, the gravely ill Muhsinzade

Mehmed agreed to a cessation of hostilities and an immediate peace conference. He died just after the 21 July 1774 treaty was signed by Ahmed Resmi and Nikolai Repnin, plenipotentiaries for the two sides, at Küçük Kaynarca, headquarters of Rumiantsev. Henceforth 'Küçük Kaynarca' became synonymous with the Eastern Question.

It was in fact a Russian triumph. Rumiantsev's army had occupied the Principalities, and penetrated deep into Ottoman territory, to the south (right) bank of the Danube, however temporarily. Russian success was predicated on the characteristics we attribute to modern armies: massive conscription, disciplined soldiers, and a unified and hierarchical chain of command. Inhospitable terrain, unpredictable rivers, and disease-ridden marshes were all much magnified by the distance from home territory and access to supplies. The Russians were well matched in the renowned Ottoman tenacity and ability to sustain sieges, but the Ottomans were operating with a confederative, temporary force that suffered from lack of expertise, and opportunism rather than sense of purpose. Russian and Ottoman soldiers were both credited with the ability to withstand hardship and deprivation, but in fortress defence the Ottomans had the edge. 'Experience has often demonstrated,' mused Vasıf, 'that 10,000 well-disciplined troops sufficed to vanquish 100,000 Muslims; but an army of 100,000 did not suffice to force 10,000 besieged Muslims.'[77] There is no way to calculate real statistics for the 1768–74 confrontations, but the combination of battle and disease may well have cost the two sides in the region of 500,000 lives. 1768 inaugurated more than 50 years of continuous fighting on the Danubian and Black Sea littoral, with accelerating damage to both the ecology and the human community.

The Ottoman supply system faltered in this war, but even so, besieged fortresses often held out because of well-supplied and well-fed troops. That contrasts with the subsistence level offered (perforce and as strategy) by the commanders of the conscript army of the Russians. Time after time in Russian battlefield accounts, they report that when the Ottomans fled, they left behind them rich treasures of arms and supplies. This is a topic waiting for significant study and analysis in this period.[78]

By the treaty stipulations, Russia retained Kilburun, Taganrog and Azov on the mouth of the Don River; Kerch and Yenikale where the Sea of Azov joins the Black Sea, and Greater and Lesser Kabarda on the eastern end of the Black Sea. Kabarda caused endless diplomatic problems, territory which the Ottomans would contest until the treaty ending two more wars over the same terrain in 1829. The Ottomans retained Ochakov, Yedisan (later Bessarabia), Akkirman, Kili, İsmail, Bender and

Hotin, but had to surrender their sovereignty over the Crimean Khanate, and grant the Tatars their independence. The Russians had a toehold on the Black Sea, cause enough for alarm bells to ring in European courts. The Principalities were restored to their former status, but the Russians assumed the right of protection over the Orthodox populations. In spite of much debate among historians about the significance of the religious clauses of the treaty which precipitated later Russian interference in Ottoman minority affairs, real Ottoman anger was reserved for the article that had caused the breakdown of negotiations in 1772: Tatar independence. It was well understood that the so-called independence was simply a step towards incorporation of the Crimea into Russian territory. While remaining ceremonially and religiously tied to their fellow Muslims by the treaty (the sultan was recognised as the Grand Caliph, a term alien to Tatar–Ottoman relations, and to Muslim law),[79] no one was to interfere with their political and civil liberty. Ottoman loss of the first major piece of Muslim-inhabited territory represented the most humiliating blow inflicted by the treaty.[80] The year 1774 was simply Round One of Ottoman–Russian treaty negotiations that continued through the Crimean War.

It was as the Russians were withdrawing troops from Bukovina and the Principalities that Austrians occupied the former, thus acquiring by stealth 10,500 square kilometres and 70,000 inhabitants, largely Romanians and Ruthenians. To prevent further conflict, the Ottomans sanctioned the acquisition in May 1775. Habsburg Emperor Joseph II, after sending 15,000 troops into Bavaria, provoked an international crisis which diverted attention from the eastern frontier and engaged both Austria and Prussia in the War of Bavarian Succession.[81]

For the next decade, the Ottomans struggled to come to terms with the loss of the Crimea. In 1777, Şahin Giray, supported by Russia, was installed as Khan at Bahçesaray, and shortly thereafter took control of all of the Crimea from his rivals. In 1778, the war party prevailed in Istanbul and sent troops and a fleet of ships in support of a revolt by claimant Devlet Giray against Şahin Giray. Repulsed by the Russians off the Crimean coast and plagued by bad weather, the fleet returned to Istanbul. French Ambassador St Priest and British Ambassador Ainslie mediated between the courts of Catherine II and Abdülhamit I (1774–89). More level heads among the Ottomans, including many veterans of the 1768–74 campaigns, prevailed. The Sultan agreed to recognise the independence of the Crimea in the Convention of Aynalıkavak (1779), even as the Russians continued their occupation of key fortresses such as Kilburun, directly across the inlet from Ochakov, and fortified their newly-acquired fortresses on the northern Black Sea coast.[82] Catherine went one step further in 1783, and annexed the Crimean Peninsula, Taman and all of the Kuban, justifying the action on the basis of the treaty of Küçük Kaynarca, and the supposed interference of the Ottomans with Tatar sovereignty, in violation of the treaty. In early 1784, they jointly signed the Convention recognising the *fait accompli*. Ochakov remained in Ottoman hands, but not for long. Border territories at each end of the northern Black Sea become the site of significant voluntary and involuntary transference of Muslim and Christian populations, causing the displacement and ethno-religious conflicts better known in the nineteenth century. The Russians frequently evoked their treaty right to protect the Christian population in the Principalities as an excuse for continued occupation.

Austro-Russian–Ottoman confrontations 1787–92

Public concern in Europe, especially Britain, was not truly engaged, however, until the events of the second Ottoman–Russian war of 1787–92, when Field Marshal Rumiantsev was eclipsed by two other major Russian figures of the period: Grigorii Potemkin and A.V. Suvorov. Catherine II had not relinquished the larger aim of her southern policy, no matter how illusionary its real prospects might remain. For one, she established a protectorate over parts of Georgia in the Caucasus in the course of 1785, while the Ottomans affected a similar policy with rival Georgians. The Russian ambassador pressed their new treaty-based privilege of establishing consulates in Ottoman cities, especially at Varna on the Black Sea. The Ottomans remained hostile to such establishments, perceiving them as centres of provocation.[83]

Catherine II's diplomatic correspondence of the period resonates with a project to restore the Greek Empire, to be called Dacia (ancient country and Roman province, contiguous with present-day Romania and Bessarabia), envisioned as a vast Orthodox imperium. She corresponded especially with Joseph II, who stood to benefit from the scheme but was reluctant to see the further dismantling of the Ottoman European territories. Lured by her own rhetoric, as well as by her favourite Potemkin, who was installed on vast estates in Poland, Catherine II engaged to tour the Crimea, especially the new Russian Black Sea ports, with Joseph II in the spring of 1787. With an international diplomatic entourage, she visited the new Russian bases at Kherson on the Dnieper and Sebastopol, passing under archways erected in her honour inscribed with 'the road to Byzantium'.[84]

In response to such provocation, the Ottomans requested mediation from the English, as their long-time allies the French were about to sign a commercial treaty with Russia (which they did in January 1787). Franco-British rivalries influenced the course of negotiations in Istanbul. Real attempts at mediation were never seriously undertaken. British Ambassador Ainslie continued to thwart the French without offering any real backing from England, while Catherine II preferred to take the belligerent road with the Ottomans, rejecting all such mediation. The Prussian ambassador, working throughout the period on a Prusso-Ottoman alliance, was equally culpable for stirring up the Ottoman officials (an alliance was in fact negotiated and signed in 1790). Public outrage and dishonour at the loss of the Crimea, and potentially the Caucasus (Georgia), however, probably had a greater influence on the final decision in Istanbul. Catherine II's trip was the *coup de grâce*. The period is characterised by court rivalry between Grand Vizier Koca Yusuf Pasha (1786–89), head of the pro-war party, and Grand Admiral Gazi Hasan Pasha, the more seasoned and cautious politician. In the event, the 'war party' prevailed and the Ottomans declared war on Russia in August 1787.[85] Austria, reluctant partner to the Russians since 1781, declared war in early 1788.

Situated at the junction of the rivers Bug and Dnieper on the northern Black Sea coast, Ochakov was the key to the entire Russian Black Sea presence and the main focus of the war. For the Ottomans, recovery of the Crimea was the main reason for the war. Ochakov was the last remaining Ottoman outpost of the Danube and Black Sea strategic line that stretched east from Belgrade. It had previously sustained sieges of a significant and bloody order, especially in the 1736–39 War. In times of peace the Ottoman garrison numbered perhaps 6,000 to 8,000 guards. During the campaigns of the eighteenth century, the number could rise to 20,000 *sipahis* and Janissaries, and more. Ochakov was difficult of access from land, and required manning and stocking from the sea, as well as investment from the sea.

In August 1787, hostilities began when an Ottoman naval detachment fired on Russian frigates off Ochakov, as part of their assault on Kilburun. Two Ottoman attempts on the Kilburun fortress, a narrow strip of land located on the Dneiper mouth, only a few hundred metres opposite Ochakov, were repulsed by Suvorov. The Russian fleet did not perform well in the confrontation, however, and later was badly damaged by a late autumn storm.

André Lafitte-Clavé, sent by the Ottomans as head of an artillery corps to prepare the Ochakov defences in April 1787, was the engineer of the assault on Kilburun. He knew the Black Sea coast, having circumnavigated the estuary of the Danube and the northern littoral in 1784. In spite of the recent *rapprochement* of France and Russia in the newly-signed commercial treaty, Lafitte-Clavé was appointed after presenting his plan for the refortification of Ochakov to Ottoman officials. Ambassador Choiseul-Gouffier expressed astonishment at the confidence the Ottoman had reposed in a Christian, '. . . un preuve qu'il n'est plus impossible d'engager les Turcs à vaincre leurs préjugés.'[86] The town resembled '. . . a long parallelogram from the crest of the hill down to the waterside, fortified with a wall of considerable thickness running around it, a double ditch . . . flanked by six bastions, a spit of sand running out from the west flank into the Liman which flanks the sea wall and terminates in a covered battery.'[87]

Lavitte-Clavé noted that the plan would have succeeded if the Janissaries had followed his advice. He envisioned an attack in stages with a landing on the very end of the sandy peninsula on which Kilburun rested. Some of the troops made an earlier assault on a beachhead closer to the fortress but it was disorderly and gave away their intentions. Lafitte laboured in vain to have the commanders recall the unruly troops, who provoked a response by the fortress guards. But to his surprise, the Russians did not press their advantage after repelling the attack. The Ottomans captured two Russian cannons, as they had been unable to unload their own artillery. The Janissary sûr'atçıs (rapid-fire artillerymen) put them to good use, even though they had to use paper fuses. Lafitte-Clavé reckoned that a disciplined force could easily outsmart the Russians, but the majority of the Ottoman troops were inexperienced and simply flew at the first engagement. He regretted the bad habit of the Turks [the term he uses throughout his narrative instead of Ottoman] of rewarding their soldiers on the spot for capturing men and property. He saw that it led to soldiers quitting the fracas (where they might be more useful), in search of a potential prize. In less than a minute, he noted, 'I saw several of them approach the commander with seven or eight Russian heads, and myself saved one of their captives by expressing the need to extract information from him.' Lafitte-Clavé estimated that the excursion cost some 250 deaths and 500–600 wounded on each side. Lafitte is at pains to assert that the Janissaries blamed themselves and assured him that they would have won if they had listened to his advice.[88]

Lafitte was among the technical advisers recalled to France at the end of 1787, and was not witness to the final Russian assault on Ochakov the following year. The consequences of the Ottoman failure to reduce Kilburun became apparent the next June, when fifteen ships of the Ottoman Black Sea fleet were destroyed while trying to escape the newly-installed Russian battery at Kilburun.

For the Habsburgs, the campaign against the Ottomans assembled the largest Austrian army to date. Joseph II's military reforms had included the introduction of a limited conscription system, which allowed for the confrontation with the Ottomans now before them. Field Marshal Franz Moritz Lacy, son of Peter Lacy of Crimean fame, and veteran of Ottoman–Austrian campaigning, proposed that six separate army corps cover the Habsburg–Ottoman line from the Adriatic to the Dniester: the main army under the emperor concentrating at Semlin opposite Belgrade; a second in Croatia; a third army corps stationed along the Sava River; another to cover the Banat/Temeşvar; a fifth protecting Transylvania and a sixth in Galicia/Bukovina. Some 245,000 troops, with 898 field guns and 252 siege guns, were initially deployed on the Ottoman frontier – a number which later rose to some 294,000 – approximating the Russian mobilisation in the Principalities and the Crimea. The plan was to capture Belgrade and secure the left bank of the Danube into Lesser Wallachia, while the Galician army pushed towards Hotin to meet with the Russian army in Moldavia.

The Russians had mobilised two armies: one under Potemkin, to capture Ochakov, the other to concentrate on the Danube, Prut and Dniester basins, after joining up with the Austrians. Slow Russian war preparations meant the Ottomans could concentrate their force on Belgrade in 1788, once Joseph II reluctantly declared war in February of that year. Austrian strategy depended on the Russian support in Moldavia, which failed to materialise until late in the 1788 campaign season. Joseph II seems to have been reluctant to confront the Ottoman army as the summer advanced. Ottoman troops reached Vidin in July 1788, crossed the Danube and broke through the Austrian defences of the Banat. The old foes found their armies ranged once again in the Mehadia passage to Temeşvar as in the 1736–39 war. Both sides were short of supplies, and by mid-July 200–300 men were falling ill every day in the Habsburg camp. Refugees, largely Serbian, perhaps as many as 50,000, flooding across the Danube caused logistical problems for the Austrian army. Joseph II moved 20,400 troops into the Banat in mid-August, while the Ottomans dug in at [Old] Orsova after destroying lives and property at will, in a scorched-earth campaign which proved detrimental to both sides. They failed to follow up on the confusion in the Habsburg army north of Mehadia. In mid-September, the Ottomans forced the blockade of the Danube, instead, while the Habsburg troops withdrew from its shores. In late October, Habsburg commanders braced for a full-scale attack by Grand Vizier Koca Yusuf Pasha at Semlin, but it failed to materialise, and by the end of October, the Ottoman army had withdrawn from the Banat into winter quarters in Sofia. Habsburg casualties (military and civilian) were estimated at 80,000.[89] The Austrians had more success in Moldavia: by September they had occupied Jassy, and with the Russians besieged Hotin, which surrendered that same month. The Ottoman forces were given ten days to evacuate the fort at Jassy, with full honours, which infuriated Field Marshal Rumiantsev, commander of the Russian army on the Prut.

Potemkin, meanwhile, continued to propose the evacuation of the Crimea altogether, while he gradually encircled Ochakov by land. A total of 50,000 troops crossed the Bug in June, and by mid-July, Potemkin had spread his forces in an arc around the town. He chose in the end to delay the final assault to mid-December. Suvorov however was more impatient, and in late July engaged with a sortie of 50 Ottoman cavalrymen, which proved disastrous when a superior force of Ottomans forced the Russians back to their lines. Suvorov himself was wounded, and 200 soldiers were killed, their heads displayed on stakes around the fortresss. Potemkin censored him for his haste, and the waste of manpower. Suvorov would figure much more prominently in the fall of the Danubian fortresses later in the war.

In mid-August, the Ottoman garrison made another sortie, during which General Mikhail Kutuzov, later commander in the 1806–12 Russo-Ottoman war and famous for the defence of Moscow against Napoleon in 1812, was wounded and blinded in one eye. Russian ships fired on the Ottomans, forcing the end of that particular engagement.[90] Part of the reason for Potemkin's delay till winter was the presence of the Ottoman fleet, which as late as October had managed to break the Russian blockade and disembark 1,500 soldiers at the fortress of Ochakov.[91] Potemkin also hoped to negotiate a surrender rather than force a bloodbath. Furthermore, with the surprising resilience of the Ottoman army on the Danube, he understood very well that the war would not end with the taking of Ochakov and was already making plans for the campaigns on the lower Danube the following year.

Waiting took its toll on the besiegers. Fresh water was scarce; winter arrived early, with temperatures of –15° Celsius, and 'the camp became "snow and shit"', making life unbearable for the soldiers in the trenches, who created burrows for themselves. In spite of conditions, and much illness, Potemkin appears not to have lost as many soldiers to disease and dysentery as Münnich or Rumiantsev, his predecessors in the Crimea, or indeed as many as Joseph II was losing daily on the upper Danube.[92]

The final assault on 16 December 1788 occurred after a month of Russian shelling from the harbour, and delays due to the severity of the winter. The barbarity on both sides was unparalleled. Creasy, writing a hundred years later, evoked the battle as follows: 'The Turks of Oczakov had before the siege surprised a Russian village in the vicinity and mercilessly slaughtered all the inhabitants. Potemkin and Suwarrow caused the Russian regiments that were there to assault the town, to be first led through this village as it lay in ashes, and with its street still red with the blood of their fellow countrymen . . . [the] Russians advanced . . . whole ranks were swept away by the fire of the besieged: but the supporting columns still came forward unflinchingly through musketry and grape; 4,000 Russians fell; but the survivors bore down all resistance, and forced their way in to the city, where for three days they revelled in murder and pillage. No mercy was shown to age or sex; and out of a population of 40,000 human beings, only a few hundred (chiefly women and children) escaped.' Potemkin himself described the Russian soldiers like a 'strong whirlwind', and the Turks 'fell in piles, over which [the Russians] trampled, their legs sinking into bleeding bodies'. The final confrontation was commemorated in a song: 'Turkish blood flowed like rivers, and the Pasha fell to his knees before Potemkin'.[93]

The new Triple Alliance of Britain, Prussia and the United Provinces (the Dutch Republic) had offered mediation in August 1788, but Catherine II rejected it. The extent to which such gestures were meaningless is borne out by the continuous Prussian thrust for an Ottoman–Prussian alliance and the significant trade in arms between the Ottomans and the British, primarily in gunpowder, but also warships.[94] Istanbul was in disarray, because of the defeats of the campaign season, and the continuing rivalry between Grand Admiral Gazi Hasan Pasha and Koca Yusuf Pasha, the latter held responsible for the fall of Ochakov and Hotin. In April of 1789, a change in sultans further complicated the scene. The new sultan, Selim III, 1789–1807, was young, idealistic and a decided francophile. He ascended to the throne at possibly the most critical moment in the entire history of the dynasty. Advised by his counsellors to settle with Austria and Russia, he chose to continue the war, with Koca Yusuf Pasha, universally despised by those he commanded, in charge.[95]

In 1789 the Austrians, under Field Marshal Laudon, 72 years old, stormed Belgrade with 62,000 troops against a garrison of 9,000 Ottomans, which capitulated in October 1789. The Austrians had occupied Wallachia by November. The Russians captured Akkirman in October and Bender in November. Cooperation between the two armies in Moldavia led to a defeat of the main Ottoman army under Grand Vizier Kethüda Hasan Pasha (1789), who had replaced KocaYusuf in May, at Martineshti on 22 September 1789.

The Austrians, increasingly reluctant eastern frontier warriors, wanted out. Financial and human costs were high: sick and wounded for one year alone (1788–89) numbered 172,000 soldiers, of whom 33,000 died.[96] Revolt in Hungary, and resistance in the Austrian Netherlands against Emperor Joseph II, was partly responsible for the poor showing of the Habsburg armies against the Turks. Indecisive leadership also played a part. Prussia looked poised to attack Austria in the north in the spring of 1790. Public opinion in Vienna opposed the continuation of the war. Once Prussia and Austria settled their differences in the Convention of Reichenbach, in 1790, the Habsburgs were free to negotiate the Austro-Ottoman treaty of Sistova in August 1791, mediated by the Triple Alliance. Old Orsova, on the Ottoman–Habsburg frontier, and a subject of dispute arising from the 1739 Belgrade treaty, was the sole, but important, fruit for Austrian efforts. Relieved of the pressure on the western front, the Ottomans still had to face the Russians entrenched along the Prut and threatening the forts on the Danube basin. The campaign of 1790 was perhaps the worst year ever for the Ottoman forces in terms of catastrophic collapse of mobilisation and logistical systems.

There were now essentially two fronts: one on the Black Sea to the east of the Crimean Peninsula at Anapa and Kerch – the latter of which the Ottoman navy aimed to recapture as part of the effort to regain the Crimea – and another focused on the forts of the Danube estuary, with Ottoman headquarters at Şumnu as in 1774. By the end of 1790, the Ottomans had to acknowledge the finality of the loss of the Crimea. Kilya, Tulcea and İsakçı also fell in October and November. İsmail, the strongest fortress of the Danubian system like Ochakov, was essential to Ottoman military and naval operations. There were probably 35,000 men with 265 guns in the fortress, and it was well supplied. (İsakçı and İsmail, it will be remembered, figured heavily in the first Russo-Ottoman War as centres of supply.) Russian forces stood at 31,000 men with 600 guns. Suvorov was put in command in early December when Potemkin despaired of the indecisiveness in the Russian command. Suvorov boasted he would storm

İsmail in five days. At his side was Kutuzov. In fact, it took even less time. On 10 December, the attack began at dawn and by 11 a.m., a number of the gates were in Russian hands. The battle took to the streets, each one of which was fought for with great ferocity, until four in the afternoon. The *Serasker* and 4,000 men defended the last bastion but were slaughtered to a man. Turkish losses stood at 26,000 dead and 9,000 prisoners.

Suvorov allowed his soldiers three days of looting, following one of the bloodiest confrontations in all Russian history, which cemented Suvorov's military fame.[97] Austrian Field Marshal Laudon, victor at Belgrade, wrote: 'It is beyond all human powers of comprehension to grasp just how strongly these places [Ottoman defensive works] are built, and just how obstinately the Turks defend them. As soon as one fortification is demolished, they merely dig themselves another one. It is easier to deal with any conventional fortress and with any other army than with the Turks when they are defending a stronghold.'[98]

Peace – lessons learned

The year 1791 was even worse for the Ottoman forces. Russian armies penetrated twice deep into Ottoman territory south of the Danube, until by July they were in control of the entire estuary. Russian guns could reputedly be heard from Istanbul in a final Russian naval victory near Varna. Peace, as in 1774, when the Ottomans finally capitulated, was arrived at with astonishing rapidity. It also closed the Potemkin phase of Catherine II's Russia, as he died in October of the same year, still hard at work attempting to modify certain unacceptable clauses of the final Jassy treaty, concluded in December 1791. By it, the Ottomans ceded Ochakov, and regained the Principalities and the strategic fortresses at the mouth of the Danube. Catherine II's aggressive focus on the southern territories had paid off, but in tremendous societal costs and financial difficulties that restricted the continued modernisation of the military. Success also bred complacency. Catherine II's successor Paul (1796–1801) proved more interested in military parades than serious reform, dropping many of the initiatives of Rumiantsev, Suvorov and Potemkin. The consequences of lack of attention to military affairs would only become apparent on the battlefields of the Crimean War. This was particularly true around the question of artillery and engineering. The training of native engineers lapsed in the second half of the century, and reverted to a reliance on foreigners. Many of the initiatives of Peter and Catherine II did not survive into Tsar Paul's reign. Paul was responsible for abolishing the

General Staff, consolidated over the previous century from a bewildering number of bodies responsible for the organisation of warfare.

Russian success in the 1768–92 era lay in its infantry organisation, and in the leadership of Potemkin and Rumiantsev, who were given full authority on the battlefront. Regular regiments relied on continuous conscription. There were 31 levies between the years 1762 and 1799, costly in human terms but effective in fighting the southern wars. Russia's expansion into the new territories in Ukraine and Belorussia was accompanied by the extension of the levy, so demand rarely outstripped availability of manpower. Rumiantsev and Suvorov were not just brilliant commanders; they were also innovators in the use of drill, discipline, and camaraderie usually attributed to Napoleon and his 'citizens' army'. As Suvorov once noted, they understood that there were 'two distinct military worlds': the linear order of the Prussians versus the irregulars of the Turks.

Under Tsar Paul, the Russian officer corps, drawn primarily from the nobility, gradually lost its status and individuality after this period, a trend observable in the evolution of western European armies of the period. By 1788, civilians were recruited into the officer corps. More foreign officers were added to the Russian army, a source of considerable resentment among native officers and rank and file alike. Real innovation in the evolution of the military lay with the organisation of the border garrison troops. Ukrainian Cossack *landmilitia*, the equivalent of the Habsburg *Militärgrenzer*, became regular army regiments under Rumiantsev in 1769. Potemkin understood the need to draw the Cossack populations of Ukraine in general, and the Crimea in particular, into the Russian system. Potemkin called the Cossacks 'the eyes and protectors of the army', and worked towards the formal integration of the informal Cossack irregulars into the Russian military until his death in 1791. As military settlers, they could be self-sufficient in peace-time; as auxiliary light cavalry, they had a significant role to play in the unconventional war that continued to unfold in Eurasia. The emphasis on mobility, self-sufficiency in logistics, and the increased use of light infantry formations presaged the tactics of the United States army against the plains Indians of the nineteenth century. Still, as with other aspects of the military, innovative approaches to geography of warfare were abandoned by Catherine's successors.[99]

While the Austrians under Joseph II had gone a long way to modernising their military machine, they never reached the stage of total war adopted by the French after 1793. Coupled with a lack of popular investment in a radical revolutionary worldview, and an ongoing crisis in command structure, the Austrians were out-manned by a two-to-one margin, once the *levée en masse* was in place in revolutionary France. The perpetuation of privilege, and the general *ancien régime* mistrust of the public in arms, temporarily in abeyance in France, continued to influence military thinking in Vienna. Universal conscription in both Austria and Hungary was only introduced in 1868.[100] Austria and Russia were caught up in the struggle with the new European leviathan, Napoleonic France, while the Ottomans entered the first phase of desperately-needed reform.

To survive the disasters of the second Russo-Turkish war, the sultans relied on the Ottoman provincial elites to mobilise and supply the fortresses. The infantry and cavalry both resembled voluntary militia, and reckoned on the rewards for individual valour rather than fear of discipline, or regimental camaraderie. Such a reliance on a confederative approach to mobilisation and supply set up serious expectations and pretensions among the provincial notables that a new political order was emerging, a more inclusive distribution of power and command. In stark contrast to the Russian use of irregulars, the Ottomans took up and discarded the irregular troops as need dictated, and were careless about the impact of such a policy on local populations. Hence there was considerable blurring of peasant and soldier, and reduction of border territories to banditry and unrest.

The events surrounding the rebuilding of the Ottoman military, after its complete disintegration in the two Russo-Ottoman Wars, is the subject of the chapters to follow. Selim III's reign is most often characterised as the period when attempts at military reform failed. I think it more correct to view the years after 1793 as the prelude to the complete overhaul of the Ottoman system of governance, not just the military. Not until the reign of Mahmud II could that sort of revolution be contemplated, much less pursued, and then not until the decade of the 1830s. The Russia of Peter the Great was invariably the reference point for the Ottoman reformers who emerged after 1792. Surely emulation of the artillery success on the battlefield would suffice to restore Ottoman greatness, they reasoned. The understanding that such piecemeal military change did not produce the modern state, but was rather the product of the entire western system, came rather more slowly. Selim III struggled throughout the Napoleonic period to maintain the integrity of the Ottoman Empire, as territories bordering the empire succumbed to the colonial powers of France, Britain, and Russia. Selim III's diplomatic manoeuvres, and his perspicacity in sending permanent Ottoman representatives to the capitals of Europe for the first time in 1793, most assuredly rest in the experience of the previous hundred years of the 'Europeanisation' of Ottoman diplomacy.

Which, is, of course, another way of returning to the Eastern Question. The 'Crisis of Ochakov', as it is sometimes called, describing the intense diplomacy arising from the fall of that fortress to the third partition of Poland in 1795, involved the cabinets of Prussia, Austria, Russia and Britain in a long and potentially bellicose debate over the significance of the collapse of the Ottoman Crimean system. Further bloodshed was finally forestalled by the third partition of Poland in 1795. Catherine II died a year later, in December 1796. On an even larger map lay the confusion and disorder of the French Revolution. Napoleon's bold thrust into the eastern Mediterranean in 1798, prelude to his imperial reign, is very often held to be the beginning of the 'modern' age in the Middle East. My sense is rather that the modern age for the Ottoman Empire began on the fields of Kartal and at the walls of Ochakov.

Notes

1 At least one history acknowledges that '[T]he sheer volume of fiscal innovation implemented by the Ottoman government during the eighteenth century belies the myth of stagnation so popular among historians until recently', Halil İnalcık and Donald Quataert, eds., *An Economic and Social History of the Ottoman Empire, 1300–1914* (Cambridge: Cambridge University Press, 1994), 710. Mehmet Genç is the acknowledged expert on the late eighteenth-century financial state of the empire: 'L'économie ottomane et la guerre au XVIIIe siècle', *Turcica* 27 (1995), 177–96. Absolutely central to the discussion of Ottoman finances in the transition period are Yılmaz Cezar's *Osmanlı Maliyesinde Bunalım ve Değişim Dönemi (XVIII.yy dan Tanzimat'a Mali Tarihi)* (Istanbul: Alan Yayıncılık, 1986), especially his description of novel tax instruments such as the *imdad-i seferiye* (special campaign tax) for the use of provincial governors, 53–70, and Ahmet Tabakoğlu's *Gerileme Dönemine Girerken Osmanlı Maliyesi* (Istanbul: Dergâh, 1985).

2 Michael Hochedlinger, *Austria's Wars of Emergence: War, State and Society in the Habsburg Monarchy, 1683–1797* (London: Longman, 2003), notes that the two decades after 1718 forced a 'revolutionary modernization from above', (205).

3 As described by Toledano, 'The emergence of Ottoman-local elites, a framework for research', in Ilan Pappé and Moshe Ma'oz, eds., *Middle East Politics and Ideas* (London and New York: Tauris, 1997), 145–62.

A definition of the characteristics of a local Ottoman elite of 1700–1900 would include: 'well-rooted loyalty to a legitimate sovereign dynasty; an elaborate set of political–social–economic networks; a widely accepted legal system and concept of state justice, etc.', 160. This is an area where new research has exploded over the last two decades, with new monographs on Aleppo by Dina Khoury, Nablus by Beshara Doumani, and Acre by Thomas Phillipp, to name just a few.

4 İnalcik and Quataert, *An Economic and Social History*, 659. The term *mutasarrıf* was most often used to designate governors of *sancaks*, or local tax officials.

5 Mehmet Genç, 'L'économie ottomane' et la guerre au XVIIIe siècle', *Turcica* 27 (1995), 177–96.

6 Evgenii Radushev, 'Les dépenses locales dans l'empire ottoman au xviiie siècle', *Études balkaniques*, 3 (1980), 74–94, discusses the full range of such taxes.

7 I was inspired here by Robert Zeidner's MA thesis, 'The army of the Tanzimat: the evolution of a new Ottoman martial tradition under the stimuli of European diplomatic and military pressures', (American University, 1961).

8 See Şevket Pamuk, *A Monetary History of the Ottoman Empire* (Cambridge: CUP, 2000), 150–71, who views the 1787 war measures such as price ceilings and coin debasement as '. . . the most comprehensive and ambitious package of intervention in the money and commodity markets that occurred during the eighteenth century'. His book includes valuable tables of exchange rates with foreign currencies.

9 Jeremy Black, *European Warfare, 1660–1815* (London: UCL Press, 1994), challenges the standard view that a 'military revolution' occurred in Europe in the century 1550–1650. For Black, the scientific advances in cannon casting and firing after 1650, as well as the widespread use of highly disciplined and mobile troops and small-calibre cannon, are more significant. Warfare then became a tool of conquest. For all the debates, including Black's challenge, see Clifford J. Rogers, *The Military Revolution Debate: Readings on the Military Transformation of Early Modern Europe* (Boulder: Westview, 1995).

10 The Charter of the Nobility was a confirmation of a manifesto originally drafted in 1762. Its significance is debated, and de Madariaga discusses both sides in *Russia in the Age of Catherine the Great* (New Haven, CT: YUP, 1981), 296–300. Fuller notes two things: the small size of the provincial administration (5,000 bureaucrats, often semi-salaried, or unpaid, for the entire empire in this period), and the fact that the peacetime military *was* that administration. William Fuller, *Strategy and Power in Russia, 1600–1914*. (New York: Free Press, 1992), 97.

11 M.S. Anderson, *Europe in the Eighteenth Century, 1713–83* (London: Longman, 1961), 182–83. Russian mercenaries were in demand in Europe for those characteristics.

12 See Robert E. Jones, 'Opposition to war and expansion in late eighteenth-century Russia', *Jahrbücher für Geschichte Osteuropas* 32 (1984), 34–51.

13 Brian L. Davies, 'Russian military power 1453–1815', in Jeremy Black, ed. *European Warfare, 1453–1815* (London: Palgrave Macmillan, 1999), 145–79; see p. 177 – 9.2 million roubles in 1762; 21 million roubles in 1796. See also John Keep, on whose work *Soldiers of the Tsar* (Oxford: Oxford University Press, 1985) much of this discussion depends.

14 Well summarised in de Madariaga, *Russia*, 587–88.

15 The struggle over baptism by Orthodox missionaries and subsequent apostasies by frontier populations played itself out well into the nineteenth century. See Michael Khodarkovsky, 'Not by word alone: missionary policies and religious conversion in early modern Russia', *Comparative Studies in Society and History* 38 (1996), 292, who notes: 'Missions in Russia were part of a concerted colonization process directed by the state and, as such, were subservient to government interests.' See also Agnès Kefeli, 'Constructing an Islamic identity: the case of Elyshevo village in the nineteenth century', in Brower and Lazzerini, eds., *Russia's Orient* (Bloomington, IN: Indiana University Press, 1997), 271–91. According to Kefeli, born Muslims who became Orthodox Tatars of Kazan collectively apostatised to Islam on numerous occasions: 1802–03, 1827–30, 1865–70 and 1905 (271). See also Emanuel Sarkisyanz, 'Russian imperialism reconsidered', in Taras Hunczak, ed. *Russian Imperialism from Ivan the Great to the Revolution* (New Brunswick, NJ: 1974), 69–70.

16 Hochedlinger, *Austria's Wars*, p. 296. He includes a list of army strength in absolute numbers from 1741–92 in tables on pp. 300–1.

17 Hochedlinger, *Austria's Wars*, p. 350.

18 Hochedlinger, *Austria's Wars*, pp. 382–96; also Scott, *The Emergence of the Eastern Powers, 1756–75* (Cambridge: Cambridge University Press, 2001), chapters 5–6.

19 See Aksan, *An Ottoman Statesman in War and Peace: Ahmed Resmi Efendi 1700–83* (Leiden: Brill, 1995), chapter 2, *passim.*

20 See Kemal Beydilli, *Büyük Friedrich ve Osmanlılar – XVIII Yüzyılda Osmanlı-Prusya Münasebetleri* (Istanbul, 1985). Aksan, *An Ottoman Statesman*, p. 78. Resmi's description notes that the Russians 'stuck their claws' into Poland in 1762, and set a Polish noble on the throne, saying 'You are King of Poland.' This episode is generally accorded little significance in western historiography concerning Poland, but the nature and

uses of such eastern embassies have been underestimated by historians, who accord sagacity to Frederick.

21 Aksan, *Ottoman Statesman*, p. 116.

22 Peter F. Sugar, *Southeastern Europe under Ottoman Rule, 1354–1804* (Seattle, WA: University of Washington, 1977), chapter 6, still one of the best. T. Stoianovich, 'The conquering Balkan orthodox merchant', *Journal of Economic History* 20 (1960): 234–313.

23 See Aksan, 'Whose territory, whose peasants? Ottoman boundaries on the Danube in the 1760s', in F.F. Anscombe, ed., *The Ottoman Balkans, 1750–1830* (Princeton: Markus Wiener, 2006), 61–86; Cengiz Orhonlu, 'The geography of Wallachia written by a Turkish politician', *Revue des études sud-est européennes* 13 (1975), pp. 447–52. The author of the report refers to the usurpers of peasant lands and goods as: contemptible rabble, spoilers, bandits, especially Albanians (*Arnavut*), fugitives, officers and *ayan* who turn a blind eye, and conniving voyvodas and boyars'. (*Eflak ve Bogdan* Topkapı ms H 445, 20b–21a.)

24 Topkapı ms H 445, 50.

25 Bulgaria State Library, Sofia S166, folios 16–22a.

26 This summary is drawn from Vasıf Efendi, court historian of the period, from a French version of his text: *Précis historique de la guerre des turcs contre les russes* (Paris, 1872), 2–11.

27 Scott, *The Emergence of the Eastern Powers*, chapter 7, but especially p. 195, n. 48.

28 The term 'Tatar' masks many rival confederations, some more settled than others. Their use as a raiding vanguard, and to harass Russian supply lines and troop movements, helped the Ottomans in the early phases of the war, but they were undisciplined, autonomous, and unreliable as a fighting force by the end of the eighteenth century. They very often precipitated the general routs that characterised the field battles of this war, the last time they played a significant military role.

29 Mustafa Kesbi, *İbretnüma-yı Devlet*, Ali Emiri ms 484. Mustafa Kesbi describes himself as a secretary in the *Defterdar's* office at the time (folio 32b). Kesbi's manuscript offers extensive information on the Polish and Tatar questions as well, and remarkable comparative financial statistics about previous campaigns. It remains, like much of the primary material for the eighteenth century, unstudied for its military information. The manuscript was published in Ottoman, with Cyrillic annotations, in Petersburg in 1881. (*Sbornik Nniekotorykh Vazhnykh Izviestii I Ofitsial'nykh Dokumentov Kasatel'no Turtsii, Rossi I Kryma. Devlet-i 'Aliye iyle Rusya Devleti ve Kırım Hakkında Baze Malumat Muhimme ve Tahrirat*

Resmiye-yi havi Mecmu'a dır, ed. V.D. Smirnov.) Two Istanbul manuscripts have been compared and transcribed into modern Turkish characters as *İbretnümâ-yı Devlet (Tahlil ve Tenkitli Metin)*, ed. Ahmet Öğreten (Ankara, 2002). See pages 74–94 for this description. It is important to remember that these were expectations of the centre, not the realities of the battlefront.

30 I have written extensively on the *levendat* mobilisation elsewhere. Here, I will try to cover some more generalised aspects of mobilising that this manuscript reveals.

31 Mustafa Kesbi, fol. 35b, and p. 87 in the 2002 edition. Tott's figures of 200,000 Tatars (100,000 under the Khan alone), suffer from the inflation disease so pervasive when speaking of the 'Turk'. As the Tatars generally travelled with multiple horses, both for their own use and to sell, it is reasonable to suppose they could fool observers as to the actual size of the fighting force. Tott was at the time serving as an *agent provocateur* for the French in the Crimea, and proved rather inept at it (Scott, *The Emergence of the Eastern Powers*, 172–73).

32 It is worth remembering that the Ottomans were also manning both the Baghdad and Caucasus frontiers, and soon to face the encroachment of the foreign powers in Egypt and Greece.

33 Gunners and armourers and waggoners were counted among the 60,000+ Janissaries above, at just over 8,000. Richard Ungermann, *Der russisch–türkisch Krieg, 1768–74* (Vienna: W. Braumüller, 1906), 33, has the estimates concerning the actual guns. Rumiantsev reported capturing 150 cannon and carriages at Kartal in August 1770 (Rumiantsev, *Sbornik dokumentov* [Collected Documents], (SIR1O), vol. 1, 344.

34 Ahmed Resmi and Enverî, eyewitnesses to the war, both remark on the paucity of soldiers in the camps, especially over the winter months, and on the necessity of using retainers to defend fortresses (see Aksan, *Ottoman Statesman*, 148).

35 Kahraman Pasha, for example, commander of the advance guard sent into the Principalities, was let out of prison to put his troops together, and later killed on the spot by the grand-vizierial entourage for subordination when he threatened his commanding officer. See Aksan, 'Mutiny and the eighteenth-century Ottoman Army', *Turkish Studies Association Bulletin* 22 (1998), 116–25, for Kahraman's complete story.

36 Aksan, *An Ottoman Statesman*, 93.

37 See V. Aksan, 'Ottoman military recruitment strategies in the late eighteenth century', in Erik J. Zürcher, ed., *Arming the State: Military Conscription in the Middle East and Central Asia 1775–1925* (London: I.B. Tauris, 1999), 21–39. This statistical summary is drawn from John Keep, *Soldiers of the*

Tsar: Army and Society in Russia 1462–1874 (Oxford, 1985); Jeremy Black, *European Warfare, 1660–1815*, and William Fuller, *Strategy and Power in Russia, 1600–1914* (New York, 1992), the latter particularly useful on the question of military defeat as an impetus for reform. Bruce Menning, 'Paul I and Catherine II's military legacy, 1762–1801', in Frederick W. Kagan and Robin Higham, eds., *The Military History of Tsarist Russia* (New York: Palgrave/St. Martin, 2002), 77–105, uses figures of 80,000 for the Golitsyn command and 40,000 under Rumiantsev.

38 Figures are drawn from Ungermann, *Der russisch–türkische Krieg, 1768–74*, 31–32. The 40,000 included 10 regiments of Ukrainian militia. Ungermann further lists two Ottoman armies: 40,000 in Moldavia/Poland, and 80,000 under the Grand Vizier in Ukraine. The Polish Bar Confederation was with the Grand Vizier (33).

39 Ibid.

40 Fuller, *Strategy and Power*, 97: 'the peacetime military was the provincial administration of Russia'. This is the period in Europe and Russia, of course, when domestic and external armies began to be differentiated, the former emerging as police forces. It was only after 1762 that Russian aristocrats even had the option 'not' to serve the state in that capacity. Foreign officers filled the gap – a surplus of demobilised men from the mid-century, world-wide conflicts. They also are some of the primary witnesses of the wars that now unfolded.

41 John T. Alexander, *Autocratic Politics in a National Crisis: The Imperial Russian Government and Pughachev's Rebellion, 1773–75* (Bloomington, IN: Indiana University Press), 14–15. In 1773, 38.5% of budget was spent on the military (Black, *European Warfare 1666–1815*, 230); in 1767, 2.3% of the budget was spent on provincial functions (Alexander, 27).

42 Aksan, 'Whatever happened to the Janissaries?' *War in History* 5 (1998), 30.

43 Fuller, *Strategy and Power*, 107.

44 Aksan, *An Ottoman Statesman*, 142.

45 Fuller, *Strategy and Power*, 108.

46 Madariaga, *Russia*, 213. Similarly, Tott reported 150,000 dead in Istanbul from plague, *Memoirs*, vol. 2, 83.

47 He was writing from headquarters on the Prut River just prior to the confrontation at Falça/Larga and Kartal in late July, early August 1770 (Rumiantsev, SIRIO, vol. 1, 145, dated 24 June 1770). All translations are by Maryna Kravets.

48 Rumiantsev was reporting in November 1770, SIRIO, vol. 1, 190. One is struck by the remarkable supply of weaponry reported in these same despatches.

49 Murphey, *Ottoman Warfare 1500–1700*, n.12, p. 235. The fleet was composed largely of frigates and riverboats (*shayka*) for transport of troops and grain, frequently captained by Christians. Murat Çızakça's recent article, 'The Kapudan Pasha and the shipowners (eighteenth and nineteenth centuries)', in *The Kapudan Pasha: His Office and His Domains*, Elizabeth Zachariadou, ed. (Rethymnon: Institute for Mediterranean Studies, 2002), 203–13, reveals the extent to which the state intervened in shipping.

50 Başbakanlik Arşivleri (BA), BSM-BNE 15970 Sept. 1768 to Apr. 1769. Similar instructions can be found in the Mustafa Kesbi manuscript.

51 BA, D.MKF 30413, actually a series of records in a large register about the late 1769, early 1770 preparations. Since my own study of biscuit ('Feeding the Ottoman troops') was published, Rhoads Murphey's *Ottoman Warfare 1500–1700* has appeared. Chapter 5 is on provisioning.

52 BA, CA 11773; also D.BSM 3913, 2–3.

53 BA, D. MKF 30413, 2–5.

54 BA, KK 2929, 5–6.

55 BA, D. MKF 30414, 14.

56 BA, CA 32630 and CA 22902. There is no indication that the *Defterdar* was executed for incompetence, but one has to wonder. Russian commanders would have been envious of the supplies here described, however scarce.

57 BA, CA 14462.

58 *Memoirs*, vol. 2, p. 136.

59 Ahmed Resmi's role in the war, and his critique, are well documented in my *An Ottoman Statesman*, chapter 3. Tott noted the absolute bleakness of the landscape after the passage of the main army in April of 1769 (*Memoirs* vol. 1, 234–35). See also Vasıf, who along with Ahmed Resmi, noted the rapacity of the Bender *Defterdar*, who was given large sums to stock the fortress but embezzled the funds, refusing the Tatar suppliers their proper due for grain and other victuals. When the main army, already suffering shortages at Hantepesi, arrived in Bender in mid-June 1769, they found empty larders. The baking ovens had not even been dug. Perhaps as many as five to six thousand soldiers deserted as a result. (Ahmed Vasıf, *Précis historique de la guerre des turcs contre les russes*. Paris, 1872, 243 – the account is almost identical to that of Ahmed Resmi).

60 Rumiantsev, SIR1O, vol. 2, 164–67.

61 I have written up this set of events in *Ottoman Statesman*, chapter 3. I summarise here the version presented in Keralio, who had access to Golitsyn's campaign journals, and the Ottoman chronicles.

62 Ahmed Resmi railed about the lack of expertise, denouncing the 'armchair heroes' who thought that going to war was a picnic, and with the exception of Grand Vizier Mehmed Muhsinzade, considered all the commanders of the war incompetent, if not corrupt: V. Aksan, *Ottoman Statesman*, 122; 130–32.

63 Rumiantsev, SIR1O, vol. 2, 164–65, and (BA) D. MKF 30413, see above.

64 Rumiantsev, SIR1O, vol. 2, #115, 224–25 and #121, 256; Vasıf, 80–84. The incident is worth noting in the context of border populations and fluctuating loyalties.

65 Vasıf, *Précis historique* 76–81. Rumiantsev, SIR1O vol. 1, #114, 223.

66 These events are covered in Yuzo Nagata, *Muhsinzade Mehmed Paşa ve Ayanlık Müessesesi* (Tokyo: 1976).

67 Rumiantsev, SIR1O vol. 1, #159, 347. He described the Janissaries as ancient in appearance and age. Perhaps this battle, not 1826, is the actual end of the last of the former Janissary corps.

68 Rumiantsev, SIR1O, vol. 1, #158, 345.

69 Alexander H. de Groot, 'The Ottoman Mediterranean since Lepanto (October seventh, 1571). Naval warfare during the seventeenth and eighteenth centuries', *Anatolica* 20 (1994), 280. See also *An Authentic Narrative of the Russian Expedition Against the Turks by Sea and Land*, compiled by an officer on board the Russian fleet (London, 1772), which is a lively version of the events in the Mediterranean in summer 1770. Isabel de Madariaga, *Russia in the Age of Catherine the Great* (London, 1981), 211–12. See also M.S. Anderson, 'Great Britain and the Russian fleet, 1769–70', *Slavonic and East European Review* 31 (1952–53), 148–63.

70 See Aksan, *Ottoman Statesman*, 157–60.

71 As discussed in John Alexander, *Autocratic Politics in a National Crisis.*

72 Rumiantsev, SIR1O, vol. 2, #317, 619.

73 Ibid., #324, 629.

74 Ibid., #333, 641–55.

75 Ibid., #347, 684.

76 Ibid., #348, 685–89, late 1772, from Focsani.

77 Vasıf, *Précis historique* 243.

78 Work is under way: see for example Gábor Ágoston, *Guns for the Sultan: Military Power and the Weapons Industry in the Ottoman Empire* (Cambridge: CUP, 2005); Veysel Şimşek, 'Ottoman military recruitment and the recruit: 1826–53', MA Dissertation, Bilkent 2005.

79 C.J. Heywood, 'Küčük Kaynardja', *EI2* CD edition.

80 M.S. Anderson, ed., *The Great Powers and the Near East, 1774–1923* (London: Arnold, 1970), 9–14.

81 Hochedlinger, *Austria's Wars*, 364–70.

82 A.I. Bağış, *Britain and the Struggle for the Integrity of the Ottoman Empire: Sir Robert Ainslie's Embassy to Istanbul, 1776–94* (Istanbul: ISIS, 1984), 9, one of the few studies of the Istanbul context from within the British, French and Ottoman archives.

83 Bağış, *Britain*, 26–27.

84 About Catherine's 'Greek Project', see Alan W. Fisher, *The Russian Annexation of the Crimea, 1772–83* (Cambridge: CUP, 1970); M.S. Anderson, *The Eastern Question, 1774–1923* (New York: St Martin's Press, 1966). The 'Chemin de Byzance' quote is from L. Pingaud, *Choiseul-Gouffier: La France en Orient sous Louis XVI* (Paris: Picard, 1887), 183–89. Ottoman diplomats were fully aware of the rumours concerning the Greek Project (Aksan, *Ottoman Statesman*, 179).

85 Bağış, *Britain*, chap. 2. Gazi Hasan Pasha was absent from Istanbul during the crucial events (41).

86 Frédéric Hitzel, 'Défense de la place turque d'Oczakow par un officier du génie français (1787)', *İkinci Tarih Boyunca Karadeniz Kongresi Bildirileri* (Samsun, Turkey: 1990), 644. Lafitte-Clavé was one of the more renowned members of the French technical mission which advised the Ottoman grand viziers. Marie-Antoinette intervened on behalf of her brother Joseph II to force the recall of all military advisers, particularly those involved with the defence of Ochakov (November 1787).

87 Simon Sebag Montefiore, *Prince of Princes: The Life of Potemkin* (London: Weidenfeld and Nicolson, 2000), 404, quoting an unpublished observation by Henry Fanshawe.

88 Hitzel, 'Defense', 646–47.

89 Matthew Z. Mayer, 'Joseph II and the campaign of 1788 against the Ottoman Turks', MA thesis (McGill University) 1997, is the source of much of the information in this paragraph. Hochedlinger, *Austria's Army*, 382–84. '36,000 civilians were said to have been killed, abducted or forced to flee', (384), in an ongoing cross-Danubian struggle resonating since 1718 – when likely as many fled the Banat into Ottoman territory.

90 Montefiore, *Prince of Princes*, 405–06.

91 de Madariaga, *Russia*, 403–05.

92 Montefiore, *Prince of Princes*, 408.

93 Edward S. Creasy, *History of the Ottoman Turks* (London, 1878), 432. The garrison was indeed slaughtered, although Creasy's description is likely hyperbolic, operating by his time on the then common assumption that Eurasian warfare was somehow more barbaric than other kinds. Montefiore, *Prince of Princes*, 412–13, estimates between 8,000 and 11,000 casualties, men, women and children. Potemkin is said to have ordered a stop to the slaughter after four hours. He reported 9,500 Turks killed, and 2,500 Russians.

94 Bağış, *Britain*, 69.

95 Mayer, 'Joseph II', using Habsburg sources, argues that Koca Yusuf Pasha was an able commander (79); V. Aksan, 'Selim III', *EI2*, CD version.

96 Hochedlinger, *Austria's Army*, 385.

97 de Madariaga, *Russia*, 415–16; also Montefiore, *Prince of Princes*, 451, and 580, footnote 2. He estimates a total of 40,000 dead in the one battle.

98 Mayer, 'Joseph II', 88–89.

99 Bruce Menning, 'Russian military innovation in the second half of the eighteenth century', *War and Society* 2 (1984), 23–41, and also his chapters in *The Military History of Tsarist Russia*.

100 Hochedlinger, *Austria's Wars*, 438–42, in a section headed 'Why the Austrians lost'.

CHAPTER FIVE

Selim III and the new order (*Nizâm-ı Cedid*)*

Introduction: crisis and reform

The next few chapters examine the consequences of the Russo-Turkish wars of 1768–92, and concentrate on the reigns of Selim III (1789–1807) and Mahmud II (1808–39). To do so requires identification of both the main hurdles facing the Ottomans, and the perceptions of what was required: in other words, the climate and articulation of reform within Ottoman society. The incorporation of new-style Ottoman elites into the circles of power in Istanbul; the regulation or 'regimentation' of army and society as a result of (or as a prelude to) significant military reforms; and the technicality (or elsewhere, rationalisation, or bureaucratisation) of the 'culture of rule', are all aspects of the reform period necessitated by the collapse of the old military system. Related to the question of acquiring control over the countryside, addressed by Selim III but only achieved with some success by Mahmud II, was the necessity of overhauling the political and social premises of Ottoman 'Muslimness', which also grew out of reforming the military. Recasting of Ottoman dynastic and religious ideology was one of the unintended results of the reform documentation, as the necessity to persuade local and international critics became part of Ottoman governance. Such changes distinguish the Russian and Ottoman 'military revolutions' from those of western Europe, where the question of the creation of modern armies was a constitutional rather than a cultural issue.[1] In all aspects of reform, Selim III encountered significant opposition, facing a multiplicity of voices and a paucity of finances. This present chapter ends with an outline of the major reforms of Selim III and his advisers.

Chapter Six takes a broader look at Selim III's challengers, both internal and external. The 1760–1830 period saw significant upheaval not just in the Ottoman military order but also in its political order, especially with the rise of provincial challengers to Ottoman hegemony, precisely in those areas in the Danubian theatre where warfare had been sustained and proved so costly. Chapter Six includes a brief return to the events of the Napoleonic period, most significantly events just prior to Selim III's fall, the treaty of Tilsit between Russia and France in 1807, and the Anglo-Russian–Ottoman naval confrontations over the Dardanelles. A description of Selim's downfall and an assessment of the achievements and failures of the *Nizâm-ı Cedid* period will conclude the chapter. As will be clear from the narrative that follows, it was the greatest moment of crisis the empire had yet faced. How it was possible for the Ottoman Empire to have survived the period in question? Part of the answer lies in Selim III's engagement in international diplomacy, as much as in the great power struggles as they were played out in Egypt and the Balkans. Part of the answer lies in the vision of a small number of bureaucrats at the centre, who counselled Selim III throughout this period.

The problem

Shall I give you some account of the troubles which occurred in the world before the Nizam-y Gedid existed . . . ? Such as . . . the calamities inflicted by the unemployed Levendis, who turned the province of Anatolia upside down; and the continued bad success which attended the arms of the followers of Islam, for the space of seven years, during the Muscovite war, which began in the year 1182 [1768–69]; the defeats which our great armies suffered every year, with the loss of so many thousand tents, such abundance of camp equipage, treasure, artillery, bombs and military stores, sufficient for the consumption of many years, and so great a loss of our troops, either taken, drowned or killed, and the capture as well of our small forts and retrenched posts, as of our large fortresses, some of which were reduced by famine, and others by force; and the impossibility of delivering so many thousand women and children whom they contained, and who, still remaining in captivity, pass their lives in tears. These are things, the bitter remembrance of which can never be erased from our hearts.

Thus Vasıf Efendi exhorted discontented troops resisting the military reforms of Selim III circa 1807. He and other Ottoman officials were members of the sultan's 'kitchen-cabinet,' the group of close advisers responsible for engineering and implementing the military and administrative reforms of the period. They were a new breed of Ottoman administrators, many of whom were experienced veterans of the battlefront, and they understood the need for the overhaul of the defeated army. They emerged during this crucial period of empire, and in a desperate attempt to remedy some of the worst of the empire's excesses, began a process of reorganisation that carries through the rest of the period covered by this book.[2]

A description of a mutiny by cavalrymen of the Ottoman army, which broke out on the battlefront just as the Russo-Ottoman truce ending the 1787–92 war was being negotiated, more than demonstrates the difficulties facing Ottoman bureaucrats and military men. Cevdet's history records two versions of events, as well as his own comments. The first version is drawn from Ahmed Vâsıf, the second from Sadullah Enverî. At the beginning of the war, while 12,000 of the palace *sipahis* were on the rolls, only 2,000 could be mustered, and those with much difficulty. By spring 1791, there were none. Replacement cavalrymen (the *Levendis* of Reşid's comments) from Rumeli and Anatolia had been hastily organised and ordered to headquarters on the Danube for the spring campaign of 1791. Some 10,000 were said to have responded to the call. Rather than have that many join the grand vizier and the main army, select members were chosen and sent to the front, the others maintained as a reserve behind the frontlines. Few of the horsemen were actually regular cavalarymen (*sipahis*), but rather rebel bands from Sivas and other areas of Anatolia, some led by leaders of provincial households of long standing, such as the Cebbarzade (also known as the Çopanŏglu) and Karaosmanoğlu. These bands had arrived at the battlefront exhausted and quarrelsome, and attacked the legitimate officers of the army who were attempting to bring some order to their ranks.

In early April, a disastrous defeat at the hands of the Russians at Maçin forced the Ottoman final capitulation to mediation of the war. Soon after the truce of July 1791, the army had been ordered to withdraw to Hirsova. The new arrivals refused to follow those orders, and determined to march on to Şumnu, the main headquarters of the army. The Grand Vizier, Koca Yusuf (1791–92), who had hastily been reappointed to the post at the beginning of the campaign season, was forced to deal with a major rebellion just as he began negotiations with the Russians. He convened an army council to discuss what to do with these rebel *sipahis*, called 'riff-raff and vermin' by Cevdet. Reluctant to attack fellow soldiers, the assembled group decided on subterfuge and execution of the instigators of the events. The Cebbarzade and Karaosmanoğlu factions were ordered to Şumnu, and followed by *Çarhacı* (Head skirmishers) Ali Pasha, who negotiated with the men and persuaded the majority to return to their bivouac in Hacıoğlu Pazarı. Upon investigation, the instigators of the rebels turned out to be part of the legitimate army corps, *Silâhdar Kâtibi* Kadri Efendi and his second-in-command, who were both executed. The account by court historian Ahmed Vâsıf ends by noting that Kadri Efendi's father had also been executed during the reign of Sultan Mustafa III, a suggestion of a lineage of revolt.

Enverî, a veteran campaigner, described the revolt in different terms. By his account, the *sipahis* had heard of the truce, and aimed for Şumnu because they knew that the grand vizier would return there with the sacred banner. Penniless and exhausted, they rushed to join him to plead their case. The Cebbarzade and Karaosmanoğlu soldiers, *along with Çarhacı* Ali Pasha, were sent after the rebels. The negotiations occurred in a small town one hour from Şumnu. The rebel *sipahis* were asked what they wanted. The fact was, they said, that the 10–15 *akçe* they were given upon mobilisation had not been sufficient. They had been forced to sell their horses and firearms to reach the front. They had come to Şumnu to demonstrate their need. In the end, they agreed to 40,000 *akçe* in lieu of official registration in the muster rolls. Enverî's emphasis on the plight of the soldiers is in distinct contrast to Vâsıf, who could only see their demands as a rebellion against the state. The two views of the nature and utility of such recruits are reflected in many of the documents of the reform era.

Cevdet, after recounting the two versions in his own history, is prompted to comment that both views had merit. He continued: 'Such is the natural consequence when law and discipline are overturned. The *sipahi* regiments were those historically given the most attention by the Ottoman state. The old order which would have kept them in place had been upset, and their wages served as a feeding trough for this one and that one . . . Even though they were useless, their wages were still paid from the treasury. Even that did them little good, as they had to share much of it with their collaborators among the officials. The state of the cavalrymen in the ranks was to be pitied.'[3]

These depictions of the events highlight not just how confusing all the military categories had become, but also the extent to which local magnates, or warlords, in this case the Cebbarzade and Karaosmanoğlu households, and their private armies, continued to play a major role in mobilising and supplying the wars the Ottomans fought after 1768. Such provincial power-brokers would become a major target of reform in the period under consideration, part of a recentralisation process that continued under sultans Selim III and Mahmud II until the middle of the nineteenth century.

Economic reforms

It is hard not to sympathise with Selim III, an isolated and insolvent sultan, who was prey to the vicissitudes of internal and international politics. One of his very first acts was to order the rebuilding of the artillery corps, after an occasion, as recorded by Cevdet, when the sultan realised the extent of the corruption represented by inaccurate registers. Looking at a recent register of 'active' artillerymen, in order to estimate the financing available for new artillerymen, Selim found that of 1,059 troops listed, 33 were wounded, 90 were assigned to the foundry, 90 to the rapid infantry corps, 76 to the fire brigade, and 770 handicapped, old, or retired, were assigned to guard duty. His advisers estimated that it would take twenty to thirty years to settle the claims and straighten out the rolls to find the salaries for the new artillerymen. Selim III is recorded as saying angrily: 'My God! What kind of situation is this? Two of the barbers who shave me say they are members of the artillery corps! If we call for soldiers, we are told "What can we do? There are no salaried soldiers to go on campaign." Let others be enrolled, we say, and we are told "There is no money in the treasury." If we say, there must be a remedy, we are told "Now is not the time to interfere with the regiments." We are not saying remove them all; rather enroll them [new recruits] as others perish.' Selim then insisted that each battalion (*orta*) of the Janissaries be equipped with ten experienced artillerymen, to be paid from the vacancies in the rolls. The new sultan's frustration was intentionally made obvious by Cevdet, most likely to emphasise Selim III's ignorance and therefore innocence regarding the extent of the corruption.[4]

The Peace of Jassy in 1792, which ended the second Russo-Ottoman war, gave Ottoman bureaucrats the opportunity to draw breath for the first time in more than two decades. Selim III's advisers outlined for him a series of proposals to reform the military that varied from restoration of the traditional to adoption of the novel. The period after 1793 saw the implementation of a number of those initiatives. Effectively bankrupt upon his accession, Selim III had first to reorganise the treasury and taxation system in order to be able to finance the needs of the military, in a systematic fashion. Hence, one of his first acts was the creation of the *İrad-i Cedid*, or New Revenue (Treasury), which was established to support the formation of a new army, the *Nizâm-ı Cedid*, or New Order, in March 1793. By 1798, the new revenues reached 32,250,000 *kuruş*, a considerable recovery based on some fairly simple fiscal reform. Mustafa Reşid was put in charge as Treasurer and Supervisor of the *Talimli Askerler*, or Trained Soldiers. The revenue of the life-term tax farms (*malikane*), and tribute from the Sacred Cities of Mecca and Medina, was assigned to the new Treasury. A serious and vigorous attempt was made to recover lost income from the essentially defunct *timariot* system, from the tax farms (*malikane*) as well as from the customs revenues, themselves very often configured as tax farms. However, opposition was vigorous from powerful land-owners, and results fairly unproductive except in the short term. It was a sign of things to come.[5]

New military organisations

Of Selim III's other significant reforms, a few with lasting repercussions stand out. First, he encouraged a climate of criticism in court politics, and involved himself in the discussions with the grand viziers and other bureaucrats about what should be done to remedy the situation of the empire. Secondly, he initiated a system of recruitment and training of soldiers that challenged the guild-like structure of the traditional army, inaugurating a transformation that reconfigured the empire by the mid-nineteenth century, and third, he continued the modernisation of the artillery corps, relying on foreign advisers, largely French, but also British and others, for technical expertise, although gradually working towards Ottoman self-sufficiency in production and deployment. The internal obstacles to such proposals must have seemed overwhelming, and defeated him personally in the end, but the initiatives of his reign were carried over into that of his successor, Mahmud II, who continued the development of the artillery corps, replaced the Janissaries altogether in 1826, but also convinced (subdued) the powerful elites in both Istanbul and the provinces of the need for a reconstituted system of governance, and enhanced bureaucracy to support the new army. Achieving such cooperation or submission of the regional power-brokers was to be Selim's singular failure.

Military reform was, in fact, pursued at the Ottoman court with considerable vigour after 1700, but only became a life-and-death matter by the end of the century. With Selim III, the nature and source of the ideas concerning the transformation of Ottoman society changed radically. There are many critical questions to consider: can it be argued that Selim III inaugurated the transformation of the empire into a Muslim absolutist state by mid-nineteenth century? Though a transitional period, and Selim III's political philosophy unarticulated, to what extent did his reformers understand the social transformation implied in the adoption of western military styles and institutions? How did Selim III's enactments of a reformed army and financial administration assist in that transformation? To answer these questions, we have to consider the nature and rationalisation of military reform prior to the Sultan's accession.

Ottoman military reform prior to Selim III

Military historians often conceive of reform as technical innovation, and are generally on the hunt for revolutionary changes in weaponry, or metallurgy, as part of the 'why didn't they' school, as in 'Why didn't the Ottomans emulate Europe?' Russia has very often come under a similar scrutiny, and serves as convenient comparison to the trajectory of Ottoman reform in the eighteenth century. Peter the Great's massive restructuring of his society began some half-century earlier than that of the Ottomans, but was continued, and in fact stalled, by many of the same cultural variables the Ottomans faced in the early 1800s, including the problem of finance, the strength of religious hierarchies, and resistance from both intellectuals and provincial elites. The comparison is relevant also because astute Ottoman statesmen explicitly drew on the Russian example to convince their sovereign and the Ottoman public about the need for restructuring the army.

Historians also tend to argue that pre-modern, agrarian empires were awakened from their supposed lethargy by teams of foreign technical advisers, who served as mediators of information, and assisted the development of indigenous industries along European models. In both instances, the Ottomans are represented as losing the arms race, with battlefield failure determined by their lack of superior weaponry. Such approaches to military history avoid the social context of resistance, preferring to mark 'revolutionary' advances or technological leaps. In the Ottoman case, particularly, fundamental social change as a product of and contributing to military modernisation has been neglected. It is important to stress the influence and limitations of particular cultural settings, while also assessing the presence or absence of an early modern military bazaar in the Ottoman context.[6]

Middle Eastern historians discuss reform in the Ottoman context by reference to series of texts, collectively called *nasihatnames*, or more recently, *ıslahatnames* (advice manuals), which invariably deal with the state of the Janissaries, who served as the barometer for the health of the empire. An aspect of Ottoman courtly culture, the political philosophy exemplified in such advice manuals is the least understood aspect of Selim III's reforms, because of the references to the 'golden age' of Süleyman the Magnificent, and calls for the restoration of past greatness, which characterise the particular examples that are regularly cited by historians to explain Ottoman backwardness and stasis. Arguably iconographic and deliberately evocative of bygone eras, such ethical critiques were conceived in a certain style and tradition, and often generated at the behest of a sultan, and thus were part of the legitimate discourse of the court. Topics of such manuals included the need to address corruption, disorder, and subordination to the sultan's justice, and restore an idealised vision of a state of equilibrium among social orders. Recommended solutions generally included purifying the ranks of the Janissaries, cleaning up the corruption among the *timariots*, regulating the appointments to both the central bureaucracy and religious class, and stabilising the economy. Süleyman the Magnificent served as the exemplar of the ideal for reform, in much the same way as Greece and Rome became the model for military organisation in early modern Europe. Most historians resist this comparison, preferring instead the obscurantist Islamic backwardness model.[7] Murphey's *Ottoman Warfare* is one of the few to engage with both these tracts and much of the official historians' literature of the seventeenth century to discuss the nature of Ottoman military culture, without pre-judging the cultural context.

Eighteenth-century examples of these advice manuals exist, but are different from their predecessors by advising directly that reform initiatives consider the successful examples of transformed (Christian) armies, particularly the Russians. The particular break with the traditional form of discourse occurs precisely as the ability to maintain or reconstruct the military forces collapses completely, as demonstrated in the previous chapter. Prior to Selim III's reign, the chief proponents of the 'modernisation' school, were İbrahim Müteferrika (d. 1745) and Ahmed Resmi Efendi (d. 1783). Müteferrika argued that the reform of the military could be accomplished by learning from other states, without jeopardising sultanic and Quranic notions of justice and order.[8] Specifically, he pointed to the disorder in strategy and logistics, to the huge size of the forces, to the total lack of discipline which overrode individual courage and wisdom, and to the need for accurate intelligence of the enemy. Ottoman bravery, agility and heroic nobility were unknown in the Christian world. All that was required was to harness and discipline that energy to triumph once again.

Müteferrika's work, published in 1732 on his own inaugural Ottoman printing press, probably reached a much larger audience than had hitherto been possible. His work enfolded the justification for military change into a larger critique of the need of the Ottoman–Muslim world to widen its intellectual horizons, and to study successful enemies, and thus joined the Muslim reformist literature which continued to punctuate reform agendas both inside and outside the Ottoman context until the fall of the empire. Such literature very often evoked the Prophet Muhammad's successful military innovations as justification for reform.[9]

Ahmed Resmi (d. 1783) represented another stage of the break with the past. An experienced diplomat, he served as deputy to Grand Vizier Muhsinzade Mehmed during the 1768–74 war, and was the unfortunate plenipotentiary who signed the humiliating Küçük Kaynarca treaty (1774). Author of *Hulâsat ül-İ'tibar*, completed shortly before his death in 1783, Ahmed Resmi was an acute critic of the state of the Ottoman armed forces and logistics of his era.[10] Of most importance is an essay he submitted to then Grand Vizier Halil Pasha while on the battlefront, sometime in late 1769. It contains a list of thirteen issues concerning mobilisation and provisioning that he thought needed attention, following the disasters of the first year of war. The *Layiha* reads as a recipe for such a reconstruction. The most significant issues include:

1) the problem of Anatolian troops, 'a filthy horde of thieves and vagabounds', who plundered their way to the battlefront – his particular targets are the governors who brought sufficient troops but could not control them, meanwhile demanding cash up front for useless soldiery.
2) the control over the army headquarters (at Şumnu) – Resmi estimated 20,000–30,000 camp followers burdened the army. He was incensed, for example, by those who pitched the tents for the troops (Druzes from Syria) who essentially went on strike if not properly compensated. Lack of control over prices for goods delivered to the camp meant extortion. Similarly, lack of control over those responsible for the delivery and maintenance of the pack animals, as well as mounts, meant considerable abuse and unnecessary waste.
3) the matter of *esame*, or the regimental pay tickets – Resmi foreshadowed Selim III's reforms by pointing to the corruption in the muster rolls, as well as suggesting that small groups of raw and untainted recruits be organised and disciplined as he had seen on

parade in Potsdam, as guest of Frederick the Great, while there as Ambassador in 1763–64.

4) the matter of supplies – the complete breakdown in shipments of flour and bread meant men were dying by eating loaves adulterated with dirt. At his angriest here, Resmi was indignant that the staff of life should thus be abused. Remedies he suggested included foresight in warehousing, a central commissary system and incorruptible officials.[11]

The *Layiha* succinctly lays out the problems Resmi alluded to in his later *Hulâsat*, itself more of a history of the war than an actual reform agenda. Resmi remained a member of the court circle of reformers until his death. He was patronised first by Koca Ragıb, Grand Vizier to Osman III and Mustafa III, Muhsinzade Mehmed, Grand Vizier and Commander-in-Chief on the Danube as noted above until his death in 1774, and then Halil Hamid, Grand Vizier to Abdülhamit I, 1782–85, all sage and loyal reformers and statesmen of the late eighteenth century.

His was not the only veteran voice of complaint in the period following 1774. Süleyman Penah Efendi (d. 1785), a seasoned bureaucrat in the military accounting offices, was the author of *Mora İhtilâli Tarihi*, a description of the massive rebellion, previously discussed, which broke out in 1770 in the Morea in the Peloponnesus. Muhsinzade Mehmed, it will be remembered, was ordered to suppress the rebellion, and he did so by diverting Albanian irregular recruits who were on their way to the Danubian front. The Albanians' disorder and ferocity caused untold suffering and flight. Süleyman Penah was an observer of these events, and his telling of the story is a fascinating glimpse into the period. As an appendix, he commented on the state of the soldiery during the events, noting three things of interest to us here: the ubiquitousness of so-called Janissaries; the widespread misuse of *esames*, and the problem of raw (Albanian) recruits. His comments are prefaced by contrasting Ottoman disorder to the regimental order of the Christians, who had become like 'a string of camels reined in and led by a camel driver'.[12]

To avoid the accusation of neglecting their duties to undertake warfare on behalf of the sultan and empire, provincial administrators created Janissaries of inhabitants both insignificant and influential, peasant and intellectual alike, without distinction of rank. With such soldiers, wrote Süleyman Penah, all order had been overturned. There was no way either to make them fight, or to keep them from flight. Such an army was worse than a rag-tag general call-to-arms (*nefer-i 'am*). Their inexperience had brought defeat, rout and a loss of zeal. Secondly, for those with official entitlements, there was no indication where they were posted, so that a Janissary could claim his right to a posting anywhere. This, Süleyman Penah continued, would be easy to fix, if a proper registration were undertaken, and the number who should be in the fortresses was well established and maintained.

His third observation concerned Albanian (*Arnavut*) irregulars. The type (*taife* – also group, sort) sent to the Morea were nothing but scum who would obey no one. Furthermore, they spoke Albanian among themselves, and did not understand Turkish, so they were untrustworthy and difficult to train. While this was a common complaint about Albanians, Penah's observations represent a different way of thinking about them. First, he compares the situation to the Spanish and their American colonies, and how they spread the use of the Spanish language by educating select individual natives. Likening it again to the Spanish example, he recommended the bringing of small groups of such troops from *sancaks* like Delvine and Avlonya to Istanbul, to train and educate them properly, and then scatter them all over the empire, with benefits both to the army and the outlying territories. He regarded the expense to the treasury as more than justified in order to improve the current system. In his view, the Morea was colonial territory, to be incorporated into the Empire by an Ottoman–Albanian civilising mission.

A third veteran of the Russo-Turkish wars, Canikli Ali Pasha, was a seasoned soldier, recognised for his contribution to the war effort in 1768–74 by being awarded much of the territory of Trabzon Province, especially the Canik *sancak* on the southern shore of the Black Sea. Likely written between 1780 and 1782, when he was Governor of Trabzon, Canikli Ali's *Nasayih al-Muluk* is an indictment, among other things, of the provincial supply system, the *mubayaa* system, based on fixed prices and state purchase. The individuals involved, he claimed, had no stake in seeing their duties carried out properly. The state made requests for provisions based on historical lists, without checking on a region's current ability to provide the grains, animals and/or other types of supplies required for the army. When a state purchaser (*mubayaacı*) was appointed, local *ayan* were happy because they could collude on the enumeration of individual holdings. For example, a peasant had one ox; it was recorded and taxed as a pair. Furthermore, the *ayan* confiscated and hoarded the grain supplies, in the hopes of a higher price and were not above selling rotted grain to the state. Even the captains of the grain transport ships in the Black Sea and on the Danube were involved – they shipped barley mixed with chaff, and wheat cut with sand. One of Canikli Ali's suggestions was to bypass the corruption, and institute a system of direct buying, half at fixed and half at market price. As with Süleyman Penah and Ahmed Resmi, Canikli Ali was calling on the sultan and his grand viziers to exert some leadership, and address the undue oppression of the peasantry.

In addition, Canikli Ali's familiarity with the Russo-Iranian frontier made him caution the grand vizier about protecting the Ottoman borders – for him, Egypt, the Black Sea and Northern Anatolia, and Baghdad were vulnerable frontiers. He especially felt that Kars, Van and Erzurum required permanent garrisons and constant vigilance. The frequency of changes of personnel, requiring constant mobility of *valis* and other officials, had a deleterious effect not just on loyalty, but on the stability of empire, when the state was relying on unknown and unreliable men. The men-of-state and the ulema were making military decisions, rather than those with the experience of warfare who should be the ones consulted to advise and restore order to the ranks of the Janissaries. Of particular interest to him were the ineligible members of the *ocak* who were enrolled as a result of the selling of appointments and the *esame* to state servants and the ulema, as well as *esnaf*. He wondered why it would not be possible to establish eligibility of each *esame*-holder on pay day by asking particular questions about the last campaign and aspects of the enemy that only a seasoned soldier could answer. His aside is particularly revelatory concerning the complete lack of control over individual soldiers. Canikli Ali was equally critical of the lack of order in the artillery corps. He devoted several pages to the need for order in the camp, and the organisation of the artillery as well as its emplacement on the battlefront. Present-day warfare, he asserted, was less a matter of raiding and swords than cannon to cannon.

Throughout Canikli Ali's writing there is a sense of injustice about the demands of Istanbul, which were largely responsible for impoverishing the provinces. During the years of war, the countryside had suffered from the deprivations caused by the raids and plunder of *kapısız levendat*, bandits (*eşkıya*), or the greed of state servants. He felt that only by reforming the government in Istanbul could one begin to tackle Rumeli and Anatolia, where major rebellions were under way and much of the population had fled.[13]

Military reforms under Selim III

As these examples have shown, serious critiques of the state of the empire had already appeared, prior to Selim III's reign which anticipated his own reform agenda. Upon his accession, Selim III, in an unusual move, called together his advisers for a large consultative assembly. Selim also solicited written recommendations about what to do, a number of which survive. Diplomats, soldiers, bureaucrats and judges alike contributed their ideas about the nature of the problem and the possible solutions. We have already noted those of court historians Ahmed Vasif and Sadullah Enverî. Two others by Grand Vizier Koca Yusuf Pasha and Rumeli Kaziasker Tatarcıkzade Abdullah Efendi were equally influential. The latter is particularly interesting, because one of his obligations was the overseeing of law and order in the Janissary ranks. Among the recommendations represented frequently in these surviving records of the early years of Selim's reign were the need to:

1) reorganise the Janissaries, by teaching them regimental order and discipline from Christian models
2) gather new troops from Anatolia, up to the number of 40,000 – introduced slowly into the existing corps
3) organise special small troops of crack infantry, drilled constantly
4) organise *sekban* (reserves – provincial militia), 15,000–20,000 strong
5) reconstitute the rapid-fire cannon corps as part of reforming the artillery corps
6) procure officers from Prussia, France and other countries to aid in the reorganisation of the army
7) translate important European texts on military science.[14]

Tatarcıkzade Abdullah's reform proposal survives in full, in multiple manuscripts, and also a printed, Ottoman version. His recommendations were far more detailed and critical than those of his contemporaries, and probably exerted the greatest influence on Selim III. Abdullah particularly noted Peter the Great's success in turning the 'beast-like' [his choice of word] tribes (Kazakh, Kalmuk, etc.) living under the hegemony of the Crimean Khanate, into well-ordered and trained soldiers. This Peter had achieved by staying abreast of all the arts of war, which the Ottomans and the Janissaries had neglected. Abdullah thought that it would be possible to build a new army by combining the best elements of the existing ortas (40 to 50 from each), with new recruits from outside the system, properly outfitted and regularly paid. Each member of the new battalion should be trained in the use of muskets; each battalion should be assigned two to four rapid-fire cannons, manned by ten sür'atçis, or rapid-fire artillerymen, like those that had been organised and trained by Baron François Tott under Mustafa III, discussed below. The new troops ought likewise to be trained with help from expert English and French officers. Such disciplined soldiers could be spread across the territories of the empire in the course of 5–10 years, reaching the number of 40,000–50,000, but the training had first of all to start in the barracks in Istanbul. Abdullah felt that similar remedies had to be undertaken for the Mortar (*Humbaracı*), Sapper (*Lağımcı*), and regular Artillery Corps (*Topçu* and *Arabacı*). He also made other recommendations about the navy and the dockyards, pointing again to the reforms that Peter had undertaken in Russia.

Such reforms in the army could not be undertaken without similar efforts being made to reform the administration and the sources of revenues, continued Abdullah. These are recommendations we have seen that Selim already implemented when he created the *İrad-i Cedid*. Significantly, Abdullah felt that the ranks of the viziers and the judges ought to be scrutinised, and their ranks thinned. He noted, again by comparison with the European context, that Ottoman administrators, especially the governors of provinces, lacked the experience and knowledge of the countryside of their European counterparts. They also needed the physical and financial security to carry out their duties effectively. He recommended longer appointments, and the abolition of official appointment fees, as a means of eliminating bribery. Abdullah was not hesitant to address the problems even in his own religious class, advising curbing the selling of offices, and restricting the appointment of judges to those who were legitimate graduates of the religious schools, and able to demonstrate their facility in the religious sciences by means of examinations.[15]

Initial reforms

In an ambitious programme to modernise the military, Selim III acted on all the proposals enumerated by Tatarcıkzade Abdullah. The reforms were both restorative, as with the existing corps and administrative systems, and innovative, especially regarding the new model army, the *Nizâm-ı Cedid*. Selim first attempted to reform the Janissary organisation, by appointing a supervisor (*nazır*) to each of the corps, weeding out incompetent officers, and setting the terms of their service at three years. He also created a hierarchy of command, dividing the troops into proper battalions (*orta*) and companies (*bölük*), with the companies assigned cannons or mortars, and each manned by master gunners. New barracks were built, with parade grounds, and drilling with arms was ordered a minimum of two days a week for the Janissary regiments; five days a week for the artillerymen. (Instituting two days a week for Janissary drills suggests that none was the standard heretofore!) An effort was made to bring order back to wages and daily rations, and establish a climate of discipline within the Janissary system, which remained self-regulating, although the supervisors were appointed from within the court bureaucracy as a kind of check on the military. Bonuses were to be handed out for special bravery, including one for information concerning violations of the new orders.[16] Selim III also set out orders for the garrisons on the Bosphorus and the Black Sea, which included regular drill with muskets.

Nor did he neglect the question of supply. The grain trade and the widespread abuses, very often the subject of complaint by the reformers, were inextricably linked to the success and failure of warfare in the Principalities, which became an especially acute problem when Russia occupied the fortresses at the mouth of the Danube after 1800. Ottoman and Russian armies on the Danube, French and British garrisons in the Mediterranean, all competed for the grain supplies. The majority of the grain (and about half of the sheep) which supplied Istanbul until the Treaty of Adrianople in 1829, was shipped down the Danube and via the Black Sea to Istanbul, or driven overland on execrable roads. Prices were generally fixed (*miri*) at untenably low levels. Adequate grain supplies were a continual preoccupation of the administration in Istanbul, as shortages and famines were certain to result in riots in the city. Selim reorganised the essential grain supply system to Istanbul, ordering the construction of granaries in the Principalities, spending some 12,500,000 *kuruş* on purchasing grain at market prices, in a deliberate attempt to restore confidence in the supply system. To oversee supply and delivery, Selim III created the Grain Administration, which regulated the cereal supply to Istanbul until 1807, when the *miri* system was reimposed after Selim III's fall from power. Selim III's efforts in this regard have been described as a 'giant economic enterprise'.[17] In 1839, the Tanzimat regulations abolished the *miri* price regulations, and grain purchases were then regulated by the market.

Gunpowder and arms

Selim also dealt with munitions, especially gunpowder manufacture, which aggravated him considerably as the state was forced to rely on foreign purchases and corrupted shipments. Demand was a consequence of the increased emphasis on artillery, as part of the modernisation process. He had the Istanbul factory re-equipped, and established another plant at Azadlı (near Küçük Çekmece outside Istanbul), close to a more convenient water supply. Together, these plants reputedly produced 10,000 *kantars* of gunpowder in 1798, of a quality equivalent to that supplied by the Netherlands and Great Britain. Azadlı's production proved sufficient for the supply, and replaced the older plants at Gallipoli, Salonika and Izmir.

This was the beginning of the recovery of the production in war material which brought the Ottomans level with the Russians within another generation, also evident in the output of the cannon works modernised by Selim III after 1793. The Tophane foundry was expanded, and the Hasköy foundry of Baron de Tott revived. Machinery imported from Britain and France, coupled with French expertise, and 70 master gunners, continued the modernisation begun by Tott, as discussed below. Saint-Denys, a French army officer, who witnessed the revolutions of 1807–8 that brought down Selim III and who traced the development of the Ottoman artillery from the empire's early days to his own, acknowledged Tott's contribution but also observed that Selim III had independently ordered the manufacture of the small-calibre cannon (4, 8, and 12 bores) which became a standard component of the Ottoman arsenal. In adopting new weapons, the Ottomans had not slavishly imitated the French, but adapted both Russian and Austrian weapons which had already been proven on the battlefield.[18]

A similar but less successful process of modernisation was begun to improve firearms. Although a new musket factory was established to serve both the Janissaries and new troops at Levend Çiftliği, under French supervision (1795–98) until the Napoleonic invasion of Egypt, incompetence and the rivalry between the English and Swedish supervisors who followed meant less success in transforming local small firearms production. Nonetheless, by 1800 the Ottomans began, according to Jonathan Grant, '. . . to fit into the general pattern of technical advances taking place in Europe.'[19]

Creation of a new army, the *Nizâm-ı Cedid*[20]

It proved impossible, however, to modernise the Janissary forces, who resisted each of the imposed reforms in the established pattern, by rebelling in the streets and demanding the heads of their officers and the court bureaucrats. The situation was so dangerous that Selim's reformers were forced to isolate the new troops, who were to be organised and disciplined explicitly along European lines, in places far afield and in garrisons separated from the Janissary and court *Sipahi* corps. By 1798, although 12,000 had been targeted as a proper starting figure, less than half that were bivouacked at new barracks in Levend Çiftliği. Recruits were drawn from the streets of Istanbul, and the provinces, in order to avoid enrolling the fractious Janissaries and their brethren, but they proved untried and equally difficult to discipline. To disguise the reorganisation, they were at first attached to the *Bostancı* (Palace Guards) and known as the *Bostancı* Musket Regiment (*Bostancı Tüfenkci Oçağı*). One regiment was to be composed of twelve companies of one hundred each, every regiment to be commanded by a *binbaşı* (colonel, later major) and made up of 1,602 officers and men, both regular infantry and gunners in considerable numbers. This was based on the regimental model which had come into fashion for European armies.[21] The new army grew very slowly, even though pay was generally better than the Janissaries', so that when Napoleon invaded Egypt in 1798, the Ottomans had scarcely begun the transformation from old to new. In fact they were maintaining two systems simultaneously, the Janissary and the *Nizâm-ı Cedid*, and relying on a third, the countryside militias, for defence of the borders against internal and external enemies.

After 1799, two more army regiments were created to accelerate the process of mobilisation, one by a different means of recruitment. First, the barracks were expanded and a second regiment was added at Levend Çiftliği. Then, the provincial governors were ordered to recruit men locally, to be trained after the new fashion, while simultaneously a central regiment was created in Istanbul, with new barracks constructed at Üsküdar on the Asian side of the Bosphorus for that purpose.

The orders for these troops were rather carefully constructed, spelling out training, when and how they would drill, daily roll call, the obligations of guard duty, maintaining the hierarchy of ranks, as well as the punishments to be meted out for dereliction of duty, failure to report, and minor infractions, all in line with the regulations of the regiments already in place in Istanbul.[22]

The new order expands beyond Istanbul

Nine governors responded to the call for provincial recruits, including Abdurrahman Pasha, Governor of Karaman, later (1801) appointed

Colonel in command of the force, and Süleyman Pasha, semi-independent Governor of Baghdad. This resembled a provincial militia: half the recruits were trained as infantry to be added to the Levend Çiftliği troops, and half as cavalry, to return to their provincial bases in support of the local governor. The recruits to be sent to Üsküdar were to be drawn from the following districts, all in Anatolia: Niğde (1,000), Beyşehir (2,000), Ankara (2,000), Kütahya (1,000), Kayseri (1,000), Kastamonu (1,000), Bolu and Viranşehir (2,000), Akşehir (1,000) and Aydın (1,000), totalling 12,000.[23] The example of Ankara may give us an indication of how many were actually recruited. Court records surviving from the years 1802–05 record that one Mes'ud Agha was charged with responsibility for recruitment in the Ankara *sancak*. He appears to have sent 750 men to Üsküdar for training. When they returned, they found they had barracks for 450 only, so they applied to the sultan for the funds to build new accommodation. Meanwhile, he continued to recruit soldiers from the surrounding area in order to increase the size of the force to 12 bölüks (1,200 men).

The new order directives also intended the abolition of the long-defunct, but unburied, *timar* system. In 1792, Selim had announced that all *timarlı* fiefs would revert to the sultan as imperial lands upon the death of the incumbent *sipahi*, in order to underwrite the new system. At the time, probably no more than 25,000 of such troops existed. They were generally ill-equipped, totally ineffective and employed as labour corps during campaigns.[24] Selim reconstituted the timariots of Bolu and Hüdevendigâr provinces as regular, salaried cavalry, and added them to the regiments stationed at Levend Çiftliği and Üsküdar. By 1806, 22,685 men and 1,590 officers formed the *Nizâm-ı Cedid* army.[25] Its significance lies not so much in performance as in composition – most of the latter recruits were Anatolian peasants, largely of Turkic stock, and Muslim. Secondly, the mobilisation of the countryside represented Selim III's attempt at establishing a relationship – a slim coalition at best – between Istanbul and the countryside, in opposition to the entrenched Janissary force in Istanbul. A further point should be made about these early efforts to recruit in the countryside: the potential for stabilising of the countryside, while obvious, was unrealised as the aim of many of the recruits appears to have been to acquire the means to go raiding in the countryside.[26] The effort to expand the base of recruitment for the new army would have disastrous consequences for Selim III's own personal rule, when he tried to establish another such army in Rumeli, but it does represent an important first stage to the later imperial consolidation under Mahmud II.

Foreign advisers and reform of the Artillery Corps

Mixed among the new recruits and advisers of the *Nizâm-ı Cedid* forces were to be found significant 'foreigners' of all stripes. Many were veterans of the long series of wars in Europe and overseas, who, upon demobilisation, sought adventure and employment as advisers and mercenaries for hire. Napoleon himself contemplated joining the French technical mission to the Ottoman Empire in 1795.[27] Others were specifically recruited and hired by the Ottomans, as noted above. Some were spies. All were suspect to the population at large, but many played important roles in the transition from the old to the new military technology.

Before and after Selim III's call for reform, foreigners were an integral part of the process. It is worth noting the significant difference between the Ottomans and the Russians as regards foreign advisers. Officers from Europe dotted the Russian army and navy, a factor which led to a swifter accommodation of western European culture of warfare, even though they were much resented by the native soldiers and officers. In the Ottoman context, foreign officers were never accorded the respect or compensation they thought they deserved.[28] Still, they were occasionally hired, and proved of considerable use. Historically, foreign adventurers brought expertise to the Ottomans concerning two of the auxiliary corps: the *Humbaracı* and the *Topçu*. This discussion will focus on the latter.

As a well-developed professional service of the infantry, the artillery corps came into its own in European armies precisely in the period under discussion. By 1750, experiments in casting and mobility had combined to produce small-calibre cannon that could be mobilised as light artillery, and incorporated into the strategy of the battlefield. Even though six- and twelve-pounder guns had been deployed to infantry regiments by Gustavus Adolphus in the Thirty Years War, their effective use as part of the course of battle awaited Frederick the Great in the Seven Years War. Frederick's organisation of the artillery assigned three- and six-pounder guns to each infantry battalion, but also kept several in the reserve artillery which could be drawn up at will in the heat of battle. By the mid-eighteenth century, each infantry battalion of the British army also had a permanent allotment of two or three light guns. The British in India developed by 1780 what were called *galloper guns*, six-pounders drawn by horses, which could support cavalry units.[29]

Technical expertise developed in Austria and France also in the eighteenth century, especially in the work of Gribeauval of France, and had considerably increased the accuracy of such cannons. By 1764, France had a superior range of four-, eight- and twelve-pounder guns. Gribeauval's experiments in casting and sighting halved the weight of the guns (a four-pounder went from 1,300 to 600 pounds, for example), and improved their accuracy, making them much more mobile. The French had professionalised other aspects of the artillery such as the train, which in both British and French armies had previously been in the hands of contractors, private teams of horses and their drivers.[30] The Russians learned from the experience of the devastating battlefields of the Seven Years War, as at least one Ottoman observer noted concerning small-calibre weapons.[31] Field Marshal Rumiantsev used mobile artillery to good effect during the 1768–74 war. He commented not so much on the Ottoman lack of artillery, as the inability to use it properly.[32]

The Ottomans lagged behind in these developments, suffering in their own artillery train from the same problems described above, but also from significant under-funding, and lack of expertise concerning metallurgy and use. The technical aspect of new field tactics and training artillerymen had been introduced to the Ottomans by Baron François de Tott, a native of Hungary in service to the French, and simultaneously the Ottomans, under Mustafa III. Tott's service is worth detailing as his observations dominate the histories of this period, and he is representative of the enlightenment views on the Ottomans. Although only connected officially with the court in 1770, Tott's experience of the Middle East was already lengthy. His memoirs begin in 1767, when he left Paris for the Crimea. His connection to Mustafa III started when he was hired as an adviser on the project to rebuild and rearm the Dardanelles fortresses after the defeat of the Ottoman navy by the Russians at Çeşme (1770), in the midst of the first Russo-Ottoman War of 1768–74. His service continued for the next five years. Tott organised and trained a number of artillerymen in Istanbul between 1770 and 1772, at the same time that he was involved with efforts to improve the casting of cannons in the foundry.[33]

Frustrated by the ignorance of the Ottoman soldiers and officers, Tott wrote: 'To the haughty ignorance of the generals was added the stupid presumption of the subalterns; and the Turks, who took the field with a prodigious train of artillery, but which consisted of pieces ill mounted, and fully as badly served, slaughtered in every action by the cannon of their enemies, could only avenge themselves for their disasters by accusing the Russians of cowardly artifice. They overpower us, said they, by the superiority of their fire, which, in fact, it is impossible to approach; but let them leave their abominable batteries, and encounter us like brave men hand to hand, and we shall soon see whether these infidels can resist the slaughtering sabre of the true-believers. This multitude of wretched fanaticks even reproached the Russians for having attacked them during the holy season of Ramadan.'[34]

Elsewhere, Tott described a demonstration of fifty gunners he organised and trained at Kağıthane (on the Golden Horn) in front of the sultan and 10,000 spectators. The artillerymen achieved a discharge of five rounds a minute, after which the following exchange took place: the *Defterdar* (Chief Financial Officer) asked Tott what the rammers were made of. (The Baron recognised a set-up.) He replied: 'Hair.' The *Defterdar* then asked 'What sort of hair?' to which Tott answered 'Pig's hair.' A huge cry arose from the crowd: 'God forbid!' Tott bested his audience by asking a painter in the crowd what they used to paint the mosques, which of course was pig bristles, which left hair on the walls. 'If then bristles do not defile your mosques, it cannot surely be improper to make use of them against your enemies!' The multitude exclaimed 'Praise be to God!'[35] Surely the reason for the popularity of the *Memoirs* was its entertainment value, as this example demonstrates. The work is full of such anecdotes, which generally credit Tott and discredit his interlocutors.[36]

Arguments about Ottoman intransigence, religious obscurantism, and general obtuseness concerning military change have remained unchallenged, chiefly because of the simplistic explanations which are continually derived from this text. The only significant critique of the *Memoirs* was that of Louis Charles de Peyssonnel, himself a French appointed consul in the Crimea and then in Izmir in the same period, and author of an important treatise on Black Sea commerce. Peysonnel wondered at Tott's assertion of Ottoman ignorance, as '. . . a prodigious number of brass pieces have been brought, and come every day from their foundery at Tophana, at Constantinople, of middling, large and enormous bores, very fine, very good, and long since brought to perfection, after the proportions and models of the European artillery. Rows of them have been continually seen all along the flat of Tophana, often two and three deep, and these sometimes of double and treble ranks; and one cannot, without injustice, accuse the Ottomans of a total ignorance in the art of founding cannon.'[37]

Fictional or not, the Tott *Memoirs* have had an enormous influence on the way historians address Ottoman reform. Other sources tell a more complex story than just obstinacy and religious obscurantism. Historian Enverî, on campaign with Muhsinzade Mehmed, south of the Danube in

PLATE 10 *Cartoon satirising the French military mission in Istanbul in 1783 (Elmar Stolpe,* Klassizismus un Krieg, Uberden Historien-Mater Jacques – Louis David, *Frankfurt and New York: Campus Verlag, 1985).*

1772, observed the discipline and training of Ottoman gunners during the extended truce in the middle of the Russo-Ottoman War 1768–74. Grand Vizier Muhsinzade Mehmed, concerned about the reduction of the numbers of trained artillerymen, organised a review. He and other high officials observed the firing of the cannon. Numerous cannon were fired three to four times, with ill-processed gunpowder and ill-mounted guns. On the fourth try, they hit the target, and were suitably rewarded by the grand vizier. He ordered daily practice, and continued to supervise it himself now and then. As a result, Enverî noted, a number of master gunners emerged.[38]

Experimentation with training as well as casting had continued throughout 1772 and 1773. During that time, Mustafa III expressed a second desire for a new Corps of Rapid-Fire Artillerymen,[39] including mobile field artillery, which now acquired the name of *Sür'atçıs*, reflected in the official order for its establishment issued in January 1774.[40] The imperial order includes 11 articles and an introductory command worth summarising. 'Even though the imperial arsenal is known for its perfection in the arts of war,' it began, '. . . in recent times, other states have invented and developed small, well crafted cannon capable of rapid fire, reaching an understanding of the science through experimentation.'[41] The orders describe the regimental organisation and pay schedules. The new troops were to be issued distinct uniforms, undergo inspections, daily roll calls, and patrol and guard duty.[42]

The *Sür'atçıs* were reorganised in January 1775. Ten guns and eighty soldiers and officers were placed under the command of Mehmed Emin Agha, a confidant of Abdülhamit I, his siladar and briefly grand vizier between 1779 and 1781.[43] The corps continued without Tott or Mehmed Emin until September 1776.[44] A large French contingent, which may have numbered as many as 300 officers and engineers during the 1780s, continued to serve the Porte in various capacities until 1788, when the mission was recalled just prior to the French revolution.[45] Precise figures of the size of the rapid-fire Artillery Corps are unreliable, but the French advisers may have been involved with training as many as 500 artillerymen before they were recalled.[46]

Under Selim III, artillery reform continued. Technical aspects of warfare were introduced in more concrete ways in this period, with the help of a new French mission invited to Istanbul by the sultan. In 1796, the new French ambassador, Albert Dubayet, presented the sultan with a complete battery of French field artillery, samples of French munitions and a number of French engineers. Juchereau de St Denys was among them, one of the few who stayed on, and he wrote extensively about the reforms. He remained in Istanbul off and on until 1844. His accounts of Selim III and Mahmud II are based on long intimacy with Istanbul and the two sultans.

The number of Artillery *ortas* under Selim III was set at 25; each *orta* was given 10 cannons, a master gunner and an assistant. By 1796, the *Topçular* included 2,875 artillerymen in 15 companies of 115 officers and men. In 1806, the number of artillerymen stood at 4,910. The Transport, Sappers (*Lağımcı*), and *Humbaracı* corps were similarly reconstituted. In 1806, the Transport Corps stood at 2,129 officers and men, the Bombadiers at 960; the *Lağımcı*, divided into sappers and military engineering, at 320 officers and men.[47]

Even after the fall of Selim III in 1807, the Artillery Corps remained the most stable of the Ottoman forces. By 1827, Mahmud II had increased the size of the force to 14,000 artillerymen and 4,414 waggoners, and continued to pay more attention to it than to the other services, which may account for the fact that the artillery corpsmen stood behind

Mahmud II in June 1826, when they turned their cannon on the Janissaries for the final confrontation.[48] While the numbers may be impressive, Ottoman artillerymen apparently still had some distance to go as a disciplined modernised force. Major Helmuth von Moltke (the Elder) observed the corps in 1828: 'Although the Turks had made great improvement in their artillery, they were still very far behind their opponents . . . The guns were 3, 6, 8, 12, and 24-pounders, roughly mounted, and the shot ill cast. The effect of their artillery could never be very great; nevertheless, as the Turks laid great stress upon this arm, it had its moral worth.'[49]

Selim III also established the Army and Navy Engineering Schools where European technical works were translated and printed on the revived press in Istanbul. Significant numbers of recruits were added to the ranks of artillerymen. Students at the new schools learned French and Arabic, mathematics, surveying, the art of warfare, and became the officers and advisers of the new army, but not of the Janissaries. Divided into four branches, the new school taught cannonry, mortars, mining and engineering. Küçük Seyyid Mustafa was a product of the new schooling. His *Diatribe de l'ingénieur Seid Moustapha sur l'état actuel de l'art mi-litaire, du génie et des sciences à Constantinople* is an enthusiastic restatement of the need to learn from Europe and the *philosophes*, and to continue the success of the *Nizâm-ı Cedid*. Seyyid Mustafa's treatise represents another of the fascinating propaganda pieces produced during Selim III's reign, published not just twice in Paris but also in Arabic in Beirut.[50]

Küçük Seyyid Mustafa began his treatise by pausing on European success in mastering military discipline and technology. This they had done by competing with one another, evolving different strategies to face the Ottoman armies, and generally pursuing military problems with energy and a willingness to profit from their mistakes. By contrast, 'our fathers, the Ottomans,' he continued, were content to rest on their successes, and furthermore, viewed military discipline and drill as if it were child's play. 'The idiotic and superstitious, of which no country is exempt,' profited from this state of affairs, and persuaded the majority that it was wrong for a Muslim to attempt to equal one's enemies. Hence, they had left us no refuge except valour and courage, denuded of all art. Sultan Selim III was determined to remedy this, Seyyid Mustafa continued, and his determination is evident in *Tableau des nouveaux reglemens*, which Selim III himself penned in French and Turkish. Nothing deterred him from his task: 'aucun de ces contretemps mis en usage par les mal-intentionnés ne parut l'ébranler un seul instant: sa carrière de restaurateur de son empire fut une fois traceé; il la suivit avec ce sang-froid inébranable qui caractérise les hommes supérieurs.'[51]

Such an encomium must have served much the same purpose as that of Vasif's lecture to the Janissaries, quoted at the beginning of the chapter, as a means of responding to the critics of the reforms. Their number increased in proportion to the loss of Ottoman territory, notably after Catherine II annexed the Crimea in 1783. Two million refugees and immigrants may be an exaggeration, but without a doubt, tremendous demographic upheaval resulting from the Russo-Ottoman wars was under way throughout the period under discussion, most on the frontiers of the empire, in the Balkans and the Caucasus. Reform authors cautioned the sultan about the need to settle such exile tribal populations, be they Georgian, Circassian, or Tatar, to name just a few, who stimulated unrest in areas such as the lands south of the Danube estuary, and who brought with them diverse views on the future of the Muslim Ottoman synthesis. Generic mountain bandits (*dağlı eşkiyalar*) came to represent both uncivilised and anti-reform forces for these authors.[52]

Some of the new immigrants found their way into the court as supporters of the reform environment; others operated from within religious organisations such as the Naqshbandi Sufis, who would have a considerable influence on court ideology over the next few decades. Selim III, for example, founded a Naqshbandi convent in his new Selimiye Barracks at Üsküdar in 1805, as part of setting a new orthodox standard for the modernised army. Among such new immigrants was Derviş Kuşmani, author of 'Zebire-i kuşmani fi ta'rif-i nizâm-ı ilhami', a dervish from Abkhazia and likely a Naqshbandi, who visited Istanbul twice in 1798 and again 1805–06, when his manuscript was produced. His work was probably the first to attack the Janissaries, not just as outmoded soldiers, but also as social pariahs, because of their allegiance to the Bektashi Sufi order.

Mehmed Emin Behiç Efendi, a childhood friend and influential associate of Alemdar Mustafa Pasha, was, like his patron, from Rusçuk. His 'Sevânih levâyih'i', written in 1802–03, is a remarkable critique of the current reform agenda, with a far-sighted view of stimulating a national economy. He too emphasised the need for instilling and spreading a Muslim orthodoxy as part of the modernising and centralising reforms necessitated by the current crisis. His proposals addressed the ignorance of the palace bureaucracy, the need to codify Islamic law and regulate the training of religious officers high and low, as well as the need to broaden the consultation base of the sultan, in an implicit criticism of the kitchen cabinet which ran affairs in secret. More than that, Behiç Efendi advocated the creation of a municipal police force to control the streets of Istanbul. His chief contribution to the reform debate was to underline the absolute necessity of creating a native mercantile class, which would counteract what he saw as the domination of non-Muslims in the luxury trades, who escaped major taxation and had little investment in the sultan's reform agenda. These latter two works were never published, likely because the revolt which would bring about the fall of Selim III began shortly thereafter. They prefigure nonetheless the reform agenda, and the changing religious climate which would unfold under his successors. The tumult which exploded in 1807 delayed further implementation of reforms for two decades.[53]

I have concentrated in this chapter on the reforms of the artillery corps because of the peculiar resilience of the view among historians that the Ottomans remained impervious to changing technology, and because it was those very reforms that saved Mahmud II's throne in later years. Selim III's vision was comprehensive and ambitious, but deeply disturbing to the traditional corps, Janissaries or otherwise, as well as to the members of the court whose wealth and/or religious standing depended on the status quo. What the reforms did not effectively address was the autonomy of the Janissary ocak, which remained, as was traditionally the right, outside any sultanic judicial system. Even the Janissary Agha, nominally appointed by the sultan, was incapable of bringing systematised order to the rank and file. Hence, the installation of western-style discipline, with significant, timely and exemplary punishment, was not even attempted there, but only in the *Nizâm-ı Cedid* corps. While ceremonial, iconographic hierarchy was always rigidly observed in the court, and hence in the Janissary corps, the hierarchy of military rank and the obligation of an officer class to obey orders, perform as instructed on the battlefield, and care for men under their command, were alien to Janissaries and court *Sipahis* of the Ottoman eighteenth century. Individual prowess was prized and bought; loyalty seldom went beyond fellow soldiers in an *orta*, or allegiance to a particular household in the court. Discipline was swift, but serendipitous, in response to extreme aggression against high state officials.[54] Command was dependent on negotiation, not automatic compliance. Admittedly, rigid discipline and rationalisation of command were not parts of military life 'native' to European contexts either, but they were integral developments of armies of the seventeenth and eighteenth centuries, which simply could not have operated otherwise, given the huge scale of military operations by the time of the Napoleonic Wars.

The Russians, with a century-long experience of conscription and consolidation of military command, were much further along the road to the 'modern' in that regard. Life-time conscription, and harsh and systematically-imposed discipline, had become part of Russian military life, and were most developed and fully articulated by officers like General Suvorov, whose ferocity, emphasis on discipline, and brilliance in battle commanded allegiance among his men, though little imitation among his commanding officers. The cost to evolving civil and constitutional rights in both the Russian and Ottoman cases was inestimable, by contrast with the European setting, where political debate allowed for public discussion of the impact of militarism on local societies, and where long-standing relationships between aristocratic families and ruling monarchies established a different kind of mutually beneficial equilibrium. As long as rigidly imposed discipline and hierarchy of command remained foreign to Ottoman military life, western-style military reform was doomed to failure, and at this late moment in imperial history, failure meant collapse of the dynasty and colonisation by the European powers. The challenges that converged on Selim III by 1800 were so great that in the end they defeated him and the first generation of reformers. Those challenges are the subject of the next chapter.

Notes

* The term *Nizâm-ı Cedid* has both a general and specific meaning: it serves as a synonym for the Selim III era, hence embodying his entire reform agenda, but it was also used to designate the new, modernised troops he attempted to introduce.

1 Marshall Poe, 'The consequences of the military revolution in Muscovy in comparative perspective', *Comparative Studies in Society and History* 38 (1996), 603–18, *passim*. I take my inspiration from his categories. His subject is the earlier period of reform in Russia, but the categories are pertinent for the eighteenth century as well. Brian M. Downing's book argues the distinctive constitutional challenges represented in European military reform. (*The Military Revolution and Political Change: Origins of Democracy and Autocracy in Early Modern Europe*. Princeton, NJ: PUP, 1992).

2 The authorship of this particular exhortation of discontented troops resisting the military reforms of Selim III has been debated for many years, but recently Kemal Beydilli has proved beyond little doubt that it is the work of Ahmed Vasıf Efendi (d. 1807), written after 1791. ('*Sekbanbaşı Risalesi'nin* müellifi Hakkında', *Türk*

Kültürü İncelemeleri Dergisi 12 (2005), 221–24.) It was originally ascribed to Mustafa Reşid Pasha (d. 1819), or Koca Sekbanbaşı as he is also known. William Wilkinson, *An Account of the Principalities of Wallachia and Moldavia* (New York: Arno Press, 1971; reprint of London, 1820 edn), Appendix no 5: 'An explanation of the Nizam-y Gedid institutions, and some curious remarks concerning it. Written by Tschelebi-Effendi . . . , and translated from the original Turkish manuscript' 222–23. In modern Turkish, the work is referred to as *Koca Sekbanbaşı Risalesi*, or *Koca Sekbanbaşı Layihası*; the actual manuscript is generally called *Hülâsat ül-Kelâm fi Redd-ül-Âvam*; Y. Hakan Erdem, 'The wise old man, propagandist and ideologist: Koca Sekbanbaşi [sic] on the Janissaries, 1807', in Kirsi Virtanen, ed., *Individual, Ideologies and Society: Tracing the Mosaic of Mediterranean History* (Tampere, Finland: Peace Research Institute, 2000), 153–77; S.J. Shaw, *Between Old and New: the Ottoman Empire Under Sultan Selim III, 1789–1807* (Cambridge, MA: Harvard University Press, 1971), pp. 87–91. The term 'kitchen-cabinet' is used by him, 87.

3 The two versions are from the two significant historians of the late eighteenth century: Ahmed Vâsıf Efendi (d. 1807) and Sadullah Enverî (d. 1795?), illustrating the diverse views in governing circles. Ahmed Cevdet, here paraphrased, used both to compile his monumental history of the Ottomans for the period 1774–1826 (*Tarih-i Cevdet*, Istanbul, 1858, vol. 5, 74–77). Piterberg argues that the debate over soldiers and rebellions, a constant to the official court narratives after 1700, is actually consolidated in the texts describing the great legitimacy crisis of the reign of Osman II (1618–22), the first regicide of Ottoman history (Gabriel Piterberg, *An Ottoman Tragedy: History and Historiography at Play* (Berkeley, CA: UC Press, 2003)).

4 Cevdet, *Tarih*, vol. 4, 265–66.

5 'By August 1793, fifty-one fiefs in Karaman alone had been seized'. Stanford J. Shaw,'The origins of Ottoman mılitary reform: the Nizâm-ı Cedıd Army of Sultan Selim III', *Journal of Modern History* 37 (1965), 296, and *Between Old and New*, 128–29, and Mahmud Raif Efendi, *Osmanlı İmparatoluğu'nda Yeni Nizamların Cedveli*, tr. and ed. by Arslan Terzioğlu and Hüsrev Hatemi, which includes the original French, *Tableau des nouveaux reglemens de l'Empire ottoman* (Istanbul, Türkiye Turing ve Otomobil Kurumu, n.d. [1798]). The *İrad-i Cedid* is described on pages 9–14 of the French version. This edition, however, has been superseded by that of Kemal Beydilli and İlhan Şahin, who have produced a far better Turkish translation and analysis of this very important document: *Mahmud Râif Efendi ve Nizâm-ı Cedîd'e Dâir Eseri* (Ankara: Türk Tarih Kurumu, 2001). It also includes the French original. The use of French to describe Selim III's ambitious plans represents a deliberate gesture of propaganda aimed at European diplomats. It was originally written in Ottoman as *Nümune-i Menazım-i Cedid-i Selim Han*. Mahmud Raif Efendi had it translated into French and printed on the press of the Imperial School of Engineering (*Mühendishane*

Matbaası) in 1798 or 1799. (See Beydilli, 'Küçük Kaynarca'dan Tanzimât'a islâhât düşünceleri', *İlmi Araştırmalar* 8 (1999)), 34.

6 Arnold Pacey has called extensive inter-exchanges concerning technology, '. . . a conversation in which incomplete information sparks new ideas and what we can call "responsive invention" ', in his *Technology of World Civilization* (Cambridge, MA: MIT Press, 1990), vii–viii.

7 A recent example, perhaps the most egregious, being Bernard Lewis's *What Went Wrong?* (Oxford, 2001).

8 See for the most recent and succinct discussion of the 'Decline and Reform School', as he puts it, Antony Black, *The History of Islamic Political Thought From the Prophet to the Present* (Edinburgh: Edinburgh University Press, 2001), 258–76. See also V. Aksan, 'Ottoman political writing, 1768–1808', *International Journal of Middle Eastern Studies* 25 (1993), 53–69.

9 Ibrahim Müteferrika, *Usul ül-Hikem fi Nizam ül-Ümem*, or *Rational Bases for the Polities of Nations*, Istanbul, 1731, translated into French in Vienna in 1769 by Baron Reviczky, with title *Traité de la tactique*. Translation used here, 40ff. Gottfried Hagen, 'The Prophet Muhammad as an exemplar in war: Ottoman views on the eve of World War I', *New Perspectives on Turkey* 22 (2000), 145–72.

10 See Aksan, *Ottoman Statesman*, throughout, but especially 186ff. More than 30 copies of the manuscript have been identified. It was also published in Istanbul in the nineteenth century, and translated into German in Halle by H.F. Diez in 1813.

11 Istanbul University MS TY 419, undated, with marginal note by Ahmed Resmi's son, concerning the occasion for its composition. See also Aksan, *Ottoman Statesman*, 188–95.

12 Written sometime after 1778 and before his death in 1785. 'Mora ihtilâli veya Penah Ef. mecmuası', *Tarih Vesikaları*, as transliterated by Aziz Berker, in four parts, in vol. 2 (1942–43), from the manuscript in the Süleymaniye Ali Emiri (Millet) Collection, 563.

13 *Nasayih al-Muluk*, Süleymaniye Library, Esad Efendi MS 1855, dated 1782. This summary is derived from a number of pages of a rather lengthy manuscript of 70 folios.

14 Drawn from a contemporary manuscript which summarised the reports, transliterated and published serially by Enver Ziya Karal, 'Nizâm-ı Cedîd'e Dâir Lâyihalar', in *Tarih Vesikaları* 1 (1941/2), 414–25, and 2 (1942/3), 104–11, 342–51, 424–32. There were some two dozen reports.

15 Besim Özcan, 'Tatarcık Abdullah Efendi ve islahatlarla ilgili lâyihasi', *Türk Kültürü Araştırmaları* 25 (1987), 55–64. Comments on the soldiers begin on 58, on the navy, on the administration, also 61. The author has used a manuscript in

the Istanbul University manuscript collection, TY 3377. The text was published in Ottoman script in *Tarih-i Osmani Encümeni Mecmuası*, no. 41, 257–84; no. 42, 321–46, no. 43, 15–34 (1916/17). Also summarised by Shaw, *Between Old and New*, 94–95.

16 Shaw, *Between Old and New*, 115–16; Raif, *Tableau*, 22–24.

17 See Salih Aynural, *İstanbul Değirmenleri ve Fırınları: Zahire Ticareti (1740–1840)*, Istanbul: Tarih Vakfı, 2001. See also Güran, 'The state role in the grain supply of Istanbul: the Grain Administration, 1793–1839', *International Journal of Turkish Studies* 3 (1984), 27–41. Also Mahmud Raif, *Tableau*, 14–16 (Beydilli ed., 171–73). See also M. Alexandrescu-Dersca, 'Contribution à l'étude de l'approvisionnement en blé de Constantinople au XVIII-e siècle', *Studia et Acta Orientalia* 1 (1957) 13–37, who lays out a very bleak picture of the consequences of the grain obligations on communities in Moldavia and Wallachia. The Principalities were also responsible for the supply of sheep, as one survey of 1818 indicated: 42% of sheep came from that region (of 147,000 head total: 49,000 from Wallachia and 14,000 from Moldavia). Anthony Greenwood, 'Istanbul's meat provisioning: a study of the Celepkeşan system', PhD (University of Chicago), 1988, 27. Merchant ships were subject to confiscation by the Russians, as happened during the 1787–92 War. (A.U. Turgay, 'An aspect of Ottoman–Russian commercial rivalry: confiscation of Ottoman merchant ships in the Black Sea', *Proceedings of the IIIrd Congress on the Social and Economic History of Turkey*, eds. Heath W. Lowry and Ralph S. Hattox (Istanbul, 1990), 59–65). Grain smuggling was a profitable enterprise, one of the chief contributions to the rise of Greek merchant families in the era.

18 A. Juchereau de Saint-Denys, *Révolutions de Constantinople en 1807 et 1808* (Paris, 1819), vol. 1, 66; Shaw, *Between Old and New*, 131.

19 Jonathan Grant, 'Rethinking the Ottoman "decline": military technology diffusion in the Ottoman Empire, fifteenth to eighteenth centuries', *Journal of World History* 10 (1999), especially 184, and 194–98. See also Mahmud Raif, 25–28, *Tableau* (Beydilli, 188–93). Still, we lack a basic history of the Janissary Artillery Corps based on the Ottoman archives. See Gábor Ágoston, 'Ottoman artillery and European military technology in the fifteenth and seventeenth centuries', *Acta Orientalia Academiae Scientiarum Hungaricae* 47 (1994), 15–48; and 'Gunpowder for the Sultan's Army: new sources on the supply of gunpowder to the Ottoman Army in the Hungarian campaigns of the sixteenth and seventeenth centuries', *Turcica* 25 (1993), 75–96, and his book *Guns for the Sultan*.

20 Here the reference is to the new troops, not to the reform era in general.

21 Shaw, *Between Old and New*, 128–33.

22 Sipahi Çataltepe, *19. Yüzyıl Başlarında Avrupa Dengesi ve Nizâm-ı Cedit Ordusu* (Istanbul: Göçebe Yayinlari, 1997), has produced a detailed picture of these troops, drawn from the Ottoman archives. Shaw's early articles on the first

reorganised troops are the source for all later descriptions: 'The origins of Ottoman military reform', as above, and his 'The established Ottoman Army Corps under Selim III', *Der Islam* 40 (1965), 142–84.

23 Küçük Seyyid Mustafa, 22–23, 1803 French ed., reproduced in Kemal Beydilli, 'İlk mühendislerimizden Seyyid Mustafa ve Nizâm-ı Cedîde dair Risalesi', *Tarih Dergisi* 13 (1983–87), 468–69. Provincial troop numbers are also reproduced in Çataltepe, *19. Yüzyıl Başlarında Avrupa Dengesi*, 149.

24 Robert Zeidner, 'The army of the Tanzimat; the evolution of a new Ottoman martial tradition under the stimuli of European diplomatic and military pressures', American University (MA) 1960, 15.

25 Shaw, *Between Old and New*, 131–32.

26 Musa Çadırcı, 'Ankara sancağında Nizâm-ı Cedîd ortasının teşkili ve *Nizâm-ı Cedîd* Askeri Kanunnamesi', *Belleten* 35 (1972), 1–13. The latter comment comes from Çataltepe, 157–58.

27 See Enver Ziya Karal, 'Osmanlı tarihine dair vesikalar', *Belleten* 4 (1940), 175–89, who reproduces a series of documents, including the one concerning Napoleon.

28 Levy, 'The military policy of Mahmud II, 1808–1839', PhD Chicago (1968): 'Tolerated by the Turks with great pain, our military mission had merely an ephemeral existence and a doubtful value'. (26, quoting Pingaud, Choiseul Gouffien, 102).

29 B.P. Hughes, *Open Fire: Artillery Tactics from Marlborough to Wellington* (Chichester: A. Bird, 1983), 38–41, 47–48.

30 Bruce McConachy, 'The roots of artillery doctrine: Napoleonic artillery tactics reconsidered', *The Journal of Military History* 65 (2001), 619–22.

31 Aksan, *Ottoman Statesman*, 127, where Resmi records a conversation with Frederick the Great while at Potsdam.

32 *Russkie polkovodtsy: Dokumenty I materialy*, vol. 2: Generalissiumus Suvorov, ed. N. Korobko (Leningrad, 1947), 59.

33 His father was a member of the Hungarian refugee community in France, who reputedly spoke French, Turkish, and Polish. Lieutenant-Colonel of the Bercheny Hussards, he served as an adviser to French Ambassador Vergennes (1755–68) at the embassy in Istanbul, where he died in 1757. Tott senior had been accompanied by his son, who later served as French consul in Ottoman territories. See my 'Enlightening the Ottomans: Tott and Mustafa III', in *Proceedings, International Congress on Learning and Education in the Ottoman World, Istanbul*, 1999 (Istanbul, 2001), 163–74; François de Tott, *Memoirs* vol. 2, 3, 114, on the 'ignorance of the Turks' casting and foundry. He knew very little about it, judging from his own comments: 'I had never seen any foundery and my taste for the arts, which I had always made my amusement, had never led me to

attempt what could amuse nobody.' In another place, he noted: 'The Memoirs of Saint Remi [sic] and the Encyclopédie were my constant guides.' He was successful at last, '. . . casting 20 pieces of cannon, with a success which surprized and enchanted the Turks . . . and astonished nobody so much as myself' (116–19).

34 Tott, vol. 2, 3, 9–10. He goes on to say that Mustafa III asked him about the different types of ordnance the Europeans were using, so 'I sent that prince the Memoirs of Saint Remy; he could only examine the plates, and these he had carried after him, when he went abroad, by one of his attendants.' He refers to the 'Bible' of the artillery corps, S. de St Rémy, *Mémorial de l'Artillerie*, 2 vols (Paris, 1693; Amsterdam, 1702).

35 Tott, vol. 2, 3, 85–90.

36 As Larry Wolff recently reminded us, the *Travels and Surprising Adventures of Baron Münchausen*, likely based on Tott, was published by Rudolf Erich Raspé in the very same year as Tott's *Memoirs*: Larry Wolff, *Inventing Eastern Europe: The Map of Civilization on the Mind of the Enlightenment* (Stanford, CA: Stanford University Press, 1994), 359 and elsewhere. The *Memoirs* were an enormous bestseller in an age when such adventures made good raw material for a rapidly-expanding reading public in Europe, hungry for the Oriental tale. Authors who wrote with greater impartiality, such as Mouradgea D'Ohsson's *Tableau général de l'Empire Othoman* (Paris 1788–1824), in seven volumes, by another long-time Ottoman observer, were generally ignored by contemporaries. Mouradgea D'Ohsson, an adviser to Selim III and servant (spy) for the Swedish at the Ottoman court, contradicted the general supposition of both French and British opinion that the sultan was the unparalleled despot of the time. Tott's *Memoirs* cap a century of fictional fascination with the east, in such works as Montesquieu's *Lettres persanes*, or the lesser-known *Letters Writ by a Turkish Spy*, which was published in 25 editions during the eighteenth century. See my 'Is there a Turk in the *Turkish Spy*?', *Eighteenth Century Fiction* 6 (1994), 201–14.

37 Peyssonnel wondered even if Tott could be the author of the work. 'I cannot believe it to be throughout, such as it is published, by Baron de Tott, because there are faults that could not have been committed by a man of his education and parts.' (*Appendix*, 118–21).

38 Sadullah Enverî, *Tarih*, Istanbul University MS T 5994, copied in 1780, covering the years 1768–74, folios 271 and following.

39 Documents reviewed in the Cevdet Askeriye (CA) collection of the Başbakanlik Arşivleri (BA) Archives in Istanbul include: CA 30301, November 1772, an order for gunpowder for the drill of cannons; CA 15612, March 1773, 50 *sür'atçıs* ordered to the battlefront, with supplies; CA 29732, May 1773, a request for supplies for 110 artillerymen remaining in Istanbul; CA 20344, July 1773, a requisition for the monthly rations money for 50 *sür'atçıs*, the preceding sample

representing only the tip of the iceberg. Tott's narration of this second initiative begins on vol. 2, 3, 136 of his *Memoirs*.

40 Text included in an account book devoted to this corps, BA, Maliyeden Müdevver Collection (MM) 4844, pp. 8–9; French translation in A. Boppe, 'La France et le "militaire turc" au XVIIIe siècle', *Feuilles d'histoire* (1912), 391–93. Tott notes that Mustafa III had already spent 25,000,000 sterling (*kuruş*) on the war, and came up with 12,500 *kuruş* for the new troops only with difficulty (*Memoirs*, vol. 2, 3, 136). As estimates about annual expenditures consider 15,000,000 *kuruş* an average before and after the war, this seems about right.

41 BA, MM 4844, 9. The Ottoman document and a transliteration are included in Mustafa Kaçar, 'Osmanlı İmparatorluğu'nda askeri teknik eğitimde modernleşme çalışmaları ve Mühendishanelerin kuruluşu (1808'e kadar)', *Osmanlı Bilimi Araştırmaları* (Istanbul, 1998), 69–137, along with a description of the Rapid-Fire Artillery Corps, 78–81.

42 Tott noted in his *Memoirs* that he received the blessing of the religious officials, encouraged payment of salaries on time, and tried to argue the rationalisation of punishment: Tott, *Memoirs*, vol. 2, 3, 136–43.

43 BA, MM 4844, 33–37.

44 Boppe, 'La France et le "militaire turc" ', *Feuilles d'Histoire* 1912, 396. Tott left Istanbul in March 1775.

45 Levy, 'The military policy of Sultan Mahmud II', 23, from L. Pingaud, *Choiseul-Gouffier*, 99ff.

46 Shaw, *Between Old and New*, 121.

47 Shaw, *Between Old and New*, 122–37, who may not have had access to MM 4844. His study is exhaustive, but lacks context. See also Raif, and Levy, 'The military policy', 30–32.

48 S.J. Shaw and Ezel Shaw, *History of the Ottoman Empire and Modern Turkey* (Cambridge: CUP, 1976–77), vol. 2, 24–27.

49 Elsewhere he noted that the artillery continued to get the attention of the sultan to the detriment of the regular infantry corps, and that '. . . the employment of Prussian officers and sergeants has led to the attainment of a perfection far in excess of anything which at Constantinople had ever been thought possible.' Helmuth von Moltke, *The Russians in Bulgaria and Rumelia in 1828 and 1829* (London: John Murray, 1854), 19. Moltke, of course, later the great Field Marshal of the Prussian army, spent a number of years in Ottoman territories as an adviser to the sultan. The comment on the Prussian officers is from *Essays, Speeches and Memoirs* (New York, 1893), 293.

50 Beydilli, 'İlk mühendisler imizden' 410. Both Ottoman and French versions are reproduced in Kemal Beydilli, 'Seyyid Mustafa'. Beydilli also reproduced the

French title page and the Turkish manuscript introduction in *Türk Bilim ve Matbaacılık Tarihinde Mühendishâne ve Kütüphânesi (1776–1826)* (Istanbul: Eren, 1995). Beydilli's work, along with the article by Mustafa Kaçar, go a long way to filling in the gaps on the history of the engineering schools, their students and publications. Shaw, *Between Old and New*, 145–49. See also Levy, 'The military policy', 34.

51 Küçük Seyyid Mustafa, 17–19, French ed., in Beydilli, 'İlk mühendisler imizden', 464–65. The authorship of Mahmud Raif's work (see above, note 5) has been conveniently attributed to Selim III here.

52 Beydilli, 'Küçük Kaynarca'dan Tanzimât'a islâhât', 26.

53 Beydilli, 'Küçük Kaynarca'dan Tanzimât'a islâhât', 34–38. See also Ali Osman Çınar, 'Sevânihü'l-Levâyih'ı ve değerlendirmesi' (MA Thesis Marmara Üniversitesi, 1992) which is an edition of a Behiç Efendi manuscript from 1802–3. My thanks to Kahraman Şakul, who brought these works to my attention, and who lent me his 2004 unpublished paper on the subject: 'A chronological categorization of the texts of Nizâm-ı Cedid: widening scope of the reform program and transformation of the reform discourse'.

54 For one example, see Aksan, 'Mutiny and the eighteenth century Ottoman army', *Turkish Studies Association Bulletin* 22 (1998), 116–25.

CHAPTER SIX

The internal and external challenges to Selim III

Introduction: provincial allies and colonial powers

We continue the story of Selim III's reign by examining the forces within and without the empire which challenged the dynasty. They include, of course, the arrival of the armies and navies of the western colonial powers, France and Britain, on the shores of the eastern Mediterranean. Before returning to the broader international setting, however, there is one aspect of Ottoman culture of the period just as important in the discussion of Selim III's fall as Napoleon: the changing nature of Istanbul–provincial relations. Responding to the opponents to his military reforms in Istanbul, Selim III had to acquire provincial allies: hence, he unintentionally recast centre–periphery politics to suit his vision of reform. He wished to extend his *Nizâm-ı Cedid* beyond the confines of Istanbul, as we have seen. In this he ran straight up against the new local Ottoman elite class, the *ayan*, who in the eighteenth century represented a burgeoning social stratum. Although the emergence of the great rural households, sometimes modelled on the imperial court structure in Istanbul and built on loyalty and clientage, was not a new phenomenon, the period from 1760 to 1830 has often been dubbed the 'age of ayan', because of their number and influence on central politics. The argument here is that they emerged precisely because of the military needs of the empire for defence as well as reform. The internal challenge was more immediately pressing to the survival of the dynasty than even the international invasions that began in 1798. On most frontiers of the empire, Selim III faced significant challenges not just to his right to rule, as head of the Ottoman dynasty, but equally importantly as the Caliph of Islam.

That was particularly true of the southern 'frontier' extending from Baghdad to Cairo, where newly emergent Wahhabi Puritanism and increasingly influential Naqshbandi Sufism stimulated debate over the legitimacy of the house of Osman. While Muslim heterodoxy was not absent on the Danubian frontiers, and especially ever-present among Albanian Bektaşis, Russian Orthodox proselytising and colonising was a far more pressing problem, even after the peace of Jassy. Pressure by French and Russian diplomats on Ottoman measures to protect their co-religionists had the unintended effect of consolidating Muslim orthodoxy within the dynasty.

The history of Mehmed Ali Pasha, first Governor and then semi-independent ruler of Egypt, is well known, as he posed one of the greatest of all the challenges of the age, and was central to Eastern Question diplomacy, the problem which Mahmud II inherited from Selim III. Less well known is the earlier eighteenth-century rise of the Qazdağlı mamluk household, from which emerged Ali Bey el-Kabir and Muhammad Bey Abu al-Dhabab, challengers to Ottoman colonialism in Cairo and predecessors to Mehmed Ali. For almost the length of Selim III's reign, Ottoman–Muslim notables such as Ali Pasha of Iannina, recently described as the 'Muslim Bonaparte', and Osman Pasvanoğlu of Vidin, alternately challenged or allied with the Ottoman dynasty and generally disturbed the balance of power in the Balkans. All played the international and ethno-religious cards as it suited them (or as necessity dictated), situated as they were on the borders most contested during the Russo-Turkish Wars, and then during the Franco-Russian–British confrontations after 1800. Indeed, both Ali Pasha and Pasvanoğlu were known to engage in international diplomacy with Napoleon.[1] Ali Bey el-Kabir succeeded in wresting semi-independence both from Istanbul and his rivals in Cairo, while challenging the commercial dominance of France in the Egyptian economy. The final and greatest challenge to Selim III's rule came not from Paris or London, or Cairo, however, but rather from the Balkans.

Local Ottoman elites: provincial loyalty and betrayal

In order to understand Selim III's downfall, we must examine the sources of support as well as opposition to his efforts in the provinces. To do so, we return to Canikli Ali Pasha of Anatolia, author of *Nasayih al-Muluk*, as an example of the new-style Anatolian local notable, and then Osman Pasvanoğlu's rise and fall in Vidin and Serbia, as the Danubian example.

The chapter ends with a discussion of events in Egypt and Syria prior to and during the invasion of Napoleon Bonaparte in 1798. Ali Pasha of Iannina and Mehmed Ali of Egypt were the most powerful challengers to the second phase of reforms under Mahmud II, and will be taken up in Chapter Seven.

Canikli Ali and Osman Pasvanoğlu, commanders of their own armies, were initially instrumental in protecting the Ottoman frontiers and fighting on behalf of the sultan. Canikli Ali was a provincial appointee, governor of Trabzon throughout most of his career. Osman Pasvanoğlu first arose as a powerful local *ayan*, and then spent his career alternately as enemy and ally of the dynasty. Their careers are emblematic of the nature of centre–periphery relationships in the late Ottoman context. Their sources of power relied on the acquisition and control of tax farms.

Caniklizades in North Anatolia

Canikli Ali, born in Istanbul in 1720/21, established his power base by becoming the tax collector (*muhassıl*) of Canik, a district stretching along the southern coast of the Black Sea. The Caniklizades controlled the life-term tax farm (*malikane*) of the *Muhassıl* of Canik from 1737 to 1808. By the end of their regime over northern Anatolia, they controlled tax-farm and custom revenues in Amasya, Trabzon, Tokat, Gümüşhane, Kastamonu, Sivas and Erzurum.[2]

Distinguished service, and recognition for raising local troops in the 1768–74 Russo-Ottoman War, guaranteed his status as a vizier, and secured additional tax farms ostensibly for having suppressed an uprising of '*kapısız levendat*', maintaining stability in the area, and facilitating as well as policing passage to the Crimea on the Black Sea. He was commended for his ability to raise troops in the area to protect the Black Sea coast. By 1772, as *Mutasarrıf* of Amasya, he was entitled to the tax revenues called imdadiye, specifically tagged for supporting war efforts.

Canikli Ali was appointed Commander of the Crimea, and Governor of Trabzon in 1773, following the Russian occupation of the Peninsula in 1771, and raised troops and supplies for the many, ultimately futile, thrusts into the Crimea. He was recognised for his ability to mobilize and command armies, and also for his adroitness in making astute alliances, and maintaining open communication and sympathies with the Crimean Tatars. Canikli's next assignment was the defence of Kars and Erzurum, as Commander of Kars and Governor respectively (1774–76). In March of 1774, he was ordered to mobilize some 15,000 soldiers that he was to recruit from the entire north-eastern sector of Anatolia. In addition, he was ordered to procure camels to be sent to Edirne for the same campaign season. As we know, such numbers of men and supplies were never achieved and the campaign year ended in failure, which led to the final capitulation of the Ottomans to the Russians.

In the five to ten years to follow, Canikli Ali was ordered twice more to the Crimea as *Serasken*, in the struggles which continued with Russia until Catherine II's unilateral occupation of the region in 1783. In 1777–78, for example, Canikli Ali, then Governor of Sivas, was appointed *Serasker* of Crimea, and ordered to assemble 40,000 Janissaries and other suitable recruits from all the districts of the southern shore of the Black Sea. He and Admiral Gazi Hasan Pasha attempted a landing at Sevastopol, but returned in disgrace. Throughout the period, he and his sons were repeatedly credited with preventing rebellions, re-establishing order, and putting down banditry in much of north-eastern Anatolia.[3] Here, it would seem, was a dedicated, loyal subject of the sultan, whose service was exemplary.

Things began to go wrong when state demands exceeded Canikli Ali's ability to fulfil them, and a rivalry developed with Mustafa Çapanoğlu (also known as Cebbarzade), also based in Sivas and Tokat. Particularly in the matter of supplies – grain and animals, desperately needed by the state for campaigns in the Danube and Caucasus regions – Canikli Ali's apparent insufficient response to the orders, coupled with the extensive and wealthy power base and independence now reported of him, made him an automatic candidate for suspicion in Istanbul. Complaints about his injustice from residents of Amasya started to make their way into courts, and after 1779 he was declared a rebel, accused of extortion and injustice. Canikli Ali marched on Çapanoglu Mustafa in an effort to defeat his rival, declaring 'either you execute him, or I'll remove him myself', and found himself an outlaw instead. While Ali remained on the run in his strongholds of Canik and Trabzon, by 1780 the state had managed to confiscate many of the supplies (tents, camp equipment and various other stuffs) he had reputedly stockpiled, which required according to one document 422 *hammals* (porters) to disembark in Istanbul. Confiscation, although no longer a standard punishment, could still be imposed as one of the routine ways the Ottomans attempted to maintain control over upstart provincial appointees.

Canikli Ali, who had taken refuge in 1781 with Sahin Giray in the Crimea, proved necessary to the state, however, and was officially forgiven in 1781, reappointed as Governor of Trabzon and survived to die in his bed. Orders from Grand Vizier Halil Hamid (1782–85) to Canikli Ali commend him for his defence of the coastline of the Black Sea, and ask for further intelligence about the Crimea and the Caucasus. This he appears to have supplied, as evident in his *Nasayih al-Muluk*, which may have been written for Halil Hamid, but more likely earlier, while he was in hiding. He died in mid-1785.[4] His is a remarkable story, especially for the evidence of the degree to which the state 'tolerated' such autonomy on the one hand, and pursued to the point of persecution emerging wealth and power on the other. Canikli Ali's story may be unique because he left us a record, but all signs indicate he was a faithful Ottoman subject, and a loyal servant to the dynasty for over a decade. His sons, Battal Hüseyin and Mikdad Ahmed, failed to supply troops for the 1787–92 campaigns, and Battal's son Tayyar Mahmud was accused of neglecting the Caucasian front. Selim III confiscated the Caniklizade wealth and executed Mikdad Ahmed, while Battal and Tayyar Mahmud escaped to Russia, where they stayed until 1799. The final chapter involves Tayyar Mahmud Pasha, who briefly figures in the intrigues around Mustafa IV in 1807, as deputy grand vizier.

This broad outline of the fortunes[5] of the Caniklizades, with different outcomes, was repeated in every region of the empire, to a lesser or greater degree, in this early phase of reforms. Istanbul was relying on hundreds of mini-despots for its provincial governance and defence. In the areas surrounding Trabzon, for example, orders to mobilise troops for the 1787–91 campaigns were addressed to 26 family names in 1788 and more than 40 in 1789. In 1796, a French expedition wishing to explore the botany of eastern Anatolia was confronted with three rulers of Trabzon: the official provincial governor, and two chiefs of local notable families: Osman Agha (Şatıroğlu) and Memiş Agha (Tuzcuoğlu), from each of whom they needed to acquire permission to explore the territory. This is only one example of a precarious style of governing which required negotiation each time a new governor was appointed.[6]

From 1807 to 1811, a period of intense political crisis in the empire, nine different governors were appointed to Trabzon, inevitably creating anarchy in the province. By the first decade of the nineteenth century, by one estimate, coalitions of local elites could mobilise 10,000–20,000 men in arms to turn against the provincial appointees. For example, when appointed Trabzon Governor in 1811, Hazinedaroğlu Süleyman Pasha tried once again to impose the *Nizâm-ı Cedid* military reorganisation on the district. His efforts engendered a multi-year rebellion which required

PLATE 11 *The Ottoman interior of the Baba Vida fortress in Vidin (author's photograph).*

tens of thousands of government troops to suppress, against regional power-broker Memiş Agha. Süleyman Pasha was an 'imperial elite' who amassed extraordinary wealth, with links to the formal state system. As a point of interest, his son, Osman Pasha, also Governor of Trabzon (1827–42), was described as having adopted western dress, living in impressive wealth with a large household, which included European doctors and experts. Osman Pasha is said to have supressed all but a handful of the leaders of the Trabzon 'social oligarchy' by the end of his period as governor.[7] The Trabzon example was replicated in smaller urban settings all over the empire.

Vidin and Osman Pasvanoğlu

Osman Pasvanoğlu, who began his career as a member of the Janissaries, like his father, ruled a wide swathe of Balkan territory from his base in the Vidin fortress on the Danube. Pasvanoğlu engineered a rebellion that engaged Selim III and his government for almost a decade, requiring at one point a massive mobilisation and general call to arms of the population to finally eliminate him. The call to arms stimulated Serbian nationalist sentiments, and Kara George (Petrovic, d. 1817), commander of the forces that eventually put Pasvanoğlu's challenge to rest, is celebrated as the founder of Serbian independence. Whether defined as freedom fighter for Bulgaria, or provoking Serbian nationalism, or as an Ottoman subject in open rebellion, Pasvanoğlu proved very costly to Selim III.

Vidin had continued to trouble Ottoman authorities, largely because of its situation on the edge of the Principalities, themselves frontier territory, but also because of the excessive number of Janissaries in the area following mobilisation and demobilisation of the army after more than two decades of war. It was here that Osman had been born in 1758, son of Ömer, a Bosnian, then one of the wealthiest aghas of the area, and a Christian mother. As with Canikli Ali, however, his wealth attracted the attention of the Commander of the Vidin garrison, and hence, of Istanbul. Ömer's story is sketchy, but expelled from Vidin, he ended up stationed with the 31st Janissary regiment near Belgrade. He and his son Osman were sentenced to death by Abdülhamid I in 1787, for extortion and rebellion. Ömer was executed, but Osman escaped into Serbia, where he distinguished himself in the war against Austria (1787–91) in Wallachia, having 'turned' Albanian and gathered around him his own band. The sultan in gratitude allowed him to return to Vidin, and reacquire part of his family property. His personal army consisted of *eşkıya*, dissatisfied Janissaries and local recruits (*yamaks*), themselves expelled from the Belgrade *Pashalık* by the Ottoman commander Ebu Bekir Pasha, when the Ottomans regained control of the city and province after the peace of Sistova in 1791.

Osman continued the pattern of raiding across the Danube into Wallachia, and eventually besieged Vidin itself, taking possession in 1792. He established an extensive and impregnable power base in Vidin, as he petitioned and was pardoned at least twice by Selim III, in documents signed by all the notables of the area. Much as with Canikli Ali, the central government could do little to actually curb such power. By 1794, Pasvanoğlu had become the most powerful *ayan* of Rumeli, especially western Bulgaria, while İsmail Tirsiniklioğlu Agha controlled the eastern half. The two clashed in 1796–97, when Pasvanoğlu destroyed many of Tirsiniklioğlu's villages strung out along the Danube to Niğbolu.

The battle then began between Belgrade Commander, Mustafa Pasha, appointed by Selim III, a committed reformer, and Pasvanoğlu. Mustafa Pasha brought order to Belgrade and attempted to appease his powerful rival in Vidin, but to no avail. Pasvanoğlu had considerable aid from the Janissary regiments stationed in Belgrade, but Mustafa Pasha repelled them in 1795, and many joined Pasvanoğlu, who had by that time assembled a militia of 12,000 troops, made up of Albanians, Bosnians, Bulgarians and Turks.[8] Mustafa Pasha worked to curb the power of the *yamaks*, and earned the sobriquet 'Serbian mother' from the inhabitants of Belgrade for the care he exhibited concerning the city.[9] Briefly appointed Governor of Rumeli, Mustafa proved unable to resist the increasing raids by Pasvanoğlu and his troops. Ali Pasha (of Iannina), *ayan* of Albania and Greece, was then called upon to rout Osman's army in 1797, incidentally one of the ways in which Ali Pasha himself began his rise. Routed but unbowed, Osman continued to harass the region, until Selim III was forced to mobilise an immense force to put an end to his reign of destruction.

In February 1798, some 40,000–50,000 (perhaps as many as 80,000) troops marched against Vidin. They included Grand Admiral and *Serasker* Küçük Hüseyin Pasha (1792–1803), already renowned for his attacks on Mediterranean piracy; Mustafa Pasha, reappointed as Governor of Rumeli, Ali Pasha himself, plus Tirsiniklioğlu İsmail Agha, Adana *Mutasarrıf* Yusuf Pasha, Tirhala *Mutasarrıf* Kürd Osman Pasha, and *ayan* from Bosnia, Şumnu, Hezargrad, Varna and Pravadi, among others such as then Aleppo Governor Mustafa Cebbarzade who joined the Commander-in-Chief's army in Edirne.[10]

Hüseyin Pasha is generally acknowledged as one of the superior admirals and commanders of Ottoman forces of the period, put in command here of both the Danubian fleet and the land forces. Still, the situation at Vidin was beyond his capacity to control. The *ayan* comprising the coalition had little control over their own forces, much less the willingness to submit to a commander from Istanbul. One such example was Kürd Alo Pasha, who, when ordered to bivouac over the winter around Vidin to keep his troops together, took off to Lom with 15,000 infantry and cavalrymen, who immediately scattered, and then opened communications with Pasvanoğlu. Lured later to the Admiral's headquarters, he was ambushed and shot by Hüseyin Pasha himself, a measure of desperation that gives a fair picture of the nature of discipline in the army of the Danube at the time.

An eight-month siege failed to bring Pasvanoğlu to heel, and the troops were withdrawn, with considerable casualties. *Serasker* Hüseyin Pasha wrote to Istanbul of the losses among his own troops: Osman Pasha and 200 of his soldiers had been killed. Soldiers from the nearby town of Lofça, often the object of raids by Pasvanoğlu's bands, were notable for their bravery in the course of the siege. Süleyman Agha, their officer, was himself wounded in the foot as he and his men attacked the trenches, and later died. (As historian Cevdet noted acidly, it was strange that

Süleyman, known previously to have recovered from more serious wounds with the help of his local surgeons, should die of gangrene in official army service.)[11] Cevdet speaks with the hindsight of the mid-nineteenth century, but his stories are indicative of the lack of trust, resources and manpower at the time, only a few years after the establishment of Selim III's new order. Furthermore, as the siege got under way, rumours of Napoleon's fleet preparations at Toulon reached both the Danube and Istanbul, causing a diplomatic flurry and redeployment of military resources.

Pasvanoğlu's power and prestige were enhanced with another pardon and more titles from Selim III when the coalition army of 1798 withdrew by the end of the year. Admiral Hüseyin Pasha's ships and soldiers were needed in the Mediterranean. Pasvanoğlu was irredeemable, however, and resumed his stimulation of local rebellion among the Janissaries in Vidin and the surrounding area. Local Serbian *knezes* (notables) appealed to Mustafa Pasha to bring order, and he did so in 1799–1800, first trying to put a stop to Janissary raids, and when that failed, by allowing the Serbian population to arm itself. No help could be expected from Istanbul, concerned with the French invasions in Egypt and occupation of the Ionian Islands. Chaos resulted in Vidin and Belgrade, as the government in Istanbul not only did not help, but insisted on the reinstatement of the Janissaries almost as soon as the new Serbian militias were able to expel them. In December 1801, with Pasvanoğlu's assistance, the Janissaries wrested control of Belgrade from Mustafa Pasha and executed him. Thus began the brief Janissary dominance of the city, the so-called reign of the *dayıs*. It was not until the winter of 1804–05 that the Serbs and loyal Ottomans combined to force the *dayıs* out of the city, and captured both Belgrade and Vidin.[12]

This tale has been told without the international context surrounding it. Pasvanoğlu is said to have negotiated with both the Russians and the French. The Russians wished to establish a consulate in Vidin, as they were already well entrenched in the Principalities, having delayed the removal of troops from the region dictated by the 1792 treaty of Jassy. They were prevented by order of the sultan. British and Russian consuls in Bucharest combined to thwart French ambitions, replicating the seething great-power diplomacy of Istanbul. The French had consulates in Bucharest and Jassy after 1795, but they were closed by the Ottomans in 1798. Napoleon sent a mission to persuade Pasvanoğlu to side with France in 1801, but neither France nor Russia appears to have aided Pasvanoğlu in any substantial way.[13]

Nonetheless, when war broke out finally between the Russians and the Ottomans in 1806, Selim III ignored Osman Pasvanoğlu, whose wings had by that time been clipped, and appointed Mustafa Alemdar Pasha (also known as Mustafa Bayraktar: both mean 'Flag-Bearer'), based in Rusçuk, and successor to İsmail Tirsiniklioğlu, as the Commander of the Ottoman forces. Alemdar proved to be supporter of the *Nizâm-ı Cedid*, or at least of Selim III.[14] We will return shortly to his role in the events surrounding Selim III's downfall.

Such is the provincial evolution between 1760 and 1830 observable in Anatolia and Rumelia as well as elsewhere in the empire, in cities like Damascus, Aleppo, Baghdad, where other lineages of local Ottoman elites can also be traced. Selim III's *Nizâm-ı Cedid* inaugurated an attempt to reimpose central order which worked itself out in extended civil unrest in such regional societies, themselves empowered from the late seventeenth century forward by the state's need for troops and supplies. The age of the *sipahis* long over, it had been replaced by the age of *ayan*, but Ottoman survival depended on curbing or engaging that power as well. The Ottomans failed to forge significant new alliances, preferring an exclusive, absolutist model that bore fruit under Mahmud II, that had unforeseen consequences for the well-being and security of the countryside. Perhaps they had little option, even less inclination. By contrast, '. . . in Europe, . . . the noble cavalryman became partly redundant but he transformed himself into the officer in his monarch's army, adding a powerful material and emotional bond to the alliance between ruler and provincial landowning nobility which was the basis of the polity both in eighteenth century Russia and elsewhere.'[15] The remarkable aspect of stories like those of Canikli Ali and Belgrade Commander Mustafa Pasha, was that there were men, willing and able, to make that same transition in the Ottoman context. They would continue to articulate the need for reform from both province and centre as the nineteenth century unfolded. Whereas such leaders might be amenable to a process of negotiation with Istanbul in 1800, a significant and well-trained officer and administrative class was many decades distant.

Osman Pasvanoğlu, however, represented the traditional order. He resolutely prohibited the imposition of the new economic and military reforms in the territories under his command, one of the chief reasons for his popularity among the peasants. Foreign observers noted that Christians (Bulgarians) as well as Muslims (Turks) sought refuge in his territories, and fought in his garrison at Vidin. He was well-informed on international affairs by a spy network, and was generally acknowledged to be an adroit manipulator of the French, Russian and Austrian diplomatic initiatives in the territory. Until the Serbian revolt of 1804, Pasvanoğlu represented the downtrodden of all ethno-religious stripes in one of the most diverse corners of the empire. After 1804, the Christian/Muslim lines started to be more firmly drawn as the Orthodox Serbian communities struggled for autonomy from the Ottomans.[16] Pasvanoğlu is also credited with creating a vibrant centre of Islamic culture in Vidin, modernising the fortress, repairing the roads, erecting fountains, a mosque, and a library, as well as a modern barracks, most of which is still standing.[17]

For Selim III and his court in Istanbul, Pasvanoğlu challenged dynastic and ideological claims, and he had to be branded both a rebel (*eşkıya*) and heretic (*Bektaşi*) in order to legitimise the attack on him as a usurper of the sultan's authority. It is possible to argue that the dynasty reasserted its Muslim orthodoxy here, as elsewhere in the period, not just as a response to the Russian and Austrian manipulation of religious identities in diplomatic instruments, but equally because challengers like Pasvanoğlu cloaked their aims in caliphal arguments. The very name Osman could be used to remind followers of the eponymous founder of the ruling house. As a threat to the new order, Osman Pasvanoğlu may be likened (only in broad terms) to the 1773–75 Pugachev rebellion in Russia, itself couched as protecting the traditional order and restoring the legitimate dynasty. Pugachev, who had seen service in a Cossack regiment in Bender in 1770, derived his support from rebellious Cossacks, and for a time threatened the social stability of Catherine II's Russia. It could be argued that Pasvanoğlu succeeded where Pughachev did not in precipitating the events which brought down Selim III.

Radical diplomacy

Quite apart from his inability to effectively curb the activities of *ayan* like Pasvanoğlu, Selim III was also frustrated by notables who had access to international diplomacy in a way he did not. Their ambivalent relationship with the state would have prevented him from acquiring accurate and speedy information on frontier and external affairs from his provincial allies, with some exceptions, as we have seen in the case of Canikli Ali. Foreign occupation, a very real, frightening possibility after 1792, and even more of a possibility by the end of Russo-Ottoman hostilities in 1812, may have contributed to Selim III's decision to establish permanent Ottoman embassies in Europe in 1793, the most significant innovation in diplomacy of the age. Deprived of the French military mission (recalled in 1788) and forced to break off relations with the French after the Revolution, the sultanate struggled to maintain Ottoman autonomy. One way to do so was to take the initiative, by embracing the European idiom of diplomatic and social communication at its source.[18]

Pre-1793 diplomacy

Between 1703 and 1774, the Ottoman–European diplomatic system was regularised by the clauses of the Karlowitz Treaty (1699), based on reciprocal rather than unilateral agreements. Most Balkan treaties thereafter used the Karlowitz treaty and subsequently that of Küçük Kaynarca in 1774 as their basis for comparison. Ottoman diplomacy functioned adequately on a significant number of occasions.[19] Each negotiated treaty required an exchange of diplomats, resulting in considerable (and expensive) traffic between the Empire and its various trading partners and erstwhile enemies. Although such embassies supplied the sultans with little significant intelligence on the courts they visited, it is precisely in the mid-eighteenth century that the aim of such embassies began to change. Notable, for example, is the report of Ahmed Resmi following his visit to Berlin in 1763–64.[20]

The Ottoman embassy to Vienna in 1791–92, to cement the newly signed Sistova treaty ending Austria's involvement in the Russo-Austrian–Ottoman war, was a temporary, purely ceremonial affair, but the ambassador was Ebubekir Ratıb, Selim III's closest confidant, whose mandate appears to have been far broader than just an exchange of treaty instruments. Ratıb submitted a massive report to Selim III upon his return to Istanbul, almost 500 manuscript pages of description of Austrian military, bureaucratic and political affairs, unique to the eighteenth-century traditional embassies. Ebubekir was present when Selim III convened his first council as sultan, and his report on Austria is acknowledged by historians to have contributed to the later reform agenda, although it is generally conceded that Tatarcık Abdullah Pasha's recommendations were those Selim III enacted. The detail of knowledge about a European court and its military secrets that is embodied in the embassy report was unprecedented in Ottoman circles. Austria, rather than France or Russia, as a model for Ottoman reform, remains a largely unexamined possibility.[21]

Ratıb's experience and advice may have precipitated Selim III's sudden decision to send permanent embassies to select capitals of Europe: London, Vienna, Berlin and later Paris (1806), in search of further such enlightenment.[22] The implication of Selim III's decision to take diplomacy to the courts of Europe has been overlooked in most discussions of his reforms, because of the embassies' relative lack of permanence, and also because the gesture was eclipsed in the chaos that engulfed Istanbul after 1798. Furthermore, the sultan did not have an independent courier system, relying on the diplomatic post of neighbours such as Austria or Prussia. Regular, European-style, permanent Ottoman embassies to foreign capitals were not in place until after 1835.

First embassies to Europe

The first, permanent Ottoman embassy was to London, an initiative perhaps conceived by Ratıb Efendi and discussed with Sir Robert Ainslie, long-time British ambassador to Istanbul. Britain's wealth and power, and potential for support against both France and Russia, must surely have swayed Selim III. Joining Ratıb in the discussions with Ainslie concerning the proposed embassy were *Reis* Mehmed Raşid Efendi and Tatarcıkzade Abdullah. The first London ambassador was Yusuf Agah Efendi, whose Chief Secretary, Mahmud Raif, wrote a report of the years in London (1793–97) in French.[23] Author of *Tableau des nouveaux règlemens de l'Empire ottoman*, discussed in the previous chapter, Raif was one of the first three men to be sent abroad by the sultan with specific instructions to undertake the study of a foreign language. He was 32 years old at the time of his appointment, and very anxious to learn French, the 'universal language'.[24] Raif is representative of the new-style bureaucrat important to Ottoman reform in the mid-nineteenth century.[25] His account includes a description of the House of Commons and Lords, and the English army and navy, with special emphasis on the ships, the arsenals and docks. Also included is an extensive description of the English economy, and of the city of London. Raif served 1800–05 as *Reis*, and was on campaign in Egypt in 1801 as part of the Ottoman–British coalition army organised to combat the French occupation. Ambassador Yusuf Agah's correspondence with Sultan Selim III, preserved in manuscript with the sultan's notes in the margin, is evidence of his keen interest in refashioning the Ottoman state on foreign models.[26] On the practical level, Ambassador Yusuf Agah sought British friendship and loans, but his official reception did not bode well for the success of the enterprise. British Prime Minister Grenville reacted indifferently to the ambassador, and quite failed to appreciate the significance of Selim III's decision.[27]

The second and third permanent ambassadors were sent to Vienna (Ibrahim Afif Efendi in 1797) and Berlin (Ali Aziz Efendi also in 1797).

Like Yusuf Agah, and his successor in London, Ismail Ferruh Efendi, these pioneers are generally portrayed in western sources at best as naive; at worst, as serving as dupes of the host governments. Generally unskilled in the languages of the country to which they were appointed, adept at ceremonial but untrained in the intricacies of European diplomacy, they were often lesser bureaucrats of the court in Istanbul, and had to rely on a staff of translators, drawn from the Greek community of the capital, the infamous 'dragomans' of all foreign observers.

Selim III's diplomatic initiative annoyed Catherine II and angered the French, the latter long considered the oldest 'friends' of the Ottomans. The Ottomans did not send a permanent ambassador to St Petersburg until 1857, a measure of the long years of hostility between the two courts.[28] Selim III had corresponded with the French court, however, before becoming sultan, and remained a Francophile throughout the period. Vergennes, former ambassador to Istanbul, and French Foreign Minister, and a long-time realistic ally of the Ottomans, had died in 1787, and with his death, the putative friendship collapsed.[29] Vergennes was responsible for the operation of the French military mission in Istanbul, which was withdrawn in 1788. It remained to be seen if Great Britain would step into the void. In the 1790s, this was by no means clear. Britain's Levant Company was close to dissolution (and actually taken over by the British government in 1821), and British officials were more engaged with India and Europe than with the Ottomans. At most, the British ambassador in Istanbul never heard more than ten times a year from his government in London. By contrast, French Ambassador Choiseul-Gouffier (1784–92) received special dispatches from Izmir on a regular basis, and was a particular source of information on international affairs for Selim III.[30] The French dominated Eastern Mediterranean trade between Marseilles and the Ottoman ports, particularly Izmır. After the French Revolution forced a breach in Ottoman–French relations, Selim III's political manoeuvring in the period 1798–1807 can be interpreted as his continued desire to re-establish the French connection. This was one of the significant reasons for his downfall.

Pressure from Ambassador Raymond Verinac (1795–97), the first fully accredited French Republican representative in Istanbul, persuaded Selim III to appoint Seyyid (Moralı) Ali Efendi as the first permanent Ambassador to Paris in 1797. The idea probably appealed to Selim III anyway. Ali Efendi was well received in Paris, in the style of Molière's 'Bourgeois Gentilhomme' to judge by one account.[31] Europeans were avid readers of the *Arabian Nights* and *The Turkish Spy*, and must have delighted in the virtualisation of their fantasies in the new ambassador.[32] Ali was accompanied by one of the more notorious of dragomans, Pangiotis Codrika, who served Talleyrand in keeping Ambassador Ali Efendi ignorant of French intentions in Alexandria. Ali Efendi was maintained as a virtual prisoner until 1801, like his counterpart in Istanbul the *chargé d'affaires* Pierre Ruffin, who was one of the last European diplomats to be locked up in the Seven Towers castle, as part of an Ottoman declaration of war. Ruffin later became one of the most influential of France's Orientalists.

Undeterred by British and Russian opposition to his moves to reconnect with France, Selim III continued to appoint ambassadors to Paris: Halet Efendi (1803–06), and Muhib Efendi (1806–11). In spite of that continued representation, Selim III was compelled, on at least three occasions, to send special plenipotentiaries to resolve some delicate negotiations. One of note is that of Mehmed Said Galib Efendi, sent to negotiate the signing of the Treaty of Paris on 25 June 1802, re-establishing Ottoman–French relations ruptured in 1799, as described below. Of most interest is the fact that Galib Efendi and Halet Efendi survived the rebellions around Selim III's demise to serve as rivals in the court of Mahmud II. Galib Efendi has acquired the title of father of modern Turkish diplomacy for negotiating the Treaty of Bucharest in 1812, while Halet has been dismissed by Bernard Lewis as 'a convinced reactionary and hater of all things Western.'[33]

While the diplomatic record of these first permanent Ottoman representatives abroad is undistinguished, their exposure to western style and culture was more significant, adding to the information slowly seeping into the upper levels of government in Istanbul, and already apparent on the streets.[34] Their experiences at foreign courts may also have contributed to Mahmud II's rejection of the system until such time (1835) as he had created a reliable diplomatic corps and trustworthy translators in Istanbul itself before appointing further permanent representatives to the courts of Europe.

European governments preferred to acquire their information on the Ottomans from the hotbed of Istanbul itself, probably the last 'European' capital where members of all states met on more or less the same terms, even as they complained of the rapacity of the system of 'gifts' required to influence Ottoman court politics. That would change with Napoleon's invasion of Alexandria, which inaugurated the colonial age in the Middle East, when spies, diplomats, translators alike were caught up in the new age of persuasion and ideology.

Napoleon's invasion of Egypt

After 1798, Franco-British commercial rivalry, played out all over the world, was intensified in the Mediterranean with warships, as European conflicts spread to the Near East. In 1797, the Habsburgs had abandoned their allies Russia and Great Britain, and signed the Treaty of Campoformio with France, one of the first pieces of news reported home from Vienna by Ambassador Ibrahim Afif Efendi. That treaty dismembered the Republic of Venice, and gave France the Ionian Islands, making France a neighbour of the Ottoman Empire for the first time and exacerbating Selim III's complicated relationships with Osman Pasvanoğlu and Ali Pasha of Iannina.

In July 1798, General Napoleon Bonaparte, largely on his own initiative, sailed for Egypt with 35,000 troops and a 500-strong team of scientists, archaeologists, linguists and scholars. Landing at Alexandria without much resistance, he marched to Cairo, where he defeated a mamluk army double in size at the Battle of the Pyramids. On 25 July, Bonaparte entered Cairo, but remained precariously perched on the mouth of the Nile. In the annals of military history, the battle stands as the great east-west encounter, with Bonaparte's hollow squares of infantry cutting down mamluk cavalry armed only with swords and bows and arrows. This was not, however, Italy, as attested by Napoleon's comments from his memoirs dictated at St Helena: 'In vain were [the French soldiers] assured that the country was the most fertile in the world, that it was even superior to Lombardy; how were they to be persuaded of this when they could get neither bread nor wine? We encamped on immense quantities of wheat, but there was neither mill nor oven in the country. The biscuit brought from Alexandria had long been exhausted; the soldiers were even reduced to bruise the wheat between two stones and to make cakes. Many parched the wheat in a pan, after which they boiled it.' Water was a problem: 'Frequently, after our most oppressive marches, nothing could be found to allay the urgent cravings of thirst but a little brackish water of the most disgusting description.' The army, almost from the day of arrival, was restive and unhappy.[35]

Napoleon's adventure, initially a private enterprise sanctioned by the Directorate, who preferred the general occupied outside Paris, proved disastrous and embroiled France and the entire region in global warfare. (The scholarly enterprise, not under discussion here, proved more lasting in the later publication of the magnificent set of documents called the *Déscription d'Egypte.*) According to his chief ideologue, Constantin

Volney, writing in the *Moniteur*, Bonaparte had achieved the creation of an Arab nation by his reforms, and aimed at 'the resurrection of former Arab glory, the destruction of the Ottomans, and the purification of Islam.'[36] By 11 September of the same year, in agreement with Russia and Britain, the Ottomans were at war with their 'oldest friends'.

Bonaparte's thrust would be short-lived, however, as the British navy under Admiral Nelson dispatched the French navy at the Battle of the Nile on 1 August 1798, surrounding and destroying most of the fleet off Aboukir. It was a blow to Bonaparte as the destruction of the ships and supplies crippled the enterprise. Great Britain signed the Anglo-Ottoman Defensive Alliance in January 1799, with mutual guarantees for their respective territories, joining the Russo-Ottoman alliance already agreed upon. Rather than becoming mired in Alexandria, Bonaparte marched into Syria in February of 1799, defeating the Ottoman garrison at Jaffa, where the French army executed 4,000 prisoners. He faced stiff resistance at Acre, and withdrew by the end of May, having failed to breach the fortress. Of 13,000 troops under Bonaparte's command at that time, 2,000 died and another 3,000 were wounded, heavy casualties even for the period.[37]

The garrison at Acre was commanded by independent *ayan* Cezzar Ahmed Pasha, another of the controversial figures of the age, who managed 35 years of rule over Syria, sometimes as Governor of Sidon (Sayda), sometimes of both Sidon and Damascus. A Bosnian native who spent time in Istanbul, Cezzar first went to Egypt in the entourage of Ali Bey Agha, who had appointed Ottoman governor in 1756. Cezzar joined the household of Salih Bey, learned Arabic, and acquired power by his associations, especially with mamluk Ali Bey el-Kabir, discussed below. After a falling out with Ali Bey, Cezzar Ahmed escaped from Egypt, only to surface again in Beirut around 1770. Much like Osman Pasvanoğlu in Vidin, the young adventurer Cezzar turned 'Albanian', to attract warriors, and created his own power base in Acre after the death of Zahir al-'Umar in 1775. His fame rests largely on his defeat of Bonaparte.[38] Cezzar Ahmed faced the French with around 700 *Nizâm-ı Cedid* troops sent from Istanbul as an advance guard among his own forces, and a British contingent led by Sir Sydney Smith. British ships lent fire support offshore, bombarding French trenches, and kept the fortress supplied throughout the siege.[39]

While events unfolded at Acre, the Ottomans made preparations for a full-scale campaign by land and by sea in cooperation with the British. Selim III's advisers in Istanbul discussed the military alternatives in Egypt, in response to Cezzar Ahmed, who had written to Selim III asking to be made *Serasker* of Egypt and to be given sufficient funds to raise troops and supplies to lead the fight against the French. The imperial council was inclined to comply, but sober thought made them realise that although he was strong in his own territories, Cezzar Pasha was insufficiently powerful to counter the large French army. One of the clinching arguments was that once Cezzar Pasha was in Egypt, who would remove him? Hence, he was ordered to prepare and defend the Syrian coastline, as he appears to have done at least at Acre. Selim III's advisors preferred to mobilise a coalition army with the British rather than empower a quasi-independent *ayan*.[40] By mid-1799, the fleet was deployed, under Grand Admiral Küçük Hüseyin Pasha (1792–1803) and the army ordered to mobilise under Grand Vizier Kör Yusuf Ziya Pasha (1798–1805; 1809–11) at Üsküdar. Quite apart from confronting the invasion, the Ottomans sought to re-establish more formal control over Egypt, which had in the previous half-century achieved a considerable degree of autonomy from Istanbul. Before continuing with a discussion of the consequences of Napoleon Bonaparte's adventure, we need to describe Egyptian affairs just prior to the invasion.

Pre-Napoleonic Egypt

Egypt and Syria had experienced much the same emergence of militarily independent-minded households as elsewhere. In Syria, as in the Balkans, the line between mercenary and bandit was almost non-existent, and the region had existed for centuries under nominal Ottoman aegis, with private armies and local recruitment. After the dissolution of the Janissaries, in 1826, themselves well assimilated into the local scene and often rivalling *yerliye*, the relationship between Istanbul and the Arab provinces became even more disconnected.

Irregular military groups grew out of many of the same cultural contexts as we have seen elsewhere: clan-based, communal warriors characterised Nablus, for example, and the Hawran of the Druze. Nomadic groups of Türkmen or Bedouins might serve as temporary local combatants, or protectors of the caravan routes. Inter-tribal warfare made that kind of force problematic for the Ottomans, and the centrally-appointed governors often found themselves pawns of local events. A third kind of local force commonly seen elsewhere were the *levends*: cavalry of mostly Anatolian and Balkan origin, who essentially became soldiers-for-hire when they were demobilised by the Ottoman army itself. They could be Albanians, Kurds, or even Maghrebins, the latter also known as *delis*, or the later *başıbozuk* cavalry. Cezzar Pasha, for example, made extensive use of the *levends*, as well as independent groups of Bosnians and Albanians. Added to the mix of foreign soldiers were the Wahhabis, who emerged as fierce anti-Ottoman Islamic warriors precisely in this period.

The city of Damascus, situated on the pilgrimage route to Mecca, enjoyed moments of stability under the al-'Azm family, members of whom were also Ottoman governors of the province at least nine times between 1725 and 1808. 'Abdullah Pasha, one of the last of the family, ruled intermittently from 1795–1807, and assisted then Governor Cezzar Pasha in the struggle against Napoleon at Acre. Subsequently he was dismissed when he could no longer control the local military factions, who could attack the annual hajj caravan at will. A period of undeniable strife and chaos unfolded in the first half of the nineteenth century in greater Syria, as in Egypt in the first decade.[41]

By the mid-eighteenth century, the Egyptian household system also resembled the Ottoman-style clientage system. The distinct style of the mamluks, however, was that membership of the households was said to be restricted to formally manumitted slaves, largely from the Caucasus. The struggle that unfolds after 1800 concerns the last of the great mamluk households, local bedouin groups, and the imported Albanian soldiers who constituted a good proportion of the Ottoman armies sent to Egypt. The Ottoman governor, an Istanbul appointee, was caught in the middle unless he was able to construct an alliance with the local brokers.

For the Ottomans, the Egyptian province was largely a colonial enterprise, governed by the *vali* who was an Istanbul appointee, and controlled by the Janissary regiments established in Cairo after the original occupation in 1517. The total paper strength of these seven regiments numbered 18,309 in 1797. The two largest were the *Mustahfizan* and the *'Azaban*.[42] The chief task of the governor was to maintain order, secure the annual tribute from Egypt, and supply 3,000 troops to imperial campaigns when called on by Istanbul. By the eighteenth century, most of the Cairo regiments had come under the control of *sancak beys*, heads of mamluk households, and were packed with local residents seeking to benefit from the salaries and protection of the corps. Hence Istanbul governors found themselves eclipsed, and subject to the rivalry among the great households for control of the beylicate, the highest of the mamluk offices, as well as appointment to the lucrative position as *amir al-hajj*, commander of the annual pilgrimage to Mecca and Medina. It is this context which made Bonaparte's landing and victory so deceptively easy. That the Ottomans

PLATE 12 *Egyptian mamluks training in Cairo (Yates, William Holt,* The Modern History and Condition of Egypt; its Climate, Diseases and Capabilities with an account of the proceedings of Mohammed Ali Pascha from 1801 to 1843. *London: Smith, Elder & Co., 1843).*

maintained 'a modicum of authority in Egypt and the flow of taxes from that province throughout most of the eighteenth century is a remarkable achievement.'[43]

Debilitating civil wars among the households had given rise to an extraordinary figure from the Qazdağlı mamluk faction, and achieved the merging of the office of governor and bey in one person by the end of the century. That person, Ali Bey el-Kabir, probably of Christian origin, was a mamluk from Abkhazia.[44] By 1768, he had ruthlessly eliminated his major rivals, disposed of possible rivals in the seven regiments, including Janissaries, and assumed the title of *shaykh al-balad*, headman of Cairo. By 1769, he had pacified most of Egypt. His tyranny extended to the foreign and minority merchant community, largely Jewish, from whom he extracted ever larger extortions, often given as one of the reasons the government in Paris was later persuaded to allow Bonaparte to attack Alexandria in 1798. Ali Bey deposed the Ottoman governor, and challenged the sultan by refusing to submit the annual tribute to Istanbul, even though he did send the 3,000-troop contingent to the campaigns unfolding in the Balkans in 1769. He reputedly had 25,000 to 40,000 troops at his disposal.[45]

As part of an ambitious foreign policy, Ali Bey sent his trusted mamluk and brother-in-law, Muhammad Bey, to the Red Sea in an attempt to re-establish Egyptian commercial control over the area, where British and French rivalries were heating up over the lucrative Yemen coffee trade. Ali Bey also had ambitions concerning Syria and Palestine. He surrounded himself with Christian advisers, and employed numerous foreigners, especially artillery experts, and beginning in November 1770, began a campaign to capture the city of Damascus. His ally was Shaykh Zahir al-'Umar, predecessor to Cezzar Pasha in Palestine. The city of Damascus, abandoned by the Ottoman governor, surrendered to Muhammad Bey, Ali's hand-picked commander after his return from the Red Sea, in June 1771. In spite of this victory, Muhammad Bey retreated precipitately to Cairo, and participated in a conspiracy to oust Ali Bey. Ali Bey, exiled from Cairo and still allied with Shaykh Zahir in Gaza, enlisted the support of the Russians in his continued assault on Syria. Negotiations with Count Orlov, already in the Aegean stirring up the Greeks of the Morea against the Ottomans, came to little other than some naval support from the Russian fleet. Rumours about an Ottoman assault on Egypt in late 1771 proved groundless. There was little the sultan could do, with Grand Vizier Muhsinzade Mehmed engaged in negotiations to end the 1768–74 war with the Russians at Babadağı on the Danube.

Mamluk household war continued in Cairo, but Muhammad Bey eventually defeated his erstwhile master at the gates of the city in May 1773, in a battle which 'would determine not only the leadership of Egypt, but the course of subsequent events in Palestine, Syria, and the Ottoman Empire.'[46] The significance of these events lies in the autonomy gained by the beylicate as opposed to the Ottoman governor. Ali Bey had defied Istanbul by deposing the Ottoman governor. Muhammad Bey restored both the Ottoman governor and the annual tribute to the sultan, suggesting his collusion with Istanbul, but achieved the same status as Ali Bey when Sultan Abdülhamid I (1774–89) gave him the rank of vizier and was set to appoint him as governor, just shortly before Muhammad Bey's sudden death in Acre in 1775, where he too was on campaign to reunite southern Syria and Egypt. Both Ali Bey and Muhammad Bey foreshadow the ambitions and achievements of Mehmed Ali by several decades. In the confusion and factional strife which followed Muhammad Bey's death, the Ottomans were able to re-establish a relationship with the Qazdağlıs which lasted until Bonaparte's invasion in 1798, but which did not include secure administrative control of such an important province. An Ottoman expeditionary force was sent to restore the sultan's control over Egypt in 1786, under the command of the well-known Grand Admiral Gazi Hasan Pasha, who had advanced as far as Cairo when he was recalled to Istanbul before he could finish the task to prepare for the campaign against Russia in 1787. As I have argued, defence of the northern arc obsessed the inner circle in Istanbul. Egypt resembled a ripe plum for the plucking, both militarily and economically, as many in Paris, such as Constantin Volney, argued.[47]

French evacuation of Egypt

By mid-1799, however, Bonaparte personally abandoned the Egyptian plum, as his project had become untenable, particularly after the retreat from Acre. Even though his army defeated some Ottoman troops disembarking at Aboukir on 11 July, Bonaparte chose to sail for France with a small escort. He left General Kléber in charge of the evacuation of the French troops. The Ottoman army had begun the advance from Jaffa through the Sinai towards Egypt. Plague and hunger posed as much a threat to the French as mamluk and Ottoman adversaries. In January 1800, persuaded by Sydney Smith, Kléber signed the Convention of El-Arish with Kör Yusuf Pasha (and Smith for the British), for the evacuation of Egypt. The convention was rejected by both Britain and Russia, who insisted on surrender, so the French were forced to dig in. In March, the French defeated the Ottoman (largely Albanian) army and Mamluk troops at Heliopolis ('Ainshams) and gained control over Upper Egypt. Returning to Cairo, which had revolted in his absence, General Kléber starved the rebels into submission, and then fell victim to an assassin in mid-June. He was replaced by General Menou.

By October 1800, British ships and troops had embarked for Egypt, in support of the Ottomans, and in March 1801, the British landed in Aboukir, shortly before Ottoman contingents under Kapudan Admiral Küçük Hüseyin Pasha also arrived by ship. The combined army numbered: 15,000 British, under command first of General Abercromby, then General Hely Hutchinson, and 7,000 Ottomans, under Kapudan Küçük Hüseyin, in Alexandria. In addition, Grand Vizier Kör Yusuf's army of 15,000 (estimated at 25,000 by the time of its arrival in Egypt) left Gaza in mid-March to cross the Sinai Desert again, arriving on the eastern bank of the Nile in mid-April. The first of 6,000 sepoys from British India, who had left Bombay in December, arrived in Qusayr on the Red Sea in May, and with great difficulty marched down the Nile, reaching Rosetta in time for the final French capitulation.[48] The East India Company used the opportunity to promote further trade in the Red Sea and the Gulf. French General Menou's ineptitude and delay permitted a disorganised Anglo-Ottoman army to isolate the last of the French army in Alexandria, forcing the French evacuation of Egypt by September 1801. Militarily insignificant, this latter campaign, and indeed the entire Napoleonic interlude, had immense social consequences for Egypt, discussed below.

Hostilities ceased with the Treaty of Amiens in June 1802, which was signed by Selim III, renewing all French capitulations and establishing French predominance in Istanbul as before the Revolution, much to the chagrin of the British, who would shortly resume the fight against Napoleon Bonaparte, in the Third Coalition with Austria and Russia (May 1803). Selim III preferred neutrality to breaking the treaty. After much wrangling the last of the British troops left Egypt in March 1803, in order to restore good relations with the sultan. The first colonial thrust into the Middle East was over in a brief three years. The consequences reverberated for another hundred.

On the surface, the Ottoman Empire, though not Selim III, escaped with little damage, or no more than the usual destruction inflicted on local populations by warfare and local oppressors, and Selim III's preference for the French connection, as well as Anglo-Russian vulnerability to Napoleonic ambitions elsewhere, restored Ottoman sovereignty to Egypt, however tenuously. He was grateful for this, and established a medal, called the Order of the Crescent, which he bestowed on the British officers 'to perpetuate the signal services rendered'.[49]

What ensued in Egypt, however, was further civil war among the mamluk households, the Ottoman governors, and remnants of the Ottoman army of liberation, mostly Albanians, who caused considerable unrest and widespread misery. A young survivor of 1801, Mehmed Ali, would bid his time, and return to threaten the new sultan Mahmud II (1808–39) in his own home territory of Anatolia. To build his modern army, Mehmed Ali relied, as we shall see, not on the British but on French ideas, technology, military models, and advisers. This was the primary consequence of the Napoleonic invasion, which produced both the chief architect of Mahmud II's reform agenda, Hüsrev Pasha, and the chief rival to the Ottoman house, Mehmed Ali. The story continues in Chapter Eight.

The audacious French thrust into the eastern Mediterranean also resulted in bringing not just the British but also the Russian fleets into fuller operation in the region. In fall 1798, the Russians moved warships through the Bosphorus Straits to support the Ottomans in the emerging struggle with the French. A Russo-Ottoman fleet drove the French from the Ionian islands, establishing the short-lived Septinsular Republic in 1802, largely to protect the inhabitants from the predations of Ali Pasha of Iannina who had captured Preveza and Butrinto from the French. While lending support to Selim III in his bid to subdue Osman Pasvanoğlu of Vidin, Ali Pasha had succeeded in establishing his own power base in Albania and Greece. He was in a position more powerful than any sultan during the first two decades of the nineteenth century, a problem greater even for Mahmud II than Selim III.

Imperial impressions

The brief Anglo-Ottoman–Russian coalition against Napoleon Bonaparte left some lasting impressions and attitudes of the Ottomans and their armies with the British officers. Used to the strict hierarchy and discipline of their own system, the British officers were appalled at the unruly disorder of the Ottoman camp, which was noted for its confusion and filth. No attention was apparently paid to the burying of dead animals, or offal. The rivalry in command between the Grand Admiral and the Grand Vizier was palpable to these eyewitnesses, and generally attributed by them as natural in Oriental court politics.

Kapudan Küçük Hüseyin Pasha, who died in Istanbul in 1803, was described as having '. . . great penetration, a marked predilection for every thing European, and a desire to better the condition of every one immediately about him, . . . [as] the best and most prominent features in his character; but to his education in the seraglio he owes the opposite and dark side of his character, profound dissimulation, and a deep spirit of intrigue.' It was noted that he was the only one who disciplined his troops properly, having under his command two good regiments of *Nizâm-ı Cedid* troops. The admiral vigorously strove to keep his troops and those of the grand vizier separate. Those reformed troops were actually under command of the above-mentioned Hüsrev Pasha, protegé of Küçük Hüseyin Pasha, who would be appointed to command Mahmud II's new army in 1827. For his efforts in Egypt, he was appointed governor. When he tried to introduce the *Nizâm-i Cedid* regime, he became embroiled in the civil strife, which led to Mehmed Ali's rise and his own defeat at Damietta in 1803. Hüsrev Pasha is, nonetheless, a towering figure in all subsequent events leading to the creation of the new army. During the Greek revolt in Peloponnesus, Hüsrev Pasha, who had been reappointed Admiral in 1822, would struggle with Ibrahim Pasha, son of Mehmed Ali, for control of the Ottoman campaign against the Greeks, and, pressured by Mehmed Ali himself, Mahmud II would dismiss him in February 1827.[50]

The gray-bearded Grand Vizier, Kör Yusuf Pasha, blind in one eye, was known for his piety and fatalism rather than his knowledge of European affairs, but was considered an able commander. His principal support came from the Qazdağlı faction under Ibrahim Bey. Of the troops commanded by the grand vizier, the Janissaries were considered dangerous; the Albanians, of the greatest number, were thought to be good light troops but prone to revolt. The cavalry numbered around 5,000 but 'they have no idea of discipline, and can never be brought to think that acting in a body is preferable to their loose mode of attack. Of course they can never make any great impression on a well disciplined enemy.'[51] British commanders generally supported the mamluk beys and Arabs over Albanians and Turks in the struggles following 1802, influenced by an increasingly widespread imaginary evocation of the Arabs as an ideal republic.[52] Local civilian populations, however, fed up with mamluk excesses, favoured the reforms of Mehmed Ali, and considerably enabled the transformation of Egyptian society.

These were the early days of British global imperialism, before the consolidation of the Raj system in British India, with all its prejudices against

'Orientals', was in place. Whether or not the French and British understood their 'imperial project' as a 'civilising mission' *at this juncture*, is very much a subject of debate. It could easily be argued that these first impressions of aspiring imperial armies with their Ottoman equivalent were pivotal to the emergence of a certain worldview, what Norman Daniel characterises as the merging of 'crusade and colony'.[53]

It could equally be argued that both the British and French transferred their American and Indian experiences and impressions to the eastern Mediterranean. A good number of the officers in Egypt were experienced East India Company army veterans, hence the celebration of the arrival of the sepoy regiments, as well as the inclination to rely on the mamluk-style warrior. The East India Company faced no significant native resistance to British rule before the 1830s, in a partnership of 'merchants, a patrician military class, and willing native rulers', until the Company fell into debt and imposed heavy local taxation to support an army that grew from roughly 90,000 in 1793 to about 230,000 in 1820, and spawned numerous local revolts.[54] The sepoys who reached Egypt in 1801 are evidence of the increasing British-style militarisation of India. British officers, such as Sir Sydney Smith, were confronted in the Ottoman command with an imperial disposition to superiority, and cultural disdain equal to their own. This infuriated him enough to refer to them as 'our ill-informed, illiberal, cruel, avaricious, perfidious allies, the Turks'. General Koehler, Commander of the British Military Mission, wrote: 'The Turks have great resources, but they are in the last stage of weakness, from the want of knowing how to call them into Action, and how to employ them with advantage afterwards, but above all from want of Patriotism or Publick Spirit, every person in every Station, and of every rank thinks no farther of this situation than as enables him to plunder the State or individuals; No person except the Emperor Himself has the good of the Emperor at Heart.'[55]

Observations here would be reinforced on the next occasion when the British and the Ottoman officer classes fought on the same side, first at Acre in 1840, and again in the Crimea in 1854–55. The gulf in the ethos of military command and the conduct of warfare between Ottoman and British remained unbridgeable for most of the nineteenth century. British military advisers, never numerous in court reform circles in Istanbul, were completely eclipsed by Prussian officers and advisers by mid-century. This is at least partially attributable to the early influence of Helmuth von Moltke, who had become chief foreign adviser to Mahmud II at the end of his reign. French influence in the capital also waned after the final defeat of Napoleon, even though the French regimental model was initially adopted for Mahmud II's new army. In the end it was the Prussian-style army, and a German officer command, that carried the Ottomans through the First World War. French influence, however, remained strong in Egypt and Greece, where the civilising mission drove much of the support for the nationalist struggles of the Greeks, and philhellenism blended past and present. 'Of course, the modern Greeks needed the guiding light of the West, which was now the repository of antiquity's legacy.'[56] Mehmed Ali made abundant use of the technical know-how cloaked in the ideology, and was consistently supported by the French in the challenges to Istanbul that lay in the future.

Abd al-Rahman al-Jabarti, Egyptian Muslim intellectual and eyewitness to the French invasion, noted that the mamluks were 'irresolute, and . . . at odds with one another, being divided in opinion, envious of each other, frightened for their lives, their well-being, and their comforts; immersed in their ignorance and self-delusion; arrogant and haughty in their attire and presumptuousness; afraid of decreasing in number, and pompous in their finery, heedless of the results of their action; contemptuous of their enemy, unbalanced in their reasoning and judgement.' By contrast, the French acted '. . . as if they were fighters in a holy war. They never considered the number of their enemy as too high, nor did they care who among them was killed. Indeed they considered anyone who fled a traitor to his community, and an apostate to his faith and creed. They follow the orders of their commander and faithfully obey their leader.'[57] His observations were probably typical of the reaction of local inhabitants, but especially representative of Ottoman thinking as regards the mamluks. The Ottomans too had their own 'imperial project' and 'civilising mission'.

Volney noted of the mamluk warriors: 'their armies are mobs, their marches ravages, their campaigns mere inroads, and their battles, bloody frays; the strongest or the most adventurous party goes in search of the other, which not unfrequently flies without offering resistance; if they stand their ground, they engage pell-mell, discharge their carbines, break their spears, and hack each other with their sabres. A panic frequently diffuses itself without cause; one party flies, the other pursues and shouts victory; the vanquished submit to the will of the conqueror, and the campaign often terminates without a battle.'[58] This is how the Ottoman reformers had begun to consider not just the mamluks, but also the auxiliary forces such the Albanians, the Kurds, and increasingly, the Janissaries.

Jabarti's likening of the French invasion to a jihad is not far off the mark. Imbued with the zeal of the revolution, which championed liberty and tolerance, Bonaparte professed friendship and brotherhood with the Muslims of Egypt, declared his wish to continue his good relationship with Selim III, and his desire to eliminate the tyranny of the mamluks. He also asserted that his landing represented a blow against British power in India, which had initially persuaded his masters in Paris. The use of ideological propaganda as a tool of warfare, if not precisely his invention, certainly matured in the era of Napoleon. Selim III prepared his own propaganda for circulation. While the Ottomans, he began, had at first hesitated to condemn the Revolution in France out of respect for an old friendship, those the revolution bought to power have subverted 'under an illusive idea of liberty . . . every established government, . . . the abolishment of all religions, the destruction of every country, the plunder of property, and the dissolution of all human society – to occupy themselves in nothing but in misleading and imposing upon the ignorant amongst the people . . . and render the government permanent in their hands . . . the French planned to divide Arabia into various Republics; to attack the whole Mahometan sect, in its religion and country; and by a gradual progression, to extirpate all Mussulmans from the face of the Earth.' Egyptians were warned that once the Frenchmen were in control, they would 'spread hatred and excite the people to revolt; ultimately to destroy the Holy Places and all the Muslims.'[59] That prediction was borne out when Bonaparte fired on rebels in Cairo in 1798, destroying parts of the al-Azhar mosque and university complex, an uprising that left 2,000–3,000 Egyptians and 300 French troops dead.[60]

While such propaganda may have circulated in Egypt to combat the French invaders, it could be and was equally used against internal, Muslim challengers such as the Wahhabis, militant purists who had already attacked the holy places in the Arabian Peninsula and would plague the Istanbul government for the first decade of the nineteenth century. The Wahhabi movement, and its militant revivalism, had an enormous impact on Middle East politics, coinciding with the age of Mehmed Ali and Mahmud II. We will return to that challenge in the next chapter.

The final challenge to Selim III 1806–07

Selim III was distracted from the chaos in his southern territories by the even more significant challenges of Pasvanoğlu and Ali Pasha described earlier. The arena of European conflict centred for the moment in the eastern half of the continent, where Napoleon was routing the Third Coalition, while religious propaganda wars moved to the Principalities. (The earliest stages of Serbian revolt have already been mentioned.) Russian frequently played the Serbian Orthodox card in gambling with Napoleon and the Ottomans between 1804 and 1812. The loser was most often the Orthodox or Muslim peasant.

Selim III's Francophilia was tested once again when he wavered over recognising Napoleon Bonaparte as Emperor of France (1804). Pressured by British and Russian diplomacy, he refused to do so. Napoleon broke off relations, and the Ottomans were forced to rely on Russia, allowing warships through the straits, and further intervention of Russia on behalf of Ottoman Orthodox subjects. Russophile princes (called *hospodars* instead of *Voyvodas*, in recognition of the Russian influence) were appointed in the Principalities as a result of the Russo-Turkish Treaty of Alliance, signed by tsar and sultan in October 1805. Two secret articles proposed by the Russians and rejected by the Ottomans continued the decades-long and unresolved problem of the Russian presence in the Principalities and the Caucasus. That problem was a sub-text of the next round of Russo-Ottoman engagements on the Danube beginning in December 1806.[61]

Selim III was emboldened by news of a massive French victory at Austerlitz (December 1805) to recognise Bonaparte as emperor (February 1806), counting on help from the French in Dalmatia (ceded by Austria at Pressburg on 25 December 1805) against the Serb uprisings. Selim III immediately sent his courtier Muhib Efendi to Paris as Ambassador (1806–11).[62] He also expelled the *hospodars* from the Principalities. That prompted the Russians to blockade the Adriatic, and send further help to the Serbs. War preparations continued through 1805–06, by the Russians in Bessarabia, by Kara George in Serbia, and the Ottomans in Bosnia and Macedonia. In mid-August 1806, under severe pressure, Selim III reversed his policies yet again and reinstalled the Russian-approved *hospodars* in the Principalities. In September 1806, worried about war between ally Russia and the Ottomans, the British ambassador demanded that the straits be reopened to the Russian warships. (They had been closed after the Franco-Ottoman rapprochement.) Selim III agreed, but news of further overwhelming French victories (over the Prussians at Jena and Auerstädt in October 1806) emboldened him to defy the Anglo-Russian request. The Russians declared war and occupied Moldavia by December 1806, initially more worried by French proximity and influence on Selim III than Ottoman belligerence.

By February 1807, a British fleet under Admiral Duckworth, in support of the Russians, dropped anchor near the Princes' Islands, after sailing through the Dardenelles. Istanbul panicked, but diplomatic dithering and adverse weather forced a British withdrawal, especially after Selim III had strengthened the fortifications and requested a formal alliance with France (by secret initiative through Muhib Efendi in Paris, which was simply ignored).[63] French Ambassador Sébastiani rounded up some 200 French officers and aides to man the batteries alongside the artillerymen. Every available weapon in the city was mobilised, and within a few days, the shores of the city were bristling with cannon. Crowds of the city's young men volunteered for service against British and Russian alike.[64] It was perhaps the last moment that Selim III enjoyed the approbation of the streets of Istanbul, as the populace waved goodbye to the British fleet on 3 March, and celebrated as the warships sailed through the Dardanelles with great difficulty. Meanwhile in Egypt, Mehmed Ali was making his bid for power. A British fleet landed briefly at Alexandria in March 1807, in an attempt to back the mamluks. A total of 5,000 British troops under General Mackenzie Frazier were ambushed and routed by the garrison at Rosetta by September 1807. While not personally commanding the garrison, Mehmed Ali's reputation grew with the victory and the withdrawal of the British fleet.

Meanwhile, Napoleon and Tsar Alexander I (1801–25) signed the Treaty of Tilsit in July, 1807, after the Russian loss at Friedland in June. The British found themselves abandoned by Russia, as were the Ottomans by France. The treaty stipulated that should negotiations to restore peace among the belligerents fail, France would join Russia in its war against the Ottomans and they would divide the Ottoman European territories among themselves. Eastern Europe would became a Russian sphere of influence. It was the final blow to Selim III's strategy to remain friendly with France.

So things stood in mid-1807, but Selim III had already been deposed, a victim of his own vacillation and of a revolt in Istanbul against the extension of his new order, indeed against his reforms in general and probably his Francophilism in particular. In 1805, enemies at court and in the Balkans had conspired to resist Selim III's lastest effort to raise *Nizâm-ı Cedid* troops in the provinces, in Edirne. A conscription system, rather than the previous voluntary sign-up, characterised his extended new order. It was an ill-advised initiative. Edirne historically served as the staging point for all Balkan campaigns, and was territory whose leaders were fiercely proud of, and profited from, its importance in the military history of the empire. They were therefore likely to be resistant to such an endeavour to replace a well-entrenched system. As we have seen, there were already a plethora of small private armies assembled by the *ayan* in the Balkans. Selim III must have been desperate for troops, imagining a new campaign against the Russians, and for the defence of his capital, but equally mistaken in his assumption that the public at large welcomed his reforms. Resistance to the conscription orders was fierce, and forced Selim III and his advisers to back down.

Balkan notables, chief among them İsmail Tirsiniklioğlu in Rusçuk and Silistre, took advantage of the resistance to mobilise what is called by historians the 'Edirne Incident'. They were aided by the sultan's opponents in Istanbul, who were led by Grand Vizier Hafız İsmail (April 1805 – September 1806). In June 1806, Selim III again ordered a *Nizâm-ı Cedid* force to be assembled in Edirne, which was to be commanded by Kadi Abdurrahman Pasha, one of the few provincial governors and commanders loyal to Selim III's reforms. Abdurrahman assembled troops intended for Rusçuk and the defence of the Danube from all over Anatolia in the new barracks in Üsküdar, on the Asian side of the Bosphorus.[65] In early June, İsmail Tirsiniklioğlu, at the grand vizier's secret urging, marched on Edirne with a force estimated at 80,000, commanded by a coalition of Balkan notables who refused to cooperate with the sultan's orders. Selim III delayed sending Abdurrahman and his loyal troops, numbering 15,000–20,000, against them. When they did march in mid-July, they were confronted outside Edirne by a force of 10,000 Janissaries. Sustaining considerable losses, the *Nizâm-ı Cedid* army retraced its route to Silivri, where they were halted by Selim III himself. Deliberately supplied misinformation about the size and strength of the resistance, and repeated demands for the disbanding of Abdurrahman's troops, led Selim III to capitulate, dismissing Abdurrahman and sending the assembled loyal troops to Anatolia. The Grand Vizier was replaced by the Agha of the Janissaries, İbrahim Hilmi Pasha (September 1806 – June 1807), which temporarily derailed further opposition from that quarter. All the rebels in Edirne were pardoned.

Selim III's vacillation clearly played a large role in his own demise. Unwilling to confront the traditional forces, he undermined his own policies by the betrayal of those loyal to him. The reputation of the new soldiers was much tarnished, and contributed to the increasing discontent in Istanbul. An *alafranga* mode had begun in Selim III's Istanbul, the city was full of foreigners, discontented Janissaries, and a plethora of unemployed religious students. The latter two were the lethal combination that had plagued urban politics in Istanbul since the early 1700s.

Economic reforms were not fully applied, and caused inflation and shortages. Rumours started to circulate about Selim III's sanity, but especially about corruption and about the sultan's credentials as a Muslim leader. For a short while, however, his capitulation to the malcontents kept the lid on the cauldron in Istanbul.[66]

Among those in Tirsiniklioğlu's entourage at Rusçuk was Alemdar Mustafa Pasha. As with the careers of the other well-known *ayan* already described, Alemdar, born in Hotin but based in Rusçuk, began his rise in Janissary service, as the flag-bearer of his regiment. Caught up in the struggles against Osman Pasvanoğlu, he served with distinction and was awarded the title of *Kapıcıbaşı*, by request of the Silistre governor and the Wallachian *Hospodar*, and made *ayan* of Hezargrad in 1803. Thereafter a lieutenant of Tirsiniklioğlu, he remained in his service until the latter was murdered in August 1806.[67] Upon the recommendation of local officials and followers of the dead chief, Alemdar was appointed *ayan* of Rusçuk, and subsequently of Silistre. Selim III resisted the accumulation of power, playing the notables one against the other, but Alemdar seems to have won the admiration not just of the local Ottoman appointees, but also of inhabitants of the Danube shores, for refortifying the towns and lifting the imposition of the much-hated *corvée*. Selim III was forced to accept him as the chief notable of the Ottoman Danubian shores east of Belgrade. Alemdar was prepared for the Russians, when in December 1806 they attacked and took Bender and Hotin. He faced the Russian penetration of Wallachia on the northern shores of the Danube in the early days of the war, and checked the Russian troops under General Michelson at Guirgevo. Cevdet records: 'The primary aim of the Russians was to reach Bucharest, there to assemble 50,000 troops, and uniting with the Serbs, to alarm the Ottoman State. Hence, a division had moved in the direction of Bucharest. This division met up with Alemdar's army. In the confrontation that followed, Alemdar triumphed, causing many losses among the Russians. The news of the victory of Muslim over Russian soldiers, with few casualties, arrived in Istanbul Alemdar and caused great joy.' Selim III appointed *Serasker* of the Danube with the rank of Vizier, and Governor of Silistre in February 1807.[68]

The brief confrontation did not keep the Russians from spreading rapidly into Wallachia and Bessarabia by mid-1807, and being greeted with joy by the inhabitants of Jassy and Bucharest, who were suffering from a century of *boyar* and *hospodar* oppression, a succession of bad harvests, and a collapsed economy. As we have seen in earlier campaigns, the Russians, who expected to be supplied locally, were forced to bring supplies across the Dniester, a factor which influenced the direction of the 1806–12 war.

The combined efforts of Grand Vizier İbrahim Hilmi Pasha and Alemdar Mustafa checked the potential Russo-Serbian offensive across the Danube. By mid-1807, the line between the Russian and Ottoman armies was drawn. The Franco-Russian rapprochement at Tilsit brought a momentary truce in August 1807 which lasted until March 1809, during which time the Russian army remained in occupation in the Principalities, and Russian policy makers started to think about annexation.[69]

The other explanation for the hiatus in military confrontations was, of course, the fall of the sultan, and the political uncertainty that followed. In June 1807, as the Grand Vizier, on the march to the front with the main Ottoman army, exchanged greetings and the gifts from Selim III with *Serasker* Alemdar Mustafa, news arrived of the rebellion and deposition of their sultan. An explosion of anti-Russian sentiment after the invasion of Wallachia, and the obvious inability of Selim III to defend the empire's borders, had blown the lid in Istanbul.

The revolt began in May 1807, when Mahmud Raif was sent to the Bosphorus fortress at Rumeli Kavağı to distribute the quarterly pay to the soldiers stationed there, and attempt to persuade them to adopt new uniforms. The immediate response was a mutiny by the *yamaks* who manned the fortresses. These were brand new recruits, numbering 2,000, mostly Albanian, or Laz from Trabzon, only recently added to the *Nizâm-ı Cedid* troops already in place, in the hopes of inspiring them with the spirit of the new order.[70] But *Kaymakam* Musa Pasha (formerly *Mutasarrıf* of Salonika), who had been called to Istanbul to substitute for Grand Vizier in his absence, and *Şeyhülislam* Ataullah, both enemies of the sultan, secretly placed well-chosen discontented Janissaries among them. The rebellion from the start was well organised and manipulated by the inner circles of the court. Once under way, however, it proved difficult to control by the instigators. Mahmud Raif was the first victim of the revolt.

Chosen by the rebels as their leader, one Kabakçı Mustafa demanded dissolution of the 'infidel innovations', and the removal of all the *Nizâm-ı Cedid* troops from guard duty. Selim III capitulated again, and the last of his loyal troops remained confined to their barracks as he was brought down. Musa Pasha ensured that all the regular army joined the rebellion, including the artillery corps, by spreading the word in the barracks. Within three days, the rebels, *yamaks* and Janissaries, assembled at the Janissary parade ground and demanded the sacrifice of the leaders of reform, with a proper *fetva* in their hands from the traitorous Ataullah. This too Selim III agreed to, sacrificing his remaining loyal advisers, secretly moving others out of his palace as he could. Of the reformers closest to Selim III, only Mustafa Reşid Efendi survived, escaping in disguise to return as one of Mahmud II's chief aides. Prince Mustafa, son of Abdülhamit I, previously prepared, was enthroned as Mustafa IV (1807–08).

On 3 June 1807, the *Nizâm-ı Cedid* and all of its associated institutions were eliminated by the new regime. Janissary Agha Pehlivan Hüseyin had already mutinied against his commander on the Danube, Grand Vizier Hilmi, who resisted the Janissary demand to return to Istanbul. Alemdar Mustafa, responding to the Grand Vizier's call for help, arrived at Giurgevo on 8 June and restored order to the army under the Grand Vizier. Expecting to be appointed Grand Vizier himself, he was forestalled by the new regime in Istanbul, which appointed Çelebi Mustafa Pasha instead (June 1807–July 1808). Furious, Alemdar withdrew to Rusçuk, taking all remaining Selim III loyalists with him. Cut off from support from both Istanbul and Rusçuk, the main army simply fell apart, and was demobilised by August, leaving the Danube defence line entirely to Alemdar Mustafa. Luckily for the Ottomans the news of the Tilsit Treaty brought respite, as we have seen.

A Russian blockade of the Dardanelles was also lifted after the Russian and Ottoman fleets engaged off Limnos, bringing much-needed grain supplies to starving Istanbul. By mid-July, when news of Tilsit arrived, initial fury at perceived French treachery changed to a desire to negotiate, and Muhib Efendi, in Paris, was empowered to do so. By August an armistice with the Russians was signed at Slobozia, negotiated by Galib Efendi, who had sought refuge with Alemdar's army at Rusçuk. The new Ottoman government accepted the armistice in September, but Tsar Alexander refused to ratify it. The Russians withdrew from Bucharest, but not from the Principalities. By the end of the next year, the abortive armistice was finished and preparations began for a resumption of hostilities in the spring. The war continued, mostly as a stand-off, until 1812, when the French threat to Moscow forced Russia to conclude hostilities in order to redeploy troops. We will finish that story in the next chapter.

The Rusçuk Committee and the end of an era

Meanwhile in Rusçuk, the surviving generation of reformers under Abdullah Ramiz Efendi, a judge and veteran of the Egyptian campaign, having served as Commander of the Mortar Corps (*Humbaracı Başı*), persuaded Alemdar of the advantage of reinstalling Selim III by force if necessary.

Istanbul needed help against the erstwhile rebels, who, under Kabakçı Mustafa, were terrorising the city. By mid-1808, the original conspirators *Kaymakam* Musa and *Şeyhülislam* Ataullah Efendi had fallen out, and administrative disorder ensued as each tried to remove the other from office. The Rusçuk Committee manipulated events to their own ends, persuading Mustafa IV to call on Alemdar and his army for support. By early May, Alemdar had joined the grand vizier in Edirne, and many of the Rusçuk Committee had been appointed or re-appointed to offices in the Istanbul bureaucracy. On 14 July, Alemdar Mustafa and Grand Vizier Çelebi Mustafa marched on Istanbul, with 15,000 troops, where *Şeyhülislam* Ataullah, still favourite of the Janissary Corps, held sway. *Kaymakam* Musa Pasha had earlier been replaced by Caniklizade Tayyar Mahmud Pasha, former Governor of Trabzon, a Russophile and known for his hatred of Selim III. Mustafa IV complicated the situation by courting Tayyar Mahmud and promising to make him grand vizier in order to rid himself of Ataullah and the two remaining princes of the Ottoman dynasty, the deposed Selim III and his cousin, Mahmud. Ataullah secured the dismissal of Tayyar Mahmud, who then joined the side of Alemdar Mustafa as *Serasker* of Varna.[71]

The Rusçuk contingent broke the back of the Bosphorus *yamak* rebellion by ordering Kabakçı Mustafa to be attacked and killed secretly. One Hacı Ali was sent with a detachment of cavalry to a village on the Bosphorus, where Kabakçı Mustafa was living, caught him asleep in his harem, and promptly beheaded him. The *yamaks*, learning the next day of the death of their leader, were encouraged to revolt by Süleyman Agha, uncle of Kabakçı Mustafa: 'These men, who have treacherously caused the death of our chief, wish for nothing more than your death and that of our glorious sultan. Let us seek vengeance for our father; defend our sovereign; support our religion and our laws; and punish these vile murderers, who, sent by that perfidious grand vizier, and the infamous Alemdar, who has declared himself the protector of our enemies, has just committed the most cowardly and horrible of crimes.' Hacı Ali and his eighty men took refuge in a lighthouse, where they were besieged for three days, until the *yamaks* grew tired of the effort and dispersed. Rumours abounded, as the noise of the cannons could be heard in the city, that 'All the brigands of Rumeli and the mountains of Albania and Rhodope had united to pillage Istanbul, and Hacı Ali was the chief of the advance guard.'[72] The rumours weren't far wrong.

By the end of July, the grand vizier Çelebi Mustafa and Alemdar Mustafa had, however, restored order to the city, and eliminated most of the opposition, largely because the Janissaries in the city were disinclined to raise their weapons against their brothers-in-arms of the same regiments who had been on campaign on the Danube. Alemdar imposed strict discipline in his camp, and restored confidence in the merchants of the city, by forcing his soldiers to pay properly for what they bought.

The rivalry between the grand vizier and Alemdar Mustafa, however, determined Selim III's fate. Alemdar hid his real aim, the assumption of the grand vizierate, until 28 July 1808, when he rose up against Çelebi Mustafa and Sultan Mustafa IV, intending to restore Selim III to the throne. The grand vizier surrendered the seal to Alemdar, who immediately marched on the palace. In the very final moments of his rule, Mustafa IV sent his own servants to kill Selim III and prince Mahmud, who escaped. Shown Selim III's body, Alemdar was overcome with grief, crying: 'Unhappy prince, what have I done? I wanted to restore you to the throne of your ancestors, and I am the cause of your death. Is this the fate reserved for one with your virtues?' The *Kapudan* Pasha, Seyyid Ali, who had joined Alemdar Mustafa, exhorted him: 'Is it proper for the Pasha of Rusçuk to weep like a woman? It is vengeance and not tears that Sultan Selim III requires of us. Let us punish his assassins. Let us especially not allow a bloody tyrant to profit from his crime, and to assure his reign by the death of his brother, Sultan Mahmud.'[73] Messengers were sent to find Mahmud, who climbed down from the roof where the Chief Imam Ahmed Efendi had hidden him. The Imam declared, 'This is Sultan Mahmud. He is next in line as Caliph. I have sworn allegiance to him.' Alemdar enthroned him forthwith as Mahmud II (1808–39).[74] So ended a revolution unprecedented in Ottoman history, where an army of occupation from the provinces replaced the ruling sultan, without (or in spite of) the ruling of *Şeyhülislam* Ataullah, who had been removed from office immediately, disregarding the will of the Janissaries of Istanbul.[75]

The death of Sultan Selim III is a natural point at which to end this chapter, although it is impossible not to view the 1793–1826 period as a continuum, with the rise and fall of sultans and grand viziers just one of the threads running through the narrative. Selim III was represented in contemporary histories as fostering decadence in Istanbul, preferring entertainments and excursions on the Bosphorus to a serious dedication to his reform programme. His reluctance to press on with it in the face of opposition undermined his initiatives. Observers like Juchereau commented that Selim III had squandered the good will of his Istanbul subjects, who had proved more than capable of saving a nation in moments of extreme crisis.[76]

Selim III's reign coincided with the initial phase of the 'Eastern Question', that interplay of great power diplomacy and the dismemberment of the Ottoman empire which dominates most histories of the period. This chapter has argued that Selim III's diplomatic initiatives played an important role in forwarding the reform movement and maintaining Ottoman autonomy and independence.

The Selim III period is also transformative in a number of ways, most notably in the proximity and communication between centre and province so evident in Alemdar Mustafa's arrival in Istanbul. It was an age when the Ottoman provincial aristocracy made its bid for the sharing of power with the bureaucracy in Istanbul. The decisive presence of Alemdar's troops in Topkapı palace certainly brought that home to Istanbul residents.

Recording the changes in power and associations so characteristic of events in 1807–08, one is struck by the small size of the ruling elite of the vast empire, a problem faced on a far greater scale by Catherine II and her successors in Russia as well. Secondly, it is possible to observe a new swiftness in trans-imperial communication. For example, the army on the Danube was aware within a few days of the Kabakçı Mustafa revolt. Similarly, the Rusçuk committee reinserted itself with fair rapidity into Istanbul politics after May 1807. Thirdly, there was a certain familiarity (and contempt for one another) among all the *ayan* that must have resulted from the mobility of armies in the period. The army that was mobilised was more often than not a coalition of provincial mini-despots, as we have seen. Although these forces could not be called swift by present-day standards, they were nonetheless more mobile in comparison with the army manoeuvres of two previous Russo-Ottoman wars, prefiguring the regimental formations of the nineteenth century. The first challenge to Mahmud II would arise from this new class of governors–soldiers–bureaucrats, and their relations with the new sultan would determine the course of events at least until the 1830s.

The private armies of the *ayan* were composed of Albanians who were everywhere, and increasingly, 'Laz' from Trabzon, as well the exiled communities from the Caucasus, Tatars and, finally, Kurds from eastern Anatolia. Such mobile soldiers and their leaders would become the officer corps and 'cannon fodder' of the reformed armies of Mehmed Ali of Egypt and Mahmud II. The military crisis created a large migrant population, cause of many woes to sedentary populations but also the carrier, not just of orthodoxies, but of heterodoxies. Ethnic categories were further complicated with Muslim, as well as non-Muslim, religious diversity. All this required a certain degree of recasting of the Muslim premises of rule, also an aspect of the strategy of Mahmud II.

Simply to polarise the events into a confrontation between Muslim reactionaries and secular reformists is to misrepresent the history of the period altogether. First and foremost, the struggle was an economic as well as a social conflict. As we saw with Canikli Ali, the provincial *ayan* described above, the wealth in Istanbul, and the hierarchies of power embodied in the imperial court and bureaucracy, were both magnets for the greedy and targets for the resentful.

There is little doubt that the 'golden age' evoked by the ideologues of reform, cast in the ethical frame of Islam, referred to powerful sultans like Süleyman the Magnificent. The Prophet Muhammad could also be used to justify both the traditional and the innovative, as exemplified in the hadiths concerning his early military prowess against his enemies, the Quraysh. In army life, moreover, the sacred banner and the appeal to fight for God and country were tropes of modern, conscripted warfare everywhere, and often were deliberately employed by commanders to stir the hearts of ignorant and unwilling foot soldiers.

But the Ottoman house by 1830 was under siege not just by external imperial and colonial powers but also by its own populations, and foreign missionaries seeking control over the souls of its Christian subjects. Orthodox, and to a lesser extent Catholic, priests under Russian, Austrian and French aegis, reinvigorated Christianity in territories with divided populations, such as Bulgaria, or the Principalities, or Syria in what became ethno-religious nationalist movements. Wahhabi fundamentalism and Naqshbandi sufism also emerged as two very important forces to challenge Ottoman legitimacy in majority Muslim territories. That struggle will unfold further in the next chapter.

Mahmud II's later achievements can be measured by an adulatory story that survives in Cevdet Pasha's history, written after the sultan's death. Alemdar Mustafa swore vengeance on those traitorous members of sultan's inner household (*Enderun*) that he held responsible for the murder of Selim III. Mahmud II, newly-made sultan, replied with his first order: 'Pasha, I will find them and send them to you. Dismiss your soldiers and have them remove their weapons. Let us then retire to the precincts of the Mantle of the Prophet.' Alemdar asked if he could continue to wear the scimitar he had at his side, as it had been a gift of Selim III. Mahmud II allowed him the privilege, an extraordinary gesture for a ruler who had just been saved from execution. As if the point was not clear enough, Cevdet continued: 'Mahmud II was just twenty-four, and had never been in such a crowd, except of his own servants.' Alemdar later commented: 'That first order of the Padishah frightened me more than any other time in my life.' Here, concluded Cevdet, with his hindsight, was the type of brave sultan the Ottoman Empire desperately needed at that moment.[77] An immense task awaited the young sultan.

Notes

1 K.E. Fleming, *The Muslim Bonaparte: Diplomacy and Orientalism in Ali Pasha's Greece* (Princeton, NJ: Princeton University Press, 1999).

2 Canay Şahin, 'The economic power of Anatolian *Ayan* in the late 18th century: the case of the Caniklizâdes', *International Journal of Turkish Studies* 11 (2005), 33.

3 One example has him sending 10,000 troops against the district of Of to put down an uprising: Michael Meeker, *A Nation of Empire: The Ottoman Legacy of Turkish Modernity* (Berkeley, CA: UC Press, 2002), 212.

4 I have extracted Canikli's tale from an archaeological exercise in the Ottoman archives and *kadi* court records undertaken by Yücel Özkaya, 'Canikli Ali Pasha', *Belleten* 144 (1972), 483–525, esp. 503, 508 and 510.

5 Şahin, 'The economic power', 33–36, reproduces documents from 1792 (BA, MAD 9720 and Cevdet Maliye 7401), recording the official value of the confiscated Caniklizade holdings, some 151,915 *kuruş*; 32,710 was cash.

6 Meeker, *A Nation of Empire*, 193–98, 202.

7 Meeker, *A Nation of Empire*, 203, 216–17, notes that residents of Of in the 1960s could still recall 22 agha families of the nineteenth century.

8 As laid out by Robert Zens in 'Pasvanoğlu Osman Pasha and the Pashalık of Belgrade, 1791–1807', *International Journal of Turkish Studies* 8 (2002), 89–104.

9 Zens, 'Pasvanoğlu', 96, from Ahmed Cevdet's *Tarih*; also M.O.H. Ursinus, 'Sirb', *EI2*, CD edition.

10 Zens, 'Pasvanoğlu', 99. Ahmed Cevdet notes that a rebellion broke in Aleppo while Cebbarzade was absent, requiring the appointment of a substitute.

11 Ahmed Cevdet, *Tarih*, vol. 7, 9–13, as quoted by Zens.

12 Zens, 'Pasvanoğlu', 102–03.

13 Robert F. Frost, 'Napoleon's diplomats in the Danubian Principalities: revolutionaries, opportunists, and spies', *Proceedings: Consortium on Revolutionary Europe* 20 (1990), 853–60.

14 Zens, 'Pasvanoğlu', 103. Pasvanoğlu died in January 1807.

15 Dominic Lieven, *Empire: The Russian Empire and its Rivals* (New Haven, CT: UP, 2000), 146–47.

16 Rossitsa Gradeva, 'Osman Pazvantoğlu of Vidin: between old and new'. In *The Ottoman Balkans*, Frederick F. Anscombe, ed. (Princeton, NJ: Markus Wiener, 2006), 115–61.

17 Svetlana Ivanova, 'Widin', *Encyclopedia of Islam*, 2nd edn vol. 9 (2001), 205–8.

18 The phrase belongs to C.A. Bayly, 'Knowing the country: Empire and information in India', *Modern Asian Studies* 27 (1993), 3–43.

19 Some 68 diplomatic instruments survive, 29 by Ahmed III (1703–30) and 30 by Mustafa III (1730–57) alone, indicating a considerable international diplomatic record for a state generally declared moribund and impenetrable; see Donald Quataert, *The Ottoman Empire, 1700–1922* (Cambridge: Cambridge University Press, 2000), 76–77.

20 See Aksan, *Ottoman Statesman.* Ahmed Resmi reported not just on his journey, but on Frederick the Great, his aspirations, and his characteristics as a commander and monarch.

21 Carter Findley has supplied us with a comprehensive summary of the contents of the report, and some very interesting questions about possible sources in Vienna of some of the text. Carter Vaughan Findley, 'Ebu Bekir Ratib Efendi's Vienna Embassy narrative: discovering Austria or propagandizing for reform in Istanbul?' *Wiener Zeitschrift für die Geschichte des Morganlandes* 85 (1995), 40–80. The text is now available in modern Turkish: V. Sema Arıkan, 'Nizâm-ı Cedit' in kaynaklarından Ebubekir Ratib Efendi'nin "Büyük Lâyiha"sı', Istanbul University (PhD), Istanbul, 1996. Both Findley and Arıkan note that there is a strong possibility that much of the report is a translation from German, perhaps at the instigation of Ignatius Mouradgea d'Ohsson, Swedish–Armenian ambassador, translator and spy, who also submitted recommendations to Sultan Selim III.

22 It remains a remarkable circumstance that the lives and embassies of most of the diplomats in question remain unstudied, with some notable exceptions.

23 Ahmet Alaaddin Yalçınkaya, 'Mahmud Raif Efendi as the Chief Secretary of Yusug [sic] Agah Efendi, the first permanent Ottoman-Turkish Ambassador to London (1793–1799)', *Ankara Üniversitesi Osmanlı Tarih Araştırma ve Uygulama Merkezi Dergisi* (1994), 385–434. Ebu Bekir returned from Vienna in late 1792. Raif Efendi's *Journal du voyage*, in manuscript only, is translated into English and included as an appendix to Yalçınkaya's article, 422–34, which is part of his PhD from the University of Birmingham, 'The first permanent Ottoman-Turkish embassy in Europe,' 1993.

24 Mahmud Raif Efendi, *Tableau des nouveaux reglemens de l'Empire Ottoman* (Constantinople, 1798), preface.

25 The other two were Mehmed Derviş and Seyyid Mahmud Tahir. Yalçınkaya, 'Mahmut Raif Efendi', 417.

26 Ibid., 420–22.

27 Allan Cunningham, 'Robert Liston at Constantinople', In Edward Ingram, ed., *Anglo-Ottoman Encounters in the Age of Revolution: Collected Essays, vol. 1* (London: Frank Cass, 1993), 58. Yalçınkaya's account suggests that all protocols were followed, but does not suggest any serious foreign policy exchanges took place between Grenville and Yusuf Agah ('Mahmud Raif Efendi', 410ff.).

28 Ercümend Kuran, *Avrupa'da Osmanlı Elçiliklerinin Kuruluşu ve Ilk Elçilerin Siyasi Faâliyetleri 1793–1821* (Ankara, 1968), 16. The death of Catherine II in 1796 threw Russian diplomacy in chaos as her successor Paul I dismissed her advisers and abandoned or reversed many of her policies.

29 Cunningham, 'The Ochakov debate', in *Anglo-Ottoman Encounters*, 3: 'the French barrier around the Ottoman Empire fell'.

30 Cunningham, 'The Ochakov debate', in *Anglo-Ottoman Encounters*, 5–6, note 11; 20. S.J. Shaw, *Between Old and New: the Ottoman Empire Under Sultan Selim III, 1789–1807* (Cambridge, MA: Harvard University Press, 1971), 91.

31 Shafik Ghorbal, 'The Missions of Ali Effendi in Paris and of Sedki Effendi in London, 1797–1811', *Bulletin of the Faculty of Arts, University of Egypt* 1 (1933), 114–29. See also Bernard Lewis, *Muslim Discovery of Europe* (New York: Norton, 1982), chapter 1.

32 The first translations of the *1001 Arabian Nights* by Galland appeared in 1797, Tott's *Memoirs* were a bestseller of the age, and *Letters Writ by a Turkish Spy*, to name just a few, had fashioned an oriental mania in the reading public. On Seyyid Moralı Ali and Muhib Efendis, see Stéphane Yerusimos, *Deux Ottomans à Paris sous le Directoire et l'empire* (Paris: Sindbad, 1998).

33 Shaw, *Between Old and New*, 189. E. Kuran, 'Mehmed Sa'īd Ghālib Pasha', *EI2* v. 6, 1003b, CD edition; B. Lewis, *Muslim Discovery*, 57.

34 Robert Liston reported a craze for all things European in the largest cities of the empire in 1796. Shaw, *Between Old and New*, 194.

35 *Napoleon in Egypt: Al-Jabarti's Chronicle on the French Occupation of Egypt, 1798*, tr. by Shmuel Moreh (Princeton, 1993), which includes parts of the chronicle, plus excerpts from the memoirs of Napoleon's private secretary, Loius Antoine Fauvelet de Bourienne. The comments here can be found on pages 119 and 162. The French lost 300 men; the mamluks some 6,000. See P.M. Holt, *Egypt and the Fertile Crescent, 1516–1922* (London: Longman, 1966) for the standard account.

36 Norman Daniel comments further: 'This was purest Bonapartism', *Islam, Europe and Empire* (Edinburgh: EUP, 1968), 98.

37 A.L. MacFie, *The Eastern Question 1774–1923*, revised edn (London: Longman, 1996), 11, and *Napoleon in Egypt*, 9.

38 Like Ali Pasha of Iannina, his ferocious reputation and ambivalent historical stature has prevented a true picture of the individual. For his biography, see Thomas Philipp, *Acre: The Rise and Fall of a Palestinian City, 1730–1931* (New York: Columbia University Press, 2001), especially 48–78. Cezzar (meaning 'butcher') Ahmed Pasha was known for his cruelty as well as his prowess. Philipp quotes an Arab source for the 4,000 soldiers executed at Jaffa. William Wittman, surgeon to the British Military Mission to the Ottomans 1799–1801, *Travels in Turkey, Asia Minor, Syria and Egypt* (London, 1803; reprint New York: Arno Press, 1971), quotes the figure 4,000, plus 500–600 of the Turkish garrison from el-Arish. He also noted that upon the retreat from Syria, the French commander ordered the poisoning of all the French sick in Jaffa (128).

39 A. Juchereau de Saint-Denys, *Révolutions de Constantinople en 1807 et 1808* (Paris: Brissot-Thivars, 1819), vol. 1, 92–93, notes that Cezzar was authorised by Istanbul to raise troops to combat the Druzes and rebellious pashas of Aleppo and Damascus. He first defeated such enemies of the government, annexed their armies, and became powerful enough to execute the imperial orders when they suited his own interest.
The Albanian and *Nizâm-i Cedid* troops resisted more than ten assaults on Acre. Ahmed Cevdet, *Tarih*, vol. 7, 50.

40 Ahmed Cevdet, *Tarih*, vol. 7, 48.

41 Dick Dowes, 'Reorganizing violence: traditional recruitment patterns and resistance against conscription in Ottoman Syria', in Erik J. Zürcher, ed., *Arming the State* (London: I.B. Tauris, 1999), 110–27. See also Linda Schatkowski Schilcher, *Families in Politics: Damascene Factions and Estates of the 18th and 19th Centuries.* (Stuttgart: Franz Steiner, 1985), chapter 2.

42 Daniel Crecelius, *The Roots of Modern Egypt: A Study of the Regimes of 'Ali Bey al-Kabir and Muhammad Bey Abu al-Dhahab, 1760–1775* (Minneapolis: Bibliotheca Islamica, 1981), 21. The number in 1664 stood at 13,673.

43 Crecelius, *The Roots of Modern Egypt*, 28.

44 Jane Hathaway, *The Politics of Households in Ottoman Egypt: The Rise of the Qazdağlıs* (Cambridge: CUP, 1997), 101–4, argues that the Qazdağlı household was largely made up of Caucasus and Anatolian stock, although she admits it was opportunistic rather than strategic, in what was a 'hodge-podge' of ethnicities. Her conclusion asks us to imagine a Georgian manumitted ethos as part of the changing cultural scene in late eighteenth-century Egypt. The impact of continuous Ottoman losses to Russia in the

Caucasus, normally the source for *mamluks*, cannot but have influenced later choices for conscription, as we will see with Mehmed Ali.

45 Crecelius, *The Roots of Modern Egypt*, 55.

46 Crecelius, *The Roots of Modern Egypt*, 102.

47 As explored by H. Laurens, *Les origines intellectuelles de l'éxpedition d'Egypte: L'orientalisme Islamisant en France (1698–1798)* (Istanbul: ISIS, 1987), 180–82. C.F.C. Volney's *Voyage en Egypte et en Syrie pendant les années 1783, 1784, 1785* (Paris: 1787), like the *Memoirs* of Baron de Tott, was one of the bestsellers of the enlightenment. Kapudan Gazi Hasan Pasha, better known as the veteran of the disastrous naval defeat at Çeşme in 1770, was then ordered to relieve the Russian siege of Ochakov, and managed to deliver supplies to the fortress but was unable to lift the siege, as we have seen. He died in 1790. (See J.H. Mordtmann, E. Kuran, 'Djezā'irli Ghāzī Hasan Pasha,' *EI2* CD edition.)

48 Thomas Walsh, *Journal of the Late Campaign in Egypt, Including Descriptions of that Country, and of Gibraltar, Minorca, Malta, Marmorice, and Macri: With an Appendix Containing Official Papers and Documents* (London, 1803), supplies the details used here. See appendix for William Whitman's description of the Ottoman march from Gaza.

49 Walsh, *Journal*, Appendix no. 35.

50 For concise biographies of both Küçük Hüseyin and his protegé, Hüsrev Pasha, see 'Khosrew Pasha', *EI2*, vol. 5 (1979), 35–36, and Münir Aktepe, 'Husayn Pasha', *EI2*, vol. 3 (1971), 627–28.

51 Walsh, *Journal*, 146–50.

52 Daniel, *Islam, Europe and Empire*, 74–75.

53 Daniel, *Islam, Europe and Empire*, 70.

54 C.A. Bayly, 'The British military-fiscal state and indigenous resistance: India 1750–1820', in Lawrence Stone, ed., *An Imperial State at War: Britain from 1689 to 1815* (London: Routledge, 1994), 224–54.

55 Daniel, *Islam, Europe and Empire*, 146–47. Sydney Smith's service in Acre was impeccable, in spite of his private observations. Koehler's superiority apparently provoked considerable ill-will among the Ottoman commanders. Both he and his wife died of malignant fevers in Jaffa (Wittman, *Travels*, 196).

56 Victor Roudometof, 'From rum *millet* to Greek nation: enlightenment, secularization, and national identity in Ottoman Balkan society, 1453–1821', *Journal of Modern Greek Studies* 16 (1998), 24.

57 *Napoleon in Egypt*, 36.

58 Cited in Crecelius, *The Roots of Modern Egypt*, 60–61.

59 Daniel, *Islam, Europe and Empire*, 91–92.

60 Darrell Dykstra, 'The French occupation of Egypt, 1798–1801', in M.W. Daly, ed., *Cambridge History of Egypt* (Cambridge: CUP, 1998), vol. 2, 123.

61 Armand Gosu, 'The third anti-Napoleonic coalition and the Sublime Porte', *International Journal of Turkish Studies* 9 (2003), 199–237.

62 Muhib Efendi was bolder than his predecessors. He recorded his conversations with Napoleon, to whom, when he referred to Cezzar Pasha, the defender of Acre, as the Butcher, Muhib replied: 'Cezzar is a man devoted to the sultan. He was raised in the shadow of the palace, and, in spite of his origins as a "montagnard kurde", he is no less worthy of the honour of being favoured by his master'. Napoleon mentioned the Serbian revolt and its possible consequences to the tranquillity of the Ottoman Empire, to which Muhib Efendi replied boldly that it never would have happened without the expedition to Egypt. Bertrand Bareilles, *Un Turc a Paris 1806–1811: relation de voyage et de mission de Mouhib Effendi* (Paris: Bossard, 1920), 19–20.

63 Ibid. Napoleon reassured Muhib Efendi that the Ottomans had nothing to fear from Russia, and entertained Muhib's secret mission to propose a Franco-Ottoman alliance. Shortly after this conversation, the Tilsit Treaty was signed.

64 Juchereau, *Révolutions de Constantinople*, vol. 2, 76.

65 Juchereau reported that the *İrad-i Cedid* raised more than 50 million *kuruş* in 1800, and 75 million in 1806. Abdurrahman Pasha had previously raised 8 regiments of *Nizâm-ı Cedid* troops in his own territories: *Révolutions de Constantinople*, vol. 2, 21–23.

66 Juchereau, *Révolutions de Constantinople*, vol. 2, 28–34. Also Cevdet, *Tarih*, vol. 8, 180.

67 Juchereau, *Révolutions de Constantinople*, says Alemdar Mustafa was responsible, carrying out the sultan's orders (vol. 2, 31). Shaw says that it was an agent of Kadi Abdurrahman (*Between Old and New*, 347).

68 'Mustafa Pasha, Alemdar', *Islam Ansiklopedisi* (Uzunçarşılı), vol. 7, 10–27. Cevdet, vol. 8, 149.

69 Shaw, *Between Old and New*, 364.

70 Juchereau, *Révolutions de Constantinople*, vol. 2, 108.

71 Shaw, *Between Old and New*, 399–400; Juchereau, *Révolutions de Constantinople*, vol. 2, 148.

72 Juchereau, *Révolutions de Constantinople*, vol. 2, 175–76, an indication of the kind of terror and its source that was probably the subject of daily alarms in Istanbul in the period.

73 Juchereau, *Révolutions de Constantinople*, vol. 2, 186–87.

74 Cevdet *Tarih*, vol. 8, 310–20.

75 Juchereau, *Révolutions de Constantinople*, 177. Shaw, *Between Old and New*, 384–407.

76 Juchereau, *Révolutions de Constantinople*, vol. 2, p. 90.

77 Cevdet, *Tarih*, vol. 8, 310–20. Cevdet blames the murder of Selim III on the *Bostancı* guard and the Chief Black Eunuch, who persuaded the naive (*saf*) sultan.

CHAPTER SEVEN

Mahmud II and the new absolutism

Thrust on the throne by Mustafa Alemdar Pasha, Mahmud II faced three significant crises which would determine whether or not the empire survived: the resumption of the Russo-Turkish war not terminated until the treaty of Bucharest in 1812; the Greek and Serbian revolts, which extended to 1830; and the invasion of greater Syria and Anatolia, 1831–40, by Mehmed Ali Pasha and his son, Ibrahim. The first two are the subject of this chapter. The reforms of Mehmed Ali, and Mahmud II's own reforms, will be the focus of the following chapter. The confrontations between the Egyptian army and the new Ottoman forces in Anatolia and Syria are described in Chapter Nine.

The young sultan inherited an empire whose terminal illness became the cliché of European diplomacy by mid-century, and his own grip on power was indeed tenuous. The geopolitics of the northern defensive border had been altered permanently by two Russo-Ottoman wars, with Russia in possession of the entire northern littoral of the Black Sea and a large part of the Caucasus. Austria had been defeated by Napoleon and Russia would shortly face the same challenge. The southern tier, especially the Egyptian–Levantine coast, had achieved an autonomy which overturned the traditional Ottoman–Mamluk balance, and carried Egypt into the world economy. The sultan could be said to be in control of very little territory except Anatolia, Bulgaria and Thrace. This too was soon to be challenged.

Selim III reorganised his court, installing what were to become regular meetings of his advisers, who counselled him on foreign affairs and the need for reform, and had attempted to extend his reform initiatives into the provinces, with some success but considerable opposition. It was his gesture to potential provincial partners that resulted in Alemdar's return to Istanbul. Discussions concerning the future of the Ottoman Empire dominated international relations. Intervention by Britain, France and Russia at pivotal moments in the civil wars and rebellions within the empire often determined the outcome. Like Selim III, Mahmud II proved resilient in manipulating, but was just as often forced to capitulate to the demands of Great Britain and Russia. Mahmud II proved more successful at dealing with the two major obstacles to a reconsolidation of imperial power. The first were the *ayan*, necessary to the survival of both Selim III and Mahmud II. The second, of course, were Janissaries. As this book has argued throughout, the two groups were often at odds, but frequently united in opposition to the dynasty. This chapter will weave the threads together in an exploration of Ottoman power politics of the 1820s. As part of the justification for the reconsolidation, 'Ottomanism' emerged as the new orthodoxy of the latter days of the empire. To what extent this represented a departure from the traditional view of Ottoman–Muslim sovereignty, and its consequences for the military establishment, are assessed at the end of the following chapter.

The over-reaching ambitions of the *Şeyhülislam* Ataullah and Istanbul rebel Kabakçı Mustafa, provoked a reaction and counter-revolution of forces under Rusçuk *ayan* Alemdar Pasha, who makes a curious figure as a counter-revolutionary. Claimed by some of our sources to be opposed to the military reforms, Alemdar was welcomed to Istanbul by many (including the Janissaries) as a restorer of the traditional order. He championed Selim III's and Mahmud II's reforms when it served his interests. Juchereau called the new grand vizier 'l'idole du jour et le but des espérances de tous les partis.'[1]

Forming part of the ceremonial coronation parade of Mahmud II, Alemdar Pasha continued to flaunt tradition by surrounding himself with an entourage of 300 Albanians bristling with muskets and sabres. The challenge to the new sultan did not go unnoticed by his enemies, who represented him as an insolent adventurer who violated the traditional laws.[2]

Persuaded by Selim III's reformers in his entourage, Alemdar had re-established the *Nizâm-ı Cedid* army out of Janissary troops, and renamed them *Sekban*, an old name for one of the earliest divisions of the Janissaries. He placed them under the command of Abdurrahman Pasha, a survivor of Selim III's court. Alemdar Pasha was also persuaded to assemble a large council of notables to agree to this reordering of the Janissaries, in order to convince Mahmud II to undertake what they considered necessary. A call was sent out to the principal governors and *ayan* of the empire. In October 1808, the chiefs of many of the great houses of Anatolia, such as the Karaosmoğlu and Çapanoğlu families and some from the Balkans, convened in Istanbul. Many of the family chieftains were accompanied by large numbers of their own troops, perhaps as many as 70,000 in total by one contemporary account.[3] Notably absent were Grand Vizier Alemdar's rivals from Bulgaria, and Ali Pasha of Iannina, although the latter sent a delegation. The governors of Kars, Erzurum, as well as the Arab provinces, in Mosul, Damascus, Baghdad, and Cairo, were too far away to make the meeting in time. In effect, the notables who gathered in 1808 represented the major regional forces whose territories would make up the rump of the empire by the end of the nineteenth century.

Sened-i İttifak – the Deed of Agreement

Twenty-five of those present, including Şeyhülislam Esad Efendi and Grand Vizier Alemdan, signed the *Sened-i İttifak*, or Deed of Agreement, which outlined the relationship and obligations between the sultan and his notables. Seven articles, with an introduction and conclusion, form the document. Article 1 reaffirmed a pledge from the *ayan* not to oppose or resist the sultan, and to come to his aid should others do so. Article 3 affirmed the commitment to the imperial provincial financial system, essentially confirming the continuation of tax farming. Article 4 asserted the obligations of the grand vizier as absolute representative (*vekalet*) of the sultan, to uphold the laws of the empire (*kanun*), and the obligation of the signatories to the document to stand as his accusers if he violated those laws. Articles 6 and 7 dealt with preventing disorder in the capital city, and protecting the empire's subjects from abuse and oppression, with the *ayan* serving as the watchdogs. Articles 2 and 5, however, are the most interesting from the military point of view. Article 2 committed the signatories to cooperate in providing the state troops (*devlet askeri*) for the benefit of the survival of the empire, and assisting the sultan against foreign and domestic enemies when required. Article 5 regulated the relationships among the *ayan*, the sultan and the central bureaucracy, on the basis of mutual guarantees. The *ayan* pledged to preserve the authority of the state by maintaining good relations with one another, and with the central state authorities, and in return were confirmed in the possession of their lands, and the rights of their heirs.[4]

However short-lived, this is still an extraordinary document, sometimes called the Ottoman *Magna Carta* and elsewhere described as the origin of public law in modern Turkey.[5] Continuing the tradition of grand vizierial councils (*meşverets*) prevalent on Ottoman battlefields and in the court of Selim III in the eighteenth century, the deed echoes the bargain made by the Janissaries with Mustafa II in 1703, described in Chapter Two. It was, nonetheless, innovative in striking a balance between the sultan and his noble, *provincial* subjects. By all accounts, the real opposition to the document lay with the sultan himself, who found his power circumscribed by an agreement negotiated between the grand vizier and the *ayan* which the sultan was invited to ratify. He did so because it was his only recourse, a measure of the degree to which the sultan's authority had been reduced. Events overtook the intentions of the assembled *ayan*, however, as a Janissary revolt, which began just a month later, killed the architect of the agreement himself.

The ostensible reason for the revolt was that those Janissaries who refused to enroll in the new disciplined forces found themselves off the rolls. Not just Janissaries, but scores of merchants, civil servants, and other beneficiaries of the Janissary payroll, themselves in possession of pay tickets (*esame*), had been deprived of a source of income. Furthermore, Grand Vizier Alemdar proved over-bearing and ostentatious, flaunting his new-found wealth, which he had largely confiscated from his enemies. It is possible to see such revolts as manifestations of a corporate will at work. This one appears to have been a genuine protest against the changes to the system, and the abuses which accompanied them, not as well organised or provoked internally as the betrayal of Selim III described earlier, but by far the most destructive. Palace intrigues lobbied tacitly for the restoration of Mustafa IV, but that did not prove of interest to the Janissaries. The new sultan no doubt also encouraged the destruction of the fragile alliance among fractious notables.

By early November, when the city erupted, most of the provincial *ayan* had left for their own territories, although some 400 of their men had been added to the new troops. Abdurrahman Pasha had assembled 3,000 musketeers from among the disbanded *Nizâm-ı Cedid* forces. Conscription began in Istanbul, and 4,000 more were added to the soldiers billeted at the barracks in Levend Çiftliği and Üsküdar, as before. Christians may have been among the conscripts, as Alemdar Mustafa was known to have Christians among his own troops. The new force was clearly to be a substitute for the Janissaries, with its own horsetail, insignia, and military band, but was initially made up of the riff-raff from the streets of Istanbul, hardly promising material for immediate action. They were divided into three divisions, white, green and red, determined by the colour of their uniforms. Alemdar had ambitiously envisioned one hundred *bölük* of 1,600 each, or 160,000 men. The army which supported him against the Janissaries, however, probably numbered no more than 25,000. Their number was even further reduced when his Istanbul enemies prompted the rival *ayan* on the Danube to threaten his main base at Rusçuk, which forced him to send troops to the defence of his provincial territories.

The coup, or, more accurately, civil war, began among junior officers of the Janissary Corps, who first attempted to enlist their commander, the Janissary Agha, who was promptly killed when he refused to join them. Alemdar was caught at the Sublime Porte, grand vizierial headquarters at the centre of the city, and reputedly himself caused the blowing up of the building during the Janissary attack. He was killed in the explosion and fire that followed. The rebels' target remained their leadership, so ordinary soldiers went unpunished, and much of Alemdar's army vanished into thin air. Still, Kapudan Ramiz Pasha elsewhere in the city put up a good fight with the *Sekban* troops, and regrouped others. The rebels were initially prevented from storming the Tophane arsenal, where the Levend Çiftlik troops were deployed. Abdurrahman's troops in Üsküdar helped to complete the circle of the city, but no one thought to fight their way through to the grand vizier. The Janissaries demanded concessions from the sultan in traditional fashion, by coercing the *ulema* (who this time at least were not directly complicit) to secure the appointment of a new agha and new grand vizier. Faced with his predecessor's fate, Mahmud II turned to the assembled reformed army for aid. On 15 November, 4,000–5,000 troops from Üsküdar and Galata were transported to Topkapı Sarayı. The new sultan had made his choice. Rebellion continued, while Mahmud II was reluctantly persuaded to execute Mustafa, his only dynastic challenger to the throne.

The revolt, called the 'Alemdar Incident' in Turkish history texts, turned into a general riot, which left 600 *Sekban*, and perhaps as many as 5,000 rebels, dead. An extensive fire destroyed many of the districts around the palace by Wednesday afternoon, 16 November. The sultan reached an accommodation with the Janissary commanders, who were ready to pledge obedience if the *Sekban* corps was dissolved and any of the reformers who escaped were punished. The Janissaries did not demand the head of those who remained in the palace. Mahmud II was the last of the Ottoman line. His death would have created even greater upheaval on the streets of Istanbul.

The rebellion received new stimulus, however, when news spread about Alemdar's death, and the seizing of the *Humbarıcı* barracks and the imperial arsenal across the Golden Horn at Hasköy the following day. By the end of Thursday, the rebels also controlled Tophane arsenal. The barracks at Selimiye and Levend Çiftlik were torched. Ramiz Pasha and Abdurrahman Pasha escaped initially, but were eventually caught and executed. At least one contemporary account, probably an exaggeration, reckoned as many as 50,000 deaths in the events of November.[6]

Mahmud II was still sultan. The *Sened-İttifak* experiment lay in ruins. The success of the Janissary revolt in 1808 indicates the spontaneous and fragile nature of the reform 'party', a cobbled-together set of ideas which attracted adherents according to the self-interests of the parties involved, of which Alemdar Mustafa is the most obvious example. A consequence of the confrontations in the streets revealed the degree to which urban/rural, centre/periphery political worldviews were antithetical, and left an indelible mark on Mahmud II, who vigorously pursued his own program of suppression of independent-minded governors, such as those in Damascus, Baghdad, Vidin and Belgrade, for the next ten years. Furthermore, in building his entourage of loyal advisers, Mahmud II came to rely on many individuals of Caucasus origin, as had Selim III to some extent, the consequences of which were to further 'Turkify' the empire in ways which will be discussed below, part of a redefinition of imperial populations along ethnic as well as religious lines. A comparison with Russia is apt for the period, especially in the Caucasus where Russian expansion was extensive, and many of the same questions of the control of violence and state loyalties arose.

More astutely than Selim III, Mahmud understood that popular politics had to be manipulated in order to effect major changes. The best way to do so was to establish successful relations with the huge class of ulema, but also to curb their claims not just on the minds, but the lands, of Ottoman subjects. In these ways, Mahmud II was more 'modern' than his predecessor, perhaps more 'western' as well. The nineteenth century is not only the great imperial age, but also a time when hitherto unregulated populations came under considerable central control, not the least of which involved the development of police forces to establish and maintain urban security. Events in Greece and Egypt, as an extended arena of the Napoleonic Wars, are immensely better known histories, and have dictated a view of Mahmud II's reign which underestimates the extent to which he managed to redraw the course of the empire long before the final confrontation with the Janissaries in the streets of Istanbul in 1826.

The real test of his early years lay in the war with Russia over the Principalities. Begun in 1806, halted briefly in 1808, it had resumed once again in 1809 when negotiations over a possible treaty broke down as Franco-Russian relations grew frosty. Uncertainties about Napoleon's potential attack on Moscow, as well as the disorder in Istanbul, delayed a resolution. Mahmud II would not entertain the surrender of all of Wallachia and Moldavia. Alexander I of Russia, similarly, required something to show for years of campaigns and occupation that had yielded little. Given the degree of destruction to the military institutions and material in Istanbul, one can only wonder at the nature of the Ottoman army called upon to defend the Danubian fortresses in 1809/10.

The state of the Russian military

The military system established by Peter the Great would continue for more than 150 years on the strength of the peasant army and the continuation of serfdom. Both would be called into question only after the disastrous performance of the Russian army in the Crimea in 1854–56. Success bred complacency, much as it had in the Ottoman context after the victory at Belgrade in 1739. Paul I (1796–1801) and Alexander I were erratic military leaders, who created divisions among their commanders. Paul's policies were disruptive of the reforms of Catherine II, and while Alexander I's resolve in the face of Napoleon's army before Moscow has salvaged his reputation, in reality, incoherence was the order of the day.[7] The Russia of Nicholas I (1825–55), however, who came to power just as Mahmud II's plans to eliminate the Janissaries came to fruition, has been described as 'a gigantic garrison state, an armed camp under the rigorous supervision of an autocratic drillmaster.'[8] Confronted by the Decembrist revolt of 1825 among progressive officers on his accession to the throne, Nicholas remained suspicious of these who suggested reforms and established the Third Section in 1826, a security force concerned with disloyalty among the officer corps and charged with 'promot[ing] correct thinking while combating dissent'.[9]

Russian recruitment, logistics and military command looked much the same in the first half of the nineteenth century as they had in the 1760s. Conscription was more than a reality, it had become a fixture in the lives of serfs. In Alexander's reign, close to two million men were to be levied from poll tax rolls, although intake may never have matched estimates, as shortfalls were characteristic of the system.[10] Exemptions were granted, dependent on pragmatic decisions, such as the need for settlers in newly-acquired territories, most of whom were involved in borderland defence anyway. Such was the case with the Cossacks in Ukraine and on the Black Sea. The heaviest levy burden fell on the peasants and the provincial estates, although communities were, by and large, able to determine whom they would send and how they would pay both the dues for the recruit, as well as support for the dependents left behind. The target of the levy became the extended household unit by the end of the eighteenth century, with economic and social consequences similar to the Ottoman context. Ottoman villagers were obligated to support militia members and pay collective taxes as part of the burden of warfare, generally proportioned by the local judge (at least in the Balkans) across the village tax-paying population. Of course, this led to the indebtedness of individuals and villages alike, and the beneficiaries were more often than not the local *ayan*.

Rights to conscript substitution in the Russian context created a system of donors (of able-bodied recruits) who could anticipate levies and ameliorate the impact on their own work force. Volunteers for service were often remunerated by other village families, one means of supporting the dependents left behind. Corruption and abuse more often led to extensive indebtedness of the less fortunate. Like the Ottoman *esame*, or certificate of entitlement, the donors' certificate of fulfilment of recruitment became a marketable 'security paper', although the abuse never exceeded ten per cent. Donors provided a cash grant and foodstuffs collected from their peasants, or fellow villagers, for the initial stages of a recruit's journey to induction. While the levy system did not lack critics, such as Field Marshal Rumiantsev himself, who advocated the creation of territorial armies, major reform would not be enacted until after the death of Nicholas I.

By contrast with the Ottoman system, Russian discipline was much more systematised, and brutal. The use of the gauntlet, an extreme punishment in other European armies, was applied frequently for minor offences, augmenting fear and tension among miserable soldiers.[11] Verses by a grenadier, one Ivan Makarov, survive from 1803:

For my country I stand on guard / Yet my back is beaten hard / The stick's the sole reward for me / Who defends us from the enemy / He who beats his men a lot / Rises straight up to the top / And is thought extremely keen / Though a devil he has been. He who fails this brutal test / Has to serve with the rest.

For his literary efforts he was forced to run the gauntlet.[12]

While Catherine II's military reforms are often lauded, discipline under Tsar Paul was tightened, which had the effect of standardising and legitimating corporal punishment. Consciousness about the plight and contribution of the common soldier to Russian victories, although dotting the military writings of the age, would not be visible as policy in military manuals until the latter half of the nineteenth century.

Soldiers' pay and rations were in theory highly regulated in the Russian army, according to a system often commended by foreign observers. *Artels*, company level cooperatives, were one way of ensuring the well-being of troops. For example, such collectives could finance the purchase of winter victuals, or horses and wagons, which gave the members of a unit solidarity and pride of ownership.[13] It was, however, an informal system and the monies were actually held by the company commander. The possibilities for abuse are obvious.

The soldiers were paid three times a year, although payments were very often in arrears, irregular, and reduced by official deductions. Pay was doubled over the century, with hussars, artillerymen and bombardiers paid double that of the land-militia in Ukraine. By the time of the war under discussion, a cavalryman in the field could expect 12 roubles a year, and an infantryman 14 roubles. Those in internal garrisons received, by contrast, 7.50 roubles. Little is actually known about the purchasing power of such a salary, but Keep's calculations argue for a deterioration of buying power over the course of the eighteenth century. In this, the Russian army looks little different from others of its age.[14]

The Janissaries maintained a well-organised system of *orta* treasuries, like the *artels*, and traditional rights as well as obligations concerning their pay. With the chaos of the period under discussion, and the destruction of so many of the Janissary records, it is hard for historians to determine how many of those rights were transferred to the new system put in place by Selim III. That the *esame* were virtually eliminated we understand, but not until the military reforms of 1843 is there a real effort to codify all aspects of military life.

What little we do know about wages is that they did not keep up with inflation, in fact, and were fixed at abysmally low levels throughout the latter half of the eighteenth century. Thornton in 1807 estimated an average wage of 30 *kuruş* every six months, then the equivalent about one-quarter of a shilling sterling per day.[15] Devaluation, and the loss of other army privileges such as the uniform allowance, caused Janissary revolts throughout the history of the corps, no less so in the period

1807–09. We have seen how cash awards for brave deeds, extraordinary services, and enemy casualty counts continued to serve as a stimulus and compensation for individual valour. This contrasts enormously with Russian military culture.

Ottoman commanders disdained (and were unable to impose) discipline. As noted in an earlier context, one Ottoman witness to Frederick the Great's army on parade in 1764, observing the punishment meted out via the gauntlet, equated the life of a soldier as worse than a slave. Some forty years later, the idea of rationalised discipline remained largely foreign to the Ottoman context, and systematic conscription regulations lay in the future: 1822 in Mehmed Ali's Egypt, and 1843 in Istanbul.

Looting was a fact of life in both armies, in part in the Russian case because of the inability or unwillingness of the government to supply the armies properly. Hunger was a constant; disease was a killer of more soldiers than bullets. Russian military strategy, especially under Nicholas I, was to run the army on as little as possible. Authorities had the luxury of seemingly inexhaustible sources of manpower, and remained impervious to the losses of men to wounds and disease at least until the 1850s.[16] On the borders, the burden of feeding the armies fell heavily on the Principalities and Bulgaria, and increasingly parts of eastern Anatolia and the Caucasus. In the post-1800 Principalities, hunger was an endemic aspect of military service. Half the Russian army died in the winter campaign of 1828, a combination of miserable weather, plague, bad hygiene, poor food supplies, and insufficient medical facilities.[17] Ottoman casualties are always accounted to be large, but fixed figures, or percentages, remain beyond recovery from surviving sources.

Religion, or religious propaganda, continued to play a part in both armies, as an ideological stimulus to ferocity on both sides, especially in the Danube and Crimean theatres of war which were notorious in Europe as bywords for the barbaric. We have already seen Suvorov and Potemkin at work at Ochakov and then Suvorov at İsmail in 1788 and 1790 respectively, when entire garrisons were slaughtered.[18] Russian soldiers expected to be badly treated if captured by the Ottomans, and often were killed by individual soldiers for head counts. Ottoman prisoners of war perished on long treks following capture, as much casualties of disease and hunger as brutality. Cossacks fighting for the Russians, and Tatars on the Ottoman side, were especially ferocious mutual enemies. Russia had embarked on a series of colonial wars in the Black Sea and the Caucasus which continued well into the nineteenth century, with consequences for the demographic displacement of Muslim populations into surviving Ottoman territories.

Still, in spite of the brutish and dismal nature of the military existence, desertion in the Russian army averaged one per cent by the end of the century, a figure which astonished foreign observers.[19] The common Russian soldier, unlike his Ottoman fellow, could be relied on to respect, even worship, able and strict commanders, on the battlefield, in folktales and songs. Peter the Great had created an officer corps in Russia, and by 1800 a single battlefield command structure was theoretically in place, but a systematic military bureaucracy was slow to develop. Imaginative thinking had to be squared with court politics, resulting in slow steps rather than drastic reform. Talented and critical foreigners, such as the Comte Louis Alexandre Andrault de Langeron, servant of four different tsars, were little appreciated by the native officer corps. Langeron's writings on the period are voluminous, and he had much to say about the disjuncture between Tsar Alexander I and his senior officers in the field, who were dealing with the shortages and lack of morale of an army that had been stationed in the Principalities since late 1806.[20]

Much has been made of the 'Prussianisation' of the Russian military, but the prime movers of military strategy concerning the Ottomans were the Russian field marshals P.A. Rumiantsev and A.V. Suvorov, both noted for their concern for the well-being of the troops under their command, who responded with immense loyalty. These two giants of the southern battlefield arena were most responsible for the Russian army's repeated success on the Danube. In 1811, it was to be the turn of General Mikhail I. Kutuzov, the less able but famed field marshal of the Russian army that faced Napoleon in the winter of 1812.

Ottoman campaign preparations

In March 1809, Tsar Alexander informed the sultan that he intended to annex Moldavia, and follow that by the annexation of Wallachia, an outcome of his agreement with Napoleon at Tilsit whereby eastern Europe came under Russian influence with French acquiescence. When Mahmud II demurred, insisting on the Dniester River as the boundary, hostilities were resumed. Istanbul, according to Cevdet, was a city of disobedient soldiers, an empty treasury, and a hotbed of French intrigue. The Janissaries (hereafter referred to as *zorbas*, here bullies, by him), stimulated by the prospect of a campaign, became intolerable, authors of shameful deeds that had shopkeepers closing their stores and people afraid to look out of their windows. The Janissaries resisted the enlargement of their numbers, or subverted the Anatolian recruits, calling them the sultan's 'flesh fortresses' (cannon fodder?) or 'brave sacrificial rams', and forcing their commanders into collusion with them.

Mahmud II implored then Aleppo Governor Kör Yusuf Pasha, veteran of Egyptian and Syrian campaigns, to assume command of the unruly mob and remove them from Istanbul as quickly as possible. In state councils it was decided to appoint commanders to the Danubian fortresses prior to the grand vizier's departure. Hurşid Ahmed Pasha, like Kör Yusuf of Georgian slave origin, was left as *Serasker* in Sofia, and Çelebi Mustafa, grand vizier, and co-conspirator with Alemdar Mustafa, remained as Commander in İsmail. Ahmed Agha was appointed Commander of İbrail.

As news arrived of a Russian siege at İbrail, and of Russian intentions to cross the Danube at Silistre, Grand Vizier Kör Yusuf assembled his troops at the parade grounds of Davut Pasha. Although only 4,000 Janissaries and 1,000 retainers, they insisted on the pay and rations for 14,000, as registered in the rolls. Stratford Canning observed the Ottoman forces at Davut Pasha as they mobilised for the campaign, and noted: 'The mass of the army was composed of levies, whether infantry or cavalry, . . . animated by a fervent spirit of loyalty and religion, but quite as formidable to the country which they traversed as to the enemy which they had to encounter. Nothing could be more picturesque to an artist's eye than their appearance under arms, nothing less satisfactory to an experienced observer than their accoutrements, maneuvers and commissariat.'[21] The army finally left Istanbul in the third week of July, and was in Edirne by mid-August 1809. Istanbul breathed a sigh of relief.[22] Russian estimates put the size and placement of Ottoman forces as follows: 12,000 at İbrail; 10,000 at İsmail, 10,000 at Guirgevo and Rusçuk; 10,000 stretched along the upper Danube to Vidin.[23]

Support from the *ayan* of the empire was reluctantly forthcoming, even when they were pressed for aid by the young sultan. In most cases they kept their real soldiers around them, and 'sent a regiment of farmers and laborers, odds bods, with string stirrups, broken saddles, guns without flints, and pistols without stocks, satisfied they had fulfilled the imperial request.'[24]

The Battle for Rusçuk and the Treaty of Bucharest, 1811–12

1806–12 was not a brilliant war for the Russians, as noted by Langeron: 'Never was there a war so poorly fought with such a lucky end.'[25] Events in Europe and relations between France and Russia, as we have seen, dictated the terms and length of engagement. A final push across the Danube was ordered for 1811 precisely because Alexander I expected Napoleon's attack on Moscow in 1812.

Russian commander Michelson had occupied the Principalities in 1806–07, but his operations were limited because of the small army under his command. The war lost its momentum in the spring of 1807, and an armistice was concluded at Slobozia, after the Franco-Russian rapprochement at Tilsit. When hostilities resumed in spring 1809, Prince Prozorovsky had more troops but failed to defeat the Ottomans, and was badly beaten at İbrail, Kladovo and Guirgevo. Russian Commander Count P.I. Bagration then captured Maçin, and Köstence at the mouth of the Danube, and besieged Silistre with an army of 30,000. Bagration took İsmail and İbrail, but had to lift the siege at Silistre and return across the Danube. Ahmed Agha, Commander at İbrail, held out for over 20 days, and victory caused rejoicing in Istanbul. In September 1809, Grand Vizier Kör Yusuf crossed the Danube towards Bucharest with an army of 40,000, and was confronted by the Comte de Langeron and his 'regiment of fever', who had assembled 6,000 reserve troops in Bucharest. (Reputedly 5,000 Russian soldiers in Bucharest and 3,000 at Craiova were incapacitated by fevers and plague.) Langeron drove the grand vizier's army back across the Danube. In the winter of 1809/10, Count Mikhail Kamensky, with Langeron under his command and over 80,000 men, broke the Danube barrier once again, and captured Silistre and Bazarcık, on the right bank of the Danube. Yet when Kamensky attacked at Şumnu, he was driven back, but not before capturing Rusçuk, Niğbolu and Sistova and retiring to Wallachia. Russian losses were large.[26]

Ottoman mobilisation orders indicate a clear reliance on the *ayan* and the populations of the Balkans for the Serbian and Wallachian fronts. Alı Pasha of Iannina, for example, then siding with the sultan, was ordered to Şumnu with 8,000 of his Albanians during the 1809 campaign. Osman Pasha, Governor of Anatolia, was appointed to face the Russians as they crossed at the mouth of the Danube. Troops were ordered to be sent to him from Karinabad, Yanbolu, Hacıoğlu Pazarı, Akyol, Kozluca and İslimiye, as well as those of the *ayan* of Elbasan and Köstendil. As in previous campaigns, reality rarely reached expectations. Osman Pasha, for example, was sent an additional 300 infantry (raised from the general population) after the first 1,000 vanished on the way to his camp at Şumnu. A total of 14,000 soldiers were to be recruited from among the Pomaks, and commanded by Pehlivan İbrahim Pasha, for the defence of Niğbolu. Some

3,000 soldiers were to be sent from Plevne to Vidin, similarly, to reinforce the garrison there. The heavy reliance on the goodwill of the local *ayan* rarely paid off in disciplined and war-ready troops, with some exceptions such as Pehlivan Pasha and Boşnak Agha, who distinguished themselves at Tatariçe, four hours from Silistre, when they were vastly outnumbered by Russian troops (Cevdet says 60,000) and forced a Russian withdrawal to Silistre and back into Wallachia at the end of 1809, as above.[27]

Russian advances were often facilitated by disloyal or vacillating *ayan*. The *ayan* of Silistre, Yıllıkoğlu Süleyman, for example, first declared his neutrality, and then surrendered the fortress to the Russians without a shot, in June 1810. Boşnak Agha, who replaced Alemdar Mustafa at Rusçuk and heroically defended the citadel against the Russians under Kamensky, stood by as the Russians defeated an Ottoman army in September of the same year.[28]

In the spring of 1811, Kutuzov was made Commander of the Moldavian army on the Danube. Kutuzov was a veteran of previous Russo-Ottoman campaigns, who had fought under both Rumiantsev at Kartal and Suvorov at Ochakov. The Russians were in control of the Danubian fortresses as far upstream as Niğbolu, but had failed to secure the right bank of the Danube in order to force an Ottoman capitulation. Negotiations stalled continuously because of Russian demands for surrender of the Principalities, and concessions on Serbian independence, which the Ottomans steadfastly refused to consider. Europe was certain that Napoleon played the Istanbul fiddle, although Ottoman obstinacy should never be underestimated in delaying the outcome. Mahmud II demonstrated the staying power that Selim III had lacked. Kutuzov was well known and respected in Ottoman circles, not just for his war experience but for participating in the treaty negotiations which had terminated the earlier conflicts.[29]

Preparations in Moscow for a French attack forced the withdrawal of five divisions from the Danube, although they remained stationed on the Prut River, leaving Kutuzov with 46,240 men, 180 field and 38 siege guns under his command in four divisions: Craiova, Niğbolu, Silistre, and the lower Danube. Of the assembled forces, a considerable number were cavalrymen: 6,539 Cossacks, and 7,605 regulars, a strategy of troop strength and distribution developed by Rumiantsev for Ottoman frontier warfare. The battle line from Belgrade to the Danube mouth was over 600 miles, and the Russian troops were thinly stretched along most of it. Kutuzov's strategy was not to engage in prolonged sieges, which he thought a difficult and wasteful exercise, but to await the grand vizier's move from his campaign headquarters at Şumnu. He anticipated crossings at Vidin, where a large flotilla of boats awaited Ottoman troop passage, and at Rusçuk. He was under immense pressure from Alexander I to finish the matter quickly.[30]

In 1811 the Sultan, encouraged by deteriorating Franco-Russian relations, decided to go on the offensive against the Russians. He replaced 80-year-old Grand Vizier Kör Yusuf Pasha with Ahmed Pasha (called Laz Ahmed, from Trabzon, Grand Vizier 1811–12), the former commander of İbrail, who had repulsed Prince Prozorovsky's assault in 1809. Under Ahmed Pasha was Boşnak Agha, hero of Rusçuk. Mobilisation was extensive and thorough; by the end of May, there were 50,000 men at Şumnu, 25,000 men at Sofia and additional troops to arrive.[31] The Sofia army, commanded by Hurşid Pasha, was assisted by the Bosnian Governor Hilmi Pasha, and aimed at keeping the Serb rebel army and the Russians separated, as in 1809 and 1810. The Russians had attacked Lofça (in Bulgaria) over the winter of 1810, but had withdrawn. The Serbian rebellion, under the leadership of Kara George, with an army of possibly 10,000, had spread into the territory around Niş.

Sofia became a strategic centre for the remainder of the war, and later as the command headquarters for the sustained offensive against both the Serbs and Ali Pasha of Iannina. Thus, there were two command centres in the Balkans, an unusual occurrence in Ottoman military history. Negotiations were restarted at Şumnu, and in earnest in Bucharest as well, where the representatives had moved at Kutuzov's invitation. Meanwhile, Grand Vizier Ahmed had arrived in Şumnu, prepared to cross the Danube at Vidin.

The bulk of Kutuzov's army awaited him at Rusçuk, to secure the passage, with smaller divisions stationed on the Danube island of Slobozia and in Craiova. He had previously ordered the destruction of the fortifications at Silistre and Niğbolu, and the withdrawal of all his troops except those at Rusçuk, under commanders Counts Langeron and Essen. He secretly opened negotiations with the Vidin Commander, Mulla Pasha, to secure 400 Danube vessels, most locally owned, for Russian use, and dispatched Cossack patrols to reconnoitre across the Danube.

Commander İsmail Bey, with 25,000 men, arrived at Vidin in early June, forestalling the treachery of Mulla Pasha. He had orders to send 100 of the boats downriver to the Olt estuary, to aid in the crossing of the Ottomans at Niğbolu. İsmail Bey was to cross with his army into Little Wallachia, towards Craiova, while Grand Vizier Ahmed moved his headquarters first from Şumnu to Razgrad, dug in there, feinted a crossing at

Rusçuk, and then marched on Niğbolu. Kutuzov was aware of the plans, and moved troops into position in April and May. Peace negotiations continued, initiated by the grand vizier and carried out in Bucharest by Hamid Efendi and former Russian Count Andrew Italinsky. They were stalled by the intransigence of sultan and tsar alike, and the delegates awaited the outcome of the summer campaign.

By mid-June, Ottoman cavalry were harassing the Russians at Rusçuk, and seizing all boats for the crossing of the Danube. The grand vizier was within a day's march of the fortress, with an estimated 60,000 men and 78 guns. Facing them were a Russian advance infantry guard, under the command of Voinov, and cavalry regiments, mainly Cossacks. The Ottomans dug in at Kadıköy, an hour's march from Rusçuk. They then marched on the fortress, and engaged with the advance guard. Upon hearing of the Ottoman attack, Kutuzov ordered the main body of troops at Rusçuk to come to the aid of the Russians troops, in what would prove to be the largest confrontation of the entire war. On 21 June the grand vizier sent troops to cut off Russian access to the Danube on the left flank. On 22 June he ordered 10,000 cavalrymen under the command of Boşnak Agha to make the assault on the fortress, while other troops attacked the Russian right flank in an effort to surround the Russian army. A day-long battle ensued, during which considerable bravery was shown on both sides, but Kutuzov prevailed with a combined infantry and cavalry assault on the Turkish forces late in the afternoon, forcing them to withdraw to Kadıköy. In spite of his success, he subsequently abandoned Rusçuk, and recrossed the Danube, arguing that it would be far more prudent to draw Ahmed Pasha into a trap on the left bank of the Danube than waste men and supplies on penetrating deep into Bulgaria. At least, that is how he later defended his decision, when subsequent events bore him out.

The Russians lost 800 killed and wounded, and the Ottomans 1,500 according to Russian statistics. Kutuzov later observed that during the numerous campaigns against the Turks, he never faced such ferocious and overwhelming attacks as by the troops of Boşnak Agha at the battle of Rusçuk. The tsar rewarded Kutuzov with his portrait. Istanbul rejoiced at the victory. Valuable fur coats were bestowed on the Ottoman messengers, and later on the grand vizier.[32]

By mid-July, Kutuzov had destroyed the fortifications at Rusçuk and retreated into Wallachia, taking with him 635 Bulgarian households.[33] The grand vizier spread his army along the Danube from Niğbolu to Silistre, and waited to hear of the outcome of İsmail Bey's thrust into Wallachia at Vidin. İsmail Bey and his troops fought through August to

PLATE 13 *Nineteenth-century print of Rusçuk* (The Danube: Its History, Scenery, and Topography. Splendidly Illustrated from sketches taken on the spot, by Abresch and drawn by W.H. Bartlett. *By William Beattie, M.D. George Virtue, London, 1844*).

establish a strong foothold on the left bank of the Danube, hampered by the marshy area west of Vidin where they tried to cross. They also established an outpost on an island at Lom Palanka to facilitate the grand vizier's troop passage using the Danube flotilla. Considerable resistance from the Russian troops at Craiova, and the duplicity of Mulla Pasha, Commander of Vidin, prevented İsmail Bey from carrying out his orders and forced his retreat to the right bank of the Danube.

In early September 1811, Ahmed Pasha faced Commander Kutuzov across the Danube. Ahmed Pasha blinked first, and began to move his army across the river. First they attacked and captured the island of Slobozia, on the night of 8 September. Then 6,000 Janissaries and supplies achieved the crossing, and began building fortifications, while a small group of soldiers feinted a crossing a few miles upstream from Giurgevo as a diversion. Within a week, and in spite of some ferocious confrontations, Ahmed Pasha had 36,000 men on the left bank of the Danube, and crossed over to join them, reputedly leaving 30,000 on the right bank. Kutuzov moved from his headquarters at Giurgevo to within two miles of the Turkish army at Rusçuk, and called up two divisions of the Russian Moldavian army that had been withdrawn from the Principalities in anticipation of the war with Napoleon. Ahmed Pasha remained inactive after his landing on the left bank, consolidating his fortifications and awaiting the outcome of his orders to İsmail Bey to attack Vidin once again. His troops were well supplied from the vital camp outside Rusçuk across the river.

Kutuzov prepared his own surprise attack on Rusçuk carefully, sending small numbers of troops to build redoubts on the right bank in anticipation of encircling the army stationed there. Meanwhile, desertions increased in the Ottoman camp – as many as 10,000 Albanians vanished as Ahmed Pasha awaited news of Vidin. On the night of 13 October Commander Markov's troops succeeded in crossing the Danube, and attacked the grand vizier's camp in Bulgaria, taking the army and bureaucrats stationed there by surprise and effectively surrounding the Ottoman army. By 14 October the grand vizier's army, 36,000 men and 56 guns, was stranded on the left bank of the Danube. After a spirited defence of Silistre, which fell to the Russians at the end of October, Ottoman troops on the right bank fled in all directions. 'Meantime, the Grand Vizier's army suffered under complete blockade. Hunger and bad weather (sleet and frost) spread diseases among the Turks and increased the death rate. Horses were either dead or eaten. Poor Turks had to eat rotten meat and had no salt. They cropped and ate the grass and roots on the territory of the camp, often paying with their lives for such terrible food and dying under the Russian artillery and musket fire. The cold October night worsened their conditions because the Turks did not have wood to warm themselves.'[34] The grand vizier himself had escaped to the right bank, and pressed for an armistice with Kutuzov in order to save his best troops so helplessly isolated. By 25 October a ceasefire had been agreed, which included biscuit and salt for the trapped troops. Negotiations began at Giurgevo, with Italinsky and Major Sabaneev for the Russians and Müftizade Selim Efendi, *Kaziasker* of Anatolia, and Galib Efendi, former *Reis* but *Kethüda* (Second-in-Command) to the grand vizier since July, for the Ottomans. Ahmed Pasha and Kutuzov were assumed to have the powers to make peace.

The condition of the Ottoman army did not improve, however, and by the end of that November, Kutuzov intervened on behalf of the surviving soldiers, and as a means of propelling the negotiations, took them under his protection, promising that he would return them in the spring. As historian Michailovsky-Danielevsky noted:

Kutusov accepted the 12,000 disarmed Turks that remained of the Vizier's army of 36,000 men that crossed the Danube earlier in the year. During its three month long captivity, [the Turkish] army lost over 20,000 men as a result of Russian artillery bombardment, wounds, starvation and cold, while some 2,000 men deserted. On the Vizier's request and to protect [Russian] troops, Kutusov transported to Ruse some 2,000 men that were ill with fever and other contagious illnesses. He instructed Count Langeron to escort the Turks to designated settlements, some 50 versts *[33 miles] behind the Russian army, and demonstrate to them that 'they are not our prisoners of war, but our guests, willingly staying here.' Leaving the campsite, the Muslims [Turks] were elated for being rescued from certain death. Capan Oghlu, courageous commander who shared the sufferings of his troops, approached Kutusov with dignity. Kutusov welcomed him and presented him with the best steed in the army.*

The Turkish camp looked like a dreadful cemetery. Numerous wounded, some of them without limbs, were lying without any medical aid among the decomposed corpses. The elite of the Turkish army had perished there. In the following years, the Porte could never assemble forces similar to the army commanded by Ahmed Bey. These troops were not familiar with the new movements that would soon convulse Turkey. They stood united in their faith and died for the glory of the Ottoman

Empire. Therefore, Kutusov was the last of the European generals who fought the Turks, while they still fanatically followed the ancient customs and traditions of their ancestors.[35]

Kutuzov bivouacked his army for the winter at the outskirts of Giurgevo and Bucharest, and moved his headquarters to Bucharest, where the Ottoman delegates joined him to negotiate the peace. The Russian negotiators began by insisting on acquiring Moldavia and Wallachia, and on some concession of the eastern Black Sea ports they had occupied in Kuban. The aim was to consolidate their holdings in the northern Caucasus, following the annexation of Georgia in 1801. The problem of Russian sovereignty over the Caucasus stemmed from the ambiguities of the 1774 Küçük Kaynarca treaty, in ambiguous wording about Kabarda in the Caucasus, and would continue until the end of the Crimean War.[36] The Russians also demanded the settlement of the Serbian problem by insisting on an amnesty for the rebels, and a guarantee of internal autonomy which the Serbs had not previously had. In spite of the humiliating position the army was in, the Ottoman negotiators would consider neither the secession of Moldavia and Wallachia nor the recent Russian acquisitions in the Caucasus. Furthermore, they consistently argued against any discussion concerning the Serbians, whom they considered part of the Ottoman *reaya* and not tributary subject peoples like those of the Principalities.

General Kutuzov and Grand Vizier Ahmed had settled on the Seret River as the new border in Moldavia, in order to begin the negotiations. By the end of November the negotiators had met twelve times and reached agreement on a number of lesser issues, such as the exchange of prisoners, and stipulations concerning the Principalities. Kutuzov broke off the sessions, however, when instructions and official powers were sent by Mahmud II and his advisers, insisting not on the Seret, but the Prut as the new border, and allowing no concessions in the Caucasus.

These treaty negotiations are usually represented, like those of the 1739 Treaty of Belgrade, as having been successfully concluded only with the intervention of international mediators, in this case the British, and more especially Stratford Canning, at the end of his first term of service as minister plenipotentiary to the Ottomans. The Ottomans signed the Treaty of the Dardanelles with the British in 1809, following laborious and difficult negotiations by British Ambassador Robert Adair, in the midst of all the dynastic upheavals and international crises previously described. Hence, the British were in a position to press the Ottoman point of view with the Russians. To view 'Ottoman fate as being decided in European chancelleries',[37] as did diplomats of the time and most later historians, however, is to lose sight of the astuteness of both Mahmud II and Galib Efendi, later briefly grand vizier (1823–24).

In January 1812 the two sides met again after the hiatus, to find the Russian treaty demands unchanged, except that they had renounced the armistice and now considered the stranded army as prisoners of war. Alarmed, Grand Vizier Ahmed Pasha wrote to Mahmud stressing that the condition of the army precluded a resumption of hostilities. Galib Efendi wrote to his sultan at this point as well, suggesting a letter from Stratford Canning as a means of persuading the Russians. Galib had previously approached Canning in 1810 and again in 1811, but Ottoman–British relations had remained strained.[38]

Mahmud II convened a general council in early February which met for three days. The council found the Russian demands for the eastern Black Sea littoral particularly onerous, clearly viewing it as establishing a Russian sovereign presence within striking distance of the heart of the empire. From both a strategic and religious point of view, such a development was completely unacceptable. They were, however, fully cognisant of the military and financial woes of the empire, and unwilling to propose a return to war. Mahmud II's directives to the council forced the decision on them: acceptance of the terms meant discussing the road to peace; rejection of the terms meant a return to war, and were that the case, discussion should focus on the means to that end. He was, he added, unprepared to accept his troops being made prisoners of war, which they took to mean his preference for renewal of hostilities. It may well have been a strategy on the part of the sultan to pacify the streets of Istanbul, demonstrate the empire's complete inability to go back to war, and secure peace. On 14 February, Mahmud informed Grand Vizier Ahmed Pasha of the general council's decision, and then proceeded to order him to reverse it, while ensuring the compliance of his own ministers in Istanbul through further discussions. As we have seen at other crucial moments in the empire's history, this involved discussions with religious officials as well as the military officers, whose agreement that a return to war was impossible was secured. Grand Vizier Ahmed Pasha was granted exclusive power to decide on war and/or peace.

Mahmud II also followed up on Galib's advice, and approved of the correspondence between Stratford Canning and Count Italinsky in Bucharest and the Duc Sierra Capriola in St Petersburg, urging the Russians to moderate their conditions. Belatedly, Napoleon appealed to

Mahmud II to join him with an army of 80,000 in Poland, in correspondence that arrived in Istanbul in early March. The appeal followed the signing of a French alliance with Austria, which included a guarantee of Ottoman integrity. Mahmud II and his grand vizier still recalled what was generally perceived as the treachery of Tilsit, and appear never to have seriously considered such an alliance.[39]

The Russians reoccupied Sistova in February 1812, but negotiations continued. The main issue remained the eastern Black Sea littoral. Kutuzov offered a Russian alliance at the end of March, to counter the French initiatives, subsequently insisting on it being part of acceptance of the treaty. Galib held out steadfastly for neutrality, as did his sultan, but in the event of a need for capitulation, Galib recommended a preference for a French alliance rather than a Russian one.[40] Both sides were anxious to conclude peace, fearing that a Russo-French war would protract negotiations and even renew the conflict, which Russia could ill afford.

At the beginning of May, the Russian demands were altered to draw the new border at the Prut River rather than the Seret, including the mouth of the Danube at Kilya. They also asked for the use of the Black Sea coast littoral on the Asian frontier. The sultan intially agreed, on condition that Kilya and İsmail remain Ottoman possessions, and that the prewar Georgian frontier be restored. What remained for settlement were the Serbian question and the restoration of certain Ottoman rights concerning the Principalities. The grand vizier insisted on the Serbian surrender of arms and renewal of loyalties to the empire. The Russians were also not to interfere with the appointment of *hospodars* in the Principalities. By the end of May a treaty was drawn up with sixteen articles, and a separate, secret treaty with two articles: one regarding the demolition of the fortresses of İsmail and Kilya, and a second allowing Russian access to the eastern Black Sea ports. Article VIII stipulated that the Serbians acquire internal autonomy, and a fixed tax rate which they themselves would collect, as part of their reconciliation with the sultan.

On three counts, Mahmud II remained adamant. First, he refused to accept the secret articles, and the proposed alliance, even at the risk of going back to war. Secondly, he and his advisers understood that having access to a port and passage at the eastern end of the Black Sea was just a means for the Russians to supply weapons to the area, as part of the Russian aim to 'conquer Georgia, Iran, Abkhazia and Circassia totally, and to execute the designs they have long harboured against the Ottoman Empire'. Finally, they understood that acceptance of Article VIII was tantamount to accepting future Russian interference in Serbia.[41]

Ahmed Pasha had already signed the definitive instruments when he received these instructions on 28 May, and had sent the text to Istanbul for ratification, where Mahmud II and his court considered it on 2 June. He and Galib Efendi pressed for an understanding with the Russians concerning Article VIII. Galib argued that the article did not force anything on the Ottomans, as the Serbians continued to be *reaya*, and any settlement of the revolt would require negotiations with the rebels on that basis. Canning was asked to press the Russians not to insist on the alliance, and to drop the secret treaty clause concerning the Caucasus, acknowledging in his dispatches, like Mahmud II, that Russian colonial aims in that region posed a serious threat. They were anticipating the 'Great Game' of the nineteenth century, shortly to become a major concern of British India.

Those initiatives were Canning's last efforts concerning the Treaty of Bucharest. By the end of June, Robert Liston, the new British ambassador, had arrived to take up his post. Similarly, Kutuzov had been recalled to Moscow and replaced by Admiral Paul Chichagov, who had presented Alexander I with an ambitious design to attack Austria and French Dalmatia with a multi-ethnic Balkan army of 150,000. The core of that army was to come from Kara George, the Serbian rebel, and his militias, and the *reaya* of the Principalities, who would spontaneously rise with their Orthodox brothers. Alexander, displeased with Kutuzov over the conduct of the war and treaty negotiations, recalled him to Moscow in mid-May, appointing Chichagov in his stead. Perhaps it was a measure of the tsar's desperation that he entertained the Chichagov plan, which fell apart as soon as its author was confronted with the realities of the Danubian battlefront. Most telling was the total lack of sympathy and refusal among the *boyars* of Wallachia and Moldavia to entertain the idea of arming the countryside, even when his aims were reduced to calling for the formation of a militia of only 20,000 men.[42] In the end, Chichagov withdrew with his army towards Moscow, and the Treaty of Bucharest stood as signed on 28 May 1812. The idea of an alliance, and the two secret articles, never ratified by the Ottomans, were dropped from Russian demands. The final exchange of documents occurred in mid-July when Napoleon was already on his way to Moscow.

The Ottomans may have escaped relatively easily, but the Principalities did not. Six years of Russian occupation and the fighting up and down the Danube reduced Wallachia and Moldavia to desperate straits. Initially, the Russians dealt with the local *divans* and the *hospodar* Constantin Ypsilanti, but in March 1808, they appointed Sergei S. Kushnikov as

President of the *divans*, whose first priority was the supplying of the army on the Danube. He achieved little in the two years he was in Bucharest, stymied by a seething centre of multi-lingual, multi-ethnic and cultural rivalries. Kushnikov pressed for order in the army supply system, and demanded such immense amounts from the countryside that much discontent erupted and peasant flight increased.[43] The Russian occupation forces were dealing with contradictory methods of operation in territories of ambiguous status. On the one hand they cooperated with the local notables, such as with the *hospodars* in Wallachia, while on the other they colonised and resettled in more direct ways. Significantly, in the territory east of the Prut, ceded in 1812 and henceforth acquiring the name of Bessarabia, the Russians took control, expelling the entire population of Tatars and sending them to the Crimea. The inhabitants of Turkish-speaking areas were sent into Russia as prisoners of war, and the land designated crown land. Refugees from across the Danube (Bulgarians) and many orthodox Greeks were encouraged to settle into the emptied lands. The legacies of the Russian occupation and use of the Principalities as a logistics base were ill-will and impoverishment. For those in Bessarabia, it meant becoming part of the Russian empire. For the rest, reverting to Ottoman tributary status did not promise any better future.

Confronting rebellion: the Serbs

European preoccupation with Napoleon, and studied neutrality on the part of Mahmud II, gave the Ottomans a respite from European affairs. For the period between 1812 and 1826, Mahmud II's determination to pacify the countryside and rid himself of 'the usurpers of legitimate authority', as the *ayan* were frequently labelled, bore fruit. He faced resistance in Serbia, Iraq, Greece and Albania, as well as Egypt.

In Serbia, Kara George continued as leader of the Serbian revolt, which had erupted in February 1804. Initially the revolt aimed at protecting the population of Belgrade against the rapacious *yamak dayıs*, and restoring the privileges the Serbian population had gained under reformers like Mustafa Pasha, who had been killed in the 1801 *yamak* coup d'état. Limited autonomy concerning bearing arms and the right to oversee the tax regime were part of the demands. By January 1811, Kara George had acquired the title of Supreme Leader and ruled Belgrade in conjunction with a Legislative Council. Article VIII of the Treaty of Bucharest recognised these developments without committing either side to full Serbian independence. There is considerable evidence that Mahmud II ordered the commanders stationed in the Danube fortresses to implement Article VIII of the treaty by declaring a general amnesty and peaceful takeover of the Serbian garrisons. Deserted by Russia in 1812, the Serbians resisted the overtures and continued fighting, with disastrous consequences. Grand Vizier Hurşid Pasha (September 1812 – April 1815), stationed in Sofia during the recently completed hostilities, attacked and defeated the Serbian rebels. By October of the following year, the Ottomans had retaken control of Belgrade. Hurşid Pasha declared a general amnesty, and called on Serbians to return home. Some 30,000 reputedly did so. The new governor of Belgrade appointed many Serbian leaders, among them Milos Obrenovic (1780–1860), to local administrative offices. This was accompanied by an attempt to pacify the region by eliminating many of the powerful local *ayan*. Harsh imposition of order had the opposite effect of creating more chaos, and precipitated the Second Serbian Uprising, in 1815, which forced the Ottomans to recognise at least part of Article VIII, concerning Serbian internal administration of the *Paşalık* of Belgrade, under Supreme Leader Milos Obrenovic (1780–1860), rival and murderer of Kara George in July 1817.

Continued Russian pressure to implement Article VIII of the Bucharest Treaty resulted in the Convention of Akkirman in 1826, specifying the limits of Serbian autonomy and reiterated in the Russo-Ottoman Treaty of Adrianople in 1829. By August 1830, Serbia was a principality within Ottoman suzerainty. Obrenovic was confirmed as Hereditary Prince. In 1833 the remaining fiscal obligations to the Ottomans, as well as the stationing of troops along the frontier, were negotiated, but it was not until April 1867 that all Ottoman troops were withdrawn from Serbian soil. Full Serbian independence was only finally enshrined in the Treaty of San Stefano in 1878.[44] Nobel prize winner Ivo Andric's *The Bridge on the Drina* imagines the social evolution as follows: 'The border between the two *pashaluks* of Bosnia and Belgrade, which passed just above the town, began in those years to become ever more sharply defined and to take on the appearance of and significance of a state frontier. . . . a whole company of Miloš' soldiers and a delegate of the Sultan, a soft, pale fellow from Stambul . . . confirmed everything. Thus, he said, it had been ordered by Imperial decree that Miloš should administer Serbia in the Sultan's name and that the frontier should be marked out, to know exactly to what point his authority stretched . . . And old Hadji-Zuko who has twice been to Mecca and is now more than ninety years old says that before a generation has passed the Turkish frontier will be withdrawn right to the Black Sea, fifteen days march away. . . . Time did its work.

. . . The Turks had now to abandon even the last towns in Serbia. One summer day the bridge at Višegrad was burdened with a pitiable procession of refugees from Užice. . . . in all there were about 120 families. . . . [One of them said] ' "You sit here at your ease and do not know what is happening behind Staniševac. Here we are fleeing into Turkish lands, but where are you to flee to when, together with us, your turn will come?" '[45]

From the point of view of Istanbul, concessions to the Serbians resulting from foreign mediation were simply unacceptable. The Ottomans had swallowed the annexation of the Crimea by Catherine II in 1783 with great difficulty. That unilateral occupation was perceived to be a humiliating loss of a Muslim frontier zone. Any further concession concerning the eastern Black Sea frontier continued to be unthinkable. The long protestation about outside interference in Serbian negotiations could be that it reflected the sultan's deep-seated commitment to the preservation of Ottoman sovereignty on the edges of Europe. In 1807, 1808, and again in April 1809, the Serbian rebels were offered amnesty and limited internal autonomy by successive Ottoman sultans. The stumbling block appeared to be the Serbian refusal to accept the offer without foreign guarantees. Preservation of Ottoman territorial sovereignty without international intervention is but one explanation for the decades-long, violent contestation of Serbian rights and privileges during the first half of the nineteenth century.

A second is obviously the geopolitical position of Serbia. Belgrade remained the pivotal end of a garrison line that stretched to the mouth of the Danube. Its loss would arguably leave much of Bulgarian territory vulnerable, even though the real weaknesses in the line were the fortresses downriver, as we have seen – at Rusçuk, for example. The Danube had become a lively commercial waterway, integral to the feeding of Istanbul, but also, since 1699, a waterway open to the markets of Europe. Loss of the Principalities threatened at any moment to cut off supplies to Istanbul, as the occupation of Bucharest by the Russians demonstrated. Belgrade (and Vidin) served as a check on those who aspired to colonise Little Wallachia in particular.

Thirdly, the initial reaction of the authorities in Istanbul to the rebellion, at least before Selim III's fall, was to support the resistance of the *reayas* to unruly Janissary *yamaks*, ultimately the source of the problem in Belgrade as in Vidin and Little Wallachia. Rebellions demanding justice or restoration of rights, as has been argued throughout this book, were often treated by applying 'friendly persuasion' to the rebels, a coaxing back into the sultanic fold, so to speak, accompanied by intimidation when necessary.[46] Such a viewpoint was in keeping with the idealised theory of sultanic justice as protector of the *reaya*, source of the wealth of empire. The forces of intimidation brought to bear, when necessary, were often neighbouring *ayan* at the head of their own troops. In this case the neighbouring *ayan* challenged Mahmud II both domestically and internationally, even as he was frequently called on to intimidate his unruly brothers – what has been called 'modulated anarchy'.[47] The Serbian rebellion crystallised for Mahmud II the problem of the *ayan* in general, and bringing them to heel would be his chief target over the next decade. There was a consistency of Ottoman policy towards the empire's Balkan subjects from 1792 to 1822, based on a strong Ottoman tradition of statecraft arguing for reassertion of central control over unruly provinces as the solution to rebellion. Mahmud II was ferocious in his pursuit of that end. In the process, by eliminating the local notables, he also uprooted and destroyed the regional networks and organisations which underwrote sultanic power in 1800. The consequences of this fiercely fought re-centralisation of power would play themselves out over the rest of the nineteenth century.[48]

Hakan Erdem has asserted that one of those consequences was in the clarification of 'Muslimness' under Mahmud II. The debate among historians over Ottoman perceptions of ethnicity and religiosity continues, often centred on the meaning of *reaya* itself. Did the Ottomans in Istanbul consider the word *reaya*, or peasant, as a marker of tax status, or as a religious (exclusively Christian) identity? Many assume the former to be the case until well into the nineteenth century. Others argue that *reaya* was consistently used as a label of exclusion and discrimination for all Christian subjects of the empire. At what point did the term *reaya* become differentiated in official policy by specific ethnic and religious categories? And to what extent were such categories created as part of the reconstruction of the army under Mahmud II? What is clear at this juncture is the sultan's alarm at the influence of Russian orthodoxy on his *reaya*, and a gradual sorting out of those Christians who were loyal, and those who were not. With the Greek revolt in 1820, public attitudes in general, and the sultanic reaction in particular, responded to such challenges by the further drawing of ethnic and religious frontiers.

Confronting rebellion: Ali Pasha of Iannina

Mahmud II continued his policy of containment of the *ayan* throughout the second decade of the century. The instrument of his will was Halet

Efendi, former Ambassador to Paris (1802–06), a man of widespread experience of the provinces. Halet Efendi joined court service in Istanbul through the patronage of Mehmed Raşid Efendi. Briefly *Reis* during Mustafa IV's short-lived reign, Halet Efendi was exiled to Kütahya for his supposed pro-British sympathies in 1808. By 1810 he had been recalled to Istanbul, and joined Mahmud II's inner circle. He was immediately sent to Baghdad to confront semi-independent Governor Kücük Süleyman Pasha, who had siezed power in 1808, and who was refusing to pay tribute to Istanbul. Halet, writing to Mahmud II from Baghdad at the time, presages the problems that would await the Ottomans two decades later. Iraq had become a stronghold for Mamluks, local strongmen who seized the governorship during the Wahhabi incursions, and their influence had extended to Mardin, Basra and Shahrizur. The mixture of populations and mentalities, Arab/Kurdish, Sunni/Shiite, and the proximity to Iran, made it imperative to install a strong governor.[49] Rebuffed, Halet Efendi recruited the militias of Abdülcelilzade Pasha, *Mutasarrıf* of Mosul, and Abdurrahman Pasha, *Mutasarrıf* of Baban, and marched on Baghdad in June of 1810 with 10,000 men to overthrow Süleyman.

His success in securing the deposition and death of Süleyman gave him considerable cachet in Istanbul. Upon his return, he was appointed as *Kethüda* (or steward) to the sultan's inner circle, the equivalent of the minister of domestic or internal affairs. For more than a decade, he exercised extraordinary influence at court, as the chief adviser to the sultan. He refused to entertain the elimination of the Janissaries, who were the basis of his support, and he had a significant influence on the direction of imperial affairs in the first decade of Mahmud II's reign. His hand seems to have been everywhere, from securing the appointment of his favourites as the *hospodars* of Wallachia and Moldavia, to the sabotaging of all efforts by the young sultan to reform the Janissary corps. Halet Efendi concentrated on subduing a large number of the *ayanlık* dynasties of Rumeli and Anatolia at the behest of the sultan.

The result was that by 1820, much of the territory in Anatolia and Rumeli and the north Syrian territories had reverted to central government control, except for Ali Pasha of Iannina, 1744–1823, whose life and adventures have influenced all accounts of Greek and Albanian national narratives. Ali Pasha was adroit at manipulating the international context, especially during the confusion in the Adriatic following Napoleon's occupation of Venetian Corfu and the Ionian islands in 1797, and the attack on Egypt in 1798. Ali Pasha was more comfortable in his dealings with his potential allies Britain and France than with his master, Mahmud II, or the Russians. He proved implacable against any and all, Greek, or Albanian, who rebelled against him in his own lands, and against whatever Ottoman force was sent against him. Celebrated by Lord Byron as the 'Muslim Bonaparte', and the subject of countless Oriental fantasies substituting as his biography, Ali Pasha's challenge to the Ottoman state has rarely received the serious attention he deserves, with some recent exceptions.[50]

By July 1820, when Mahmud II sent an army against his rebellious subject, Ali Pasha controlled most of Greece and southern Albania, and had his sons, Mukhtar and Veli Pashas, installed as governors of Avlonya and Morea in the Peloponnesus respectively. We have already seen Ali Pasha in his relations with the Ottoman government, as when he joined the Danubian military coalition against rival Osman Pasvanoğlu, and against the Russians in the 1806–12 war, where the performance of his troops was effective and distinguished. By such service, Ali accrued coveted tax-collecting offices and extended his power, often brutally, over much of the fiercely independent Albanian and Greek territories. Danish traveller Peter Brønsted, one of many diplomats, spies and intellectuals of the late eighteenth century who paid the 'Diamond of Iannina' a visit, said of Ali: 'He had nothing of that white and delicate complexion, which is so common amongst the Turkish seignors; his own was brown and manly; his manners, concise and lively, had absolutely nothing of that slowness and gravity, which is the *bon ton* with the Turks of high rank. Ali, as the Greeks told me, even frequently mocked such a theatrical sort of gravity. He spoke quickly, moved with rapidity, and in one word, had nothing of an old Turk, except his dress; – his body was that of a Greek, and his soul that of an Albanian.'[51]

Ali Pasha first derived his power from his appointment as *Derbend Başbuğu* in 1786, an important frontier position which he held until 1820. It was precisely his own ties to Albanian chieftains (Tosks) in Tepedelen, as well as the increasing use of such warrior stock for the late eighteenth-century Ottoman armies, which gave him his leverage. Albanian recruits pressed into Ottoman service as *levends* were diverted to the Morea in 1770/71, when Grand Vizier Mehmed Muhsinzade used them to suppress a rebellion in the midst of the 1768–74 war. This inaugurated a cycle of violence between the two communities that Ali Pasha capitalised on adroitly, by coupling the animosities of rival chieftains, Greek and Albanian, with his own particular brand of diplomacy, which was to cultivate, marry his children into, and then eliminate his Ottoman rival appointees in the various *sancaks* of his own and surrounding areas.

In 1810 his reach extended into northern Albania, when he added the *sancaks* Ohri and Elbasan to his holdings. Mahmud II watched this process helplessly. Ali did respond to the sultan's call for soldiers in 1809, as we have seen, sending Mukhtar and Veli Pasha with their armies to the Danubian front. By 1819, Ali Pasha reputedly could muster 40,000 men if need be.[52]

In 1819, however, Ali Pasha fell foul of Halet Efendi, who turned the sultan's attention to Ali Pasha's disloyalty. Mahmud II had succeeded by that time in removing or de-fanging most of the lesser *ayan* of the Balkans, replacing them with more amenable local governors, or central appointees. Dismissed from his position in early 1820, Ali Pasha was ordered to withdraw all troops to his seat of power, the *sancak* and city of Iannina. Hurşid Pasha, veteran of Sofia command and the Serbian rebellion, was appointed Governor of Morea, and ordered to march against Ali Pasha with the pashas of adjacent provinces and their personal armies. One of those who joined Hurşid Pasha was Buşatlı Mustafa Pasha, Governor of İşkodra in northern Albania, who had recently been under attack himself by Ali Pasha. Ottoman ships were dispatched to the Albanian coast. Ali Pasha's response was to stoke an uprising among the Greek rebels in his territories, by signing a mutual aid pact with their leaders. Besieged in Iannina, Ali Pasha held out for two years, but was fatally wounded in fighting with the Ottoman troops and died on 24 January 1822. Halet Efendi, blamed for the Greek rebellion which broke out and subsequently engulfed Greece, Moldavia, and much of the rest of Rumeli, was himself dismissed, sent into exile, and finally beheaded in Konya in November 1822.[53]

The Greek Revolt 1821–27

The decision to confront Ali Pasha had not been taken easily, according to historian Cevdet. Opponents to Halet Efendi, such as *Reis* Canib Mehmed Efendi and others, were fully aware of the potential for rebellion in the Peloponnesus. Ironically, by subduing the strongest of the *ayan* in the Balkans, perhaps the only one who might have prevented the revolt, Mahmud II unleashed the greatest threat to the dynasty to date. The Greek uprising that began in Moldavia in March 1821 had its antecedents in Ali Pasha's resistance after 1820, and more importantly, went largely unchecked for more than a year while the Ottoman army was engaged in the siege of Iannina.

Furthermore, the ethno-religious group that rebelled against the Ottoman state was significantly embedded in the central administration of the dynasty, in both commercial and diplomatic spheres. There was not, however, unanimity of aims among the various Greek populations of the empire (perhaps as much as one-quarter of the inhabitants of Ottoman territories) who found themselves caught up in the rebellion. Four distinct areas: the Peloponnesus and the Greek islands of the Adriatic and the Aegean, Wallachia and Moldavia, Bessarabia and the new Russian city of Odessa, and Istanbul constituted areas of great concentrations of Greek-speaking Orthodox Christians. The Peloponnesus was largely self-governing, but also split among the educated classes: priests, large land-owners and merchants, and a large peasant population, the latter slow to respond to the call to revolt. Inhabitants of the area were well acquainted with Ottoman retaliatory measures, such as the crushing of the previous uprising in 1770/71, and more recently the heavy hand of Ali Pasha. The Principalities, as we have seen, were administered by Ottoman–Greek boyars, with attachments to the Phanariot families of Istanbul, who had a stake in continuing as Ottoman clients. Moldavia was different even from Wallachia, in that Russian occupation of Jassy had been more or less continuous since the 1760s, and Russification was farther advanced than in Wallachia, where Hellenisation had a stronger hold. In Bessarabia, a policy of resettlement was instituted by the Russians after its acquisition by treaty in 1812. Greeks were numerous among the new settlers, enticed by the commercial possibilities of the frontier territory, Danube trade and the new Black Sea port in Odessa.

The Philike Hetairia Greek liberation secret society was established in 1814 in Odessa. Its members originally envisioned the restoration of a greater Byzantium, with Istanbul as its capital, rather than the establishment of a modern nation-state. In its first two years the society attracted 30 members, but by 1821 numbered closer to 1,000. The leadership of the organisation eventually fell to Alexander Ypsilanti, son of former Wallachian *Hospodar* Constantine Ypsilanti. The son was a general in the Russian army and aide-de-camp of Tsar Alexander I. Ypsilanti may or may not have had the tacit approval of Alexander I for his activities, but he hinted at the tsar's support and attracted important members of Greek elite groups to the fledgling society. More than half of the membership was of the merchant class. Plans for an uprising continued, with the Peloponnesus and the Principalities targeted as the strategic centres. Serbian leader Milos Obrenovic, having established his own peace with the Ottomans, maintained neutrality throughout. In Wallachia, however, Tudor Vladimirescu, a former member of the legally sanctioned (by the Ottomans) Serbian militia, the *pandours*, was simultaneously organising a Romanian revolt in Oltenia, believing that Russian support would follow. In January 1821, Vladimirescu inaugurated a social revolution in Oltenia, with the following declaration: 'Brothers living in Wallachia, whatever your nationality, no law prevents a man from meeting evil with evil . . . How long shall we suffer the dragons that swallow us alive, those above us, both clergy and politicians, to suck our blood? How long shall we be enslaved? . . . Neither God nor the Sultan approves of such treatment of their faithful.' Vladimirescu's call to arms, and promise of Russian support, set off a peasant uprising against the *boyars*, not against the Ottoman dynasty. Importantly, those who joined Vladimirescu called on Istanbul to investigate the injustices in Wallachia and restore the principalities to pre-Phanariot days.[54]

In March 1821, Ypsilanti crossed over into Moldavia, seizing the government and implying that a Russian army would soon arrive. On 17 March, however, official repudiation and denial of support from both Alexander I and the Orthodox patriarch in Istanbul arrived in Bucharest. Russian consuls in Bucharest and Jassy were recalled. Russian Foreign Minister Ioannis Capodistrias, himself a native of Corfu, and likely co-conspirator with Ypsilanti, was later dismissed by the tsar. Vladimirescu, with 65,000 men, arrived in Bucharest in early April. Ypsilanti, with no more than 5,000 rag-tag troops, arrived shortly thereafter. Their occupation of the Principalities had resulted in the usual denuding of the countryside we have seen, endemic to Romania of the period. Without Russian support, the two revolutionary armies were stranded. Furthermore, atrocities against Muslim residents in Jassy and Galati, in the early days of the occupation of Moldavia, promised reprisals from Ottoman armies. The chaos had engendered native flight, with one estimate that 12,000 *boyars* and merchants fled to Brasov and 17,000 to Sibiu in Transylvania.[55]

On 6 April, revolt had erupted in the Peloponnesus. The sultan could no longer avoid such assaults on his territories and subjects, and prepared to send armies to both the Peloponnesus and the Principalities. At peace with Russia, however, Mahmud II was hesitant to offer Alexander I a cause for a renewal of hostilities. The commander of Vidin, Derviş Mustafa Pasha, wrote to the sultan shortly after the outbreak of the revolt, asking for new soldiers and supplies. Mahmud II and his advisers were hesitant to violate the Bucharest Treaty, which prohibited the Ottomans from sending troops into the Principalities. Derviş Mustafa warned, in fact, that the peasants were afraid of just such an eventuality, and taking to flight.[56] Nevertheless, by 13 May Ottoman forces had crossed the Danube. They encountered a dissolving militia, inexperienced military commanders, and discord between the two leaders. Ypsilanti's followers captured and executed Vladimirescu on the night of 8 June, which sealed their fate. On 19 June, at the Battle of Dragatsani, Ypsilanti's forces were easily overrun by the Ottoman forces, and the leader of the revolt himself escaped into Transylvania, where he was arrested, and he died in an Austrian prison seven years later.[57] Among the fallen was a group of young Greeks, called the Sacred Band, who unlike the majority of Ypsilanti's troops, reportedly stood firm under the Ottoman assault: '. . . educated in Europe, and for the most part students or merchants' clerks, who had repaired to the standard of Ipsilanti from Russia and Germany. They wore a uniform of black, as an emblem of mourning for the sufferings of their country, to whose service they had devoted their fortunes and lives . . . These young men, animated by the spirit which taught their ancestors to perish at Thermopylae, preferred a glorious death to flight or dishonour.' Four to five hundred of them stood in the open plain against the Turkish attack, 'at once the flower and hope of their country. The heroism displayed on this occasion will bear an advantageous comparison with the best days of Grecian history.' Such hyperbole filled the gazettes and journals of London, Paris and St Petersburg, stimulating many more *philhellenes* to volunteer.[58]

Mahmud II had escaped the threat of all-out war with Russia for the moment. In April 1822, delegates from the two territories met with Ottoman and Russian delegates to re-establish order. The sultan granted the restoration of native rule, the removal of Greek clergy, to be replaced by native clergy, and the creation of a Romanian militia in place of the Albanian guard. Ottoman–Russian disagreements arose over the old issues: the status of the Principalities, and the right of Russian intervention in the appointment of the *hospodars*. With the extension of the revolt to the Peloponnesus, relations between the two powers would enter another period of crisis resulting in war in 1828.

Revolt in the Peloponnesus and Ionian Islands

The revolt in the Peloponnesus proved more resilient than that of the Principalities. Organised by militia bands (*klefts*) and their commanders, in concert with local notables and wealthy shipping magnates, it began as a spillover from Ali Pasha's resistance as described above, and was exacerbated by Ottoman attempts to coerce the Orthodox leaders into submission. What started as sporadic resistance coalesced into a general revolt in April 1821. Muslim residents were the targets of local anger. Perhaps as many as 15,000 of an estimated 40,000 Muslim residents in the Peloponnesus may have been killed in the course of events. Such atrocities were matched by the ferocity of Ottoman Muslim subjects and soldiers alike. The massacre of Christians on Chios in 1821 resulted in thousands of deaths, and awakened Europe to the fierce struggle unfolding in the eastern Mediterranean.[59]

Istanbul's population was traumatised by the outbreak. Mahmud II was forced on the one hand to take measures to protect the large Greek population of the city, while on the other, to secure Muslim residents from the potential violent acts of the Philike Hetairia society, which had relocated its headquarters from Bessarabia in 1818. Reprisals in Istanbul included the public execution of Patriarch Gregory V in April 1821. His death reverberated throughout the Orthodox Christian world. Among other victims was the chief court translator, or dragoman, Constantine Mourouzis, symbol of an office dominated by Greek Orthodox families and considered the most obvious symbol of Greek treason.[60] The outbreak of full rebellion in the Peloponnesus, the internationalisation of the cause, and the factionalism which developed among the Greek parties to liberation, meant the eclipse of the Philike Hetairia and the construction of a fledgling national Greek government.

Ottoman commanders had great difficulty controlling their soldiers, as Grand Vizier Salih Pasha reported concerning Chios. There, he said, 'so-called' soldiers were plundering pardoned *reaya* villages and enslaving the inhabitants. The *Şeyhülislam* condemned the enslavement of pardoned villagers, who were lawfully considered free men and women. Such was the situation that the grand vizier suggested that the commander on Chios exercise caution with the unruly troops, informing the miscreants of the illegality of their actions, but stopping short of alienating them. His words indicate the depth of the problem of discipline.

Marines guarding Ottoman merchant ships, in another example, were castigated for attacking peaceful *reaya* upon landing in the Dardanelle ports. Commanders in the area warned the sultan of the possibility of massacre. Ordered to discipline the violators, the Janissary Agha refused to assume responsibility, saying they were irregulars, *miri levendat*.[61] Mahmud II was furious, equating the plunderous marines with the Greek rebels and berating his grand vizier for seeking the consent of the Janissary commanders. Such episodes were endemic during the conflict, and incapable of control from Istanbul. Repeated instances of plunder and massacres, rife on both sides, must have hardened the resolve of Mahmud II to rid himself of the Janissaries. Doubtless, the shock of the revolt prepared public opinion in Istanbul for the series of reforms initiated by the sultan in the middle of the Greek war.

After the death of Ali of Iannina, Ottoman armies were able to reassert control over the northern territories of Epirus, Thessaly and Macedonia. They concentrated their efforts on retaking the fortresses in the Peloponnesus, as well as the cities of Missolonghi, Thebes and Athens on the mainland. They faced considerable resistance both from entrenched local armies and from Greek merchant piracy at sea, which crippled supply lines from Istanbul. They were, in fact, losing control of the situation.

The continuing success of the Greek side was largely due to the *klefts*, renowned for indiscriminate banditry throughout and adept at piracy at sea. Their style of warfare suited the terrain, and the political environment. They made fierce opponents to the Albanian warrior bands constituting the bulk of the Ottoman forces. The *klefts* became the symbol of resistance to the tyranny of the Ottomans, and inspired the *philhellenes* of Europe, caught up in the enlightenment celebration of classical learning and style. Volunteers, including George Finlay, later historian of the events, travelled to Missolonghi to be with Lord Byron, whose verses stirred the enthusiasm of intellectuals as well as military adventurers. Greek Committees were formed in all the major cities of Europe, raising funds and volunteers for the cause. They joined groups of Greeks, educated abroad, who were imbued with French revolutionary fervour but riven by factionalism. Guerrilla warfare complicated the war of liberation, even as it continued to be supported by the *philhellenes*. Checked, the Ottomans withdrew to the large fortresses where they were besieged. As always, hunger and illness often claimed more victims than actual battles.

Unable to suppress the rebellion, Mahmud II turned to his Governor of Egypt, Mehmed Ali, in late 1824. By that time, Mehmed Ali's reputation was more than equal to that of Ali Pasha of Iannina. Veteran of domestic and international conflicts, Mehmed Ali had preceded Mahmud II's army reforms by a decade, and had disciplined troops at his command. Mahmud called on Mehmed Ali to assist him in ending the Greek rebellion. Mehmed Ali agreed to participate only on the condition that his son, Ibrahim, would be awarded the Peloponnesus, while he reserved Crete for himself. In February 1825, after retaking Crete, the Egyptian army, commanded by Mehmed Ali's son Ibrahim, entered the Peloponnesus. He arrived with 5,000 men, and was soon joined by another 7,000 from

Crete, eventually assembling 17,000 troops in the Morea.[62] Ibrahim Pasha relieved Ottoman forces besieged in Patras and by May had captured Navarino.[63] The reformed Egyptian troops, drilled by French advisers, proved more than a match for the Greek regulars and militias. One Greek observer noted that 'these Arabs make war in a manner that no one has seen before; they advance in regular squares, and standing upright as if a bullet could not harm them, they then rush the Greeks with bayonets stuck on their Tophaics [*tüfenk*, *or* musket] so long, and what soldiers in the world could be supposed to endure that?'[64] The Greek army was demoralised, and as ill-equipped as their Ottoman opponents. Though Greek ships continued to interrupt Ottoman supply lines, their efficacy was checked by the reformed Egyptian navy.

Missolonghi and Athens

By April 1826, Ibrahim Pasha and Rumeli Governor Mehmed Reşid Pasha, *Serasker* of the Ottoman army in northern Greece, had joined forces and retaken Missolonghi on the Greek mainland after a year-long siege. Just prior to the fall of the fortress, a hastily assembled Greek provisional government committee petitioned Europe with an appeal to 'all ye who have Greek blood in your veins, to prepare your sinewy arms against the barbarous enemy of Christendom.'[65] The Ottoman victory was accomplished with significant bloodshed, perhaps as many as 3,000 of the defenders dead, with another 5,000 women and children enslaved.[66] Ambassador Stratford Canning, Ambassador to Istanbul, was invited to Topkapı palace in June 1826 to view the 5,498 human ears from the victory at Missolonghi, and was the only diplomat present who refused to congratulate Mahmud II on his victory. Stratford was equally dismissive about the Greeks, however, as were many of the *philhellenes* when confronted with the realities of the Morea. They were 'Unfit for a state of complete independence . . . the powers of Europe would dread the establishment in their immediate vicinity of a new state so utterly unprepared for the arduous duties of self-government.'[67]

By 15 August 1826, the joint Ottoman–Egyptian army under Mehmed Reşid Pasha, and Ibrahim Pasha, had retaken Athens, and laid siege to the Acropolis. Admiral Thomas Cochrane, commander of a fleet organised by the London Greek Committee, after more than two years of dallying, arrived in March 1827 to assist the Greeks besieged in the Acropolis. The combined forces, when added to the Greeks commanded by Karaiskakis, numbered 11,000. They first confronted the Ottomans at St Spiridon in

Piraeus in an attempt to relieve Athens from the south. The 300 Albanian troops surrendered the monastery, but were slaughtered by enraged Greeks, an act which had most of the foreign officers quitting in disgust. Heavy fighting in open country ensued, ending in the death of Karaiskakis. By the time of the attack to relieve Athens, 6 May, the operation was in disarray, and in the battle which followed, some 900 Greeks, and many of the remaining *philhellenes*, had been killed. On 5 June, the besieged Acropolis surrendered and was reinvested by the Ottomans.[68] One participant, Thomas Whitcombe, described the style of fighting:

> *the followers of Panayèa [the Virgin – the Greeks] and the Prophet [the Muslims] . . . are trained to a unique mode of tactics – for 'in joyless union wedded to the dust,' they lie at full stretch on the ground, and pop up and grin at, and vituperate one another, past all bearing. They throw up a small heap of stones before them – where no barrier ready made to their hands presents itself – so that little more than the head is visible to an opponent, and that only for the second or two they raise themselves for the purpose of firing; then they bob down again and reload. They are advocates for long shots, and have little sympathy for advancing unless their enemy retire. Sometimes, several of them fire from the same cover; and, again, many others act quite independent of, and at considerable distances from, each other; – straggling in such various directions that one stands almost as good a chance of being shot by a friend as by an enemy.*

In describing the enemy, Whitcombe noted 'the Turkish camp at Athens as exhibiting a scene of the greatest confusion; the Pasha's troops – especially the Albanian and Delhi [*deli*] cavalry – being worked up to a state of complete mutiny owing to the scarcity of provisions and the withholding of their pay.'[69]

For this particular campaign in the Morea, we have a unique manuscript of a soldier, son of a standard-bearer in the Ottoman military, whose rather curious narration has recently been translated. Mustafa Vasfi Efendi of Kabud, a village near Tokat in Anatolia, spent the years from 1801–33 as a soldier for hire, first in Erzurum, as part of the troops under Dramalı Mahmud Pasha. He later travelled to Ağrıboz, where he signed on with *Çarhacı* Ali Pasha and then Ömer Pasha. Vasfı Efendi described battles, sieges, looting and pillaging, without batting an eye. The following passage is typical, and is evocative of the life of an Ottoman irregular, just before Mahmud II's reforms. The events described near Ağriboz:

'The Janissaries, because they were on foot, soon fell behind. We, who had good horses, went on ahead. We were altogether eighteen horsemen. Anyway, we went off and arrived in an infidel village.' They sat down under two mulberry trees, whereupon some local inhabitants approached them, and said: 'We are afraid of you. We have wives and daughters on that mountain over there. If you give us protection, we will come down. We said: the pasha has sent us and we have orders to protect you. The infidels were extremely glad, went away and brought lambs and bread to us. About twenty to thirty women and girls came with them.' The cavalrymen were afraid of being outnumbered, and isolated for the night, when they assumed the infidels would slay them, so '[we] took the infidels, cut off their heads, captured these thirty women and girls', and took off. They came upon a church, captured the infidels who were inside the church, cut off their heads and hid in the church for the night. They found 5,000 sheep beside the church the next day, and with sheep and captives, returned towards their camp. It is at this point that they encountered Janissaries, who stripped them of their captives and booty at gunpoint. 'I had a girl and woman with me, and two mules. They arrived, plundered all my possessions. I remained behind as a simple foot soldier.' Then, he ran into other Janissaries. Pretending to be a Janissary himself, 'he complained of his treatment at the hands of other members of the *ocak*'. A Kurdish servant of the Janissaries addressed those who had abused him: 'You have taken this man's possessions, slave girls, and severed infidels' heads. Things like this do not befit our corps. Now give this man his belongings.' Vasfı Efendi thus retrieved his booty, and returned to camp. His commander rewarded him with two coins for the heads, but chastised him: 'No-one of the deli horsemen should advance on the main army corps.' Henceforth, Vasfi Efendi noted, they remained with the pasha.[70]

Of the *philhellenes*, their fellow American Samuel Howe wrote satirically: 'But this was no Greek scheme [referring to the bungled confrontation of the Ottomans in open country beneath Athens], Karaiskakis [Greek commander] would not approve of it; but could a rude uneducated mountain Chief oppose the opinions – of my Lord Cochrane, than whom a better *sailor* existed not; of his Excellency, Sir Richard Church [Generalissimo], whose skill at the toilet was indisputable; – of the Bavarian Colonel Heideck, whose string of orders, and ribbons, and medals, proclaimed what he ought to have been; and who besides could draw horses to such perfection; – of the crowd of newly arrived foreigners from all nations, who, in more tongues than were heard at Babel, spouted their own titles, the valour of their own countrymen, and breathed vengeance upon the Turks? Surely all these men must be better qualified to judge, than ignorant Greeks.'[71]

With the help of Mehmed Ali and his sons, the Ottomans appeared poised to re-establish imperial order in the remaining rebellious territories. The revolutionary forces were in disarray. By 1827, the Greek rebels retained control over Nauplion and Corinth only. The Orthodox Patriarch in Istanbul was seen presenting the sultan with petitions for pardon and amnesty from his followers in many parts of Greece, requesting reinstatement as *reaya*.[72] Two events altered the outcome of the conflict. First, Mahmud II had found an opportune moment to eliminate the Janissary corps in Istanbul, on 15 June 1826, and did not press the Ottoman advantage in the Peloponnesus. The purge and reconstruction that followed his destruction of the last Janissaries left him reliant on the proxy army of Mehmed Ali in the Greek conflict, and particularly vulnerable in the Balkans.

Secondly, Britain, France and Russia intervened in the conflict. Although the three powers pursued a policy of neutrality in the Greek question until 1825, events in the Principalities brought the Russians and Ottomans perilously close to war once more, and the prospect alarmed British and French diplomats, who were increasingly sensitive to Russian imperial aims in the Caucasus and the Black Sea. The disruption of trade was a major cause of concern, as Greek pirates showed little discrimination in their choice of targets.[73] The fall of Missolonghi and Athens raised a public outcry in Europe. Tsar Alexander I died in 1825, and was succeeded by Nicholas I (1825–55), who proved of much firmer resolve concerning the Ottomans. In April 1826, the Duke of Wellington, attending the coronation of Nicholas I, concluded the Convention of St Petersburg with the newly-installed Tsar, by which they agreed to mediate in the Greek conflict. On 6 April 1826, the new tsar presented an ultimatum to the Ottomans concerning the Principalities and the Caucasus, which insisted on implementation of the clauses of the Treaty of Bucharest. Wellington erroneously assumed that would end Russian ambitions in Greece. After stalling during the first months of military reform in Istanbul, Mahmud II concluded the Convention of Akkirman with Russia in October 1826, which temporarily settled the unresolved conflicts from the treaty of Bucharest, calling for the withdrawal of Ottoman troops from the Principalities, and asserting Russian rights at the eastern end of the Black Sea. This gave Russia unparalleled access to the Bosphorus. Mahmud II too may have considered it a means of removing

Russia from the Mediterranean equation, or as a delaying tactic until he had his new army ready.

Nonetheless, the Greek conflict continued to fester, and as noted above, looked particularly bleak for the rebels in June 1827. On 6 July of that year, the Treaty of London, the official acknowledgment of the Convention of St Petersburg, was signed. By that time France, Austria and Prussia had been added. Unanimity of internal pressure was brought to bear on the Ottomans in a proposed armistice and European mediation to settle the Greek question. The European powers envisioned the initial creation of a vassal state in Greece much like the Principalities, but the sultan had already handed the Ottoman ultimatum to the diplomatic representatives in Istanbul, strongly emphasising that the Greek question was his to solve, and rejecting intervention and mediation by any foreign power.

In the *firman*, the sultan reminded the international powers that he had absolute authority over the nations subject to his dominion. Each independent power 'possesses also institutions and relations which concern only itself and its internal state, which are the offspring of its legislation and form of government.' He denounced the revolt of the Greeks, who, 'from generation to generation, have been tributary subjects to the sublime Porte, . . . have been treated like Mussulman in every respect, and as to everything which regards their property, the maintenance of their personal security, and the defence of their honour; that they have been, particularly, under the glorious reign of the present sovereign, loaded with benefits far exceeding those which their ancestors enjoyed.' The Ottomans had never refused to pardon and reinstate rebels, the statement continued, and insisted on non-intervention by foreign powers, and respect for Ottoman sovereign law. The sultan felt it necessary to reiterate that the 'Greek question' be recognised as belonging to the internal affairs of the sublime Porte. Although he recognised mediation between two independent nations, 'a reconciliation may be brought about by the interference of a third friendly power: it is the same in respect to armistices and treaties of peace. But the Sublime Porte being engaged in punishing in its own territory, and in conformity with its sacred law, such of its turbulent subjects as have revolted, how can this cause ever be made applicable to its situation? And must not the Ottoman government attribute to those who advance such propositions, views tending to give consequence to a troop of brigands?'

From this declaration, then, it is clear that Istanbul perceived the Greek Revolt as a *fitne*, a rebellion against the state, rather than as the rise of a Christian nation, at least for public consumption. Mahmud II saw restoring order (*nizam*) as his absolute right, without outside interference. For the Ottomans, the Greeks had violated the *zimmi* pact that regulated relations between Muslims and non-Muslims. Mahmud's statement of June 1827 clearly articulated the revolt in the language of the shar'ia, and warned against outside interference. In that framework, rebel non-Muslim subjects could be declared *harbis*, warring non-Muslims, and the sultan could call for a general mobilisation, *nefir-i am*, to confront a (theoretical) attack on the state.[74]

When the Ottomans refused mediation, an allied naval blockade was imposed in the eastern Mediterranean to prevent men and supplies from Egypt reaching the Peloponnesus. In October 1827, a fleet of allied ships entered Navarino Bay, where a combined Ottoman–Egyptian fleet lay anchored. On 20 October, after a fierce battle, the entire Ottoman and most of the Egyptian fleets were destroyed. A total of 57 ships sank, and 8,000 men drowned. In little over a month, Mahmud II had repudiated the Akkirman Convention, closed the Dardanelles straits to shipping, and declared war against Russia. By August 1828, Mehmed Ali had deserted the sultan, signed a convention with the allies, and withdrawn all forces from the Peloponnesus, to be replaced with French troops of occupation.

Between 1774 and 1830, the particular combination of 'nationalism, romanticism and populism – fueled by the fires of the French Revolutionary period' joined the international relations arena to bring about '. . . a new and secularized concept of crusade' which was 'pouring out of Europe. The first and most consistent recipient of this inundation was the Middle East.'[75] With Navarino, a new phase of the Eastern Question was about to unfold. It was preceded by the Ottoman response to the Christian crusade, the jihad against Russia of 1828.

Notes

1 Juchereau, de Saint-Denys, *Révolutions de Constantinople – en 1807 et 1808* (Paris: Brissot-Thivars, 1819), vol. 2, 190.

2 Juchereau, *Révolutions de Constantinople*, vol. 2, 195.

3 Avidor Levy, 'Military policy of Sultan Mahmud II, 1808–1839' (PhD, Harvard, 1968), 52.

4 B. Lewis, 'Dustūr', *EI2* 1999 CD edition; Levy, 'Military policy', 54–56.

5 Levy, 'Military policy,' 53.

6 I have followed Levy's summary of the events here, as he consulted all the contemporary accounts. Levy, 'Military policy', 65–83.

7 William C. Fuller, *Strategy and Power in Russia 1600–1914* (New York: Free Press, 1992), 204.

8 Fuller, *Strategy and Power*, 238.

9 Fuller, *Strategy and Power*, 228.

10 John L.H. Keep, *Soldiers of the Tsar: Army and Society in Russia, 1462–1871* (Oxford: Clarendon, 1985), 145. To be precise, 1,933,608. For the previous period, 1725–1801, Keep proposes the figure of 2,150,000 men.

11 Keep, *Soldiers of the Tsar*, 145–67.

12 Keep, *Soldiers of the Tsar*, 168.

13 Keep, *Soldiers of the Tsar*, 172.

14 Keep, *Soldiers of the Tsar*, 182.

15 Thomas Thornton, *The Present State of Turkey* (London, 1807), 173–74; see also Virginia H. Aksan, 'Whatever happened to the Janissaries?'

16 Nicholas I had a standing army of close to a million regulars and irregulars by the 1850s. Fuller, *Strategy and Power*, 239.

17 Fuller, *Strategy and Power*, 253.

18 Keep, *Soldiers of the Tsar*, 216–18. Suvorov had an exceptional reputation for sanguinary tactics, even for the times.

19 Fuller, *Strategy and Power*, 167. European armies had a habit of disintegration, based on the preference for large numbers of mercenaries, before the development of 'native' conscript armies.

20 George Jewsbury, 'Comte de Langeron and his role in Russia's armies against the Porte and France, 1805–14', *Proceedings, The Consortium on Revolutionary Europe 1750–1850*, vol. 1, 222–30 (Athens, GA, 1980).

21 Stanley Lane-Poole, *The Life of the Right Honourable Stratford Canning* (London: Longmans, Green, 1888) vol. I, 55–56, as quoted in Levy, 'Military policy', 112. This was during Canning's first term as the British plenipotentiary to Istanbul (1810–12). He returned as ambassador in 1825, when he was to play a large role in events leading up to and during the Crimean War.

22 Extracted from Cevdet, *Tarih*, vol. 9, 132–42. Cevdet takes pains to note that the inhabitants who had sided with the Janissary rebels got what they deserved. The number of troops can only be one estimate, probably deliberately exaggerated by Cevdet for effect.

23 Alexander Mikhailovsky-Danilevsky, *Russo-Turkish War of 1806–1812*, trans. by Alexander Mikaberidze (West Chester, OH: Nafziger Collection, 2002), vol. 1, 96.

24 Cevdet, vol. 9, 142. Cevdet is obviously creating an effect to heighten the later success of Mahmud II in subduing both Janissaries and *ayan*.

25 George F. Jewsbury, *The Russian Annexation of the Bessarabia: 1774–1818: A Study of Imperial Expansion* (Boulder, CO: Westview Press, 1976), 32.

26 Mikhailovsky-Danilevsky, *Russo-Turkish War*, vol. 2, chap. 13–14, *passim*. The plague numbers come from Jewsbury, 'Comte de Langeron', 226. Cevdet notes the brave action of Ahmed Agha, *Tarih*, vol. 9, 135–7.

27 Pehlivan's celebrated bravery and self-sacrifice earned him the sobriquet Baba Pasha (Father Pasha). The sultan awarded him with the rank of *vezir*, as Governor of Rakka. Cevdet, *Tarih*, vol. 9, 160–68. I reproduce the language of his text for effect.

28 Avigdor Levy, 'Ottoman attitudes to the rise of Balkan nationalism', in Bela Király and Gunter E. Rothenberg, eds, *War and Society in East Central Europe, vol. 1: Special Topics and Generalizations on the 18th and 19th Centuries* (New York: Columbia University Press, 1979), 333–34.

29 For example, in 1792, when he was sent to Istanbul by Catherine II to strengthen Russo-Ottoman relations and counter French influence. Constantin de Grunwald, 'Ambassade russe a Constantinople au XVIIIe siècle', *Miroir de l'Histoire* 7 (1956), 491–99.

30 Mikhailovsky-Danilevsky, *Russo-Turkish War*, vol. 2, 97–98.

31 Mikhailovsky-Danilevsky, *Russo-Turkish War*, vol. 2, 99–101.

32 Mikhailovsky-Danilevsky, *Russo-Turkish War*, vol. 2, 104–5; Cevdet, *Tarih*, vol. 9, 328–29.

33 Mikhailovsky-Danilevsky, *Russo-Turkish War*, vol. 2, 106.

34 Mikhailovsky-Danilevsky, *Russo-Turkish War*, vol. 2, 109–10, 120.

35 Mikhailovsky-Danilevsky, *Russo-Turkish War*, vol. 2, 126, drawn from Kutuzov's dispatches. We have already seen other death sentences applied to the Ottoman army, such as Rumiantsev's concerning the battle of Kartal in 1770.

36 Sean Pollock, 'Empire invitation? Russian empire-building in the reign of Catherine II', PhD (Harvard, 2006), chapter 2.

37 Cunningham, 'Stratford Canning and the Treaty of Bucharest', in *Anglo-Ottoman Encounters in the Age of Revolution: Collected Essays, vol. 1* (London: Frank Cass, 1993), 147. Canning was much more interested in procuring timber, masts, hemp, tallow, etc. for the British navy. Cunningham describes his intervention as 'useful but not decisive' (150). He was also awaiting the new ambassador, Robert Liston, who arrived in June 1812.

38 F. Ismail, 'The making of the Treaty of Bucharest, 1811–12', *Middle Eastern Studies* 15 (1979), p. 171; Allan Cunningham, 'Stratford Canning and the

Treaty of Bucharest': in the year 1811, Stratford received only four dispatches from the Foreign Office in London (146). As Cunningham concludes, 'All the evidence suggests that there would have been a treaty without him, at about the time and on precisely the terms agreed' (183).

39 Ismail, 'The making of the Treaty of Bucharest', 172–75.

40 Ismail, 'The making of the Treaty of Bucharest', 179.

41 Ismail, 'The making of the Treaty of Bucharest', 180–81.

42 Jewsbury, *Russian Annexation*, 35–37. Chichagov's idea may not have been so far-fetched considering that the French army was only half-French. The army marching on Russia was one of the largest ever assembled: 368,000 infantry, 80,000 cavalry and 1,000 guns, with troops from Holland, Italy, Poland, Spain, Austria and Germany (Fuller, *Strategy and Power*, 186). The Russians called Napoleon's Moscow campaign 'the invasion of the twelve nations', referring to the number participating in Napoleon's army. (Mikhailovsky-Danilevsky, vol. 2, 121, n 147).

43 Jewsbury, *Russian Annexation*, 48–51. As examples, from a Russian source, the income for Wallachia in 1808 was 2,737,809 *lei*, of which 1,969,968 *lei* were spent on the army of occupation. In Moldavia in 1809, of an income of 2,561,866 *lei*, 1,569,720 were spent on the army. Such statistics are simply one measure of the imposition.

44 Ursinus, 'Sirb', *EI2*, CD edition.

45 Ivo Andrić, *The Bridge on the Drina* (Chicago, IL: University of Chicago Press, 1977), 94–99, one of the most remarkable portrayals of the complexity of Bosnian life.

46 Levy, 'Ottoman attitudes', 332–36. Specifically the Ottoman governor tried to disarm the Serbian population, which touched off the second rebellion.

47 Deena R. Sadat, 'Rumeli ayanlari: the eighteenth century', *Journal of Modern History* 44 (1972), 354.

48 Levy, 'Ottoman attitudes', 340–41.

49 BA, Hatt-ı Humayun 20896, dated 1225 (1810). My thanks to Dina Khoury for drawing this document to my attention. There is a confused account of events in Tom Nieuwenhuis, *Politics and Society in Early Modern Iraq: Mamluk Pashas, Tribal Shayks and Local Rule Between 1802 and 1831* (The Hague: Martinus Nijhoffs 1982), 62–68.

50 K.E. Fleming, *The Muslim Bonaparte: Diplomacy and Orientalism in Ali Pasha's Greece* (Princeton, NJ: PUP, 1999), and E. Prevelakis, K. Kalliataki Mertikopoulou, eds *Epirus, Ali Pasha and the Greek Revolution: Consular Reports of William Meyer from Preveza* (Athens: Academy of Athens, 1996), as well as Peter Oluf Brøndsted, *Interview with Ali Pacha of Joanina in the autumn of 1812* (Athens, 1999).

51 Brøndsted, *Interview*, p. 60. The description says more about the views of the traveller than about Ali. Brøndsted was astonished at the loyalty Ali Pasha engendered among the inhabitants of the area, Muslim and non-Muslim alike. Among others appointed as envoys, D.R. Morier and William Martin Leake actually took up residence as special British envoys in 1804 and 1809 respectively, one measure of the degree of the British concern about Ali Pasha's increasing power.

52 Levy, 'Ottoman attitudes', 336.

53 Outline of story from H. Bowen, "Ali Pasha Tepedelenli', *EI2*, CD edition, and the other sources cited above. Ironically, both towering figures were connected to the Mevlevi order of dervishes, Ali Pasha by descent, Halet Efendi by affiliation with the Mevlevis of Istanbul. Mahmud II, standing in front of his tombstone in the garden of the famous Galata Mevlevi Monastery, reputedly asked *Şeyh* Kudretullah Dede: 'Efendim, what do you make of this, *our* Halet Efendi?' The response: 'Efendim, now *he* was *some thing*, but he [it] has gone [passed]!' 'Şeyhim, şu bizim Hâlet'e ne dersin? Efendim, o da bir hâlet idi, geçti!' Emphasis and translation mine – the *Şeyh* makes a play on the meaning of 'hâlet', meaning 'condition, situation, aspect', and in 'hâlet-i ruhiye', meaning 'mental attitude'. The effect is that Halet was 'some piece of work'. Quoted in M. Şihâbeddin Tekindağ, 'Hâlet Efendi', *Islam Ansiklopedisi* 5, 123–25.

54 Barbara Jelavich, *History of the Balkans: Eighteenth and Nineteenth Centuries* (Cambridge: CUP, 1983), 204–34, quoted on 210.

55 Jelavich, *History of the Balkans*, 212.

56 Hakan Erdem, 'Do not think of the Greeks as agricultural labourers: Ottoman responses to the Greek war of independence,' in Thalia G. Dragonas and Faruk Birtek, eds, *Citizenship and the Nation-State in Greece and Turkey* (London: Routledge, 2005) 67–84.

57 Jelavich, *History of the Balkans*, 212–13.

58 John L. Comstock, *History of the Greek Revolution; Compiled from Official Documents* (New York, 1828), 150–51, this excerpt taken from Edward Blaquière, noted *Philhellene* (*Letters From Greece*, London, 1828). *Philhellenism* is liberally and vaguely applied to a general enthusiasm for all things classical in Europe, but more specifically to those who volunteered in the Peloponnesus, at the siege of Missolonghi in 1826, and of Athens in May–June 1827.

59 Jelavich, *History of the Balkans*, 217.

60 Richard Clogg, *A Short History of Greece* (Cambridge: Cambridge University Press, 1992), 36–37. The classic work on the revolution is George Finlay, *History of the Greek Revolution and the Reign of King Otho*, 2 vol. in 1.

(Oxford: Clarendon, 1877, reprint London: Zeno, 1971). On the death of the patriach, see vol. 1, 187–88; on Chios see vol. 2, 254–56.

61 Erdem, 'Do not think', 67–84.

62 Khaled Fahmy, 'The era of Muhammad 'Ali Pasha', in *Cambridge History of Egypt* (Cambridge: CUP, 1998), vol. 2, 139.

63 Howard Alexander Reed, 'The destruction of the Janissaries by Mahmud II in June 1826' (PhD, Princeton, 1951) , 86–7.

64 J.L. Comstock, *History*, 372.

65 Allan Cunningham, 'Stratford Canning, Mahmud II, and Greece', in *Anglo-Ottoman Encounters in the Age of Revolution: Collected Essays*, vol. 1 (London: Frank Cass, 1993), 289.

66 Comstock, *History*, 386. Mehmed Ali contributed 50,000 troops to the enterprise in the three years Egypt was in the Morea (396). Byron died at Missolonghi on 19 April 1824.

67 Cunningham, 'Stratford Canning, Mahmud II, and Greece', 290. A total of 3,500 of Missolonghi survivors were sent to Egypt, to be redeemed by European merchants and 'Turkish grandees' (298); Finlay, *History*, vol. 1, 394–97.

68 Thomas Douglas Whitcombe, *Campaign of the Falieri and Piraeus in the Year 1827* (Princeton, NJ: Princeton University Press, 1992), 4–5, included various estimates of dead and imprisoned. Most of those taken prisoner were beheaded.

69 Whitcombe, *Campaign*, 103, 105. The *Deli* here refers to the cavalry of that name, who took the most serious risks, but it was also applied to irregular cavalry in general.

70 Vasfı Efendi appears to be Kurdish himself. Hardly a polished piece of literature, it remains a rare bit of autobiography of a common Ottoman professional soldier. His escapade was a private enterprise, and evoked no discipline other than a scolding. Jan Schmidt, 'The adventure of an Ottoman horseman: the autobiography of Kabudlı Vasfī Efendi, 1800–1825', in *The Joys of Philology: Studies in Ottoman Literature, History and Orientalism (1500–1923)* (Istanbul: Isis Press, 2002), vol. 1, 284–5.

71 Whitcombe, *Campaign*, 7; Finlay, *History*, vol. 1, 431–35, commended the restraint shown by Reşid Pasha after the fall of Athens.

72 Comstock, *History*, 405. Cunningham, 'Stratford Canning, Mahmud II, and Greece', 312.

73 Afaf Lutfi al-Sayyid Marsot, *Egypt in the Reign of Muhammad Ali.* (Cambridge: Cambridge University Press, 1984), 208.

74 The translation is from Comstock, *History*, 406–14, dated here 9–10 June 1827, but paraphrased by me. It is clear that Mahmud II considered the British largely to blame for supporting the insurgents, and the continuation of resistance.

75 L. Carl Brown, *International Politics and the Middle East: Old Rules, Dangerous Game* (Princeton, NJ: Princeton University Press, 1984), 29.

CHAPTER EIGHT

The Auspicious Occasion and its consequences

Before continuing the story of the transformation of the Ottoman military, we have to return to Egypt, and understand how Mehmed Ali Pasha (1760s–1849) came to challenge his own sultan on the plains of Anatolia. That will be followed by a description of the events around the 'Auspicious Occasion', which was Mahmud II's final confrontation with the Janissaries on the streets of Istanbul, and his general reforms that unfolded thereafter. A discussion of the Russian–Ottoman war of 1828–29, the first test of the new military organisation, will be followed by a brief assessment of the state of the empire before we move on to a description of the important Treaty of Adrianople, and the two campaigns against Mehmed Ali in Syria and Anatolia in the following chapter. The contest of the two titans of reform, one in Istanbul and one in Cairo, engendered a particular style of international intervention in Middle Eastern affairs that has persisted to the present.

Mehmed Ali Pasha and the transformation of Egypt

Mehmed Ali rose to power as a participant in the defence of Alexandria and Cairo against the French after 1798. Born in the late 1760s, at Kavala in Macedonia, Mehmed Ali was the son of an Albanian Ottoman soldier. In 1801, he was appointed deputy commander of the Kavala contingent to the army of Egypt. By mid-1803, he was commander of his own Albanian army, part of the forces assembled for the 1801–02 campaign. While the newly-appointed Governor of Egypt, Hüsrev Pasha, contended with his rival Tahir Pasha, commander of the combined Ottoman Albanian force, Mehmed Ali consolidated his own power and prepared to confront the mamluk households. In 1803, a riot over pay broke out among the Ottoman Albanian troops, who drove Hüsrev out of Cairo. Tahir Pasha was soon assassinated by the mamluks he had called on for support. Mehmed Ali then assumed command of the Ottoman Albanians as well, defeated Hüsrev Pasha's army at Damietta, and was appointed governor himself in 1805, following unsuccessful attempts by Istanbul to impose another Ottoman governor on Egypt.[1] Selim III acquiesced to the *fait accompli*. It was his recognition of the coalition Mehmed Ali had constructed with the backing of the military and the ulema, as well as the merchants of Cairo. It could not have been easy in the chaos of military occupation and reoccupation by both French and Ottoman forces, a period recognised by all who study it as the nadir of modern Egyptian history. Mehmed Ali's reputation was further bolstered when the British forces landed at Alexandria in March 1807 and were trapped there after their attempt to occupy Rosetta.[2] Selim III's deposition and death allowed Mehmed Ali a respite from internal challengers. He consolidated power while awaiting his moment, which came in March 1811 when he confronted and eliminated a number of the mamluk Beys in Cairo, by which he acquired independence of action in Egypt. Survivors were incorporated into Mehmed Ali's new forces. The process of suppression of the traditional forces, and the support of the residents of the city, presaged the events which would unfold in Istanbul in 1826. Mehmed Ali was particularly aided by Umar Makram, the *Nakib al-Ashraf* (Chief of the Muslim Notables), who was persuaded to return from exile in Palestine to assist in the elimination of the last mamluk chieftains.[3]

Between 1811 and 1824, when called upon by the sultan to come to his aid against the Greeks in Morea, Mehmed Ali undertook a series of striking and innovative reforms that completely reorganised Egyptian society. This has led some historians to see him as the founder of the modern nation of Egypt.[4] Centralising finances, introducing conscription, monopolising export and import trade, and experimenting with industrialisation were all part of his programme of consolidation and reform. In 1809, he experimented by taxing *rakf* lands, challenging his erstwhile allies among the ulema, and exiling his former supporter Umar Makram. He began as early as 1810 to reorganise the agricultural system by introducing direct taxation and administration at the provincial level. Between 1812 and 1815, Mehmed Ali revoked the tax farms (*iltizam*), and increased taxes on *rakf* lands. By 1821, a cadastral survey, registration and the imposition of the new system of taxation had been achieved. Sixty per cent of the territory of Egypt was brought under the new tax regime by these efforts. Mehmed Ali created government monopolies of all of Egypt's cash crops: grains, rice, sugar, and notably cotton. War in Europe proved a boon to his agricultural reforms, as the demand for grain during the Napoleonic wars made Mehmed Ali a wealthy man. He also introduced a number of long-staple varieties of cotton, increasing the crop yield. In this, as with his other reforms, he worked by negotiation and building coalitions with trusted associates who formed the core groups of Ottoman–Egyptian elites of the nineteenth century. The idea of extending his power to Syria was hatched in this early period as well. As has been argued in previous chapters, control of greater Syria was a preoccupation of Egyptian rulers. Mehmed Ali would prove the most able to carry out his expansionist schemes.

Factories were also built in this period, as a means of achieving self-sufficiency in certain goods as well as supplying the army. Arsenals and textile factories, sugar refineries, rice mills and tanneries were all established, starting in 1815. Within two decades, however, most of them were closed or had collapsed because of technical and operational difficulties. The net effect was to facilitate Egypt's integration into the world market as a supplier of raw cotton to the textile mills of Britain and the United States.[5] Nonetheless, by 1820 Mehmed Ali can be said to have achieved a robust economic base, and sufficient state revenues to underwrite his next series of adventures.

The Wahhabi challenge and Mehmed Ali's new army

Mehmed Ali's reforms of the Egyptian military system remain the most important to this narrative. His first step was to eliminate the mamluk leadership, as described above, in early 1811. In this, he was following in the footsteps of the Ottoman Egyptian governor-appointees, who, on numerous previous occasions, had attempted to eliminate the well-entrenched organisation. Mahmud II, the new sultan, ordered his governor that same year to subdue the Wahhabis in Arabia, who had seized the holy cities of Medina and Mecca.[6] The Wahhabis, named for Muhammad ibn 'Abd al-Wahhāb (1703–92), followed the Hanbali school of Muslim practice, 'Abd al-Wahhāb further advocated that all non-conformists to the shar'ia, such as Sufis or Shiites, were heretics, and open to attack. This included the Ottoman dynasty, considered illegitimate by 'Abd al-Wahhāb and his successors. Selim III and Mahmud II both were confronted with the vigorous revolt of the Wahhabis and their adherents, the Arab tribes of the Najd and Hijaz, especially the family of Ibn Saud, who established a theocracy in Diriyah, near present-day Riyadh, that proved hard to eradicate.[7] The contest in Syria centred on imperial Muslim legitimacy, and the promotion of the Hanafi school of law preferred by the Ottomans. One of the main tenets of Ottoman legitimacy was the support and protection of the annual pilgrimage caravan from Damascus to Mecca and Medina, also a source of revenue for province and sultan alike. This ideology was severely tested by the Wahhabis, when the pilgrimage caravan and holy cities alike came under attack.

Once Mehmed Ali was confirmed as the legitimate governor of Egypt, he was obligated to serve the sultan in putting an end to the Wahhabi threat in the Hijaz. Without a doubt, it also served his own purposes. Sending his second son, Tusun Pasha, as commander of the assembled forces, Mehmed Ali recaptured both Mecca and Medina. Over the next two decades, his sons Tusun and İbrahim, especially the latter, would lead military expeditions to the Hijaz, Sudan, Crete, the Morea, and greater Syria.

The Wahhabis, however, proved a stubborn foe. Tusun was initially bested by the Wahhabis, but Mecca and Medina were regained by 1813, at a cost of 8,000 men, 25,000 animals and 170,000 purses. When the keys of both cities were presented to the sultan in 1813, Mahmud II and Istanbul rejoiced. Mehmed Ali's campaign to subdue the Wahhabis completely, however, would not be accomplished until 1818, when his elder son Ibrahim Pasha's army defeated them in Najd. Perhaps his father's later ambitions regarding expansion into Syria arose as a result of that expedition. The Wahhabi leader Abdullah was sent to Istanbul where he was promptly executed, and Diriyah was razed. Ibrahim was appointed governor of the Hijaz.[8] Although the Hijaz remained under Ottoman hegemony, Najd was reoccupied by the Wahhabis in 1824. Wahhabism continued to influence the religious debates of greater Syria, stimulating many to question the authority of the sultan, and allowing in part for the ease with which Mehmed Ali and his son would conquer and occupy the Ottoman territories there after 1830. Mehmed Ali held Yemen and the Hijaz until 1840.

Mehmed Ali's achievements are remarkable, and no more so in the striking reorganisation of the military, an idea generally acknowledged to have been his. All his other reforms were subservient to the creation of a conscription army and a navy in the Mediterranean and the Black Sea, which, in Fahmy's view, had made a garrison state of Egypt by the time of his confrontation with the Ottoman army in Anatolia in 1831. Perhaps it was the experience of seeing French soldiers on Egyptian soil that originally stimulated Mehmed Ali's interest, or his knowledge of the *Nizâm-ı Cedid* reforms of Selim III. He had likely seen Hüsrev Pasha, part of the Anglo-Ottoman coalition against Napoleon in 1801/02, commanding a *Nizâm-ı Cedid* regiment. Mehmed Ali demonstrated such knowledge in his treatment of his own Albanian forces, who had repeatedly proven their unreliability and inability to maintain order and discipline in Cairo. In 1815, Mehmed Ali explicitly ordered the unruly Albanians to begin target practice and drill in the manner of the *Nizâm-ı Cedid*. His orders precipitated a revolt which he had great difficulty bringing under control. He rid himself of the nuisance by repeatedly sending Albanian troops to the Arabian desert in the seven-year struggle with the Wahhabis.[9] By 1822, the date of the first conscription order, Mehmed Ali was instructing his advisers to draft new organisational plans on the model of Selim's *Nizâm-ı Cedid*. (Universal conscription eluded the Ottomans until the detailed regulations of 1846.)[10] Mehmed Ali's decision to turn to the conscription of Egyptian peasants rested on the experience of his army in Sudan from 1820–22.

His fame having grown as a result of the restoration of the holy cities to the sultan, Mehmed Ali had turned his attention to the conquest of the Sudan, where he intended the elimination of a few surviving mamluks, and the consolidation of his control. Two aspects of this campaign influenced later developments. First, Mahmud II imposed an embargo on the shipment of mercenaries and slaves to Egypt after 1820, which likely had an influence on his ability to re-man his depleted armies.[11] Second, Mehmed Ali's Turkish-speaking forces deserted in droves from the intolerable conditions of the Sudan. Hence, a manpower crisis was likely important to Mehmed Ali's decision to mobilise a Sudanese slave army, some 6,000 strong, following the occupation of Semnar, Kurdufan and Dongola, the latter the stronghold of a number of mamluk emirs.

There are important differences between the army sent to the Hijaz to quell the Wahhabis, and the army that resulted from two expeditions to the Sudan in the summer of 1820. The former was contractual and confederative, as were the expeditions in 1820, made up of Turkish, Arab and Albanian cavalry and infantry mercenaries and commanded by their own warlords. The new standing infantry army was to be paid from government revenues, through direct taxation, and commanded by non-native, i.e. non-Egyptian (or Arab-speaking) officers, trained in European style. This initiative was paralleled and supported by the construction of a

PLATE 14 *Mehmed Ali's reformed army uniforms (Yates, William Holt,* The Modern History and Condition of Egypt, its Climate, Diseases and Capabilities with an account of the proceedings of Mohammed Ali Pascha from 1801 to 1843. *London: Smith, Elder & Co., 1843).*

new navy. By 1820, shipyards all over the Mediterranean were supplying Mehmed Ali with his new vessels.[12]

The expeditions to the Sudan proved an unmitigated disaster. A major revolt in 1822 resulted in the death of Commander İsmail Pasha, Mehmed Ali's third son. The Sudanese population proved unable to survive enslavement and transportation. After four years in the Sudan, of 20,000 Sudanese troops collected by the army, only 3,000 were still alive.[13] The expeditionary armies themselves fell ill of dysentery and other fevers, and grew mutinous. As a result, the conscription of the peasants (the *fellahin*) of Upper Egypt began, although the Sudanese continued to form the mainstay of the army in training in Aswan in 1822 under Mehmed Bey Lazoğlu. A second training camp was created outside Farshut in northern Egypt, with 4,000 recruits from among the *fellahin*. Service was set at three years, and replacements were to be drawn from village census rolls resulting from the new land surveys and tax registers. In reality, conscription was brutal and haphazard, and the length of service was set at 15 years only in 1835.[14]

Simultaneously, Mehmed Ali organised an officer training school of his own slaves and Turkish-speaking soldiers of the old army, requesting teachers from his agent in Istanbul, Necib Efendi. He also sought the advice and assistance of the French, notably Colonel Sèves, known as Süleyman Pasha, who rose to rank of chief-of-staff to İbrahim Pasha's campaign in Syria in 1831.[15] As early as 1816, Sèves began training 500 selected youths at a barracks in Aswan.[16] Initially, however, all non-foreign commanders were appointed from among ethnic Turks. Arabs could not rise above the rank of captain (*yüzbaşı*). As Mehmed Ali once explained: 'I have not done in Egypt anything except what the British are doing in India: they have an army composed of Indians and ruled by British officers, and I have an army composed of Arabs ruled by Turkish officers . . . The Turk makes a better officer, since he knows that he is entitled to rule, while the Arab feels that the Turk is better than him in that respect.'[17] Rank-and-file conscripts resented the discrimination bitterly. Native officer training was escalated in 1827, under the direction of the French mission. Only in 1828, when Mehmed Ali openly defied Mahmud II by not sending the troops to the Danubian front as requested, and dismissed some 700–800 Turkish-speaking officers, were new native graduates of the school commissioned.[18] Still, the highest ranking officers remained Turkish-speaking even during the Syrian campaign.

Mehmed Ali's militarisation of Egyptian society did not go uncontested. In 1823, a large revolt against the new taxes and conscription broke out in lower Egypt, which was only defeated by troops led by the Pasha himself. In 1824, some 30,000 men and women participated in a similar rebellion in upper Egypt. Mehmed Ali sent the new troops to quell the revolt successfully after only two weeks, leaving behind 4,000 casualties. He also used them against Wahhabi resurgence in the Hijaz, again with some success. In March 1824, a powder magazine exploded in the Cairo citadel, killing some 4,000 people. A battalion of the new troops put down the putative rebellion, rumoured to be the work of disgruntled Albanians. Hence, Mehmed Ali and his new army were already tested as a domestic force before the call came from Mahmud II.[19] We will return to the state of Mehmed Ali's army in 1831–40, during the campaigns of Mehmed Ali's greatest aggression against the sultan.

In 1824, however, a desperate Mahmud II, facing defeat in the Peloponnesus, offered Mehmed Ali Crete, Cyprus and the Morea if he would eliminate the Greek threat to the Ottomans, the outcome of which was described in the previous chapter.[20] Mehmed Ali's intervention in Greece gave Mahmud II the respite he needed to transform his own army. Given the extent of the threat to the dynasty in the latter days of 1826, it is remarkable that the sultan moved forward with the reform programme at all.

The Auspicious Occasion, 1826

Charles Macfarlane, resident in Istanbul in these crucial years, provided a description of Mahmud II shortly after the elimination of the Janissaries:

Instead of melancholy and a doomed man, I remarked an expression of firmness and self-confidence, and of haughtiness not unmixed with a degree of ferocity. His lift and orientally arched eye-brows, his large coal-black eyes, (which are habitually however rather heavy than otherwise,) his thick black beard and mustachoes, which completely veil the expression of the lower features, the lordly carriage of his head, are all calculated to strike, and coincide perfectly, with our picturesque idea of an eastern despot. He had become fond of military training exercises, and commanded his own squadron of horse. He was renowned for his prowess with the bow and arrow. His transition from 'the habitual life of the sultan to the life of a sort of Frederick the Great', has rather improved his general health than otherwise. In the summer of 1827, he was continually out with his tacticoes at Daut Pasha.[21]

It was but a brief moment of triumph. Mahmud II had successfully eliminated the gravest of his internal threats, the Janissary Corps, and temporarily checked the Greek Revolution, but ahead lay the disaster of Navarino, in October 1827, when the entire Ottoman–Egyptian fleet sank to the bottom of the Mediterranean. Ahead also lay the Russo-Ottoman war over the Principalities, 1828–29, and the first major confrontation with Mehmed Ali's son, İbrahim Pasha, in the plains of Anatolia in 1831. Both would severely test the newly formed army, the *Asakir-i Mansure-yi Muhammadiye*, on the battlefield. How did Mahmud II manage it?

PLATE 15 *Sultan Mahmud II, 1829 (Macfarlane, Charles,* Constantinople in 1828: A residence of sixteen months in the Turkish capital and provinces, with an appendix to the autumn of 1829. *2nd edn. 2 vols. London: Saunders and Otley, 1829).*

Halet Efendi, engineer of the savage repression of the countryside following 1810, was blamed for causing the Greek revolt in 1821, and dismissed and banished by 1822. With Halet's subsequent execution, the sultan ended what might be construed as a regency, and came into his own. From 1822 forward, Mahmud II appointed the most able men around him as his advisers, but kept a tight rein on state affairs, removing rapidly those who opposed him or proved unable to carry out his orders. When the Janissaries challenged the sultan in the streets of Istanbul, he matched violence for violence. Mini-revolts became an excuse for the shakedown of senior officers of the corps, and the dismissal of those who opposed the sultan's designs. Mahmud II removed many of the trouble-makers in the city by first exiling them one by one to a distant province, and subsequently ordering a secret execution of the offender by the provincial governor. By these ruthless means he contained much of the corruption and disorder, and may have produced a degree of loyalty to his reform mission, but he also eliminated most of the experienced cadre of officers around him, with grave consequences on the battlefield. Moreover, the policy did not achieve its aim of controlling the Janissaries. Thinning the ranks did produce Hüseyin Agha, who rose through the hierarchy to be appointed as Commander of the Janissaries in February 1823. He proved to have the mettle in the struggle with the corps almost immediately, removing suspected adversaries from the rank and file and banishing them in large numbers. For his efforts, he was awarded the honorary rank of vizier. Hüseyin hence came to be known as Agha Pasha. When his harsh measures provoked a reaction in the corps, Mahmud II removed him from the post, but made him commander of the Bosphorus forts, as well as Governor of Bursa and İzmit, close by when required. When two grand viziers, Deli Abdullah (also known as Hamdullah Pasha), and Silahdar Ali Pasha, proved incapable of disciplining the Janissaries, they were relieved of command in quick succession. In December 1823, Mahmud II turned to his reliable servant, the aged Mehmed Said Galib Pasha, and appointed him Grand Vizier. Galib, it will be remembered, had survived the riots and purges around Selim III and Alemdar Mustafa, and returned to help to negotiate the 1812 Treaty of Bucharest. Galib was secretly informed of Mahmud's plans for the elimination of the Janissaries, and recommended the appointment of a more seasoned soldier for the task than himself, alighting on Benderli Mehmed Selim Pasha, then Governor of Silistre. Mahmud sent for him, and Mehmed Selim arrived in Istanbul disguised as a Tatar messenger. He was appointed Grand Vizier on 14 September, 1824.[22]

While searching for the appropriate instrument for the destruction of the troublesome Janissaries, Mahmud II continued focusing his attention on his artillery, doubling the size of the corps in Istanbul to 10,000 gunners and 4,400 drivers by 1826. Within the larger organisation, a small mobile artillery regiment, which had been almost annihilated in the

1806–12 war, was reorganised and continued its training on western models. After the war, that regiment's strength was increased to around 1,000 officers and men with 70 light field pieces.[23] The artillery forces were well paid, and allowed long leaves home, all part of Mahmud II's effort to earn their loyalty.[24] These precautions of the sultan bore fruit, as the regiment stood by him in 1826.

Hüsrev Pasha rebuilt the Danube flotilla while Admiral of the Black Sea fleet from 1811–18. It was composed of 20 sloops and 80 large *sandals* (rowboats) divided into four equal forces and moored at İbrail, Silistre, Rusçuk and Vidin, with 400 men attached to each fortress, for a total of 1,600 men. After 1821, when Greek sailors were eliminated from the Ottoman fleet, Muslims from North Africa and Christian Cossacks from Silistre, some 500 of them, were enlisted and brought to Istanbul in 1825 upon the suggestion of Grand Vizier Benderli Mehmed. These initiatives would prove of importance to the war effort in 1828 and 1829 against Russia.

The sultan also directed the strengthening of the fortification system after 1812, on the Danube, Bosphorus and the Dardanelles. The military arsenals were reorganised, gunpowder production stepped up, and large quantities of arms, such as 50,000 muskets from Liège, were imported from Europe in early 1826. Thus, when June arrived, Mahmud II had shored up his defences and secured a cadre of loyal officers and bureaucrats in order to see him through the final confrontation with the Janissaries. They included Celaluddin Mehmed Agha, appointed Commander of the Janissaries and the last to hold that office, in September 1825. Admiral Hüsrev Pasha, implacable foe of Halet Efendi, alike, was reappointed in 1822, after Halet Efendi's exile, and detailed to the Morea upon the outbreak of the Greek revolt. Dede Mustafa Agha, known to be steadfast and capable, was appointed in February 1826 as Commander of the Bombardiers.[25]

By the spring of 1826, public opinion was in favour of reform as well. The successful siege of Missolonghi had proven the mettle of Egyptian Governor Mehmed Ali's reformed troops, as contrasted with the rag-tag Ottoman army. Many observers, native and foreign alike, asserted that it was clear that discipline and training gave the Egyptian army an edge. The population of Istanbul must have been heartily sick of the constant upheavals in the streets of the city. In 1821, for example, when the leader of the Greek revolt, Ypsilanti, invaded Moldavia, 5,000 Janissaries were recruited, and sent to the Danubian Principalities to put down the uprising. They started their 'campaign' before they had left the Istanbul district of Beşiktaş, destroying the houses of Christians on their march out of the city.[26]

Thus, Mahmud II had carefully paved the way for the confrontation by staffing key positions with his most trusted officers. The Janissaries themselves had alienated most of their potential loyal backers in Istanbul. Still, Mahmud II had to acquire the approbation of the ulema, notably of the office of the *Şeyhülislam*. In November 1825, Mahmud II appointed Mehmed Tahir Efendi to the post, known to favour the reforms the sultan intended. Earlier in the year, an influential Muslim official in Istanbul, by the name of Halim Efendi, began speaking out against the disorderly Janissaries. Once word of his disapproval reached the Corps, his banishment was demanded. Mahmud II acquiesced in his banishment, but as Cevdet commented: '. . . there were many who, in their hearts, were pleased . . . [as] everyone had become exasperated with the evil state and disorder of the corps.'[27]

Nor did Mahmud II neglect the propaganda necessary to convince the public at large about the necessity of his reforms. Mehmed Esad Efendi, himself a member of the ulema, was appointed as official historian in September 1825 and later commissioned to write a history called *Üss-ü Zafer* (*The Foundation of Victory*), which was published shortly following the events of 15 June 1826. It serves as one of the primary sources concerning the final destruction of the Janissaries.[28] Mahmud II also ordered the writing and distribution of books on the life of the Prophet Muhammad, which may have been intended to appease the ulema as well as the public.[29]

Formation of the Eşkinci Corps, May 1826

Ambassador Stratford Canning witnessed the last quarterly payment ceremony of the Janissaries on 30 March 1826. Power in the Istanbul barracks still lay with those who controlled the pay tickets of their fellow Istanbul officers and men. Plans were very carefully laid by the sultan to forestall the kind of resistance which had overcome his predecessor Selim III. At the end of May, Grand Vizier Mehmed Selim Pasha, *Şeyhülislam* Mehmed Tahir Efendi, and Janissary Agha Mehmed Celaluddin assembled prominent members of the ulema, and presented the sultan's programme of reform to them. The arguments focused on the disorder and lack of discipline in the Janissaries. Apart from the decay of the corps itself, they argued, foreign spies had infiltrated their ranks. They spread the 'disease' of sedition, and the 'fever' of anarchy. Many previous sultans, they continued, had endeavoured to restore discipline to the infantry, notably Mahmud I (1730–54), who died before he could finish his undertaking, and Abdülhamid I (1774–89) who was forced to confront internal disorder in Egypt, but also that engendered by the unruly *levends* in Morea. Selim III, who set about his reforms too hastily, saw his new troops defeated by the Janissaries. Hence, they concluded, the only way to succeed was to begin the reorganisation from within the existing corps. They proposed the formation of squadrons of soldiers, called *Eşkinci*, from within the Janissaries themselves. The assembled ulema agreed to the scheme, unanimously approving the action by ruling that it was incumbent upon Muslims to study military science. The officers of the Janissaries pledged the cooperation of the members of the *ocak*.[30]

All was in readiness to enlarge the circle of participants. On 28 May 1826, 66 Ottoman officials and ulema, as well as the grand vizier and other members of the bureaucracy, assembled at the headquarters of the *Şeyhülislam*. Of those present, 34 were prominent religious officials. The grand vizier addressed the assembly concerning the necessity for reform. Among his arguments: (1) that Muslim soldiers had fallen into a state of permanent revolt; (2) that they fled in the face of the enemy, bringing shame and dishonour upon the community; (3) that the corps had been infiltrated by all kinds of impostors, especially those so-called *reaya*, the Greek rebels (*eşkıya-yi Rum*). They had spread the rumour that the war against the rebels was intended to suppress the corps rather than suppress a revolt of the *reaya*. Such rumours had paralysed the zeal of the Janissaries. (4) The weakness of the religious spirit and inattention to the ancient military laws had led to a decadence within the corps. (5) Muslim soldiers no longer paid any attention to religious duties, to the merit of being a 'warrior for the faith', or to the obligation to obey their superiors. The grand vizier concluded by exhorting those assembled to choose the methods to redeem the glory of the empire. The *Reisülküttab* spoke of the external threat to the empire on the Danube, and the unjust demands of the foreign powers vis-à-vis the Greek insurrection. The ulema followed one by one, reiterating their approbation of the study of military sciences, and the necessity of obeying the sultan. Finally, Hasan Agha, Janissary *Kul Kahyası* (Lieutenant General), exhorted his assembled cohort of his readiness to accept the reforms. They too acclaimed the remedy. Once the approval had been secured on all sides, the new *Eşkinci* regulations were read out loud. This was followed by drawing up an affidavit of commitment to the reforms, as well as a *fetva* submitted by the *Şeyhülislam*.[31] Signed by all present, it was then carried to Janissary headquarters. There, all three documents were read again, and further signatures were attached, some 209 in all. Approximately two-thirds of the signatories were Janissary officers.[32]

The document itself is very carefully prefaced to legitimise reform and military discipline as a religious duty, and as conforming with the Qur'an. Then a long, rambling text refers to the new regulations. The *Eşkincis* were to be drawn from the existing *ortas* of the Janissaries: 150 men from each of the 51 *ortas* in Istanbul. Hence, 51 new companies (or 7,650 men) were to be established, each with seven officers, listed here in rank order, beginning with a Captain (*Çorbacı*), a Lieutenant (*Odabaşı*), Paymaster, Standard Bearer, Commissary officer, Top or Master Sergeant (*Başkarakollukcu*), and a Sergeant. For each of the companies, 15 corporals (*karakollukcus*) would be selected from among the 150, in other words, one for every nine enlisted men. Each company was also assigned a surgeon and a prayer leader. Salaries ranged from 66–67 *kuruş* per quarter year for an enlisted man to 750 *kuruş* per quarter for the Captain. The Janissary high command structure remained in place, but a Corps Superintendant (*Nazır*) was to oversee the *Eşkincis*. Rank and promotion were determined not by merit so much as by vacancies, and movement through the hierarchy described. Captains who did not move on into the high command would be retired after six years. Pensions were assigned to retirees, the ill and the wounded.

Each soldier was to be issued a musket and sabre, traditional weapons. The corporals were to be issued a crimson waistcoat, baggy trousers and a pair of red gaiters, as well as a fur cap. Other uniforms were not regulated. Rations were also regulated, in an attempt to remove the monopolies and irregularities of the traditional corps. The trade in pay tickets was strictly forbidden. The document is clear on the intention to round up the tickets of retired soldiers, while acknowledging that those of active Janissaries would continue to be paid. The *Eşkincis* were to be paid in the presence of their commanding officers, but only after a roll call.

Drill was regulated in the document. The main intention in establishing the troops, according to the text, was to 'train them in the art of war, requiring regular and constant practice.'[33] Companies were to drill in rotation, and target practice was also mandated. Discipline was to be strictly enforced. The men were to be present in their barracks at all times. Leaves were also to be strictly regulated, and those who did not return were to be reported and sought. Guard duty was mandated and specified for each of the new companies. Considerable detail was included on the daily routine of the new trainees, obviously one of the sources of discontent with the Janissaries.[34]

Missing from these regulations are the details about the specific responsibilities of the various officers, and any definition of the kind and extent of punishments to be meted out for violation of the new regulations. Most historians agree that the *eşkinci* reform project was an overhaul of the Janissary regulations rather than a radical break with the past, but it is worth remembering how consensual this process was, with Mahmud II insistent on acquiring the approval of large numbers of both military and religious officials. Even so, on the documents submitted to him on 28 May, he noted that the potential for rebellion still existed.[35]

In spite of the careful preparations, rumours circulated widely among the Janissaries that Mahmud II sought their destruction. They appear to have been especially unhappy about the new regulations concerning the pay tickets, as well they might, for by one estimate some 135,000–145,000 names were on the rolls, while not more than 10,000 were on duty in Istanbul.[36] On 30 May 1826, orders were sent to the Janissary Agha to organise the reformed troops, and within a few days, their number had climbed to 5,000 men. On 12 June 1826, a small number of the new troops, among them the captains and senior officers of the new organisation, demonstrated the drill in front of Ottoman high officials and members of the ulema.

But all was not well with the Janissaries, including one or two of the officers supposedly convinced to join the reforms, who were among the instigators to rebel in the corps. The government, aware of the unrest, hastened to assure existing Janissaries that their pay tickets would continue to be honoured. On Wednesday, 14 June 1826, Janissaries began to assemble in Et Meydanı, the traditional square for signalling a rebellion. They first attacked and burned the Janissary Agha's headquarters. According to one source, their numbers grew to 20,000 or 25,000, suggesting that parts of the population joined in. The residence of Necib Efendi, Mehmed Ali's representative in Istanbul and a known reformer, was attacked, as was that of the Grand Vizier, the Sublime Porte. Both men escaped. Loyal auxiliary corps, artillery, bombardiers, sappers and marines, combat ready, were ordered to the palace. The ulema sanctioned force for putting down the rebellion. Mahmud II picked up the sacred banner, wishing to lead the troops in person, but was persuaded otherwise. In the event, the sultan, having prepared the ground well, bested his opponents for the moral high ground and found large parts of the city's inhabitants willing to fight with him. The assembled Janissaries found themselves cornered in their barracks in Et Meydanı, as troops of mounted artillery under Hüseyin Agha Pasha approached from the south via Divan Yolu, and another column, under İzzet Mehmed Pasha, advanced from the northwest. A citizens' group was thrown together and sent as a third column. After a fierce confrontation, and adroit use of the artillery, the barracks were invested, and began to burn. By mid-afternoon 15 June, the rebellion was over. The Sultan had managed to eliminate the dreaded rabble on a day which became known as the 'Blessed Affair' or 'Auspicious Occasion' (*Vak'a-yi hayriye*).[37]

On 16 June 1826, when Sultan Mahmud attended Friday prayers, he had already replaced his Janissary escort with a special guard of the new troops, which signalled to all his commitment to the new regime. His careful, studied approach had achieved the opening for a complete transformation of the Ottoman order. When Selim III had tried the same thing with his *Nizâm-ı Cedid* troops, his action contributed to the revolt which brought down his reign.[38] But Ottoman officials wasted no time in eliminating possible pockets of resistance. That same day, two courts were set up to interrogate the surviving Janissary prisoners, one by the Grand Vizier in the Hippodrome, and the other by the Janissary Agha at his headquarters and 300 were summarily executed. The day after that, 17 June 1826, a great council was assembled to formally abolish the Janissary Corps and replace it with the *Muallem Asakir-i Mansure-yi Muhammadiye*, the Trained Triumphant Soldiers of Muhammad, hereafter *Asakir-i Mansure*.[39] Henceforth the name Janissary was to be abolished, and the former Janissary Agha became the *Serasker*, or Commander-in-Chief. This was accomplished by the reading of a public edict by Esad Efendi, who later published the text in full.[40] He reported that he 'kissed the precious banner of the Prophet with trembling lips, retired respectfully a few steps down the pulpit staircase, faced the immense crowd of Muslims and read aloud the *firman* ordaining the destruction of the Janissary corps, their name and insignia, whose existence had desecrated the temple of Islam for too long a period.'

'As all Muslim people know, the Ottoman state owes its creation and its late conquest of the East and the West solely to the powerful influence of Islamic religious spirit, Muslim law, the sword of *jihad* and the *gazi* spirit,' the document began. It continued by reviewing the history of the reforms, from the time of Selim III, casting the blame on the Janissaries who refused to submit. 'This had given the enemies of the Ottoman state the opportunity to advance against us . . . The Janissaries had revolted one last time: They shouted, "We don't want any drill," and, in spite of the holy sanctions, the proclamation, their sworn allegiances, and the sacred *fetvas*, they turned the weapons given them by the state against the government of the Muslim people and broke out into rebellion against their legitimate sultan. Today they are nothing more than a useless and insubordinate body which has become the asylum of the spirit of unrest and seditions in which the number of evil men have outgrown the number of good ones . . . Among those who have just been executed were found some Janissaries who bore tattooed on their arm along with the insignia of the 75th [Janissary] mess, the cross of the ghiaurs. This simply proves that infidel traitors, parading in the disguise of Muslims, have for a long time been using the Janissary corps to further their own nefarious ends by spreading false rumours . . . Hence, let all the congregation of Muslim people, and the small and great officials of Islam and the ulema, and members of other military formations, and all the common folk, be of one body. Let them look upon each other as brethren in faith. Let there be no difference and contrariness between you. Let the great ones among you look with a merciful and compassionate eye upon the little ones, and let the minor ones, moreover, in every instance be obedient and submissive.' The document ended by exhorting the various officials and imams to spread the word and instruct the true believers to thank the state for its beneficent intentions on this occasion, and to live quietly and contentedly in the shadow of the power of the sultan."[41] Thus, the corps was wiped from the pages of Ottoman history, the reformers pausing only to blame the 'enemy-in-our-midst', the infidel infiltrators. We will examine the implications of that for the reorganised army in due course.

Elimination of the Bektashis

The proclamation made for wonderful theatre, and in many ways presaged the Gülhane rescript text inaugurating the Tanzimat thirteen years later. It is, however, also a straightforward statement of Mahmud II's Ottoman-Muslim absolutism, no matter how much it was a gesture to reassure the people of Istanbul of the legitimacy of his actions. The document also shows the influence of new strains of religious thought in Ottoman ruling circles, one of numerous claims that accompanied diverse peoples exiled into central Ottoman territories, particularly from the Caucasus. Especially influential were the teachings of the Naqshbandi *tekkes* (lodges), whose sheikhs first made a significant appearance in elite ranks in the late eighteenth century, and who had already become advisers to Selim III. By the 1820s, their teachings had become part of popular discourse in Istanbul, and were extremely influential in the latter half of the nineteenth century especially in the inner circles of the sultans following Mahmud II. The rather more strictly defined Muslim orthodoxy of the Naqshbandis clashed with more tolerant (liberal or heterodox) Sufi organisations such as the Bektashi, installed as the chaplains of the Janissary Corps in the sixteenth century, or the Mevlevis, followers of Rumi's teachings, who served as spiritual advisers to many of the sultans from the seventeenth century forward. Sultans Abdülmecid (1837–61) and Abdülaziz (1861–76) are said to have embraced an ideology based on an Ottoman/Naqshbandi synthesis which emerged first during the reign of Mahmud II. The passage into court circles was via influential court imams but also as part of the household slave system, which produced individuals such as the ubiquitous Hüsrev Pasha, himself from Abkhazia, and his protegés, many of whom came from the 'Turkic' Caucasus and later became the prominent statesmen of the Tanzimat.[42]

The teachings that spread to the *tekkes* of Istanbul were likely those of Shaykh Khalid. They advocated a strict emphasis on the *sunna* and the shar'ia, the supremacy of the shar'ia in the life of the community and state, hostility to Shiites, and especially to non-Muslims. Among those with links to the order in the inner circle of Mahmud II were Şeyhülislam Mustafa Asım Efendi, the aforementioned Hüsrev Pasha, Pertev Pasha, and Mustafa Reşid Pasha, later grand vizier, and chief architect of many of the Tanzimat reforms.[43] The prominence of Naqshbandi thought probably exerted an influence on the banning of the Bektashi order, which occurred three weeks after the elimination of the Janissaries. Hajji Bektash was the patron saint of the corps; new inductees to Janissary regiments were required to swear by the twelve imams, and on the virtue of 'Ali as well as the light of the Prophet. Closely intertwined as they were with the Janissaries, they were no doubt doomed from the day the decision was made to eliminate the Janissaries themselves.[44]

Şeyhülislam Tahir Efendi, Mahmud II's chosen enforcer, convened a meeting of top ulema and sheikhs of eleven Sufi orders: Naqshbandi, Qadiri, Halveti and Mevlevi among them, but not the Bektashi. He opened the meeting with a lengthy diatribe against the Bektashi, and asked for evidence of their heresy, in what was essentially an inquisition. Those in attendance were reluctant to give evidence one way or the other, so encountering no significant objection, Tahir Efendi proceeded to supply the Sultan with a report that the *tekkes* should be shut down, and some 25 Bektashis should be banished to Anatolia. Three main Bektashi leaders were executed, accused of fomenting schism and heresy, of being libertarians and heretics. Their tolerance of religious diversity in light of the Greek revolt must have been extremely irksome. The extremity of the language argues for the reiteration of orthodox ideology in Istanbul and the palace.[45]

Istanbul was also purged of potential troublemakers among the *yamaks*, who were stationed in the Bosphorus. They were replaced with loyal artillerymen. Other units such as the standing cavalry were disbanded and reformed. The former Janissary corps who remained loyal, such as the artillerymen (*topçu*), the bombardiers, armourers (*çebeci*), etc., were not demolished but reorganised. Many loyal Janissary officers were rewarded for their support of the sultan, but not given commissions in the new army.[46] Boatmen, watchmen, and the corps of firemen, traditional supporters of Janissary uprisings, were also disbanded and exiled from the city. By the order of the sultan, the Armenian patriarch was to assemble 10,000 of his community to serve as firemen (the *Tulumbacıs*). As early as 26 June, 2,000 of them had been organised, and replaced the Janissaries hitherto assigned.[47] Given also the upheavals surrounding the Greek community in Istanbul following the 1821 uprising, it must have been an extraordinary period of re-sorting of ethnic categories in the capital. Nor were residents spared the usual fires. On 31 August 1826, the worst fire in a century broke out in the city. The blaze, likely arson set by disgruntled ex-corpsmen, destroyed private and public buildings alike, with losses estimated at three million pounds by Stratford Canning.[48]

Systematic obliteration of sedition continued throughout Ottoman territories. Mahmud's preparations were well laid. Special couriers were sent in advance to governors of provinces with large Janissary contingents, but they appear to have capitulated to the new regime with little difficulty, with some exceptions, such as Bosnia.[49] In major concentrations of Janissaries such as Vidin, Edirne, and Erzurum, resistance was subdued with banishment or execution of a few key figures. The venerable Galib Pasha, then commander of the Eastern frontier, reputedly addressed the troops at Erzurum saying that 'he did not wish to subdue them by force, that he would require at least 10,000 men to do this, and that he had but 500; that if they did not want him to carry out his orders, he was ready to retire; that poor, infirm, devoted to his master, he did not dread his vengeance, but that he was much more concerned about the fate of a province he had administered, and which, if it should resist, would have to combat all the forces of the empire at a time when it was already menaced by a war in progress on its frontiers.'[50] Surprise and speed were instrumental in establishing the new order, but could not remove widespread resentment, which influenced the ability to conscript troops for the new army. Such purges, however limited, also further reduced the experienced manpower base.

One provincial reaction, by a soldier and Ottoman official in Damascus named Hasan Agha, illustrates the success of the sultan's message: 'It was reported to us that there was a great incident in Istanbul wherein approximately twenty thousand people perished. The disbanding of the *ocak* was a great event because many rulers and sultans had wanted to do this but were not able . . . God granted the great Sultan, king of the kings of the Arabs and non-Arabs, the just ruler Mahmud Khan . . . the capacity to disband them, because they had rebelled against kings and viziers and the lords of the state to the point where they were feared by the old and the young, had come to hold decisive power, and the Sultan had become their ward. They did as they pleased and the wealth of the Ottoman lands became theirs. The fact is that the disbanding of this corps was a mercy from God to the country and to the people. The janissary corps did no good, but only harm.' Hasan Agha may have once been a Janissary himself.[51]

The realities of reconstruction

There are two aspects of these preliminary stages of the transformation which merit comment. One is the financial state of the empire; the other, the question of leadership and Mahmud II's understanding of it for the new army. Without cash, rebuilding an effective army would prove exceedingly difficult. Şevket Pamuk has recently asserted that Mahmud II's era experienced a steep rise in inflation, greater than any other in the empire's history. He has calculated that prices increased 12–15 times between the late eighteenth century and 1850, attributing it to the state's debasement policies. After the elimination of the Janissaries, the most formidable obstacle to devaluation of their fixed salaries, debasement was applied liberally, 'reducing the specie content of the *kuruş* by 79 per cent within a period of four years.'[52] Furthermore, in spite of pledging to abolish the practice of confiscation of the estates of deceased officials (known as *müsâdere*), Mahmud II did not hesitate to execute three of Istanbul's wealthiest Jewish bankers, who were moneylenders and suppliers to the Janissaries, and seize their estates, worth an estimated 108 million *kuruş*, at a time when annual revenue by one estimate was some 200 million *kuruş*.[53]

The practice of devaluation of coinage, coupled with unjust confiscation of wealth and new taxes, had the net effect of further crippling the economy and alienating generations of Ottoman subjects. The cycle of indebtedness led to foreign intervention in Ottoman fiscal affairs by mid-century. Efforts were initially made to recognise the *esame* of surviving Janissaries and officials, but the scrutiny and cross-examination by Hüseyin Agha Pasha was so severe that many thought it prudent simply to surrender them, hence saving the government more money.[54] An export tax of 12% on silk, and a stamp duty of 2.5% on imported manufactures, were other means Mahmud II employed. Some effort was made to attack the problem of tax farming, and to confiscate *waqfs*, the charitable foundations (tax havens) of the ulema, but it was still insufficient. As Lieven puts it: 'More than four-fifths of the state's revenue in one way or another [stuck] to the fingers of those who collected it.'[55]

The second major problem was the lack of foresight about the formation of the new army. There are several explanations for that: the speed of the transformation, for one. Secondly, unlike Mehmed Ali, Mahmud II resisted the use of foreign advisers. Early on, in August 1826, he requested Mehmed Ali to send twelve officers to Istanbul to aid in the training of the new army. His governor, whose own forces were engaged in the Peloponnesus, and whose own reforms had not been underway for long, refused the appeal. Mahmud II could not, and just as likely did not want to, pay what Mehmed Ali did for his French officers and technical advisers. Colonel Sèves received 17,500 *kuruş* a month from Mehmed Ali. By contrast, another Frenchman, Gaillard, veteran of Napoleon's armies, was paid 1,200 *kuruş* per month as Mahmud's chief infantry officer. Others were paid even less, obviously a disincentive for particularly bright and ambitious military minds. Those who were employed were of lower ranks, non-commissioned officers, even privates, with little experience of command. They proved particularly susceptible to manipulation by the Ottoman pashas, where the real problem of command lay.[56]

The new regulations of 7 July, described below, specified that officers be appointed based on ability and seniority, rather than favouritism and venality, which had been the practice with the Janissaries. In practice, favouritism continued, in large part because of the household system I have alluded to previously, but also because that is the way Mahmud II understood military command. Senior officers were picked from a select social group, the imperial household. The imperial household was staffed, as we saw with Hüsrev Pasha, by *kuls*, brought up in their patrons' households, and groomed for palace and imperial service. Although his was an exceptional household because of its size (of 50 male *kuls*, 30 became pashas, grand viziers and ministers of the late Ottoman state), it was typical.[57] These vast organisations maintained ties throughout the empire, and their structures were imitated in the main provincial cities of the empire. The difference between the sultan's household and the courts of St Petersburg and Vienna, where favouritism and privilege were not in short supply, was in the ferocity of the factionalism among the large households, and the vulnerability of service, subject to the imperial will.

Mahmud II hence assumed that the officer corps would emerge from his own court. Members of the households, such as that of Hüsrev Pasha, were given senior commissions in the new army. Junior officers, on the other hand, were randomly picked from among new recruits, and put into service with little training. Immediately after the elimination of the old corps, Mahmud II set up the Court Battalion of the Aghas of the Inner Service, which was an attempt at an officer training school. Slaves and sons of courtiers served together, trained by the Italian Calosso, who was paid 2,000 *kuruş* per month, the largest salary given a foreigner by Mahmud.[58] By fall 1827, the battalion reached its maximum size of 400, half infantry and half cavalry. These are the troops that Macfarlane observed. The system did produce officers with a certain uniformity, with loyalty to the sultan, but promotion was rapid, and military tactics and strategy were absent from the curriculum. In 1830, Mahmud II abolished the court battalion, irritated because the courtier–trainees proved difficult to discipline. After repeated battlefield failures of military command, an effort was made to create a more professional cadre of officers, including examinations for promotion, and a system of merit which presupposed that officers began service as privates. Mahmud II himself violated his own principles, however, by appointing his own courtier, and son-in-law, Said Mehmed Efendi, to the rank of brigadier-general in 1833. The sultan was not above, perhaps was even a master of, the fractional politics of his court.

A French observer noted the casual training of the officer class in 1834, characterised by little instruction and high levels of illiteracy. In that same year, however, an officers' training school was established, the School for Military Science (*Mekteb-i Ulum-i Harbiye*), with cadets numbering around 400 by 1839. Regular classes only got fully under way in 1836, but apparently the curriculum remained restricted to Arabic, French, history, a bit of arithmetic, and infantry drill. Namık Pasha, Ottoman official and important reformer under four sultans, was the founder of the school. In fact, schools for naval and land engineering, emerging from Selim III's School of Mathematics, and a medical school for army surgeons (1827) were already in operation. The new officer school quickly created an officer caste, as senior officers were given honours and governorships, as well as pensions and benefits for surviving family members. Officer training would ultimately serve as the single most important modernising force of the empire.[59] In this period, however, rivalries and factionalism remained the order of the day, and are generally accorded to be the real cause for the later failure of the *Asakir-i Mansure* army against Mehmed Ali's army in Anatolia.

Asakir-i Mansure regulations 7 July 1826

After what has been said about speed and the lack of preparation, it will come as no surprise that the first detailed list of regulations for the new army replicated, in many respects, the *Nizâm-ı Cedid* of Selim III. The basic organisation of the corps looked much the same: 12,000 troops, divided into eight regiments (*tertib*, later *alay*), each with a *binbaşı*. Each regiment was to have 1,527 men (as contrasted with the 1,602 of Selim III), that is, divided into twelve musket companies with twelve cannons. Six of such companies constituted what normally amounts to a battalion, called here an arm, or wing (*kol*), and commanded by a *kol aghası* (major), two for each regiment, the left and right. The senior officer (*baş binbaşı*) of the 12,000-man force was paid 1,500 *kuruş* a month. He reported to the *Serasker*, who was paid 25,000, and to the *Nazir*, paid 7,500 *kuruş*. Among other officers and enlisted men, a *binbaşı* was paid 750; a *kol aghası* 400; a *yüzbaşı* (captain) of a company 180; a corporal 30, and a private 20 *kuruş*. Salaries were to be paid monthly (every three months in the provinces), with regular updating of the muster rolls. Recruits were given two daily meals of soup and bread or stew, with pilav on Sunday and Thursday nights. Uniforms, to be manufactured locally, were distributed once yearly, with shoes.[60] By May 1827, the fez, with express approval of the ulema, became the standard headgear. The crescent and star were adopted as the insignia of rank at the same time. Muskets, bayonets and cartridge boxes became standard issue, with the sabre reserved for officers. Recruitment age was 15–30 years, with twelve years as the term for pensionable service.

The regulations emphasised discipline, and prevention of desertion. Enlisted men were allowed home leave, or to perform the hajj; officers were not. Imams were appointed as chaplains to each regiment to lead prayers, and a mekteb was established to teach the basics of Islam. The critical shortage of surgeons and physicians, and of money to pay Christian doctors sufficiently, forced the establishment of the medical school mentioned above. A similar problem existed with the engineering schools, those with the longest history. It would be a number of years before the first graduates could be enrolled in the army. Each regiment was assigned craftsmen such as blacksmith, gunsmith, saddlemaker, six in all, who were considered military personnel. The army bands were reorganised in March 1827 to include clarinets, and in September 1828, Giuseppe Donizetti Pasha, brother to the better-known composer, began his 28 years of service with the Ottomans.[61] Hence the surprising western flavour to many of the military tunes of the transformed army.

Two of the new *Asakir-i Mansure* regiments remained at *Serasker* headquarters, and also served as Istanbul police; the other six were to be housed at new barracks at Üsküdar and Davut Paşa. The Superintendent's staff combined the duties of commissariat, munitions and paymaster combined. It was quickly reorganised in 1827 to resemble a more modern division of labour, with eight separate offices to cover a range of duties, such as military supply, provisions, fodder, the *yoklama* (or muster) office and a general treasury. Oversight of these operations was included as part of the regulations.[62]

The first provincial regiment was established in Edirne by order from Istanbul on 4 August 1826. It was based on the Istanbul model. Recruiting was to be local; officers sent from Istanbul would command. Plans were made for five regiments (of 1,527 men as above) in the Balkans: two in Edirne, and one each in Silistre, Vidin and Salonika. Another five were to be established in Anatolia: Erzurum, Çankırı, Bolu, Kütahya and Izmir. The colonel of the provincial regiments was under the command of the military offices in Istanbul, but also obligated to the governor in matters of local policy. Roll call was monthly; payments were directly in cash, sent from Istanbul, or written off provincial tax revenues.[63]

Further regulations 1826–27

The 7 July 1826 regulations, modelled on Selim's *Nızäm-ı Cedid*, resembled neither Mehmed Ali's new army formations nor those of any other European nation. Yet it is important to note that these regulations, ideas carried forward from previous attempts and promulgated in a period of intense crisis, served as the backbone of all military regulations thereafter. While Hüseyin Agha had loyally carried out the sultan's orders, he was essentially a Janissary, and the press of time and disorder, coupled with lack of imagination, had forced the creation of an outmoded hybrid infantry. Admiral Hüsrev Pasha, however, was an experienced reformer of the Ottoman navy, who had recently been dismissed from command in the Peloponnesus by Mahmud II at the insistence of Mehmed Ali. Hüsrev and Ibrahim Pasha had clashed constantly over command of the naval forces in the Mediterranean. Hence, in February 1827 Hüsrev was in Istanbul, temporarily unemployed but accompanied by his infantry battalion, which had been organised and trained by Gaillard, mentioned above, along French and Egyptian lines. On 8 May 1827, Hüsrev was appointed as *Serasker*, as a recognition of his abilities as much as a signal to an increasingly presumptuous Mehmed Ali.

The new regimental structure was organised around the French battalion (called *tabur*), much smaller than the company unit of the 7 July regulations. Napoleon's formations were small and mobile, with infantry and artillery separate. Each battalion had around 800, and was divided into eight musket companies, now called *bölüks*. Companies were divided into small squads, again eight in number. More officers were added, and a European-style pyramid of command instituted. Three battalions formed one regiment, now called an *alay*, commanded by a colonel (*miralay*) and a lieutenant colonel (*kaymakam*). Such was Hüsrev's influence on these developments, that three-line firing drill, already in place in the 7 July regulations, became known as 'the drill of Hüsrev's men'.[64]

The reform was accomplished rapidly, and without the necessary approbation of the ulema. In Istanbul ten new battalions resulted, after removing artillery and transport elements. In the provinces 21 new *Asakir-i Mansure* battalions were raised. Galib Pasha produced another one in Erzurum at his own request in late September 1827. The aim was to have a core force of about 25,000 men, but eventually to assemble an army of 100,000. By October of that same year, a regiment of the Hüsrev reorganisation performed the drill on the parade grounds of Davut Paşa. Full regimental organisation, however, was not completely in place until 1831. The army which marched to Şumnu in spring 1829 was arranged in four regiments of five (rather than three) battalions each, largely because of the shortage of competent officers. Notably, three of the four newly-appointed colonels were former *kuls* of Hüsrev Pasha.[65]

To concentrate on the infantry reforms is to neglect the other branches of the military system. Briefly, the *Bostancı* palace guards were abolished on 31 August 1826, and replaced with a European-style guard called the *Hassa*. In February 1829, they numbered 3,500 officers and men and remained in Istanbul during the 1828–29 war. The palace cavalry proved too costly to reform initially, so a provincial *Asakir-i Mansure* regiment was established at Silistre in November 1826, for a number of reasons: it was less expensive; fodder and horses were more accessible; the fortress was on a very sensitive military frontier, and Dobruja was full of noted horsemen – Tatar and Turkish tribesmen and Christian Zaporozhian

Cossacks, who had settled there after 1775. They were already irregulars for the Ottomans, performing guard duty and supplying men and animals. Several hundred served in the Danube flotilla. The inspiration for the move came from Grand Vizier Benderli Mehmed Selim, familiar with the region as a former governor of Silistre. The officers, in contrast to the infantry regiments, were to be recruited locally from the ethnic groups listed above. The troops were organised into companies of Turks or Cossacks, respectively, on the left wing, while the Tatars served on the right wing. Each wing was commanded by its ethnic officers, with a special additional office of *Kazak Başı* for a Cossack officer. The regiment numbers stood at 1,323 men, of which 109 were officers. Of the total number, 329 were to be Cossacks. Daily rations were specified for bread, meat, barley and straw, with pay much the same as with the new infantry companies. Both priests and imams were assigned to the companies. Restrictions on the Cossacks initially included barring them from rising above the rank of troop captain, and from commanding mixed formations. They were generally paid less, received half the rations and inferior uniforms, but were allowed to wear their distinctive headgear, the *kalpak*. The Muslim cavalrymen wore a *şubara*, a rounded-crown cap, as did their brothers in the infantry until the adoption of the fez in 1828.

Horses were a particular problem as they were so expensive: Cossack and Turkic militia communities received a one-time payment of 150 *kuruş* for every horse and saddle set, and thereafter were expected to supply the others without compensation. Tatars were expected to supply their own horses. As a saddle set alone cost 125 *kuruş* in Istanbul, it is easy to see why the government felt it prudent to start the reorganisation on the frontier. These companies were asked to serve four months, and to be combat ready for the rest of the year. When the Hüsrev reforms came into effect, the Silistre regiment was likewise reformed, and Christian and Muslim compensation equalised. In anticipation of the war, the size increased to two *tabur* of 884 officers and men each. By 1830 a cavalry battalion (*tabur*) became a regiment (*alay*) with the 884 officers and men divided into six troops. The commander became a *miralay*. These troops participated in the Danube fighting in 1828–29, with some defections to the Russians. After 1829, there appears to have been a decline in the numbers of Cossack soldiers, perhaps also based on the demographic changes in Dobruja, although such mixed regiments continued to serve throughout the 1830s as regular full-time cavalry attached to the infantry. Cossacks continued intermittently in the Ottoman army, but were disbanded under Abdülhamid II. The organisation of the Cossacks remains an exception to the general rule in the attempt to create a modernised, religiously and ethnically mixed cavalry force much as the Russians and Austrians had achieved.[66]

In February 1827, a cavalry regiment was organised in Istanbul at the time of the Hüsrev reforms. The new regiment was attached to the *Asakir-i Mansure*, under administration of the *Serasker*, and assigned 24 artillerymen and 16 craftsmen. It was expected that the regiment would number 1,582 officers and men. Salaries were set higher than for the infantry, because of the trooper's responsibility for his horse, and service ten years rather twelve. A cavalryman earned 24 *kuruş* per month, as opposed to the 20 *kuruş* for the infantryman.[67] Further reforms and expansion of the cavalry would wait until after the 1829 Treaty of Adrianople was signed.

Mahmud II was especially careful with his artillerymen, the *topçus* and *arabacıs* (gunners and waggoners), who were his primary reason for being alive, and continued to perform police duties for the sultan in an unstable moment of transition. At the time of the reforms of Hüsrev in July 1827, they stood at 82 batteries of artillery and 22 transport companies, the 20,000 mentioned above, of which half were in Istanbul. Mahmud had also raised a regiment of mounted, light artillery, of 70 mounted field pieces and 60 teams of transport men. Rather than attack the loyal organisations head-on, Mahmud II at first gradually removed the cannoneers from Tophane to other locations. In August 1826, he had replaced them with an *Asakir-i Mansure* infantry regiment, to which was assigned an artillery and transport unit: 122 artillery and 62 transport officers and men with 12 cannons. Hence he effectively merged the two branches of the military under the *Serasker*, but left the old corps intact. The Superintendent of the combined corps was responsible for day-to-day management with the commanders of each corps. The officer staff retained former positions, but their authority was reduced, as was the size of the total force, from 82 to 60 batteries and from 32 to 30 transport companies. The 60 artillery batteries were divided into two brigades: a left and a right wing. Much of the former system was preserved, including promotion by seniority rather than merit. The two new artillery brigades, of 30 batteries each, totalled 7,244, with one battery composed of 120 men and two cannons.

The mounted artillery, much the most modern of Mahmud II's military, were reorganised as two battalions, with transport and cannoneers merged as 3.5 gun companies, and 3 transport companies, and made to resemble the *Asakir-i Mansure* organisation. Drilling and training were mandatory for all, as with the infantry. A gunner received 40 *kuruş* a month. Much the same reorganisation occurred in the transport companies, with the estimated total force of all, including the mounted artillery at 12,924.[68]

It was deemed prudent to concentrate men and cannons in strategic forts, so many of the artillerymen were consolidated in strongholds on the Danube and the Aegean. A new armoury and munitions organisation grew out of the old *cebehane*, and many of the auxiliary service corps underwent some form of overhaul. Here as elsewhere, Mahmud II was thorough, but prevented from truly extensive reforms because of lack of money. In late November 1827, a month after the naval disaster at Navarino and not quite two years from the 'Auspicious Occasion', 51,000 was reported as actual total empire-wide troop strength, of which 29,000 were *Asakir-i Mansure* regiments. These included the regiments raised in the countryside, in Anatolia especially, where recruitment seems to have gone most smoothly. Even as late of 1829, the second year of campaigning against Russia, the complement of the *Asakir-i Mansure* never topped 40,000.[69]

Most foreign eye-witnesses of the period, admiring the fortitude of Mahmud II, were contemptuous nonetheless of the condition of the military, especially the artillery as we have already seen, and even more especially of the fortress organisation. Reverend Walsh, pastor of the British Embassy in Istanbul, writing in late 1827, noted that 'the Turkish empire seems just now in a perilous state of imbecility. The old military destroyed, the new unorganised; their courage subdued, their attachment alienated.'[70]

Moltke's well-known description of the new army bears repeating: 'men disciplined after the European fashion wearing Russian jackets and Turkish trowsers; with Tartar saddles, and French stirrups, and English sabres; it consisted of Timariots, or troops giving feudal service [*sipahis*]; troops of the line, whose service was for life; and of militia, who served only a term of years, of whom the leaders were recruits and the recruits mere children. The system of organisation was French, and the instructors were men from all parts of Europe. The splendid appearance, the beautiful arms, the reckless bravery of the former Moslem horde, had disappeared; but yet this new army had one quality which placed it above the numerous host which in former times the Porte could summon to the field – it obeyed.'[71]

Another foreign observer, with the troops at Rusçuk in April 1827, noted: 'The Turks showed the greatest curiosity respecting our arms; and one or two of them, taking hold of a rifle . . . went through the manual and platoon exercise as taught by Europeans. They then asked me to do the same, which I did, to their great satisfaction. They next begged me to show them how to face and wheel . . . and [I] put them through various maneuvers to their great delight. I was much surprised with the quickness they showed in learning this; for at first, they did not know how to face or wheel, but stepping backwards or forwards too far, always came violently in contact with each other. The aptitude they showed for martial exercises, made me tremble for the success of the cause in the Morea [Greek revolution]; for should the Turks once take cheerfully to the new system of organization, Europe will find them more formidable than she had ever known them to be.'[72]

In describing the garrison line on the Danube, Moltke remarked acutely that apart from the large towns, such as Vidin, Rusçuk, Silistre or İbrail, the majority of the fortresses were in very bad repair, relying on the local inhabitants for defence. 'They make up for want of outworks by a skilful use of the dry ditch, and their most vigorous defence commonly begins at the point where with European troops it usually ends – from the moment when a practicable breach has been effected. With us a large number of wealthy householders are a serious impediment to a protracted defence of a fortress; but in Turkey it is quite the reverse; every man capable of bearing arms is a soldier, and makes his appearance on the walls daily.'[73]

As I have argued previously, such territories were, in fact, defended when necessary not just by the inhabitants, but by regional armies commanded by local strongmen who participated at will and abandoned the battlefront equally wilfully. That system remained untouched by the new regulations, and in fact may have stimulated its continued existence, as Mahmud had to rely on provincial irregulars to support the new army in the new campaign against Russia. Reform of the provincial military forces had to wait until 1834, with the establishment of the *redifs*.

Mahmud II had first tried to recapture the fiefs, but the impending war forced him to reverse the policy. He retained the *timarlı sipahis*, as semi-regular cavalry, called the *Asakir-i Mansure-yi Muhammadiye Suvarisi*, by a February 1828 regulation which envisioned a battalion of 889 peacetime officers and men. For the war with Russia, battalions were increased in size to 1,087 officers and men by the end of the war. The troopers would continue to be assigned fiefs, but based on the modernised organisation model and applied to all fief holders in Anatolia and Rumeli. Fiefs were held to a maximum annual income of 3,500 *kuruş*, and all deeds/licenses (*berats*) had to be approved by the new *Asakir-i Mansure*

PLATE 16 *Uniforms of the* Asakir-i Mansure *of Mahmud II after the adoption of trousers (Chevket Pacha (Mahmond)*, L'organisation et ces uniformes de l'amée Ottomanne (Depuis sa création jusqu'à nos jours) *Vol. 1, 1900).*

administration in Istanbul. An enlisted trooper earned (or was entitled to) 750 *kuruş* per month. Initially introduced in the Balkans and Anatolia, it was extended to Damascus, Aleppo, Maraş and Malatya in late 1828, with Aleppo the headquarters and officers appointed from the Court Battalion in Istanbul. During the 1828–29 war, the Ottomans could not reorganise and register more than ten per cent of the estimated 30,000 feudatories.

For the Balkans, attempts were made to revive the *Evlad-i Fatihan* (Sons of the Conquerors) and the *Subaşı* organisations. The former, first established in the seventeenth century and based largely in Macedonia, was restored by order in March 1828. Previously an estimated 1,113 *Evlad* infantrymen could be called to arms, and the community escaped taxation during wartime but otherwise paid a collective substitute tax in peacetime, some 133,018.5 *kuruş*. The new regulations called for four battalions of 814 each, or 1,200,000 *kuruş*, tripling the community burden. They were exempted from most other taxes. An individual *evlad* earned 15 *kuruş* a month. Here, as elsewhere, a combination of new models and traditional style leadership and organisation pertained. In spite of the increased manpower and financial burdens, the Ottoman had more success with this organisation – two of the battalions served on the Russian front, and two others remained in Macedonia.

Similarly, the *Subaşıs* were reorganised in January 1828. Formerly officers of *timarlı* horsemen, the new *Subaşıs* served as police for local districts in many of the Danube provinces, and were supported by a household tax – every 50 houses to one *subaşı*. The governor of the province of Silistre organised his own force of 890 men on that basis; Vidin similarly supplied 600 men from the districts of Vidin and Niğbolu. There was no apparent restriction by religion.[74]

By late 1827, the sultan's ambitious reform agenda had resulted in a remarkable reordering of society, but was still at the elementary stage which had only begun to have any significant impact outside western Anatolia and the Balkans. In reality Mahmud II and his advisers had just scratched the surface, as would become painfully clear when the Ottomans forced a confrontation with the Russians, the subject of the next chapter.

Notes

1 One of the extraordinary sub-plots of this period of Ottoman history is the rivalry of Mehmed Ali and Hüsrev Pasha, the latter so important to

the reforms in Istanbul. They were still rivals in the final confrontation between Mehmed Ali and Mahmud II in 1839–41.

2 William Holt Yates, *The Modern History and Condition of Egypt; Its Climate, Diseases and Capabilities with an Account of the Proceedings of Mohammed Ali Pascha from 1801 to 1843, 2 vols* (London: Smith Elder and Co., 1843), 172.

3 The story of the massacre has been embellished out of all proportion, as is typical of the Oriental tales that became a favourite of the European literary marketplace. Mehmed Ali reportedly sent the ears and heads of 64 individuals to the sultan. (Afat Lutfial al-Sayyid, Marsot, *Egypt in the Reign of Muhammad Ali*, Cambridge: Cambridge University Press, 2007, 72.) By another estimate, Mamluk power had been reduced by death and flight from 10,000–12,000 fighters to 1,200 by 1802, and a further 450 *emirs* were killed by Mehmed Ali in 1811 (Darrell Dykstra, 'The French occupation of Egypt', in M.W. Daly, ed. *Cambridge History of Egypt* (Cambridge: Cambridge University Press, 1988), vol. 2, 136).

4 Khalid Fahmy, *All the Pasha's Men: Mehmed Ali, His Army and the Making of Modern Egypt* (Cambridge: CUP, 1997), 12–25, offers a critique of the historiography. His view is that Mehmed Ali was a vassal to the Ottoman sultan, and negotiated his acquisition and maintenance of power in those terms. Mehmed Ali's life and record suffer from the 'Oriental despot' stereotypes – he is known as 'the old Spider in his Den', from an 1836 description by a visitor (Fahmy, *All the Pasha's Men*, 3–6). Perhaps the most startling aspect of Mehmed Ali is that he apparently did not learn to read and write until he was about the age of 50 (Fahmy, *All the Pasha's Men*, 77). Egypt under Mehmed Ali is the subject of multiple studies, notably that of Marsot, and Fred H. Lawson, *The Social Origins of Egyptian Expansionism During the Muhammad 'Ali Period* (New York: Columbia University Press, 1992).

5 E.R. Toledano, 'Muhammad 'Ali Pasha', *EI2*, CD edition, tells the full story in greater detail.

6 Toledano, *EI2*, CD edition.

7 H. Laoust, 'Ibn 'Abd al-Wahhāb', *EI2*, CD edn, notes the oath of loyalty between Muhammad ibn Su'ūd and 'Abd al-Wahhāb in 1744, which established the link between the dynasty and the particular Muslim sect of present-day Saudi Arabia.

8 Marsot, *Egypt*, 200.

9 Fahmy, *All the Pasha's Men*, 76–86.

10 Erik J. Zürcher, 'The Ottoman conscription system in theory and practice, 1844–1918', in E.J. Zürcher, ed., *Arming the State: Military Conscription in*

The Middle East and Central Asia, 1775–1918 (London: I.B. Tauris, 1999), 79–94. He actually has the date 1848; the law dates from 1846.

11 Marsot, *Egypt*, 205. For a full description of the campaigns, see P.M. Holt, *A Modern History of the Sudan* (London: Weidenfeld and Nicholson, 1967), 35–48.

12 Lawson, *Social Origins*, 102–03.

13 Khaled Fahmy, 'The nation and its deserters: conscription in Mehmed Ali's Egypt', in Zürcher, *Arming the State*, 64; As another example, slightly over half of some 2,400 slaves shipped from Aswan arrived in Cairo alive: Fahmy, *All the Pasha's Men*, 86–111.

14 Khaled Fahmy, 'The era of Muhammad 'Ali Pasha, 1805–48', in M.W. Daly, ed., *Cambridge History of Egypt*, vol. 2, 163.

15 Toledano, 'Muhammad Ali', 424–27. Marsot, *Egypt*, 78, notes the presence also of Greek merchants and Italian doctors in the list of Mehmed Ali's advisers.

16 Avigdor Levy, 'The officer corps in Sultan Mahmud II's new army, 1826–39', *International Journal of Middle East Studies* 2 (1971), 21.

17 Fahmy, 'Conscription in Mehmed Ali's Egypt', 61.

18 Marsot, *Egypt*, 221.

19 Fahmy, 'Conscription in Mehmed Ali's Egypt', 65–67.

20 The recommendation may have come from Necib Efendi, Mehmed Ali's eyes in Istanbul and superintendent of the gunpowder factory of the arsenal. See Howard Alexander Reed, 'The destruction of the Janissaries by Mahmud II in June, 1826', PhD (Princeton, 1951), 85.

21 Charles Macfarlane, *Constantinople in 1828: A Residence of Sixteen Months in the Turkish Capital and Provinces*, 2nd edn (London: Saunders and Otley, 1829), vol. 1, 500–5. This observation was made after 1826, when the sultan came into his own as a ruler.

22 Some versions report that Galib Pasha opposed Mahmud II's plans and was replaced. The story of the destruction of the Janissaries has been told many times over, but seldom with much detail or analysis. My source here is Avigdor Levy, 'Military policy of Sultan Mahmud II, 1808–39', PhD (Harvard) 1968, 108–21.

23 Levy, 'Military policy', 102.

24 Reed, 'Destruction', 105 – 5,000 piastres (*kuruş*) was distributed to each company in April 1826.

25 Levy, 'Military policy', 107–11. Elsewhere, Levy says that some 20,000 men composed the artillery corps (15,000 artillerymen; 5,000 in transport companies in 1827, 268–70).

26 Cevdet, vol. 11 (1st edn), 211, recorded in Levy, 'Military policy', 114.

27 Reed, 'Destruction', 92, quotes the Cevdet passage.

28 Mehmed Esad Efendi, *Üss-ü Zafer* [Istanbul], 1827, published later in French translation, called *Précis historique de la destruction du corps des Janissaires par le sultan Mahmoud, en 1826*, traduit du turc par A.P. Caussin de Perceval (Paris, 1833). It is to be read as it was intended: as a piece of justificatory propaganda. The events are described as well in Mehmed Esad's history, *Vak'a-Nüvis Es'ad Efendi Tarihi, 1821–26* (Istanbul, 2002), which includes appendices written by Bâhir Efendi. The title *Üss-ü Zafer* itself is a chronogram for the year 1241/1826.

29 Reed, 'Destruction', 97.

30 Reed, 'Destruction', 108–9. The reference to Egypt is to the revolt of Ali Bey el-Kebir, predecessor to Mehmed Ali. The Morea events unfolded in the midst of the 1768–74 war when Albanian *levends* were used to suppress a revolt.

31 Levy, 'Military policy', 125–31; Reed, 'Destruction', 125–29; Esad, *Üss-ü Zafer*, 14–22 transl. Caussin, ch. 2. The events of 28 May 1826 are best summarised in Levy, 'Military policy', 129–31. The *hüccet* is reproduced in İ. Hakkı Uzunçarşılı, *Osmanlı Devleti Teşkilâtından Kapukulu Ocakları* (Ankara: Türk Tarih Kurumu, 1944), vol. II, plates 35 and 36. The document deliberately refers to *Eşkinci* as a former term for 'active Janissaries'.

32 Reed, 'Destruction', 155; Levy, 'Military policy', 131.

33 Reed, 'Destruction', 134.

34 Levy, 'Military policy', 132–42.

35 Levy, 'Military policy', 143.

36 Reed, 'Destruction', 183. The existing Janissary organisation reputedly had 196 companies of various sizes, with an average size of 400, of which half might have been active, on-duty forces. (Levy, 'Military policy', 133.) That would suggest fewer than 10,000 in Istanbul, where only 51 companies were stationed.

37 This particular description from Esad Efendi's *Üss-ü Zafer* has the feel of imperial hyperbole, especially the sultan's role. I have followed Levy's description ('Military policy', 148–54). As always, the question of the number of victims of the purge remains open. Ottoman historians generally accept 6,000 total casualties among Janissaries for the period, but somewhere in the range of 500–1,000 for Istanbul. Esad Efendi simply noted it 'exceeded thousands'. European historians then and later were wont to exaggerate excessively, as high as 30,000 (Levy, 'Military policy', 162–63). Adolphus Slade's estimate is 20,000–25,000 across the empire, likely an exaggeration.

(See *Record of Travels in Turkey, Greece, etc*, New edn. (London: Saunders & Otley, 1854), 137).

38 Reed, 'Destruction', 238.

39 More correctly, they are the 'trained soldiers granted victory by Muhammad'.

40 Esad Efendi, *Üss-ü Zafer*, 103ff.

41 Reed, 'Destruction', 245 – his translation of Esad. The text evoked tears in the eyes of the listeners.

42 Dror Ze'evi, '*Kul* and getting cooler: the dissolution of elite collective identity and the formation of official nationalism in the Ottoman Empire', *Mediterranean Historical Review* 11 (1996), 177–95.

43 Butrus Abu-Manneh, 'The Naqshbandi-Mujaddidi and the Bektashi Orders in 1826', in *Studies on Islam and the Ottoman Empire in the 19th Century* (1826–76) (Istanbul, 2001), 64.

44 Reed, 'Destruction', 323, reproduces the text of a certificate of Janissary induction from 1819.

45 Abu-Manneh, 'The Naqshbandi-Mujaddidi', 68–69. Esad Efendi, *Üss-ü Zafer*, 208–10. Hülya Küçük contends that the persecution of the order continued after the ban, including purging the new army of pro-Bektashi officers and men 1827–30. She also downplays the Naqshbandi influence in this period, noting the continuing role the Mevlevis played as advisers to the palace. (*The Role of the Bektāshīs in Turkey's National Struggle*. Leiden: Brill, 2002), 39.

46 Levy, 'Military policy', 166.

47 Reed, 'Destruction', 295–96.

48 Reed, 'Destruction', 340–41.

49 Bosnia revolted and held out for two years, and a Bosnian regiment only joined the *redifs* (reserves) in 1836. The strength of the resistance in Bosnia was likely based on the century-long privileges of a warrior caste that were threatened by the new regime. Odile Moreau, 'Bosnian resistance against conscription in the nineteenth century', in Zürcher, *Arming the State*, 129–37. Slade witnessed the execution of a conspirator to restore the Janissaries in late 1829 (Slade, *Record*, 235).

50 Reed, 'Destruction', 268. Galib is referring to the ongoing Russo-Persian conflict, which would soon turn into the Russo-Ottoman conflict. Reed notes that Galib was popular enough to have previously been made an honorary member of one of the Janissary messes in Erzurum.

51 Dana Sajdi, 'Peripheral visions: the worlds and worldviews of commoner chronicles in the 18th century Ottoman Levant', PhD (Columbia, 2002), chapter 1, 95. Hasan Agha al-'Abd had written (ca. 1806) that the

so-called Janissaries in Damascus had been corrupted by the haphazard enlistment of people from the city and surrounding areas: some 3,000 of them (also 95). Sajdi presents Hasan as an example of a hometown boy who aimed at inclusion in the imperial system, and achieved the status of *Mütesellim*, or district representative of the state.

52 Şevket Pamuk, 'Prices in the Ottoman Empire, 1469–1914', *International Journal of Middle East Studies* 36 (2004), 463. He corroborates the work of Mehmet Genç, 'L'économie ottomane et la guerre all XVIIIe siècle', *Turcica* 27 (1995): 177–96.

53 Reed, 'Destruction', 335. Reed adds, 'this loot amounted to over 15,000,000 dollars.' The sultan's action stunned the business community of the city. Avigdor Levy, 'The contribution of Zaporozhian Cossacks to Ottoman military reform: documents and notes', *Harvard Ukrainian Studies* 6 (1982), 372, is the source for the revenue figures he approximates to 3.5 million pounds sterling. Caroline Finkel lists the Adjiman, Carmona and Gabbai familes (*Osman's Dream*, (London: John Murray, 2003), 438). In 1828–29, some 20,000 Catholic Armenians were banished and sent to Erivan, and their estates confiscated (Slade, *Record*, 150–54.)

54 Reed, 'Destruction', 336. Hüseyin Agha apparently had large sums of money deposited with one of the bankers, which he recovered before the confiscation.

55 Dominic Lieven, *Empire: The Russian Empire and its Rivals* (New Haven CT: Yale University Press), 140, contrasts it to the to 25 per cent in Russia, then considered the most inefficient system of collection.

56 Levy, 'The officer corps', 24–25.

57 Dror Ze'evi, '*Kul*, and getting cooler', 188.

58 Macfarlane supplies a most amusing portrait of the gentleman from Piedmont, *Constantinople in 1828*, vol. 2, 174–83. Recommended to the sultan for his riding skill, Calosso saved Mahmud II from a fall, thereafter achieved a degree of favour unrivalled, and rose to the rank of Colonel.

59 Levy, 'Officer corps', 32–39 *passim*.

60 On the factories established in the 1830s to support the military, see Önder Küçükerman, *Türk giyim sanayii tarihindeki ünlü fabrika 'Feshane': Defterdar Fabrikası* (Ankara: Sümerbank, 1988). The textile factory in the Golden Horn produced military apparel well into the twentieth century.

61 Levy, 'Military policy', 217–24.

62 Levy, 'Military policy', 184–99, has complete description. The *Serasker's* income was derived from the revenues of the *sancaks* of Izmit and Bursa, then under Hüseyin Agha's administration.

63 Levy, 'Military policy', 208–12.

64 Levy, 'Military policy', 235.

65 Levy, 'Military policy', 236–40.

66 Avigdor Levy, 'The contribution of Zaporozhian Cossacks to Ottoman military reform: documents and notes', *Harvard Ukranian Studies* 6 (1982), 372–413 *passim*, tells this story very well, and includes the translation of the official documents creating the force (385).

67 Levy, 'Military policy', 265–66.

68 Levy, 'Military policy', 292.

69 Levy, 'Military policy', 371–74. Levy dismisses the Moltke figure of 60,000 infantry, which he claims was probably based on Russian sources (376). Helmuth von Moltke, *The Russians in Bulgaria and Rumelia in 1828–29*, tr. by Marshal Diebitch (London: John Murray, 1854), 16.

70 Quoted in Levy, 'Military policy', 404.

71 Moltke, *The Russians*, 13.

72 Quoted in Reed, 'Destruction', 359–60.

73 Moltke, *The Russians*, 44.

74 Levy, 'Military policy', 323–45, as elsewhere full of very interesting detail.

CHAPTER NINE

An Empire returns to war

Russo-Ottoman conflict resumes on the Danube, 1828–29

What possessed Mahmud II to go to war, unprepared as the empire was for another confrontation on the Danube? Most contemporaries concluded that there was little else he could do, as the empire was faced with dissolution on all sides, by an aggressive challenger in Egypt, by a cacophony of voices of resistance from his Muslim subjects, and by the continuous Russian encroachment in both the Caucasus and the Balkans, and even in the Mediterranean, where a Russian blockade of the Dardanelles began with the declaration of war. The empire was stunned and the sultan humiliated by the sinking of the fleet at Navarino earlier that same year. In retaliation, Mahmud II closed the straits to foreign shipping. On 20 December 1827, Mahmud II declared war on Russia. He presented the case to his own public, justifying the new conflict as a jihad. The text of his declaration is full of evidence of the impact the Greek rebellion had made on his thinking. In summary form: All infidels, but especially the Russians, the declaration began, were mortal enemies of 'the Islamic *millet* and the Sublime State of Muhammad'. The elimination of the Janissary *eşkıya* had allowed the Russians to invade and conquer Islamic territory as they had for the past fifty years. They had provoked their co-religionists into treason, aiming at the destruction of the Ottoman state. While the rebels of the mainland had generally been repressed, the bandits of the Morea continued their rebellion, calling themselves, 'with unspeakable nonsense', the Greek government (*Yunan hükümeti*). Then the Russians, English and French entered the fray, with the aim of separating the Greek *millet* from Ottoman subjecthood. Their offer of mediation in order to achieve autonomy for the Greek *millet* was unacceptable, even after they attacked and destroyed the Ottoman navy at Navarino. The sultan called upon all Muslim subjects to unite in this war, which was not only about land and boundaries. The goal of the infidels was to eradicate the Muslim *millet* from the face of the earth.[1] For Mahmud II, known to all as the 'infidel sultan', this was clearly a piece of propaganda which aimed at arousing the public and swelling the ranks of his new army. His declaration may have been an attempt to pre-empt resistance, for, while the sultan experienced some success at raising new troops, Mahmud II found both his administrators and his subjects deeply divided on the question of whether or not to go to war with Russia in 1828.

In spite of the initial success in raising troops for the *Asakir-i Mansure*, perhaps as many as 10,000 infantrymen in Istanbul in six months, the rush to join slowed thereafter, undoubtedly the result of increased taxation to pay for the reforms, but also because of the reordering of lines of authority and communal relationships implicit in the new order. Pockets of resistance emerged, and purges of the *Asakir-i Mansure* rolls, as well as coercive forms of conscription, became the order of the day. Reaction to the new army units ranged from the violent, especially against conscription, to the derisive. Macfarlane records the comments of a 'true Turk' in Izmir, while observing a review of the new troops. 'And what in Allah's name can the sultan expect to do with these beardless, puny boys, with their little shining muskets? Why, they have not a yataghan among them! The yataghan is the arm of Mahomet and his people and not that chibouque-wire I see stuck at the end of their guns. Masallah! And what sort of monkey's dress is that? What sort of ugly-faced, shrivelled, puling dogs are these? Why, they don't look like Osmanlis!'[2]

Mahmud II had assumed the enormous risk of hollowing out the core of his society, and waited to see the results of his actions. The population of Istanbul suspended judgement, but made fun of the new recruits, who were ridiculed for taking twelve days to march from Edirne to Filibe (Plovdiv), a march that the Janissaries were reputed to have accomplished in 32 hours. Pay for the army fell into arrears, and horses for the cavalry were raised by forceful pressure on local pashas and aghas to 'donate' them to the cause, only part of the exactions that the new military required. Rapid promotion initially proved enticing to recruits, but by the outbreak of hostilities in mid-1828, the leadership had returned to the old order of favouritism and factionalism over merit.[3]

Bosnia, home of loyal soldiers of old, proved particularly stubborn, with a popular rebellion led by ex-janissaries against the governor and the new uniforms. Mahmud II appointed a new governor who was able to settle the province, but no soldiers from Bosnia signed on for the 1828 campaign. Albanians were reluctant as well, although 10,000 of them did join the grand vizier at Şumnu in mid-July, under the command of one Ömer Vrione Pasha, who would distinguish himself at Varna. Mehmed Ali refused to send the Egyptian troops requested by Mahmud, largely because his own army was caught in the Greek struggle, which had become an international issue. Arab Syria resisted sufficiently to halt conscription efforts for fear of general uprisings. The Kurds, another backbone of the Ottoman military, largely chose not to participate, but were caught up on the eastern frontier, as we shall see. Conscription failed altogether – recruits that were enrolled were often youths aged twelve or thirteen, chained together to get them to the barracks and keep them from escaping: echoes of the Russian practice.[4]

Contemporary sources described the forces available to the sultan across the empire, which include the provincial irregulars, as well as the reformed and semi-reformed units as follows:

Istanbul and the Bosphorus:	30,000 (incl. ten *Asakir-i Mansure* battalions)
Dardanelles:	7,000
Thessaly:	10,000 (incl. 8 or 9 *Asakir-i Mansure* battalions)
Edirne:	30,000 (reserve army)
Şumnu:	25,000 (incl. ten *Asakir-i Mansure* battalions)
Varna:	10,000 (irregulars and Albanians)
Dobruja and Danube	30,000 (irregulars and Cossacks)
Other European fortresses	10,000
Asia (Anatolia and Caucasus):	30,000 (incl. two *Asakir-i Mansure* battalions)
Total:	182,000[5]

These are obviously estimates, reproduced here to reiterate the inflation of Ottoman military power which is standard in accounts of the period, but also to emphasise the degree to which regional and private armies were crucial to the survival of the empire. If the figures for the Ottoman forces are cut in half, to 90,000, the Russian army they were about to face on the Danube, on paper at least, was three times that size. In the areas where the real confrontations took place, around Şumnu, İbrail, Silistre, and Varna, they were probably numerically evenly matched. The Russian military could count on a depth of experienced reserves that simply did not yet exist for Mahmud II. The degree to which the Ottomans left the Caucasus undefended is noteworthy, a subject to which we will return in the next chapter – the concentration around Istanbul, Edirne and Şumnu is testimony to the sultan's understanding about the vulnerability of Istanbul.

Russian tsar Nicholas I insisted on accompanying his army to the Danube, and his caution often represented a serious obstacle to the conduct of operations.[6] Furthermore, the Napoleonic Wars demonstrated the vulnerability of the western frontiers of the empire. Henceforth a majority of the military forces available to the tsar were spread along the frontiers of Europe. Of 800,000 troops, only 100,000–150,000 were deployed against the Ottomans in 1829. The rest were stretched along the Austrian and Prussian borders, the Baltic and Black Sea littoral, in Siberia, and in the Caucasus.[7]

The ostensible Russian reason for going to war was to enforce the Ottoman understanding of existing treaties, especially Bucharest, as we have seen, on the question of the Principalities and on the fortresses and frontiers in the Caucasus. Caution about the international impact kept the Russians from responding to the Ottoman declaration of war until late April 1828, at the same time as their troop movements into Ottoman territories began. By Moltke's account, 120,000 men of the second army were deployed along the Danube, under Field-Marshal Count Wittgenstein, composed of 88 squadrons of cavalry, 96 battalions of infantry and 31 batteries of artillery. A total of 45,000 men were sent to the Caucasus, commanded by General Paskevich. Realistically, the army which crossed the Danube, in mid-1828, probably numbered in the area of 65,000 fighting men, which included 4,000 Cossacks.[8] Russian cavalry was estimated as being too heavy, and too few in number, as contrasted to the lightly-armed but quick and well-mounted *sipahis*. That was an old problem, which the Cossack units had been organised to address, but they also were too few in number, and had adopted the same large horses preferred by the Russians, a hindrance in the Balkan terrain. As Moltke noted, the Russian soldier had been reduced to a machine, but 'the infantry went through all their evolutions with perfect order, and were perfectly steady under fire.'[9]

What beat the Russians in the latter half of 1828 and 1829 was not the *Asakir-i Mansure* troops but rather terrain, an inadequate commissariat, and a particularly bad plague year. The Russians were over-confident in their ability to overwhelm the Ottoman forces, but previous campaigns in the area made them cautious about problems with transport, foraging and supplies. When the Russians occupied Bucharest in mid-May 1828, Count Pahlen, appointed Governor-General of the Principalities, was ordered to supply the Russian second army with '250,000 measures of grain, 400,000 loads of hay, 50,000 kilderkins of brandy and 23,000 oxen,' let alone press 16,000 peasants into planting and harvesting hay, and supplying untold wagons and horses. The problem was that the Ottomans, and especially the Albanians, had already been there. The Ottomans had requested '20,000 loads of corn, 10,000 head of cattle, 30,000 sheep and 1,000,000 *kuruş*' from the *Hospodars*, of which less than half had been collected. What the Russians found was devastation and desertion on all sides. Mahmud II had ordered the evacuation of the left bank before the campaign began, except for İbrail and few minor bridgeheads. İbrail was the last fortress in Ottoman hands on that side of the Danube.[10]

The Russian plan was to deploy troops in three directions: against the Vidin–Rusçuk line, against Silistre, and against İbrail, the latter representing the gateway to the road to Varna and Istanbul. In the event, most of the engagements of late 1828 occurred on the mouth of the Danube, at Varna and at Şumnu. The Ottomans were typically slow to mobilise. The new 'independent commander' of the army, *Müstakil Serasker* Hüseyin Agha Pasha, did not leave Istanbul with his army until late May. The unpreparedness of the reformed army may have been one reason for delay. MacFarlane, eyewitness to the departure, notes: 'The levies from Asia Minor, armed and equipped, . . . were the most numerous part of the motley *corps d'armée*, and mixed with these were above five hundred irregulars, on horseback, mounted generally on miserable hacks, and armed in the strangest and most varied manner. The scimitar, which they were once so famed for, was a rare weapon; but twenty or thirty of them bore long spears in their hands. The tacticoes [*Asakir-i Mansure* troops] were for the greater part mere striplings, and so ragged and dirty that they looked as if they were returning from a campaign rather than starting for one; their muskets, of inferior French manufacture, were dirtily kept, furnished with bayonets of different lengths. All those on foot, regulars or irregulars, seemed to eye with particular envy their companions on horseback.'[11]

A further ostensible reason for the late start was the debates around the sultan. All of his senior administrators opposed the war, and one at least, İzzet Molla, was exiled for writing and speaking out publicly. General morale was exceedingly low, as the well-known (and increasingly well-founded) superstition that the Russians were destined to be the conquerors of Istanbul circulated widely. And finally, a bitter rivalry between *Serasker* Hüsrev and Grand Vizier Mehmed Selim kept the two best commanders of Ottoman troops close to the sultan in Istanbul. The grand vizier finally left Istanbul for Edirne at the beginning of August. Suffice it to say that the rivalry was a contributory factor to the lack of proper command of the two years of this particular war.

The Russians first crossed at Satunovo (by lake Kartal, the site of the Ottoman 1770 debacle) in early May, as it was practically the only conceivable place to cross on the marshy territory of the lower Danube. It required the building of a dike some 7,000 paces long, with a special passage of wood of 3,000 paces. Wood was difficult to find, as the Ottomans could attest from previous campaigns. The Russians who were building the dike were fired upon by an Ottoman entrenchment, perhaps as many as 10,000–12,000 men. The first actual crossing of Russian troops was accomplished with the help of the Cossacks living in northern Dobruja, formerly loyal to the sultan, who defected with their *Hetman* and transported a Russian battalion across in their light boats. The Ottoman forces, mostly irregular cavalry, were taken by surprise, and fled to neighbouring fortresses. The wooden and canvas pontoon bridge was completed in the next two days. Thereafter, some 30,000 Russian troops made the crossing. Tsar Nicholas I joined them in early June.[12]

The siege of İbrail began a week later. In contrast to many of the smaller fortresses on the left bank, İbrail was large, and had a commanding location some eighty feet above the Danube. It had 24,000 inhabitants at the time of siege; probably only 7,000–8,000 bore arms under Commander Süleyman Pasha. Well stocked, and well armed, with 278 guns and mortars, İbrail had shown itself capable of withstanding sustained sieges before, especially when the Ottoman Danube flotilla was able to keep it supplied. Russian forces, commanded by Grand Duke Michael, totalled in the neighbourhood of 16,000–18,000 men. Defence was spirited. The troops fired on the Russians with considerable accuracy, and occasionally engaged in ferocious sorties. Hüseyin Agha Pasha had reached Şumnu with the main army by early June, but was unable to send aid from Şumnu, or anywhere else, to lift the siege, which by then had been under way for over two weeks. Mines were eventually laid by the Russians, and set off on 7 June, but a Russian attempt to storm the walls resulted in loss of some 120 volunteers when the anticipated breach did not appear in the walls. Such persistence intimidated the Ottoman garrison. Süleyman Pasha continued the siege for ten more days, but surrendered in mid-June after Maçin had capitulated; the Ottoman Danube fleet lost 13 gunboats to Russian flotilla fire, and was forced to retire to Rusçuk. Süleyman and his men were allowed to retire to Silistre, leaving behind considerable supplies and munitions. A plethora of various artillery pieces were left behind, the most effective of the guns being a small, movable seven-pounder Coehorn, with which the Ottomans took precise aim with high elevation for short distances. Still, they were unable to take advantage of opportunities that such accuracy and plenitude gave them. Shortly thereafter, Hirsova, İsakcı, Tulcea and Köstence also capitulated. In abandoning the fortifications, the Ottomans unwittingly supplied their enemy with much-needed stores. By early July, the Russians were in command of the mouth of the Danube. Silistre had not been besieged, however, and its garrison was strengthened by the troops retiring from other fortresses. The main Russian army proceeded against Varna via Bazarcık instead, without back-up from the rear.[13]

Standoff at Şumnu and the siege of Varna

Meanwhile, Hüseyin Pasha had continued to fortify Şumnu, and ordered an 8,000-man *Asakir-i Mansure* unit, mostly cavalry, to defend Bazarcık against a Russian advance. On 7 July the Russians saw the new cavalry for the first time, and were impressed with the discipline and order. After driving the Ottomans back, the Russian forces assembled at Bazarcık, some 24,000 men of which 2,500 were cavalry. An imprudent, early thrust on Varna was repulsed by a strong Ottoman garrison. Varna by that time had 14,000–15,000 men, so could not easily be invested. The Russians turned to Şumnu instead. By 20 July, the tsar marched at the head of his entire army on the road to Şumnu, from Yenibazar. Şumnu, it will be remembered, was where Grand Vizier Muhsinzade Mehmed surrendered in 1774, but the fortified town was not entered then or later by foreign troops. Nor would it be now. The fortress was well situated on a flat plateau with steep heights and precipitous walls rising above the Bulgarian plain about 600 or 800 feet. There were at the time about 40,000 inhabitants, three-quarters of them Muslims. Here the Ottomans built outer works and redoubts, in contrast to many of the fortresses on the Danube. The main road to Silistre and Rusçuk guaranteed the arrival of provisions for the Ottoman army from territories not occupied by the enemy. *Serasker* Hüseyin had 40,000 troops at his command, including the 10,000 Albanians under Ömer Vrione. The Russians proceeded to entrench themselves around Şumnu. Hüseyin Pasha stayed put, sending sorties of cavalry to attack Russian emplacements but refusing to engage fully. By early August, that standoff had become untenable. Tsar Nicholas left with a regiment of horse and foot to relieve the troops before Varna and press the siege. By late August, *Serasker* Hüseyin, apprised of the vulnerability of the troops left behind, ordered a night attack on a Russian redoubt, in anticipation of an even greater attack on the remaining battalions of Russians entrenched around the fortress. At dawn on 27 August, two columns of the *Asakir-i Mansure* infantry stormed the right wing of the Russian camp. A similar attack was made on the left wing. The Ottoman troops were driven back into the fortress, but had acquitted themselves well, though still unequal to the Russian order in the open plain. The Russians suffered tremendously from hunger and exhaustion in the late summer heat. Horses, lacking forage which had become impossible to find, began to die at the rate of 100 to 150 a day. As the Ottomans gained the upper hand, news arrived that Grand Vizier Mehmed Selim was on the march from Aydos, and that the Russian siege of Varna had begun. Ömer Vrione and his 14,000 Albanians were detailed to join the grand vizier there.[14]

Varna's situation on the Black Sea meant that both Russian and Ottoman fleets could be called into action. The Ottomans, of course, were in the process of rebuilding a fleet, but still managed to engage the Russians at least once early in September. Two Ottoman ships were captured, and a third sunk by Admiral Grieg, then part of the Russian navy.

By late July the garrison numbered 15,000, of which 7,000 were Albanians, and was fortified with some 162 guns. The Russian fleet reached Varna the first week of August. It included eight ships of the line, five frigates, and a number of other gunboats and transport, as well as troops. The land forces were under the command of Prince Menshikov. With the arrival of the troops by ship and the troops under Nicholas I from Şumnu, Russian forces investing Varna now numbered about 9,000, still insufficient to force capitulation. The tsar left for Odessa, where he remained until early September, awaiting reinforcements. A month of emplacements, sorties, and skirmishes followed. Late in August, the besieged garrison vigorously attacked the Russian redoubts by night, with some success. By mid-September, however, with the arrival of Russian reinforcements, the besieging force rose to 18,000–20,000 men, which allowed the Russians to invest the northern side of the fortress and begin digging mineshafts. An explosion, and partial breach, was followed by proposals for surrender sent the following day to Garrison Commander Yusuf Pasha. The breach appeared to strengthen the resolve of the garrison, however, which held out for close to another month before surrendering. Ömer Vrione arrived at Varna, having increased his army to 25,000 by reinforcements sent him by the grand vizier. His defence of the southern flank of the fortress is one reason the besieged were able to withstand the three-month siege. The Russians spent much of September repositioning troops to confront the new arrivals. At the battle of Kurt-Tepe, Ömer Vrione's relief army, perhaps as many as 20,000, was checked by resolute Russian troops, but held firm on 30 September and forced a Russian withdrawal. Ömer Pasha then sat for three days without pressing his advantage against a weakened Russian side, and going to the aid of the besieged fortress, which, owing to the premature capitulation of Yusuf Pasha, surrendered to the Russians on 10 October. Throughout all of this, the grand vizier himself remained idle, most likely because he was without troops. Ömer Vrione retreated precipitately upon learning of the capitulation of Varna, and on 15 October was driven further back with the grand vizier to Aydos.[15]

The remaining strategic target, Silistre, was not well situated for fortification, and had already been the scene of conflict and destruction in 1810. Located on the shores of the Danube, it too could be served by the fleets of the opposing forces. In October 1828, there were likely some 24,000 inhabitants, and a strong garrison, in the neighbourhood of 10,000–15,000. Most of the summer had passed with skirmishes, small engagements and waiting. In mid-August 36 vessels of the Russian Danube fleet arrived, but the conflict remained desultory. One explanation could be disease: European newspaper accounts from early November reported that 500 men had died in two days behind Russian lines. Hüseyin Pasha remained passively at Şumnu after the siege at Varna got under way. In mid-October, the Russian troops at Şumnu marched on Silistre, in the hope of forcing the capitulation. Ottoman garrison strength, weather, disease, and exhaustion prevented the completion of the siege, however, and it was lifted on 10 November. The new grand vizier, İzzet Mehmed Pasha, made a brief attempt to retake Varna, but lacked the troops to press the Russians. All the irregulars left the battlefield for the winter. The Russians withdrew into the area around and north of Varna, where they remained for the winter.

Campaigns on the upper Danube, continuing into early 1829, in the main left the Russian army in charge of the Danube littoral. Similarly, the army in the Caucasus had made further inroads, capturing Anapa, a subject for later discussion. Russian steadfastness had gained the day, but the human costs were terrible. Half of the Russian troops died, most from some form of illness. Typhus, fevers, dysentery, scurvy, and inflammatory disorders, killed ten times as many soldiers as did plague. Russian hospital records from the period indicate that 210,108 were treated with lesser and greater illnesses from May 1828 to February 1829, twice the troop strength on campaign! By May 1829, plague had spread to Varna, and 50–80 new cases were brought in daily in June. In the five months of March–July 1829, 28,746 died in hospital, all due to sickness and not bullets. The plague proved particularly virulent at Edirne, which was crippled with the number of cases of dysentery and diarrhoea. The stench must have been frightful.[16]

The campaign of 1829 and Ottoman collapse

Mahmud II's new organisation had distinguished itself not by initiative, but rather by perseverance and traditional Ottoman defensive strategies. Nonetheless, confidence had temporarily been restored; Macfarlane noted: 'the Turks, lambs at the commencement of the campaign [of 1828], were lions at its end.'[17] No help was forthcoming from either France or Britain, both waiting for Mahmud II's acquiescence to the Greek question. French forces had replaced Mehmed Ali in the Morea in the late fall, and could hardly be thought to come to the aid of the Ottomans. Mehmed Ali, as we have seen, sent money rather than troops. Hence the two sides were left to face one another again on the Danube in spring 1829. Hüseyin Pasha was reassigned to the troops at Aydos. İzzet Mehmed Pasha, who had defended Varna till the last moment in the face of the betrayal by Yusuf Pasha, was appointed grand vizier in place of Mehmed Selim, as we have seen. Reşid Mehmed, the *kul* of Hüsrev, and veteran of Greek campaigns with İbrahim Pasha, was appointed commander of the army in the field. About 50,000 of the *Asakir-i Mansure* troops were sent to Şumnu, largely mobilised by impressment, rather than systematic conscription, 'the most unwarlike, the poorest and worst class of the Osmanlis.'[18] The Bosnians continued to resist enlistment. The Albanians refused aid until they were sufficiently paid. The *Asakir-i Mansure* suffered throughout the period from serious manpower shortages.

While Nicholas I might have preferred peace and a heavy indemnity, he had to force the issue for a second year to protect Russian interests in the Dardanelles. Replacing men and material, however, proved difficult. Most of the horses had perished, and plague was decimating veterans and new arrivals alike. The Tsar refrained from joining the troops in the second year of campaigning, and replaced Wittgenstein with Count

Diebitsch, whose able command improved the lot of the common Russian soldier to the degree possible under the circumstances. Camels were imported from the steppes to supplement scarce horses, and used for the baggage trains. The morale of the soldiers remained low, however, and with it their increased adherence to religious rituals. Troop strength stood at 68,000 men by the first engagement.

The Russian navy seized the initiative in early 1829, establishing a base for supplies and men below Varna at Sizeboli on the Black Sea, south of Burgas, where they scattered the small garrison. Mahmud II ordered Hüseyin Pasha to march on Sizeboli from Aydos, three days away, but apparently it took Hüseyin seven weeks to carry out the order. A vigorous Ottoman attack on the garrison was repelled, and not repeated. The small Ottoman fleet made a brief appearance before Sizeboli, but did not attack the Russian ships anchored there. Consequently, Russian fleets were able to establish blockades of the Bosphorus as well as the Dardanelles, further increasing shortages in Istanbul.

The Russians under General Diebitsch laid siege to Silistre in mid-May, delaying the opening of the campaign until a bridge of rafts could be built to ensure the connection to the right bank of the Danube. Material collected at Bucharest, and assembled at Oltenitza, was floated downriver to the spot selected for the crossing, and was briefly, ineffectively, harrassed by the Ottoman Danube flotilla. The Russians had 14,000–15,000 men and 56 field guns. By Russian estimates, the Ottomans had 13,000 soldiers and 8,000 armed inhabitants at Silistre, under the command of Sert Mehmed Pasha. The siege continued throughout June, but after six weeks, the defenders capitulated. Some 9,000 men, including three regiments of *Asakir-i Mansure* troops, surrendered. Perhaps as many as 7,000 others died in the siege. The Russians found 230 pieces of artillery on the walls, and 31 on board gunboats in the Danube, although more important to this particular siege were the considerable earthwork and mining operations on both sides.

Meanwhile, in early May, Grand Vizier Reşid Mehmed had marched out of Şumnu against the Russians, towards Pravadi, at the head of not more than 15,000–20,000 combatants. With Hüseyin Pasha's troops, the Ottomans probably had 60,000 in the field. The Russians had available only 12,000–14,000 to set against them. Two columns advanced on the Russian positions around Varna. The 10,000 troops under the grand vizier encountered the Russian entrenchments before Pravadi, and inflicted considerable losses, but failed to seize the day, and fell back on Şumnu. Still, the determination of the troops and their commanders –

both Reşid Mehmed and Halil Pasha, commander of the other column, were wounded in the attack – bode well for the reformed military organisation. By the end of May, Hüseyin Pasha, at the head of a second army, marched from Rusçuk, where he had been sent to raise another force, to join up with the grand vizier, but they were unable to coordinate that effort. On 28 May, Reşid Mehmed and 40,000 men advanced on the Russian position at Kozluca. Twenty regiments of infantry and five of cavalry – all *Asakir-i Mansure* – formed the core of this army. A Russian relief corps of 15,000 under General Pahlen left the siege at Silistre in early June and after a forced march of five days covered the sixty miles to join up with the Russian forces at Pravadi. On 10 June, the grand vizier might have been able to force the issue against the spread-out and much-taxed Russian side, but General Diebitsch took the initiative on 11 June instead, barring the grand vizier's passage on the Pravadi road back to Şumnu. This proved to be the defining battle of the war. The Ottoman soldiers, charging impetuously at first, soon dispersed, and turned the battle into a rout. Hüseyin's impassivity the previous year had allowed the Russians to recuperate over the winter; Reşid Mehmed's impetuosity cost him his army, and this battle of Kulewtscha cost Mahmud II his war.[19]

Too weak to follow up on the victory, General Diebitsch and his army debated strategy, redeployed troops, and began negotiations with Reşid Mehmed. Much skirmishing continued in the area around Şumnu, and the Russians faced hostility and resistance from armed local inhabitants. *Asakir-i Mansure* troops were observed exercising in camp as negotiations continued. After nearly four weeks before Şumnu, the Russian forces were considerably increased by the arrival of the troops and baggage train from Silistre. The Russians decided to bypass Şumnu, cross the Balkan mountain range and aim for Edirne. Aydos, Misivri, Burgas all fell to the Russians, and in nine days, the Russian forces had covered one hundred miles almost unchecked. The grand vizier did send 10,000–12,000 men to Aydos, but they were scattered by a brigade of Russians under General Rüdiger. Aydos became Russian headquarters. 'The filth of the camp, and of the town itself, was frightful: hundreds of dead bodies of men, horses, and camels lay festering in the streets and courts. The atmosphere was poisoned to such a degree that the seeds of the diseases which henceforth raged among the Russian troops were probably sown during their stay in Aydos.'[20] About 25,000 men gathered at Aydos; the grand vizier is thought to have had only 15,000 at Şumnu. Each side continued to overestimate the threat of the other. The Russians especially faced a completely collapsed military system, demoralised soldiers, desertion and plundering on all sides. South of the Balkan range, the Christian population welcomed the conquering army. Still, on 31 July in the march on Yanbolu (Jamboli), the Russians encountered a regiment of the new Ottoman troops, who attacked their batteries steadfastly, inflicted considerable losses, but retreated in the night towards Edirne. The Russians found huge stores of biscuit and cartridges.[21]

By mid-August, Diebitsch was within a day's march or two of Edirne, with 20,000 troops, when the grand vizier proposed an armistice. At a three-hour battle at Slivno on 12 August, the Russians cut off the grand vizier's route to Edirne. By the nineteenth of the same month, four weeks after crossing the Balkans, the Russians camped before Edirne, a town of some 80,000 inhabitants. It was estimated in the Russian camp that at least 20,000 combatants were in Edirne, under Halil and İbrahim Pashas. In the event, the Ottoman commanders negotiated the immediate capitulation of the city, and were allowed to retire unmolested. The Russians had still to fear the forces of late-comer Buşatlı Mustafa Pasha of İşkodra, who had served the sultan in helping to defeat Ali Pasha of Iannina, and also in the Greek revolt, but remained passive in the war with the Russians. He was said to have had 30,000 Albanians and Bosnians under his command in Sofia, with another 10,000 advance troops as near as Filibe. The 30,000 troops that were reputedly sent by Mahmud II in spring 1829 to man the fortresses of Istanbul, the Bosphorus, and, as we have seen, the coastal areas below Varna, had performed poorly. Şumnu, though under-strength, still remained unconquered, but the will to fight had left the imperial Ottoman forces and panic reigned in Istanbul.

So it was with relief on both sides that negotiations began between the two belligerents, urged on by the return of French and British ambassadors to Istanbul, and enabled by Prussian mediation. The Ottomans admitted defeat in mid-August by acquiescing to the Treaty of London (1827) and the Convention of Akkirman (1826). The plenipotentiaries first met on 1 September. The sultan was prepared to accept all the territorial cessions in Europe but initially resisted ceding those in the Caucasus, and was in despair over the size of the indemnity, originally set at 11,500,000 ducats or 400 million *kuruş*. An additional show of force by Diebitsch, by marching his men to within 60 miles of Istanbul, prompted British Ambassador Gordon and French Ambassador Guilleminot to intercede on behalf of the sultan: 'The Sublime Porte has formally declared to us, and we do not hesitate to attest to the truth of the declaration, that [by a march on Istanbul], it will cease to exist; and in annihilating its power the most terrible anarchy will strike indiscriminately, without means of defence, at the Christian and Muslim populations of the Empire.'[22] The Treaty of Adrianople was signed by the two sides on 14 September 1829.

Buşatlı Mustafa Pasha was not yet finished: in spite of the sultan's insistence that he lay down arms, he refused, on the grounds that he had no money with which to dismiss his Albanians. He informed General Diebitsch that he would arrive in Edirne on 10 October, to take up winter quarters. The Russians sent a small detachment to face him down, and routed the Pasha's forces. He and his largely Albanian irregulars would spend the winter laying waste to the country around Filibe, much to the chagrin of the sultan, who was later forced to send an army against him in mountainous Albania. The Russians evacuated Edirne on 20 November 1829, after Nicholas I ratified the treaty on 3 November. Istanbul rejoiced.[23]

Assessing the new army

Before turning to the implications of the treaty, and the next phase of the Eastern Question, a few comments are in order about the state of the sultan's new army. Mahmud II had broken the spine of the traditional order, but barely begun the inculcation of systematic discipline and hierarchical command characteristic of the European armies. We have already seen some evidence of the new style of control on the field, especially of heretofore unruly cavalry, of the use of night attacks and concerted sorties that observers remarked as part of the new order. Steadfastness in sieges, long an Ottoman speciality, was accompanied by an increasingly sophisticated use of artillery against besiegers. Russian officer Count Pozzo di Borgo wrote in June 1829: 'The Emperor has put the Turkish system to the proof, and His Majesty has found it to possess a commencement of physical and moral organisation which it hitherto had not. If the sultan has been enabled to offer us a more determined and regular resistance whilst he had scarcely assembled together the elements of his new plan of reform and ameliorations, how formidable should we have found him had he had time to give it more solidity and to render the barrier impenetrable which we find so much difficulty in surmounting.'[24]

In spite of shortages in 1828–29, the new Ottoman soldier could still count on being better fed than his Russian opposite number. Turkish soldiers were heard to say, as they threw out their offal, that it was 'for the dogs and the Russians'. By contrast, as we have seen, the Russian soldier lived on daily rations of black bread, a bit of salt, and a portion of spirits. Urquhart, admittedly a notorious Russophobe, noted: 'the steam of the Nizam kettles and the sight of its Pilaf is more alarming to [Russia] than all the phrases that ever were penned on the rights of man.'[25]

Further aspects of military discipline make their startling appearance in 1828. Russian prisoners were neither automatically killed, nor made into slaves. Instead, Mahmud II had them paraded through the streets of the city during the winter of 1828/29. Body parts, in return for cash payments, were forbidden during this war, a striking change from the Greek rebellion, when ear counts were routinely used to proclaim victory. Street reaction to that was caustic: 'Very pretty this! We are not to cut the *pezvanks*' throats when we have an opportunity, and we must not make them slaves; no, though the blessed prophet himself has authorised us so to do, and has declared the captive of the sword to be the property of the captor.' 'No,' another reportedly said, 'if you see an unclean Muscove kill your very child, or your brother in battle, and you should afterwards make him prisoner, you are to put your yataghan into your girdle, and kindly tell him to walk this way – talk not of blood for blood, and the ties of kindred and affection, you must not even slit the *Karata*'s ears.' 'And when they come to Stambool as prisoners of the sultan,' his companion rejoined, 'I understand that as the bagnio is not good enough for them, they are to be lodged in serais and fed on pilaff and kibaubs.'[26]

The newly 'civilised' army had serious shortcomings, but the main problem for the sultan was recruitment, for he had not convinced his subjects of the legitimacy of his project to transform the empire. For the next two years, serious revolts, such as that of Mustafa Pasha of İşkodra, continued across the sultan's territories, as if the Treaty of Adrianople signalled the nail in the coffin of empire. The revolts were undoubtedly stimulated by the chaos, shortages and exactions of the two years of war, but were also instigated by opponents of the new order, and by legitimate fears of what conscription meant. The sultan turned his new troops against those internal revolts, and the new army honed its skills against its fellows, as Mehmed Ali was forced to do in the initial stages of building of his new Egyptian army.

Who were the new recruits? By sultanic decree, they were Muslims, but the ethnic core of the new, 'docile' army was made up of 'Turks', the *Türk uşağı*' ('Turkish lads') which Ottoman commanders increasingly saw as the most reliable, most malleable cannon fodder. The word 'Turk' here is as much an ethnic stereotype as 'Albanians', or 'Kurds', or, as we have seen, 'Greeks'. The difficulties of recruiting other populations were not just a result of Mahmud II's choice to reassert the primacy of Muslim orthodoxy and shar'ia as part of his imperial image, but also because of his sense of betrayal by his Greek subjects, hence the categories of 'disloyal' or 'unsuited' (for conscription or recruitment) in public documents. Suspect groups included converts, heterodox Muslims, Greeks (*Yunanlı*), Albanians (*Arnavut*, virtually synonymous with Bektashis and classed together with Greeks), Kurds, Bosnians, and Arabs, to name those most frequently evoked. Converts were explicitly barred from enlistment. Such categories were but vague generalisations based on military prowess and faithfulness to the dynasty. The attitudes so expressed prevented a systematic rationalisation of conscription, which, coupled with fiscal difficulties and lack of accurate census information, crippled the Ottoman military system for the next decade. When the sultan attempted to extend the *Asakir-i Mansure* regime to Damascus, for example, he envisioned sending troops from Istanbul or from Sivas and Adana to replace the locals. Local recruits were not to be considered, as they were too attached in the old system, and proved more than able to resist conscription.[27]

During the Greek revolt, Ottoman commanders in the Balkan fortresses had repeatedly requested recruits from Turkic populations, such as the *Evlad-i Fatihan* mentioned previously, as antidotes to the Albanians. Greeks and Albanians occasionally collaborated, as when they served under Ali Pasha, but more often fought one another. Albanians were perceived to be of a rebellious nature even when they were not. Albanian territories were Bektashi territories, and hence banned from the new army, but Albanians were also fiercely independent mercenaries. In 1828–29, with the exception of Ömer Vrione's troops, they opted out. After the 1828–29 war, the regiments of *Fatihan* re-established by Mahmud II became regular regiments of the *Asakir-i Mansure*, and were instrumental in the pacification of northern Albania during Mustafa Pasha's revolt.

A further comment seems in order about Bosnia and northern Albania. Here, rebellion was undoubtedly driven by an intense sense of betrayal. After all, Bosnians and Albanians had heroic histories as part of the imperial military system. Some have argued that old ethnic rivalries, in this case Balkan versus Caucasus (Circassian, Georgian) military elites (the latter dominating Mahmud II's circle at the time), contributed to the contempt for the '*Arnavut*' repeatedly expressed by officers of the new army. A more compelling argument is that the radical reordering of the premises of local rule and military contributions had a profound effect on local economies and traditional relationships. From the military reformers' points of view, such independent warrior bands represented an affront to the push to centralised order. Still, Reşid Mehmed seems to have been particularly ruthless in his campaign against the once-loyal Mustafa Pasha. Mahmud II prepared the field by simply removing Mustafa Pasha's right to administer taxes. Mustafa responded by a call to the Muslims of Albania to take arms against an infidel: 'As everyone knows, Reşid Mehmed Pasha is the enemy of the entire Muslim community and is in particular the deadly enemy of Arnavudluk, and it is obvious from his character and actions that he intends to destroy us', a message he sent not only as the governor of İşkodra but on behalf of the 'Allied Muslims' (*muttefikan-i müsliman*).

It seemed particularly galling to Muslims in general that Mahmud failed to defeat the Christian armies, and aimed his wrath at those of his subjects who were most loyal.[28] Mustafa Pasha marched against Reşid Pasha in Monastir at the head of 10,000 Albanians in 1830–31, and was barely defeated. Bosnian rebels, led by *Kapudan* Husayn Gradasevic, attacked Kosovo but agreed to negotiations once Mustafa Pasha had been defeated. Rivalries among the powerful the Muslim landowners kept the rebellion from succeeding, in spite of military successes against the Ottoman army sent into the area. The insurrection was finally put down in 1832. Ali Pasha (Rıdvanoğlu), *Kapudan* Husayn's chief rival, was appointed as Governor of Herzegovina. Mehmed Ali's attack on Syria delayed further imposition of the reforms, as the immediate concern became raising an army to confront the Egyptian army of İbrahim Pasha in Anatolia. Reşid Mehmed was charged by Mahmud II to do just that. Many of the rebels were forced into the army that marched into Anatolia.

Such instances of massive rebellion by potential recruits only reinforced the presumed unsuitability for the *Asakir-i Mansure* of men other than of 'Turkic' stock. Ibrahim Pasha, the Commander of the Vidin fortress in 1828, for example, wrote to ask for Turkish recruits from Hüdavendigar, Kocaeli, Aydın and Sarıhan *sancaks* (all in Anatolia) to be sent to counterbalance the Albanians, who could only be controlled if dominated by soldiers of another sort (*sair cins asker*).[29] Nevertheless, the sultan could not survive with an army of recruits from Turkic stock only, as they were a minority population in the empire overall, and conscription exhaustion of Turkic Anatolia quickly set in. Albanian soldiers continued to play a role as part of the forces, as we shall see. By the end of his reign, Mahmud and his advisers were considering how to implement universal conscription. The basic conundrum for the military system was not the question of ethnicity, but the question of religion. Mahmud's views are contradictory in that regard, for he is often viewed as more liberal than his successors regarding non-Muslim rights. There are indications that Ottoman officials began to reiterate the official policy of *zimmi* tolerance, precisely in response to the rebellions erupting with increasing frequency in largely Christian territories.

As early as 1809, Alemdar Mustafa Pasha decreed that Muslims should not use the word infidel (*gâvur*) to refer to Christians, as they served the same God.[30] These ideas are reflected in surviving *firmans* of Mahmud II, such as that of July 1829, addressed to the Greeks of Morea: 'There will in the future be no distinctions made between Muslims and raya [reaya] and everybody will be ensured the inviolability of his property, life and honour by a sacred law and my sublime patronage.'[31] Mahmud II repeated his promises in public on an official tour of Bulgaria in the late 1830s, the first time a sultan had left greater Istanbul in over a century, one of five such journeys he made.

Ostensibly he undertook military reviews of the new *redif* militia organisations, to be discussed below. The trips were, however, an intensive effort to capture the loyalty of his non-Muslim subjects. He moved among them, dressed in western garb, gave speeches, and curtailed excessive ceremony. More importantly, he routinely gave money for the repair of churches and synagogues, and gathered children of all religious communities to stress the importance of education. During a meeting with the leaders of the Principalities in 1837, for example, he said: 'It is our wish to ensure the peace and security of all inhabitants of our God-protected great states, both Muslim and raya. In spite of all difficulties we are determined to secure the flourishing of the state and the population under our protection.' Later the same year, he referred to his subjects as his children whom he treated equally, 'the only difference perceived among them being of a purely religious nature.' Or at Şumnu, 'Your faith is different, but all of you equally guard the law and my Emperor's will.'[32]

Moltke, with Mahmud II in Şumnu, had this to say about the sultan: 'In this country where the peasant is accustomed to doing everything gratis, in the name of corvée, for the lord, the sultan pays to cover the cost of his voyage. I was told that he carried on him 2.5 million florins, and a mass of precious objects; we do not pass any destitute or disabled person that the sultan does not order that a gold piece be given to him. At his departure, he left 10,000 florins for the poor of Şumnu, and took specific measures to insure that the money arrived at its destination and that too much not stay in the fingers of the distributors. The imams were charged with reporting to him. During the voyage, I always saw groups of women holding their supplications above their heads. An officer approached, collected the petitions, and passed them to the almoner. Later, the sultan was promenading in a four-horse phaeton which he himself drove handily; a poor woman lifted a paper high on a cane, but as the procession was passing very rapidly, no one noticed her. Except the sultan, who stopped the horses, despatched an officer and continued on his way.'[33] These gestures of inclusion and *noblesse oblige* predate by only a few years the Gülhane rescript of 1839, which promulgated equality of citizenship and empire-wide conscription in the army. The problem of mobilisation of able manpower would become abundantly apparent in the contest over Syria between Mehmed Ali and Mahmud which unfolded after 1831.[34]

The Treaty of Adrianople, 1829

By the Treaty of Adrianople, Russia restored all the cities, harbours, fortresses and districts south of the Danube to the Ottomans. The fortresses of Turnu, Kaleh and Giurgevo, on the left bank of the Danube, were to be razed, and their Ottoman garrisons removed from the Principalities. İbrail, Tulcea, İsakçı, Maçin and Köstence fortresses were also razed. By Article III, the Prut remained the Ottoman–Russian frontier, but Russia acquired all the islands of the very mouth of the Danube. Article IV aimed at settling at last the territorial disputes in the Caucasus, as Russia had penetrated as far as Erzurum in the course of this war. By it, Georgia, Imertia, Minrelia, Guria, and the provinces ceded by Persia in the treaty of Türkmançay in February 1828, remained in Russian hands, and the border ran along the edge of the *pashalıks* of Akhaltzikhe, Trabzon, Kars and Erzurum. We will return to these developments in our final chapter.

Article V clearly set out the rights of Wallachia and Moldavia, which remained under the nominal suzerainty of the Porte, but the language emphasised their freedom of religion and the right to an independent national administration, guaranteeing the further influence of Russia. The regulations which had been referred to in the Convention of Akkirman became the 'Organic Regulations', a document closely identified with the period of Russian domination before Romanian emancipation after the Crimean War. Articles IV and V of the treaty were those concessions that Mahmud II so vigorously resisted in the 1812 Treaty of Bucharest, and delayed signing until absolutely forced to do so in the Convention of Akkirman in 1826. Effectively, the sultan had only the right to invest the *hospodars*, and maintain a fiction that the Principalities were still an

PLATE 17 *Column commemorating Mahmud II's visit to Tutrekan in 1837, now standing in the garden of the local Historical Society (author's photograph).*

integral part of the Ottoman system. Article VI restored six districts to Serbia, and guaranteed its freedom, another of the much-contested concessions. Finally, Article VII allowed complete freedom of trade in the Black Sea, and passage of merchant ships through both the Bosphorus and the Dardanelles.

It is hard to imagine which of these articles most offended Mahmud II, but the indemnity, later reduced to 250,000,000 *kuruş*, was guaranteed to slow the sultan's further reform agenda. It also contributed to the increased use of debasement, as we have seen, and disastrous financial strategies, such as monopolies, which crippled already over-burdened populations and threatened international commercial interests.

In addition, the Ottoman first line of defence became the Balkan Mountains, rather than the Danube itself. Russian troops remained in Silistre and the Principalities until 1835, when Mahmud finally paid 800,000 pounds sterling, or 3,000,000 *kuruş*, to invalidate Russian excuses against the evacuation. At that time, Russia still had 11,000 troops in the Principalities and 17,000 in Silistre. In 1833, when Mehmed Ali and son İbrahim threatened to storm Istanbul from Kütahya, in Anatolia, Mahmud called on those same Russian troops to prevent the imminent collapse of his empire, an astounding reversal that awoke Great Britain to the real dangers of Ottoman collapse. We will come back to that agreement, known as the Treaty of Hünkar İskelesi (1833), and the events leading up to it, shortly.[35]

One of the results of these two years of war was the decision taken almost immediately by Nicholas I to work in concert with the other European powers concerning the future disposition of the Ottoman Empire. He had concluded that the elimination of an Ottoman Europe or Istanbul was not in the best interests of the tsar. This, and the continued Russian occupation along the littoral, offered the sultan an interlude from maintaining vigilance on the Danube, and an opportunity to consolidate his hold on his remaining territories.

The Syrian Question 1831–33

The Treaty of Adrianople also laid the Greek question to rest for the moment, although Greek independence had already been determined at Navarino and by the evacuation of the Egyptian army, completed as the French occupation troops arrived in August 1828 to oversee the transformation to statehood. The drawing of the borders of the new Greek Republic only served to bring into stark contrast the fact that more potential 'Greek' nations remained outside those borders than within, a problem for future generations.

After Navarino and the surrender of the Morea to the European powers, Mehmed Ali contemplated further expansion, independent of his master in Istanbul. To the west, the French talked of joint operations against Tripoli, Tunis and Algiers, but then occupied Algeria themselves in 1830. The more logical move for Mehmed Ali was to occupy greater Syria, historically seen as an extension of Egypt. Syria represented a buffer zone, a source of manpower, and natural resources such as iron, wood and coal, as well as an opportunity for Mehmed Ali to broaden his tax base. Both Mehmed Ali and İbrahim coveted the Damascus *pashalık*, a centre of potential wealth, because the city served as the hub of pilgrimage and trade routes. Mehmed Ali had ignored Mahmud's request for troops in the 1828–29 war against Russia, sending money instead. He ignored the sultan again, when requested to send troops to the fight against the Albanians led by Buşatlı Mustafa Pasha. Mehmed Ali had spent some 20–25 million *riyals* on the invasion of the Morea,[36] and was rebuffed when Mahmud II only offered him control of Crete, rather than his coveted Syria. Mehmed Ali did not regard Crete, also the seat of Greek rebellion, as adequate compensation. Whether or not Mehmed Ali contemplated dynastic independence, or was simply behaving in the time-tested Ottoman *ayan/vali* fashion, seems a moot point. By 1828 expansion was clearly on the Egyptian agenda, and Syria was the target.[37]

What does set Mehmed Ali aside from the other notables of his era was his perspicacity in economics and the military. Honed in the Morea, the reorganised army and the financial base upon which it stood would truly be tested in the Syrian campaign, which amounts to a prolonged civil war internationalised after 1833.[38] From 1828–31, intensive preparations were under way in Cairo and Alexandria. The navy was rebuilt, the army reorganised, and conscription intensified. European observers speculated at the aim of such intensive preparations. By the end of 1831, it became clear as troops and ships left Egypt for Sidon.

After subduing Mustafa Pasha in the mountains of Albania, Mahmud II was in a position to turn his attention to consolidating the Arab provinces, which continued to resist his economic and military reforms. His preoccupation in the north, and with rebellions in the Balkans and Baghdad, had led to neglect of the territories of Syria and Palestine. Moreover, the region had long resisted full integration into the empire, as Ottoman authorities were but one of the local power brokers in the region, who formed coalitions to side against centrally-imposed governors.[39] The Acre–Sidon–Damascus triangle had undergone a decade of disorder and chaos before Mehmed Ali chose to invade, much of it driven by the relentless Istanbul reform agenda, as well as the increasing integration of the region into the world economy. In June 1831, in fact, a major rebellion broke out in Damascus. The governor, Selim Pasha, had imposed a new surtax on the merchants of the city, which engendered a revolt. The rebels, fired upon by the governor's troops, besieged the citadel, burned the government buildings, and killed the governor, and many of the 5,000 troops who accompanied him. Elders of the city established an interim government while they awaited the arrival of the new Ottoman appointee. This must have simplified İbrahim's occupation of the city in June 1832, just as the rebellion was losing force.[40]

İbrahim Pasha, universally admired as a commander of troops, is generally credited with moderation in the early days of the occupation. He would prove adroit at exploiting the simmering discontent among inhabitants of greater Syria, Muslim and non-Muslim alike. To the Christians, he promised tax reform and exemptions. To the Muslims, he (and his father) argued the sultan's unsuitability as caliph, appealing to the divergent voices already raised in protest against the reform regime. Mehmed Ali was just as willing as Mahmud II to using religious symbols to stimulate enthusiasm, hence his army was known as the *Cihadiye* (the Jihadists), certainly another explanation for the early success of the invasion.

For the next decade, Mehmed Ali and Mahmud II were rivals in Syria not just for territory and taxes, but also for adherents to their expansionist and reformist aims, and for manpower for their armies. By design, Mehmed Ali had insisted on his senior officer class being Turkish-speaking, while the enlisted men were Arabic-speaking. Were his commanding officers to be Arabs, he foresaw that bonds would be enforced with the enlisted men that would challenge his own authority. Hence, 'Arabs' could be promoted only as far as the rank of lieutenant. Such a practice had led to the elision of distinctions under the rubric 'Turkish', leading to the absorption of numerous ethnic categories which had made up the 'Mamluks', for example Circassians or Georgians, into the Turkish-speaking ranks, and the creation of an officer corps extremely loyal to Mehmed Ali. It also meant a constant flow of able officer stock from the *Asakir-i Mansure* to the *Cihadiye* camp. Such defectors joined European officers of equal rank, who were paid very well, such as the previously mentioned Süleyman Pasha (Colonel Sèves), İbrahim Pasha's second-in-command. In more than one instance, Ottoman prisoners who spoke Turkish were appointed as high-ranking officers over İbrahim's troops, and even commanded their own regiments of Ottoman prisoners.[41] It is not hard to imagine the resentment of enlisted men discovering they were under the command of privileged, but vanquished, 'Turks'. The most notorious Ottoman defection was in July 1839, when Grand Admiral Ahmed Fevzi surrendered the entire fleet to Mehmed Ali in Alexandria.

While the Turk–Arab rivalry is fairly well understood, I think military historians have long underestimated the role religion plays in this confrontation at the Taurus Mountains. The anti-reform movement which gained momentum in the 1830s was undoubtedly stimulated by Mahmud's relentless intervention in the countryside, and his overturning of the traditional order. Less well understood is the degree to which Mehmed Ali in these few years came to represent a new order, an alternative to the infidel sultan, who in the minds of many, had violated the shar'ia. After 1831, even the ulema in Istanbul began inciting the public from their pulpits. Insurrections in Albania, Bosnia, Macedonia and Baghdad, as we have seen, were couched in terms of the sultan's violations of the traditional law. Unrest spread into heartland Anatolia: to Ankara, Kayseri, Aydın and Tokat. In August, 1831, a fire in Pera, the European quarter of Istanbul, destroyed 10,000 homes. Public expressions of resistance to obvious signs of westernisation, such as the sultan's portrait and wardrobe, escalated. Mahmud was openly called 'infidel ruler' as he ceremoniously crossed the new Galata bridge over the Golden Horn in 1837: 'Infidel sultan, God will demand an accounting for your blasphemy. You are destroying Islam and drawing down upon us all the curse of the Prophet.'[42]

Such demonstrations were a prevalent expression of social identity which cannot simply be dismissed as religious obscurantism. It appeared to many across the empire that the caliph in Istanbul was at war not just with external enemies, but with his Muslim subjects, most especially the traditional political and social hierarchy of tribal and household patriarchs such as Busatlı Mustafa Pasha in Albania. Equally important, the resistance to his rule coincided with the rise of militant Muslim voices, such as the Wahhabis, as well as the quietest Naqshbandis, calling for reform from within the faith itself.

The Egyptian Governor's aggressive expansion, and initially successful campaign, however, was also engendered by increasing problems at home in Egypt. Conscription of the *fellahin* had produced some 130,000 recruits by the time of the invasion of Syria. In under ten years, Mehmed Ali had not only transformed a traditional, agricultural society into a plantation economy, he had radicalised the relationship between a government and its people. Logistics, discipline and troop welfare all appear to have been successfully coordinated in the initial stages of the new expansionist enterprise. Yet desertion remained appallingly high, and the death toll from plague and other diseases insupportable. Rural disorder was on the increase, as banditry became an essential form of collective political action. Piracy and theft from state warehouses forced patrol of Egyptian harbours by gunboats in 1831. In June 1832, press gangs posted simultaneously in every quarter of Cairo rounded up the male population for the ongoing campaigns, a measure of the strength of resistance to conscription. Paying and feeding an army of the magnitude of 130,000–150,000 led to increasing debts and restive populations.

These problems would only be exacerbated as an invasion army became a ten-year army of occupation, more especially as Egyptian exports came into intense competition from European markets, and growth slowed. The globalisation of trade was a major catalyst for many of the events of this ten-year period, and crucial to the decisions made by European allies concerning Mehmed Ali. The contest was as much about a share of the international market as about caliphal leadership. There is considerable logic to Lawson's assertion that the invasion of Syria was prompted by a need to expand the cash crop base, as the contest over shrinking agricultural surpluses intensified. The would-be beneficiaries of such an invasion were Cairene merchant families with Damascene connections.[43]

İbrahim Pasha and the siege of Acre

Mehmed Ali's ostensible ally in Acre was Abdullah Pasha, Governor of Sidon, whom he had helped restore to favour in Mahmud II's court by lending him 11,000 purses in 1822, but who now stood in his way as the sultan's protector of the Levant coast in Acre.[44] The official reason for the invasion was to punish Abdullah for sheltering some 6,000 *fellahin* recruits who had fled the Egyptian delta for Sidon. On 2 November 1831, İbrahim Pasha left Egypt at the head of a fleet of 16 ships of war, and 17 transport vessels, carrying his staff officers, 40 small cannons and several siege cannons, along with supplies for an army of 30,000. The troops, meanwhile, had begun the march overland via al-Arish to Jaffa. Their departure had been delayed until late autumn as the result of an outbreak of cholera which killed as many as 5,000 soldiers and 150,000 civilians in just over a month, perhaps as much as three per cent of the population.[45] Contemporary estimates of the Egyptian commitment to Syria ranged from 25,000–100,000 per year over ten years, if irregulars are included in the figures.[46]

In mid-November 1831, İbrahim's army entered Jaffa and Haifa without resistance. By December, Sidon, Tyre, Beirut, Tripoli, Latakia, Jerusalem and Nablus swore allegiance to İbrahim Pasha, and he could

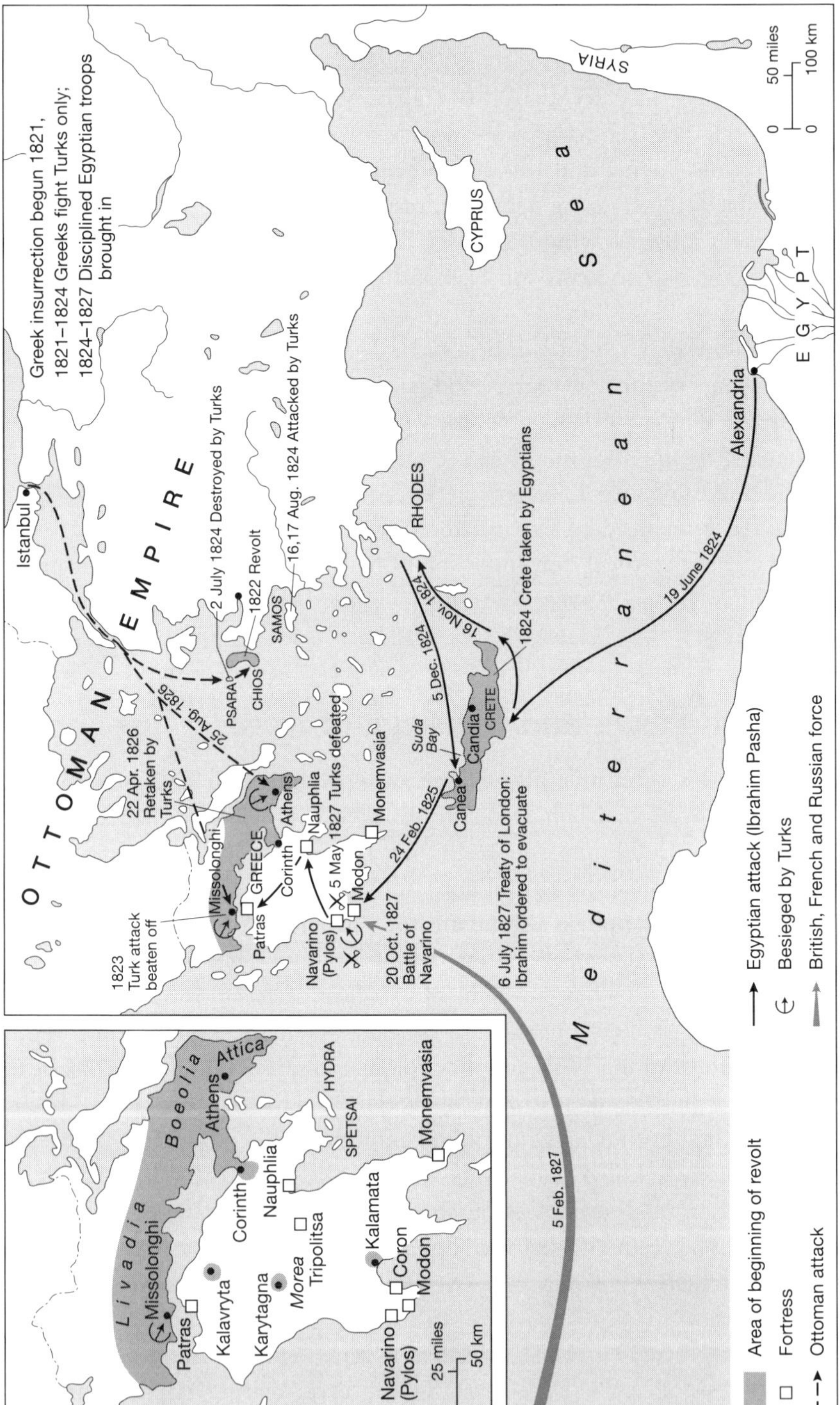
Greek insurrection begun 1821,
1821–1824 Greeks fight Turks only;
1824–1827 Disciplined Egyptian troops brought in
OTTOMAN EMPIRE
Istanbul
2 July 1824 Destroyed by Turks
1822 Revolt
16,17 Aug. 1824 Attacked by Turks
PSARA
CHIOS
SAMOS
RHODES
1824 Crete taken by Egyptians
16 Nov. 1824
5 Dec. 1824
CRETE
Candia
Canea
Suda Bay
24 Feb. 1825
19 June 1824
Alexandria
EGYPT
SYRIA
CYPRUS
Mediterranean Sea
22 Apr. 1826 Retaken by Turks
25 Aug. 1826
1823 Turk attack beaten off
Missolonghi
GREECE
Corinth
Athens
Patras
Nauphlia
5 May 1827 Turks defeated
Monemvasia
Navarino (Pylos)
Modon
20 Oct. 1827 Battle of Navarino
6 July 1827 Treaty of London: Ibrahim ordered to evacuate
5 Feb. 1827
0 50 miles
0 100 km
Livadia
Boeolia
Attica
Athens
HYDRA
SPETSAI
Monemvasia
Nauphlia
Corinth
Missolonghi
Patras
Kalavryta
Karytagna
Morea
Tripolitsa
Kalamata
Coron
Modon
Navarino (Pylos)
0 25 miles
0 50 km
Area of beginning of revolt
Fortress
Ottoman attack
Egyptian attack (Ibrahim Pasha)
Besieged by Turks
British, French and Russian force

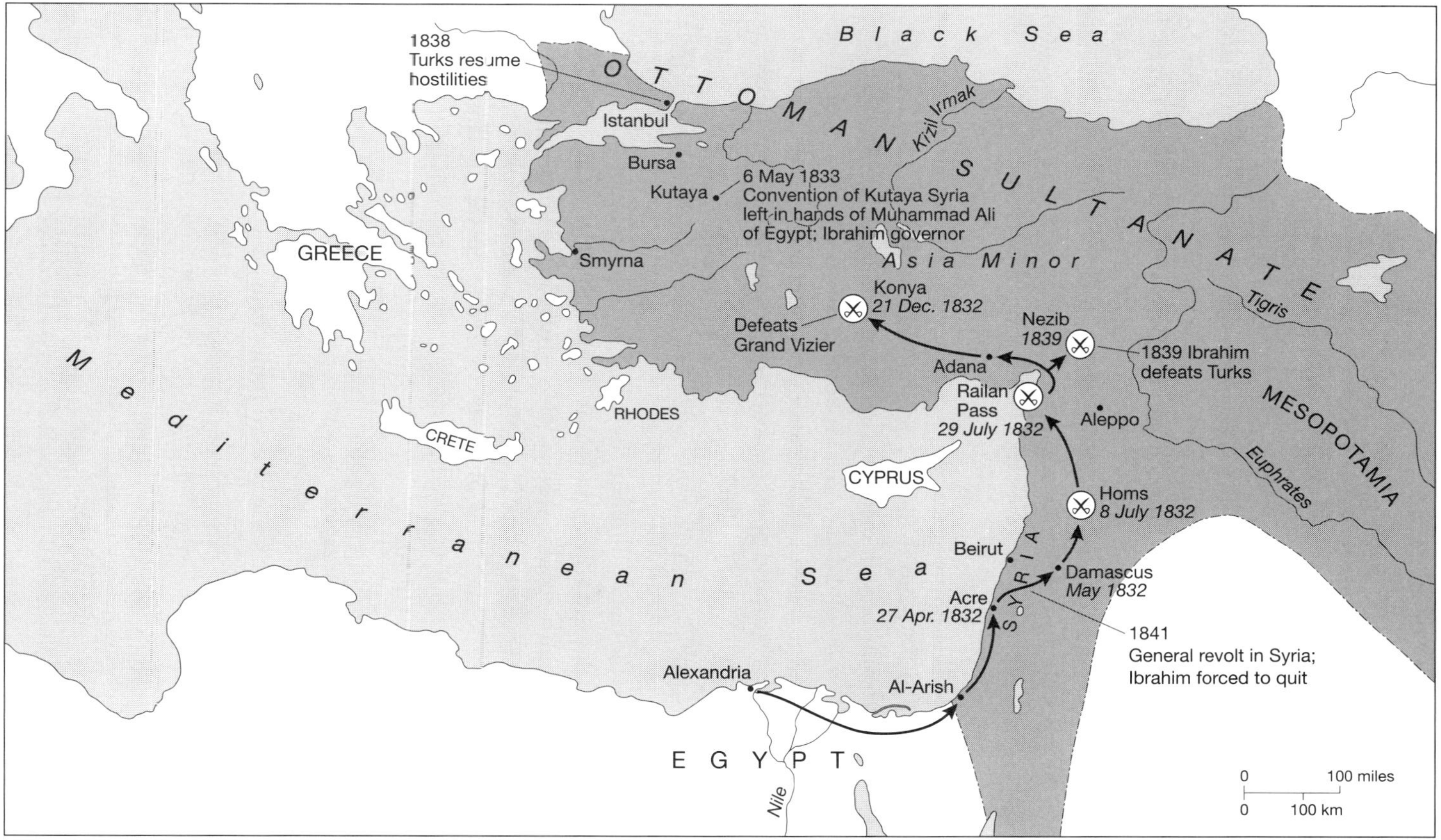

PLATE 18 *Campaigns of Mehmed Ali and his son İbrahim in Greece, 1824–33.*

concentrate on the fortress of Acre. İbrahim commanded 30,000. Abdullah was said to have 2,500 men and a year's supplies. A six-month siege ensued, during which İbrahim consolidated his position in the surrounding area. The Maronites of Mt Lebanon, under Emir Bashir II (Sihab), at first remained neutral but were persuaded by Mehmed Ali himself to join İbrahim, which gave him the command of Syria from Damascus south.[47]

As the siege of Acre continued, more troops were sent from Egypt. In January 1832, news of the imminent arrival of an Ottoman relief army kept the besieged from surrendering. Meanwhile, Mahmud II responded to the invasion first by sending an emissary to Alexandria in December, threatening reprisals. In February, he appointed Mehmed Pasha as the Governor of Aleppo, and *Serasker* of Syria and Arabia. Mahmud II ordered the other governors of the area to send troops to Aleppo, from all over Anatolia.[48] Mehmed Ali and his son were deliberately left out of the annual Ottoman appointment register (*tevcihat*) of 1832. Agha Hüseyin Pasha, Governor of Edirne, was appointed to the *pashalıks* of Jiddah, Crete and Egypt instead. A blockade was imposed on Egyptian ports. Hüseyin Pasha was also appointed *Serasker* of the Anatolian forces, and ordered to join Mehmed Pasha in Aleppo. By March 1832, Mehmed Pasha and Osman Pasha of Tripoli had gathered a large, mostly irregular force, some 18,000–20,000, at Homs. Leaving only 5,000–6,000 troops to carry on the siege, İbrahim hastened to confront the Ottomans at Homs. After a brief confrontation at Tripoli, Osman Pasha was beaten back by İbrahim's regular troops. He retreated to Homs, where a brief battle on 14 April resulted in 300 casualties among Osman's troops. İbrahim gathered his forces at Baalbeck, awaiting the outcome of the siege. In April a *fetva* was issued in Istanbul, declaring both father and son rebels, and enemies of Islam.

On 27 May 1832, a final assault on Acre forced the surrender of Abdullah Pasha who was sent as a prisoner of war to Egypt. Six months of bombing had destroyed the city, and the siege represents the beginning of the end of the once-vibrant port. İbrahim Pasha lost 4,500 soldiers. His victory freed him to march on Damascus, which surrendered with little fight on 16 June. Ali Pasha, commandant at Damascus, retired to Homs, where he joined Mehmed Pasha, with a combined force amounting to 30,000–35,000 men, again half of them irregulars (four infantry regiments, three cavalry and 15,000 irregulars).[49] Hüseyin Pasha's Anatolian army was still in Antioch. İbrahim Pasha pushed north from Damascus with 30,000 troops and arrayed his army before Homs in a form that

Marshal Marmont described as so 'judicious and complete as to leave no opening for criticism'. The army comprised seven regiments of infantry, each of four battalions; six regiments of cavalry, divided into 24 squadrons; seven batteries of artillery, and two howitzers, formed into three lines. The Ottoman army was deployed in two lines, with the cavalry placed on the flanks, and the artillery distributed among the infantry, one gun per battalion. The Ottomans were not well placed, however, and after fierce fighting, the Egyptians advanced on the centre of the Ottoman line, which broke and fled in disarray. Some 2,000 were killed and 2,500 prisoners captured. On 8 July İbrahim entered Homs, and captured an additional fifteen hundred soldiers and 24 cannons.[50] The remaining Ottoman army fled to join Hüseyin Pasha at Antakya. Aleppo shut its gates to Hüseyin, and submitted to İbrahim on 15 July. On 29 July, 16,000 Egyptian troops confronted around 20,000 Ottomans of the combined armies of Mehmed Pasha and Hüseyin Pasha in front of the village of Beylan. After three hours of heavy fighting, İbrahim's army was victorious, and his losses slight: 102 men and 172 horses, and 162 men wounded. Ottoman losses stood at 1,000, with 1,900 prisoners. Hüseyin's troops abandoned the field, camp, baggage and 25 guns.[51]

So far, the performance of the Ottoman armies had been absymal, but success of the Egyptian troops was also due to their training, to İbrahim Pasha's experienced command, and to the voluntary submission of most of Syria and Palestine, and then Aleppo, to his authority. By 31 July 1832, İbrahim controlled Tarsus and Adana. By October, and the arrival of new Egyptian recruits, İbrahim Pasha consolidated control of the region by occupying Urfa and Maraş, to maintain strategic access to Erzurum and points east. He camped in Adana, and awaited orders from his father. His primary communication with Egypt was by sea, well patrolled by the Egyptian navy.

Mehmed Ali, at least until the early months of 1832, continued to insist that his invasion was a punitive campaign against Abdullah Pasha. At the end of August of that same year, however, Mehmed Ali wrote the sultan to begin negotiations, but Mahmud II delayed responding, as he awaited the fruition of two initiatives: a diplomatic appeal to Britain to intervene, and orders to Grand Vizier and *Serasker* Reşid Mehmed Pasha, veteran of campaigns against Bosnian and Albanian rebellions, to mobilise a large army from all over the empire. Between July and October 1832, four armies were raised, for a total of some 80,000 regulars and irregulars: 20,000 Albanian and *Asakir-i Mansure* troops under his command, who were stationed in Scutari [Albania]; similarly, 20,000 from territories around Erzurum under Osman Pasha, Governor of Trabzon; a third group, of 10,000 troops, stationed north of İbrahim Pasha's army in Adana; and finally, another 20,000–30,000 stationed near Konya, survivors of the previous confrontations, under the command of (Mehmed Emin) Rauf Pasha, Governor of Karahisar.[52]

Raising troops from recently subdued Albanian and Bosnian communities seems a particular act of desperation for Reşid Mehmed and Mahmud II, can only have been accomplished with considerable force, and likely fell back on the old habits of requesting local officials to raise troops from the villages.[53] The census of 1831 established quotas for each governor across Anatolia and the Balkans, but in actuality, arbitrary numbers were assigned each town and village, forcing undue hardships on Muslim households. Enlistment was technically for twelve years by that time, but, as with the Egyptian army, and the Russian armies for that matter, conscription was a death sentence. Massive resistance and desertion, or self-mutilation to prevent recruitment, appears to have been rife in both the Ottoman and Egyptian armies. Desertion statistics averaged 10–25 per cent in the *Cihadiye* early on, but six years after the campaign began, Mehmed Ali received a report that said that 60,000 soldiers had deserted the army, a loss of one in every two conscripts.[54] By one account for the Ottoman side, from 1837, 500,000 soldiers had been conscripted since the beginning of the new order ten years previously. One in four was deserting (especially among Kurds). Death and desertion may have reached 45,000 men a year over the ten years.[55]

When the two armies finally met on the plains north of Konya, the Ottomans outnumbered the Egyptians three to one: 50,000 under the *Serasker* to the 15,000 commanded by İbrahim Pasha. Rauf Pasha had retreated from Konya to Akşehir on 17 November 1832, as İbrahim Pasha approached the city. On 18 December, the Egyptians defeated a small vanguard of Reşid's army, mostly Albanians, many of whom defected to the Egyptians.[56] The main confrontation occurred on 21 December, in a battle which began at noon with Ottoman artillery fire. A young lieutenant, Zarif Mustafa, was among those with Reşid Mehmed Pasha at Konya. This brief excerpt from his memoirs begins the night before the battle: 'Other than Reşid and Ahmed Pashas, no one else had a tent. I dug a hole in the snow, and entered it. That night, everyone stayed that way. I slept a little, and my boots froze to my feet. I cut the boots and took them off my feet. Next day, the battle began, and I had no official duty in the fight. Then, the lines began to open up, and the cannons started firing, and many shells fell on our left front regiment, which became like garden posts. Their cavalry attacked. Ahmed Pasha attacked with three companies, and I was behind him. The enemy kept coming, and we turned back. Ahmed Pasha fled, and one of the Arabs struck me with a bayonet, but it had little effect. In three to five days, the wound was fine.'[57]

A dense fog had made visibility difficult on the battlefield. The Ottoman side manoeuvred badly, allowing the Egyptian army the opportunity of isolating and attacking the left flank, largely cavalry. The Ottomans began firing at 640 yards, thereby marking their position even in the fog. İbrahim urged cavalry and reserves, backed by heavy Egyptian artillery fire, towards the Ottoman cavalry on the left flank, forcing a retreat. As Reşid Mehmed tried to regroup, he lost his way in the fog, and infantry and cavalry of centre and left flank were forced to surrender, including Reşid. The right flank had meanwhile surrounded the Egyptians on their left flank, and engaged with the battalion's square formation in a battle which lasted seven hours. A group of 2,500 Ottoman cavalry broke through the line but kept going, and combined with the capture of Reşid Pasha, precipitated the retreat of the rest of the right wing. İbrahim had his greatest victory yet at the battle of Konya. The Ottoman forces lost 92 cannons, 3,000 casualties and 10,000 prisoners. The Egyptians, by contrast, suffered 262 casualities and 530 wounded.

By the beginning of 1833, inhabitants of Anatolia from Konya to Kayseri were inclining to İbrahim, whose adroit use of real and ideological pressure attracted adherents and volunteers for his army. Marmont continues: 'Had he marched on Constantinople on the day succeeding . . . he might have arrived opposite that capital, without difficulty, by the third or fourth of January, at which period the Russian artillery corps had not reached the Bosphorus. The appearance of the Egyptians at Scutari [Üsküdar] would have produced a revolution at Constantinople, and overturned the Turkish Government, crushing the Sultan in its fall, or compelling him to act as an instrument of İbrahim's power. . . . A new Mussulman Empire might then have arisen, which with Egypt for its basis would have been guaranteed from external dangers . . . in the possession of a brilliant reputation [in İbrahim] and would have been respected by European nations, from the successful adoption of their military institutions.'[58] İbrahim must have felt much the same way, and continued his movement of troops until he stood at Kütahya, where his father ordered him to halt by the end of January 1833, but sent him further relief troops.

On hearing of the latest defeat, Mahmud II acquiesced to Mehmed Ali's request for negotiations, and sent Grand Admiral Halil Pasha to Egypt for that purpose in mid-January 1833. Halil Pasha was instructed to reinstate Mehmed Ali as Governor of Egypt, Crete, Jiddah, Sidon, Tripoli, Jerusalem and Nablus, but not Aleppo and Damascus. Tortuous negotiations, mostly facilitated by the Russians and the French, were carried on between Alexandria and Istanbul until May 1833 when the convention of Kütahya was actually signed. By March, Mehmed Ali wanted all of Syria and Adana, İçel, Alaiye and Silifke in Anatolia. İbrahim Pasha pressed his father repeatedly to ask for independence from the sultan, a stance he apparently took from the beginning of the invasion, but Mehmed Ali, at this stage at least, did not envisage independence as an option.[59] Mahmud II finally acceded to the appointment of İbrahim as governor of Damascus, Aleppo and Adana in a verbal commitment on 5 April. Still, giving away Adana, heartland Anatolia, must have rankled with the sultan, because further negotiations turned on İbrahim's appointment there. Mahmud II finally conferred on İbrahim an appointment as *Muhassil* (Tax Collector) of the district of Adana in early May 1833, with the proviso that he immediately begin withdrawing his troops. İbrahim left garrisons in Urfa, Maraş and Antep, outside the Adana jurisdiction. As of 10 May 1834, Mehmed Ali was to begin paying a tribute of 30,000 purses to the sultan for the newly-acquired territories.[60]

The European powers and the Treaty of Hünkâr İskelesi

No one in either of the courts, or in diplomatic circles, expected the Convention of Kütahya to be the end of the matter. Indeed, the situation had become rather delicate, once Mahmud II requested Russian support in early 1833. Two envoys to Palmerston's government in London had failed to generate the British naval blockade of Alexandria requested by the sultan. Envoy Namık Pasha left London for Istanbul on 17 March 1833, empty-handed. The French offered to mediate, as they sided with Mehmed Ali throughout the period. The Russians offered military assistance, as both naval and land forces stood a few days' distance of Istanbul. On 2 February, Mahmud II made a formal request to the Russian Minister in Istanbul for military support. By 20 February Russian ships with troops on board anchored in the Bosphorus, at Beykoz and Büyükdere. The troops camped at Hünkâr İskelesi on the Asian shore. Meanwhile the French continued unsuccessfully to act as brokers. As we have seen, Mehmed Ali's demands were enlarged by early March, and he sent Istanbul his ultimatum. Mahmud II counter-offered to İbrahim Pasha directly in Kütahya in early April, as described above. On 5 April, a second contingent of 5,000 troops arrived on more Russian ships. France and Great Britain reacted by sending squadrons to Besika Bay just beyond the Dardanelles. Still, relations between the two great powers was at that moment not sufficiently cordial to force the issue with Russia. In the end the Russians left, only after they had secured the Treaty of Hünkâr İskelesi, an eight-year mutual defensive alliance, with pledges to consult with one another in matters of security (8 July 1833). A famous 'secret' article was also included, by which the Ottomans guaranteed to close the Dardanelles to all foreign warships in the event of an attack on Russia.[61] Simultaneously, Tsar Nicholas I was negotiating the Münchengrätz Convention with Prussia and Austria, signed in September, 1833. That agreement undertook to oppose further Mehmed Ali expansion, and maintain Ottoman integrity. This, as we have seen, was to be the new Russian policy towards Istanbul.[62]

The British spent the rest of the decade re-establishing their relationship with the Ottomans, somewhat alarmed at the influence the Russians had acquired with the sultan. With the Kütahya agreement, Mehmed Ali had nominal authority over the two routes to India, the Suez–Red Sea, and the Euphrates–Basra in Iraq, the latter of which, coincidentally, British explorer Captain Chesney would successfully navigate with a steamer in 1836–37. Lord Palmerston, both Foreign Minister (1830–34) and then Prime Minister (1835–41), who once viewed the collapse of the empire as imminent, did an abrupt about-face. 'Because British policy was Russophobe and Egyptophobe, it now became strongly Turcophile . . . The Ottoman Empire was to be preserved, supported, reformed, and strengthened. . . . In the 1830s Britain prescribed a regimen of what might be called trade and aid for the Ottoman Empire.'[63] Palmerston gave immediate orders that the Levant fleet be strengthened and stand ready to serve the sultan if Mehmed Ali threatened further action, upon authorisation of the British ambassador in Istanbul. It was a stand-off that never really threatened war, but presaged further crises in the decade to come. The next phase of the Eastern Question had begun.

Palmerston offered both military and naval missions to Mahmud II as a gesture to redress Britain's former neglect. The naval mission had considerable success, and was instrumental in the final showdown with Mehmed Ali, 1839–41. The military mission was a failure, however, for many of the reasons we have seen before. Palmerston sent a Polish officer in British service, by the name of Chrzanowski, to Istanbul, suggesting he be added to the sultan's staff. He was joined by a Captain DuPlat and Lieutenant Colonel Considine, but they were not received well, and abandoned the effort by 1837.[64] Moltke and the Prussians who entered Turkish service about the same time fared better, largely because they advised without insisting on Ottoman military rank. Only with the Crimean War was that pattern broken – especially, as we will see, in the Caucasus.

Ottoman reforms 1834–39

Mahmud II pressed forward with his reform agenda, touching on nearly every aspect of Ottoman society. In 1835, three independent branches of government were reorganised: the military (*seyfiye*), the judiciary (*ilmiye*), and the civil bureaucracy (*kalemiye*). This elevated the *Serasker* to equality with the *şeyhülislam* and grand vizier, and reduced the status of the latter, heretofore known as the 'absolute deputy' of the sultan. It also marked the beginning of a true civil service, although well-trained candidates for offices were several decades away. Mahmud II understandably had to concentrate on the military, so it should come as no surprise that one of the long-term consequences of these reforms was the valuation of the military over the civilian career. Salaries for the officer corps and related staffs grew tremendously in an era of great fiscal crisis. Educational institutions were revived or inaugurated. The naval and military engineering schools continued to function. A medical school to train doctors for the army had been opened in 1827, but was reorganised in 1838, when it was transferred to Galatasaray, the venerable palace training school. An imperial music school followed by 1834, as well as the School of Military Sciences (*Mekteb-i Ulum-i Harbiye*) mentioned in the previous chapter. By 1838, the sultan was contemplating a system of secular elementary and secondary schools, a few of which were opened in his lifetime, and which he celebrated on his tours of Thrace and Bulgaria, as we have seen.[65]

Major initiatives around diplomatic practice also occurred at this time. Mahmud II had attempted to send numerous young men overseas for education in 1827, much as had Mehmed Ali in 1826, with only limited success. The language barrier was an acute one, which Mahmud addressed in 1833 by creating a Translation Bureau in Istanbul to train local Muslims in the language of diplomacy. It grew to a staff of 30 by 1841. In 1834, embassies were reopened in major European capitals, and the young diplomats and translators sent to Paris, Vienna and London formed the nucleus of the Tanzimat leadership for the following decades. Among them were Mustafa Reşid, Ali and Fuad Pashas, all future grand viziers, and protegés of *Reis* Pertev Pasha, himself chief opponent of Hüsrev Pasha until the latter engineered Pertev's downfall in 1837.[66]

The language of governance also changed at this point. Government departments were reorganised as ministries. For example, in 1835–36, the former offices of the *Kahya* and *Reis*, both under the grand vizier at the Sublime Porte, were renamed the Ministries of Civil and Foreign Affairs. By Mahmud II's death in 1839, Civil Affairs had become the Ministry of the Interior. In 1838, the office of Grand Vizier briefly became the office of the Prime Minister (*Başvekil*). These changes reflected the increasing importance of negotiations with the European powers, but also a slow rationalisation of the tasks of a modern bureaucracy.

In 1831, to promote the progress of reform, the sultan agreed to the publication of *Le moniteur Ottoman* by Alexandre Blacque, founder of the *Spectateur de l'Orient* (later *Courier de Smyrne*) in Izmir. A year later, a Turkish version, *Takvim-i Vekayi*, started publication. Both served initially as official gazettes. The introductory article of *Le moniteur* made an explicit reference to the important (and traditional) role of historiographers and annalists in the empire, and the contents of both presses were regularly reviewed by the sultan himself. A more independent newspaper, the *Ceride-i Havadis*, started publication in Istanbul in 1840 after Mahmud II's death, but a critical and more independent press began only after the Crimean War.[67]

Major changes continued to unfold in the military, but many of the reforms were blocked by the continued influence of Hüsrev Pasha. After the 1828–29 war, Hüsrev was regarded as the most powerful Ottoman after the sultan. When Moltke saw him in 1836, Hüsrev was more than 80 years old, having served in various posts for over 33 years, and survived (and contributed to) some of the most convulsive decades of the empire. He had managed, by the end of Mahmud II's reign, to combine in himself the functions of the office of *Serasker*, commander of the *Asakir-i Mansure* troops and all other aspects of the military: Commandant of Istanbul, Chief of the General Staff, and Minister of War. Chief of Police of Istanbul was an appropriate position, continued Moltke, 'for an individual who controlled vast wealth, had spies everywhere, and without whom it was impossible to operate. His sole passion was for power, and woe to those who stood in his way!'[68] He and his protegés literally controlled almost all the offices of the empire.

Reform of the designations and assignments of the officer hierarchy had been under way for some time, nonetheless, corresponding to the changes in organisation of the army. In 1832, brigades (*liva*) were established, and brigade commander, *Mirliva* (Brigadier General), was added to the ranks of *Müşir* (Marshal, rank of Vizier, later Field Marshal), and

Ferik (General). By 1837, there were 64 officers designated as *Ferik*. The problem of too many chiefs and no subalterns remained an acute one, however, a situation in the Ottoman command structure which continued until after the Crimean War. A true general staff was very slow to emerge. Literacy and training remained a significant part of the problem, of course, but household politics, and especially the extraordinary reach of Hüsrev Pasha, dominated all else as the primary obstacle.

To illustrate that point, let us take the example of *Müşir* Zarif Mustafa Pasha (1816–61), participant at Konya in 1832 above, and later Commander of the Anatolia army in the spring of 1854. His training appears to have been scribal, for he joined the *Hassa* regiment as a scribe at the age of 14. 'Hamdi Bey [unknown] came into the office, and began talking to his acquaintances, and I was among them. I've been made a major, he said, and I am looking for a secretary to replace me. Is there any one having such a wish, he began by asking. Me, I said. Let me see your writing, he said. Is your father willing? Whether he is or not, I want to be a soldier, I said. Well, let's go, he said. From there, we mounted up and went straight to the barracks. The late Sultan Mahmud II was there. Hamdi Bey took me directly to the office of Ahmed [Fevzi Pasha, later Grand Admiral]. Here is my replacement, he said. This is his handwriting. My son, do you want to be a soldier? he asked. Yes I do, sir. Come, follow me, he ordered.' He was given a uniform and a sword. Once in a while, they were drilled. He was assigned as *mülazimlik* (Lieutenant), and a substitute for a Captain of the *Hassa*. His first action was a year later in the entourage of Reşid Mehmed's campaign in the far reaches of Albania (1831). It was his first journey outside Istanbul. He next fought in the campaign against İbrahim, in 1832, as we have seen. In 1845, at the age of 30, he was made a *Ferik*. All those around him, with few exceptions, were raised and promoted this way. The striking thing about the Konya campaigns is the youth of all the commanders.[69]

The *Hassa* was the imperial guard which replaced the old *Bostanci* palace guards and remained under the particular attention of the sultan. By 1835 there were 11,000 men in the *Hassa*, and it was there that the sultan set up the School of Military Sciences for officer training beyond Hüsrev's reach. After 1834, during the lull in the conflict with Mehmed Ali, Mahmud II used these reforms as a means of curbing the power of Hüsrev. After Grand Vizier and *Serasker* Reşid Mehmed, Hüsrev's protégé, was captured and sent to Egypt, Mahmud II appointed (Mehmed Emin) Rauf Pasha to the post. Rauf Pasha, five times grand vizier in this period, was known to be an independent. Ahmed Fevzi, another enemy of

Hüsrev, commander of the *Hassa*, was raised first to *Ferik* and then to *Müşir*, and in 1836 to Grand Admiral. Mahmud II appointed his own courtier Said Mehmed to the post of *Müşir* of the *Asakir-i Mansure* in 1836, instead of Hüsrev's preferred candidate, Halil Rıfat Pasha. Both Said and Halil had married daughters of the sultan. Hüsrev Pasha was dismissed as *Serasker* in early 1837.[70]

Immediately thereafter, Mahmud established the first of two important councils: the *Şura-yi Askeriye* (Military Council), inaugurated in July 1837, composed of twelve army officers from all the branches and two civilians, one of whom was a jurist. The role of the new Council was to advise and oversee the office of *Serasker*. The orders for its foundation afford an interesting glimpse into the degree to which Mahmud II supervised the reforms: the new council regulations included instructions about using a round table and chairs, rather than the traditional divan-style low couches and pillows. Coffee and smoking were prohibited in the room. Attending servants were forbidden entrance, and were not allowed to loiter or make noise outside the room. Junior officers were to be encouraged to speak first, so as not to be intimidated.

Of most interest is the fact that the very first topic was the reduction of the term of military service to five years, a decision incorporated in the Gülhane decree two years later. The system of councils proved popular and proliferated across the military system, prompting the sultan to introduce further such consultative bodies at the top of the hierarchy: a Privy Council, the *Meclis-i Valâ-yi Ahkâm-i Adliye* (High Council of Judicial Ordinances), and an equivalent for the Sublime Porte, the *Şura-yi Bab-i Âli* (Council of the Sublime Porte). Both were in place by March 1838, and rather astonishingly, Mahmud II chose Hüsrev Pasha as chairman of the Privy Council. It must be said that regardless of his motivations, Hüsrev Pasha remained steadfast in his support of the sultan's reforms; indeed we will find him reponsible for organising the *redif* below.[71]

In ten years, every aspect of the Ottoman administration had been touched by the reforming hand of the sultan. Most of the changes at the centre remained tentative, but the seeds of new forms of governance had been planted which would allow the following generation of reforming bureaucrats to continue the process, and dominate the first two sultan successors to Mahmud II. Two other problems plagued the reform agenda: fiscal crises and the acute lack of manpower.

Devaluation, taxes on luxury items, the conversion of *timars* to crown lands, and the confiscation of estates of 'enemies' have already been described as methods used by Mahmud II to finance his reforms, even though he is also credited with eliminating the *müsadere*. The recovered land leases of Ali Pasha of Iannina, for example, some 1,158,000 *kuruş*, were the main revenues for a new Army Finance Secretariat established in 1827. Mahmud II had also attempted to divert the revenues of the charitable endowments by creating a Directorate for that purpose, which worked with some success under his successors and undermined the power of the ulema. Other measures included a surtax on the poll tax on non-Muslims, specifically for the army, and the increased use of *Rusûmat-i Cihadiye*, or Holy War duties. A real attempt was made to instill honesty and regularity in accounting methods, and payments of salary, but the conflict with Mehmed Ali derailed the efforts under way. In early 1835, the Sultan was spending 2,000,000 *kuruş* a month on the army; by summer 1835, it was 4,500,000, and by early 1839, the amount rose to 18,000,000 *kuruş* a month, when annual revenues have been estimated to have been at somewhere between 300,000,000 to 800,000,000 a year.[72]

We have a snapshot of the army in 1841 which illustrates the continuing burden of the military expenses. For Istanbul alone, the cost of maintaining close to 40,000 soldiers annually was 69,285,826 *kuruş*. (It must be remembered this was a year when the war with Mehmed Ali was winding down.) Some 65,093 soldiers were scattered in fortresses across the empire, at a cost of 88,357,237 *kuruş*. Discussions at the *Meclis-i Vâlâ* were estimating the cost at 197,383,000, which did not even include such expenses as munitions or tents.[73]

As one solution to the failure of conscription, and to try to minimise expenses, Mahmud and his advisers revived the idea of a national militia, first floated under Selim III. The new military organisation, known as the *Redif* (*Asakir-i Redife-i Mansure*, or Victorious Reserve Soldiers, also known as *ihtiyat*, yet later *yedek*), would survive as an organisation with various reforms until 1912, when it was merged with the regular army. It deserves particular attention for the lasting impact it would have on all of Ottoman society. It is also one of the better-documented reforms of the period because all the regulations were published in the new *Takvim-i Vekayi*.

A new national reserve system

The primary reason for establishing such a system must have been the need to supply reinforcements to the armies of Anatolia where Egyptian occupation had considerably reduced the areas of the empire available to Ottoman recruitment. A secondary aim was to create a means of screening young recruits, and to ease the burden of rapid levies on the countryside.

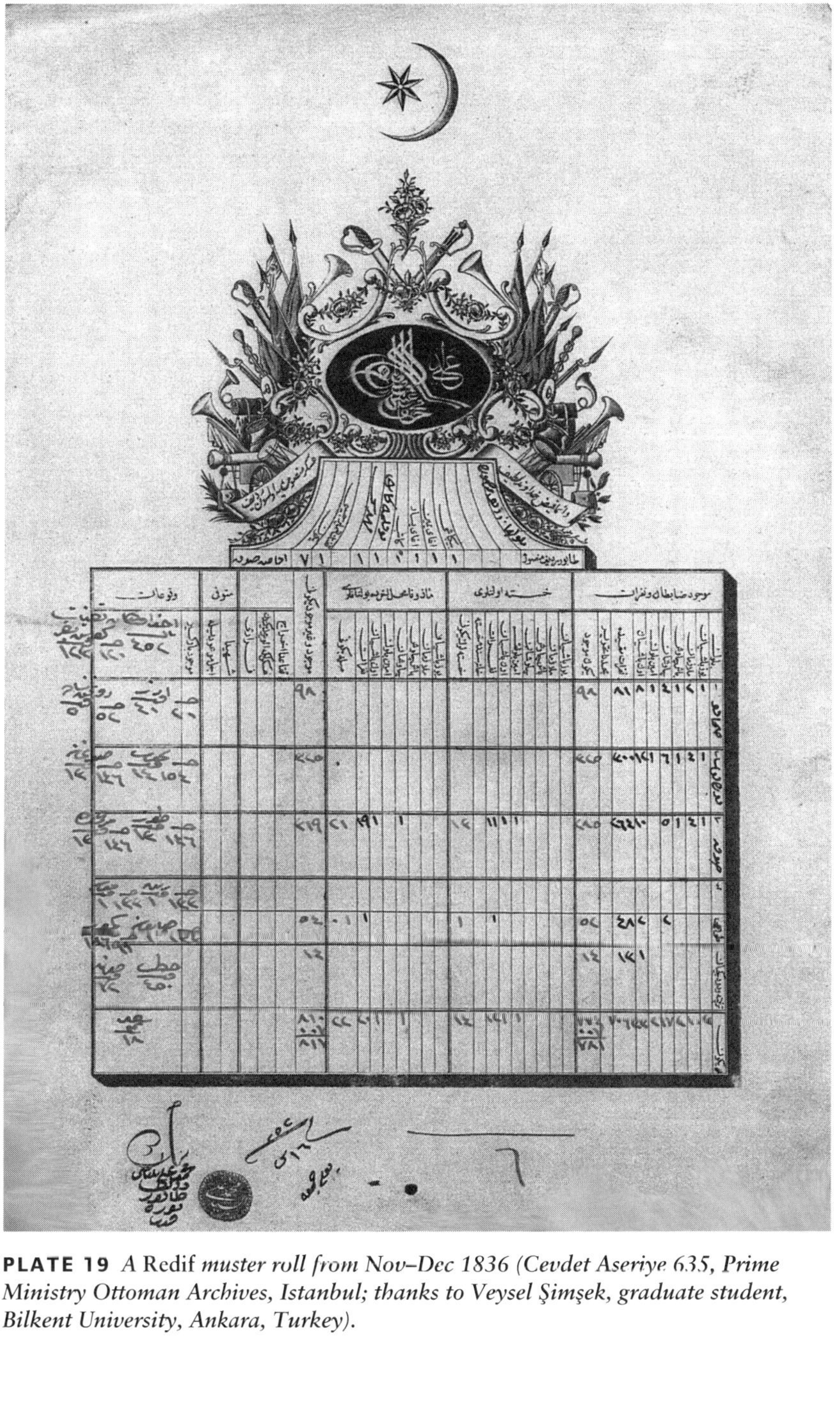

PLATE 19 A Redif *muster roll from Nov–Dec 1836 (Cevdet Aseriye 635, Prime Ministry Ottoman Archives, Istanbul; thanks to Veysel Şimşek, graduate student, Bilkent University, Ankara, Turkey).*

In May of 1834, Mahmud II assembled a large council of prominent statesmen and soldiers, coincidentally in Istanbul as his daughter's wedding guests, to discuss the creation of an empire-wide auxiliary force, likened to the Prussian *Landwehr*. Among those present were Hüsrev Pasha, and the first *Serasker* Agha Hüseyin Pasha, as well as former grand vizier İzzet Mehmed Pasha.

The regulation was published in *Takvim-i Vekayi* that August. The reserves, as the council defined them, were to be established to ensure provincial security, and to guarantee a rapid means of levying troops in war. In peacetime, such troops would continue as agriculturalists. In war, they could be called to arms. Every province of the empire was to have a battalion of four large musket companies, with a complement of 330 officers and men. Each battalion would also have four pieces of artillery with 108 officers and cannoneers. The total came to 1,432 officers and men. The men were to be paid 5 *kuruş* a month, one-quarter of the pay of a regular soldier, while the major (*binbaşı*) received 187.5 *kuruş*. What distinguished these troops was that the age was higher than the regular infantry at 23 to 32 years old; they could be married, and they did not have to serve outside their province in wartime. Those who retired from or did not re-enlist in the regular army after five years' service were eligible for this militia service. The officer corps was to be recruited from local notable families, who would be trained in Istanbul as well as their home provinces. Twice a year, the troops were to be assembled and drilled. Officers were issued uniforms and fezes, and required to wear them at all times. The enlisted men need only wear their uniforms for the twice-yearly drills. Some forty *redif* battalions were mandated, for a total of 57,000 men. Officer trainees were immediately brought to Istanbul, and regular officers were sent to help train the new units. The very first battalions were established in Ankara, Çankırı and Karahisar-ı Sâhib, and the others followed, notably in the same areas where *Asakir-i Mansure* had been most successful, in Anatolia. By early 1836, the *Takvim* announced, 33 of the new battalions had been established. The following year, battalions could be found from Cyprus to Urfa to Üsküb.

The costs per battalion of maintaining the corps were set at 250,000 *kuruş*, and deducted from provincial taxes. Small wonder that some provincial governors embraced the new system. A new treasury, named after the *redif* force, was established in the capital. A new tax, called the *iâne-i cihâdiye* (holy war donation) was introduced to further defray the expense of the new system and collected twice a year, in spring and fall, from the population at large. In March 1837, new regulations were issued, after the possibilities for corruption and abuse in the system had become apparent, and Mahmud II attempted to impose more centralised control over the battalions. An entire *redif* structure resulted, patterned on the regular army, both the troops of the line and the guards, with three smaller battalions, 2,500 officers and men making up *redif* regiments, for a total of around 100,000 in 120 battalions. These were to be commanded by nine *Müşirs* (Marshals), located in Bursa, Konya, Ankara, Aydın, Erzurum and Sivas, Edirne, and subsequently Niş and Şumnu. Small cavalry squadrons of *redif* were also established. Regular officers trained in Istanbul began to replace the children of provincial families. Three months' annual training in rotation substituted for the annual assembly, which became a fourth assembly but of the full complement. By the end of 1837, 90 such new battalions were reported established, but they remained woefully under-equipped, poorly educated and trained. In spite of the optimism of the *Takvim* report, corruption and desertion continued at all levels. Although its initial effectiveness was questionable, the *redif* organisation undoubtedly had the potential for exerting a standardisation in military training and presence in the provinces.[74]

The status quo and negotiations

Meanwhile, tension between Cairo and Istanbul escalated. The dispute with Mehmed Ali simmered around a number of unresolved questions: early on, Mehmed Ali was still concerned with the amount of tribute he owed the sultan, and the removal of Hüsrev Pasha, his old rival, from negotiations. Mahmud II demanded the tribute, and the evacuation of Urfa, which lay outside the jurisdication of İbrahim in Anatolia.

But the *status quo* was not going well for either side. Maintaining the army in Syria proved to be costly in military and economic terms for Mehmed Ali and his son. İbrahim, although initially welcomed into Greater Syria and southern Anatolia, soon found himself at odds with the population as Syria began to experience the 'new order' Egyptian style. İbrahim established a new town council, the *Majlis al-Shūrā*, the first in Damascus in 1832, comprising prominent local citizens, to deal with municipal questions beyond the jurisdiction of the shar'ia. The system was spread to every town of over 2,000 inhabitants. While self-governance was likely welcomed, increased taxation and conscription were not. In early 1834, Mehmed Ali, increasingly pressed at home for new recruits and new sources of revenue, insisted that his son impose a head tax, the *ferde*, on every male in Syria between the ages of 12 and 60. Already burdened by a series of extraordinary taxes, the population resisted when that was followed in mid-1834 by conscription and *corvée*. As taxes and the ill-paid forced labour of men and animals increased, so too did the resentment of the local populations.

As İbrahim began to impose conscription across Syria, rebellions broke out in northern Syria, Lebanon and Palestine, with the most severe in Jerusalem. By mid-June 1834, Jerusalem came back under Egyptian control, but İbrahim found himself unable to proceed further and called for reinforcements. Mehmed Ali himself, at the head of 11,500 infantry and cavalry, marched to Jaffa to the aid of his son. One of the results of putting down the rebellion was the disarming of much of the population of Palestine, Damascus and Aleppo, but pockets of resistance, in hard-to-reach regions such as Hawran, remained unbowed. Partial disarmament made conscription somewhat easier, although the old culture resurfaced once the Egyptians departed. The Ottomans began systematic application of both disarmament and conscription only after 1845, when regiments of the regular army began to be stationed in the provinces. Even then, it was sporadic and was resisted vigorously. Ottoman authorities were constantly forced to ameliorate the conscription regulations in Syria, especially as they failed to disarm the population after the Egyptian withdrawal in 1840. In urban areas they ultimately gained the upper hand over the remnants of the old para-military order; in the countryside, it took much longer.[75]

At the beginning of 1834, İbrahim maintained headquarters in Antioch, with troops in Urfa, while Ottoman army headquarters continued at Sivas under Reşid Mehmed Pasha. In August, Reşid Mehmed advanced to Malatya, north of Urfa, with 11,000 men and artillery. Mahmud II simultaneously proposed intervention while threatening naval action against the Egyptian fleet. The British fleet in the Mediterranean was placed on alert. As the stand-off continued, Mehmed Ali began to hint at independence in his negotiations with Great Britain and France, but found both powers intent at this point on preservation of the empire. To forestall the possibility of independence, Mahmud II agreed to settle the tribute question if Urfa were to be evacuated. Mehmed Ali opted for the compromise. The potentially explosive moment was avoided.

The next two years were spent in mediation, appeals to the European powers for support, offers and counter-offers, and diplomatic missions to and from Istanbul and Cairo. Mahmud II took the international community by surprise by sending an expeditionary force to Tripoli (Libya) in May 1835, ending 200 years of semi-independence of the Karamanllı family. By April 1837, Mahmud II began negotiating with Mehmed

Ali for his hereditary rule of Egypt and Syria, and Mahmud II appeared willing to concede the dynastic principle but not Damascus, Aleppo or Adana. This proved unacceptable to Egypt. In September 1837, Mehmed Ali captured Aden in the Gulf, extending his power over Arabia from Suez to Mukha on the Red Sea and threatening the British steam trade routes from Egypt to India. But İbrahim faced a sustained revolt of the Druze in Hawran at the end of 1837. An army of 20,000 finally defeated the Druze only in June 1838. In the negotiations which followed, İbrahim granted the Druzes considerable autonomy and exemption from conscription. By that time, Mahmud II's commanders had amassed 50,000 troops at Urfa, while Mehmed Ali accelerated his build-up in Syria in response. Another international crisis loomed.

Redifs and Kurds

Most of the *redif* recruits in Urfa were Kurds, as Reşid Mehmed Pasha's responsibilites there had included surveillance of Egyptian troop movements, as well as pacifying uprisings in territories around the city of Diyarbakır. Diyarbakır had an even longer history of semi-autonomous rule than Albania or Bosnia, situated as it was along the fault lines of the early Ottoman state, and Kurds, like Albanians, proved able but independent warriors. The history of the province followed many of the trajectories of resistance to centralisation and reform that we have seen elsewhere.

Kurds, like Albanians, spoke a different language from Turkish, actually several different languages. Also, like Albanians, many were putative Shiites, *alevis*. Kurdish tribal leaders, *mirs*, acquired tax-collection rights in exchange for military manpower. By the eighteenth century, much of the territory functioned as separate emirates, with official appointments from the Istanbul government such as the *mütesellim*, or tax collectors, who generally came from the local families. The governor of Diyarbakır, as with other provinces, was an Istanbul appointee. The population was extremely diverse: not only Kurds but Jews, Armenians, Türkmen and Bedouins in large numbers were native, or migratory, to the area. Isolation, mountainous and barren territory made incorporating the territory difficult – that is, until the time of Mahmud II.

In 1819, the city of Diyarbakır itself revolted against a particularly ruthless Ottoman Governor, Behram Pasha, Mahmud II's appointee to subdue the powerful merchant family elites. Behram Pasha was besieged in the fortress at Diyarbakır for close to three months, before the superior fire of his forces demolished the rebellion. Revolts began in the 1830s under the leadership of Said Bey, Ismail Bey, Vede Khan Bey, Bedr Khan of Cizre, of the Botan emirate, and Mir Mehmed Pasha of Rawandiz in present-day Iraq. The latter revolt was particularly vigorous, but successfully suppressed by Reşid Mehmed Pasha in 1835. For another decade, the territory was engaged in a long resistance to conscription and incorporation into the reformed empire. In 1838 the Ottoman forces finally defeated Bedr Khan of Cizre, but he emerged again after the Ottoman defeat at Nizib, in 1843, and with Nurullah in Hakkari, threatened the new sultan sufficiently to become a cause of international concern, especially after the massacre of local communities of Nestorian Christians. This last revolt could only be contained in 1847 by mobilising a large Ottoman army in the east under Ömer Pasha of later Crimean fame. Following its suppression, the Ottomans redrew the lines of the province of Diyarbakır, and stationed state troops on the Iranian/Russian border, arguably one of the last phases of the Ottoman pacification and incorporation of this particular frontier territory but hardly the last of its resistance. We will return to the discussion of this terrain, the Caucasus, Erzurum and Kars in the last chapter.[76]

Thus, after 1834 the Ottoman military was engaged in simultaneously quelling significant centres of revolt, and impressing large swathes of the Kurdish, Muslim population, much as they had in Albania. Zarif Mustafa Pasha was in Siverek, east of Urfa, in August 1834. 'When we arrived in Siverek, the Milli tribe was there, which had taken control of the desert, and its leader was known as the Desert Padişah. They were just three hours from us. We needed to befriend them; because they were very numerous, and had so many camels, it was impossible to estimate their number. *Miralay* Rüstem requested a volunteer, and I said yes. As I approached the tents, the Arabs, seeing me, sent up a cry, and jumped up. He was taken to the tent of one Eyup Bey. Everyone was looking at him because of his silver embossed jacket [epaulets]. They greeted one another courteously, whereupon Eyup Bey asked: "Why are you feeding your animals my grass?" "We know we are, but you are the servant of God and the Padişah. They, too, are the Padişah's animals," I replied. "Your Padişah does not interfere with the desert, but you have come as a guest, and they sent you here. If you had not come to explain, I would have destroyed you all. Do the Padişah's soldiers always go around dressed like this?" he asked. "Officers dress like this, but regular soldiers dress in a different fashion," I answered. "Aren't you at all ashamed? Such an outfit shows everything" – at which he called for a robe which they placed over my shoulders. If you come here again, do not come so naked, come with the robe, he said, and do not send your animals any further ahead [i.e. into his territory].'[77]

Helmuth von Moltke was assigned to Hafız Mehmed Pasha, Governor of Sivas and Diyarbakır and Commander of the troops in Anatolia, in the spring of 1838. Hafız Pasha was subduing the last hold-outs of the Kurdish rebellion in the mountains between Muş and Hazo, west of Bitlis. Moltke and two other Prussian officers were charged with mapping the territory, and reconnaissance. Moltke, at Hüsrev Pasha's request, had already seen to the translation of Prussian descriptions of the Landwehr organisation, and remained an adviser concerning recruitment and deployment of the *redif*.[78] Including the regiment of infantry, two regiments of Hassa cavalry, several hundred sipahis (one supposes irregulars), and three cannons, Hafız Pasha had 3,000 men. He had also called on loyal Kurdish Beys, Vede-Khan now among them, to join him. The passage across the valleys running parallel to the Euphrates tributaries was very difficult, Moltke observed. Ten horses were attached to each artillery piece in order to transport it over the rocks of the peaks. It was difficult even to find a place to pitch a tent. Passing through friendly Kurdish villages, the Pasha himself often remained in a village until the convoy passed, in order to prevent pillage. The next day, they camped next to a hostile village, whose inhabitants had not fled. Putting up a brief resistance, the villagers were surrounded and subdued. The soldiers then began pillaging the houses for booty, but the shooting continued until the last house, of the chief of the village, submitted. Trophies and prisoners were led before Hafız Pasha: men and women with bloody wounds; children of all ages; cut ears and heads. All such bounty was recompensed with 50 to 100 *kuruş*, a situation that he deplored, but acknowledged that Hafız had done his best to prohibit to the extent possible among the *başıbozuks*. Losses among the Ottomans were large, but after a day of rest, the campaign continued, amassing a mob of prisoners. By the time he was writing his letter, in mid-1838, the rebellions among the Kurds had for the most part ceased.[79]

To Moltke, the Ottomans were engaged in no less than 'brigandage' of the population. All the reinforcements for the *Asakir-i Mansure* army came from villages which resisted the government. Moltke gives one example: the town of Siirt, after being subdued by Reşid Mehmed Pasha, was enumerated as having 600 Muslim and 200 non-Muslim families. From the Muslim families, 200 recruits were impressed. When Moltke was writing his letter, the army was there again for another 200 recruits, but only the very old and very young were to be seen on the streets. The young men had fled to the mountains. In addition, disease and desertion simply carried off as many of those new recruits: in twelve months, some regiments were cut in half by deaths, even in a year of peace.[80]

For close to three years, Mahmud II had maintained large numbers of troops in Diyarbakır. As long as the *status quo* continued, however, quite apart from the added expenses, and the quartering of thousands of troops on a restive countryside, Mahmud II was simply unable to push through more of his provincial reform. By 1838 the Ottomans were clearly seeking a way to end the dispute, and so were amenable to British interests in a trade treaty, perhaps only as a means to that end. On 25 May 1838, Mehmed Ali announced his intention to declare independence to the European representatives in Cairo. The British and French counselled caution to both sides; the Russians offered military support to Mahmud II. Alarmed, the British pursued the means for reducing the influence of Mehmed Ali, and eliminating the Russian influence. The 1838 Anglo-Ottoman Convention, which resulted from this tense international crisis, satisfied two British demands: abolition of trading monopolies throughout the empire, including Egypt, and abolition of all internal tariffs. The former capitulations were renewed, but the tax structure, fixed for generations at three per cent import and export, was changed. Export duties on Ottoman goods were set at twelve per cent. Import duties were raised from three to five per cent. For Mahmud II, the attraction of such an agreement may have been entirely political, as it would deal a mortal blow to Mehmed Ali, whose economy was built on state monopolies, and perhaps presage further British naval intervention on behalf of the Ottomans. What it amounted to was a free trade treaty, which made of Britain the most-favoured nation, neutralised the Russian influence in Istanbul, and henceforth facilitated British informal colonialism in the Arab provinces. Ottoman trade had been bolstered; the British Mediterranean routes to India were secured. As a follow-up, in November 1839, Foreign Minister Mustafa Reşid Pasha was despatched to London with a request that Britain join the Ottomans in an attack on Mehmed Ali. Negotiations went on from November to March, before Palmerston's government offered a draft treaty of alliance, entirely defensive and hence of no interest to Mahmud II.

The Battle of Nizib, 1839

In the event, it remained a moot point. At the end of 1838, Mehmed Ali was known to be building up his forces around Adana and Aleppo. In

January 1839, Mahmud II called a Grand Council on the question of war with Mehmed Ali, but discussions were inconclusive, awaiting the outcome of Mustafa Reşid's mission to London. In February, Hafız Pasha assured his sovereign that his army was prepared to take Syria, and that the population was ready to rise up against the Egyptian occupation.[81] Against the advice of his court, and the foreign community in Istanbul, Mahmud II ordered a massive Ottoman build-up in Diyarbakır, Malatya, and Birecik. Some 3,000 horses, averaging 1,000 *kuruş* apiece, were purchased from the countryside around Diyarbakır; 40,000 *redifs* were assembled in Konya, Ankara, Kayseri and Malatya. By late April, the advance army under Hafız Pasha was approaching Birecik, southwest of Urfa, on the Euphrates River. İbrahim Pasha was similarly engaged in reinforcing the Egyptian army at Aleppo, and the assumption was that he would press the attack against Urfa and Diyarbakır. His passage to Adana could be blocked by the troops of Hacı Ali Pasha, mobilising to retake Konya. İzzet Pasha was mobilising in Kayseri. The sultan, cautioned by all sides to desist, would agree only if the foreign powers intervened to restore Adana, Aleppo, Jerusalem, Nablus and Sidon to him, and forced a reduction of Mehmed Ali's forces. Hafız Pasha's couriers to the sultan noted that the Egyptian army had not been paid for fifteen months, and victory was in his grasp. By May, three corps of 80,000 effectives, the majority fresh levies, *redifs*, stood under arms on the Syrian frontier, but not under a single command. Half of the regular army were recruits. Most were Kurds, treated as prisoners because of desertion, who were unable to communicate with their officers. Kurdish recruits, Moltke observed, in spite of generally being well fed, and well dressed, never lasted more than two years under arms. In Moltke's camp at Birecik, by June, 1,000 *kuruş* was being offered for the return of deserters, and still they ran. Irregulars arrived daily: Arab tribesmen; *başıbozuk* from Diyarbakır; and Kurdish allies such as Said Bey, recently defeated, with his entourage. They received a ration of wheat, 100,000 kilos of which had been requisitioned, but for all else had to live off the countryside. By Moltke's account, money and rations were not in short supply. Organisation and leadership were.[82]

On the Egyptian side, some 60,000 were said to be stationed between Aleppo and Antep, with 20,000 more on the way, but İbrahim was very short of money so there was misery in the Egyptian camp. Officers and soldiers deserting to the Ottoman side in early June reported eighteen months arrears in pay, and rations in very short supply. The Egyptian army was surviving on one third of the rations of the Ottomans. By this time, the mobilisation had been under way on both sides for over three months. Even so, the other two Ottoman corps had not yet arrived with reinforcements – indeed, had been very slow to follow orders, and the Ottomans were outnumbered, but probably not by more than 10,000.

In late May, Hafız Pasha, anxious for war, left his camp at Birecik, crossed the Euphrates into Egyptian territory towards Nizib, and began to erect fortifications. His plan was to stimulate revolt around Antep, and he was encouraged when the garrison there, having not been paid for eighteen months, surrendered after a small confrontation, and attached itself to the Ottoman army. At that provocation, İbrahim moved his entire army, commanded by Süleyman Pasha to within five and a half miles of the Ottoman troops. It was clear that the Egyptians were stronger and better situated than the Ottomans. The Ottoman line was too thinly deployed.

Frantic warnings and exchanges in the diplomatic community in Istanbul did not sway the sultan. On 9 June, Mahmud II ordered his navy to the coast of Syria. On 10 June Ibrahim led his entire army to the plains between Nizib and Birecik. On the evening of 21 June, Moltke advised leaving the first line of troops in place, to allow the rest of the camp to rest, but Hafız kept the entire force on alert for close to three days and nights, so quite apart from being exhausted, the troops were then ordered to confront the Egyptians in the open plain, against the specific orders of the sultan and the advice of the Prussian officers in his camp. Moltke had recommended that Hafız Pasha pull back to Birecik, a better position with which to confront the Egyptian army and a means of gaining time for the reinforcements to arrive. Moltke continued that while Hafız appeared poised to agree, he was dissuaded from such a course by a number of religious officials in the camp, who apparently persuaded the commander-in-chief that İbrahim intended to retreat towards Aleppo in the morning, and that 'the cause of the sultan was just; Allah will come to his aid, and all retreat would be a dishonour.'

On the morning of 24 June, İbrahim deployed his troops in three columns between Nizib and Birecik, so that he was situated between the Ottoman army and its stores. There was no place for retreat. The Ottomans had to regroup their lines to face the Egyptians, who advanced in admirable order. Moltke continued: 'the depth of their reserve [troops] was probably around three-quarters of a league . . . They arranged the artillery wisely, and used their cavalry to cover the artillery . . . Hence, while our rapid fire scattered over a large space, without reaching the reserves, their guns covered our entire line with their shot. In a few minutes, we had hardly any battalions where courage was not shaken by losses. Seven-eighths of these men had never heard the whistle of a gunshot; when a howitzer shell fell by chance on a column and exploded, entire companies began to disband. A short time later, almost all the battalions were praying, their hands to the sky . . . Once morale was lost, the battle was lost.' The army began to melt away; the Kurdish recruits became the enemy. Some fired on their own officers and comrades. Others simply threw off their uniforms and fled. Hafız and his remaining army fell back on Malatya, leaving behind artillery and supplies for the Egyptians. According to Moltke, one-sixth of the army remained when they reached Malatya: the *redif* forces simply went home *en masse*. İbrahim proceeded to occupy Antep, Maraş, and Urfa by 28 June, and there he stopped, persuaded by his father's orders not to cross the Taurus Mountains. He too had watched his army melt away, which might also explain his reluctance to follow up the resounding victory.[83]

Sultan Mahmud II died of tuberculosis on 30 June 1839, without hearing the news of the latest failure of his reformed army. On 14 July the entire Ottoman fleet surrendered to Mehmed Ali, largely because Admiral Ahmed Fevzi Pasha was fearful that the new sultan (and his new Grand Vizier, implacable enemy Hüsrev Pasha) would surrender the navy to the Russians for a joint action against Mehmed Ali. By 27 July, the five Great Powers (France, Britain, Austria, Germany and Russia), through their diplomatic representatives in Istanbul, had indicated their willingness to help the new sultan end the conflict with Mehmed Ali. It seems an appropriate place to pause, as the events mark a significant turning point in the fortunes of the Ottoman Empire.

It is difficult to assess the contribution of Mahmud, whose ferocity was matched by a vision not unlike that of Peter the Great. He inaugurated a process of reform and centralisation that would continue into the next two reigns, but did so by violently attacking his own citizens, gutting local economies, and laying the empire open to economic colonialism. During his reign, Greece, Serbia, Moldavia and Wallachia broke free; Egypt, Syria, Crete, Adana and Arabia were up in arms. The Russians acquired Bessarabia and large parts of the Caucasus. Algeria was occupied by France; Tunisia achieved virtual independence. Bosnia, Albania and Tripoli obeyed in name only, and the Russians passed the Balkan range. An infidel army had to protect the sultan in his own home against a Muslim army. Finally, Mahmud drew a line in the sand. With his journeys around the Balkans, the increasing reliance on 'Turks', and his refusal to countenance Mehmed Ali's army in Anatolia, he defended the very territories that Mustafa Kemal claimed for the Republic after 1918. He also created a template for the modern army which would allow it to happen. On his premature death, he left an empire perilously close to implosion.

Notes

1 Hakan Erdem, ' "Do think of them as not agricultural labourers": Ottoman reponses to the Greek war of independence', in Thalia G. Dragonas and Faruk Birtek, eds., *Citizenship and the Nation-State in Greece and Turkey* (London: Routledge, 2005), 67–85.

2 Charles Macfarlane, *Constantinople in 1828: A Residence of Sixteen Months in the Turkish Capital and Provinces, with an Appendix to the Autumn of 1829*. 2nd edn, 2 vols. (London: Saunders and Otley, 1829), vol. 1, 59–60.

3 Cunningham, 'Stratford Canning, Mahmud II, and Greece', in A. Cunningham, *Anglo-Ottoman Encounters in the Age of Revolution: Collected Essays vol. 1*, ed. by Edward Ingram (London: Frank Cass, 1993), 306–7.

4 Contemporaries, foreign and Ottoman alike, noted that the young boys were sometimes considered cadets, and enrolled in schools and apprenticed to artisans, but there is no doubt that youths appeared in the muster rolls, part of what Slade thought was Mahmud II's strategy to keep former Janissaries from the new ranks. (As discussed in Veysel Şimşek, 'Ottoman recruitment and the recruit 1826–53', MA Bilkent 2005, 42–3.)

5 Avigdor Levy, 'Military policy of Sultan Mahmud II, 1808–39', PhD, Harvard 1968, 405–06; Helmut von Moltke, *The Russians in Bulgaria and Rumelia in 1826 and 1829*, tr. by Marshal Diebitch (London: John Murray, 1854), 19, produces the same total, with a slightly different distribution, and estimates one-third of the force as cavalry, a high ratio for European armies of the period.

6 Levy, 'Military policy', 417, quoting Adolphus Slade.

7 Frederick W. Kagan, 'Russia's small wars', in F.W. Kagan and Robin Higham, eds, *The Military History of Tsarist Russia* (New York: Palgrave, 2002), 128.

8 Moltke, *The Russians*, 25–26. Over the course of the 1828 campaign, a total of 100,000 combatants were utilised.

9 Moltke, *The Russians*, 29. Levy, 'Military policy', ch. 7.

10 Moltke, *The Russians*, 58–59. Levy, 'Military policy', 418. It is hard to imagine Wallachia, in particular, producing these quantities at this time. The Ottoman figures represent what they expected from their arrangement with the Principalities. One has to assume these figures are expectations. This is

virgin research territory, for which there is tremendous evidence in the Başbakanlik.

11 Charles Macfarlane, *Constantinople in 1828*, vol. 2, 7–9.

12 Moltke, *The Russians*, 60–67.

13 Moltke, *The Russians*, pt. 1, ch. 5. The story is told just as thoroughly by Colonel F.R. Chesney, who arrived in 1829, in *The Russo-Turkish Campaigns of 1828 and 1829 with a View of the Present State of Affairs in the East* (New York: Redfield, 1854). They are curiously similar – Chesney quotes Moltke.

14 Moltke, *The Russians*, pt. 1, ch. 6–7.

15 Moltke, *The Russians*, pt. 1, ch. 8–9.

16 Moltke, *The Russians*, pt. 1, ch. 9. This was the era when quarantine regulations had begun making inroads into the spread of plague, but naturally isolation was hard to impose on campaign. (Moltke, *The Russians, Appendix*, 461–76). Mahmud II established quarantine and plague hospitals in 1836 (Philip Mansel, *Constantinople: City of the World's Desire, 1453–1924*. New York: St. Martin's Press, 1995, 256.) The habit of seasonal campaigning, and abandoning the battlefront in winter, a feature of Janissary history, may actually have inhibited such occurrences in former days.

17 Macfarlane, *Constantinople*, vol. 2, 308.

18 Moltke, *The Russians*, 268. Slade estimated that Reşid Mehmed never had more than 35,000 troops at his disposal (Adolphus Slade, *Record of Travels in Turkey, Greece, etc.* New edn. (London: Saunders & Otley, 1854), 160).

19 Moltke, *The Russians*, pt. 2, ch. 1–2.

20 Moltke, *The Russians*, 391. That must also have been the case for the Ottoman army, although personal cleanliness and long familiarity with the local diseases may have made the Ottoman losses lighter. Chesney's estimates of total losses reinforce that argument: 16,000 Ottomans and 50,000 Russians died from wounds and disease alike. The Russians lost 20,000 horses (Chesney, *Russo-Turkish*, 199).

21 Moltke, *The Russians*, pt. 2, ch. 3–4.

22 Vernon John Puryear, *France and the Levant: From Bourbon Restoration to the Peace of Kutiah* (Berkeley: UC Press, 1941; reprint Archon Books 1968), 85. Hüsrev Pasha reputedly told the ambassadors: 'we were so beaten, we cannot be beaten more; resistance is useless' (Slade, *Record*, 202).

23 Moltke, *The Russians*, pt. 2, ch. 5–6. On 'Mustafā Pasha Bushatlı (1797–1860)', see *EI2* CD edition, by A. Popovic. Mustafa survived defeat, and was later sent on to serve as governor of a number of provinces, under Abdülmecid. He is portrayed as the first Albanian nationalist elsewhere.

24 David Urquhart, *The Military Strength of Turkey* (London: Diplomatic Review Office, 1868), 110. The text was originally written just prior to the Crimean War.

25 Urquhart, *Military Strength*, 117–18.

26 Moltke, *The Russians*, 262. Macfarlane, *Constantinople*, vol. 2, 46–51, mused that neither of these denizens of the streets of Istanbul is likely to be a great cutter of ears. *Pezevenk*, from Persian, means 'pimp' in modern Turkish, while *karata*, from Greek, originally 'cuckold,' now has the force of 'rascal'.

27 Hakan Erdem, 'Recuitment of the "victorious soldiers of Muhammad" in the Arab provinces, 1826–28', in Gershoni, Erdem and Woköck, eds, *Histories of the Modern Middle East: New Directions* (Boulder, CO: Lynne Riener, 2002), 198–99. In fact, it is more than likely that many of the former Janissaries had been incorporated into the local forces of cities such as Damascus.

28 I am deeply indebted to Fred Anscombe for this quote, taken from his unpublished paper, 'Reform, resistance and Islam in the Ottoman Empire'. The quote is from BA, HH 21173-D, 15 April 1831.

29 Hakan Erdem, 'Recruitment', 198–99. Charles Macfarlane, returning to Turkey in 1847 and 1848, noted: 'conscription, as I have repeatedly observed, is eating up the remnant of the Mussulman people and consuming the heart's core of the Empire' (*Turkey and its Destiny: the Results of Journeys Made in 1847 and 1848 to Examine into the State of that Country*, (Philadelphia: Lea & Blanchard, 1850)), 327.

30 Ruben Safrastjan, 'Ottomanism, in Turkey in the epoch of reforms in XIX c.: ideology and policy 1', *Études Balkaniques* 24 (1988), 81–82.

31 Safrastjan, 'Ottomanism', 74–75. Viorel Panaite, 'The *Re'ayas* of tributary-protected Principalities: sixteenth to eighteenth century', *International Journal of Turkish Studies* 9 (2003), 83.

32 Safrastjan, 'Ottomanism', 75, footnote 7. The imperial tours are described in Abdülkadir Özcan, 'II. Mahmud'un memleket gezileri', in *Prof. Dr. Bekir Kütükoğlu'na Armagan*. İstanbul: İstanbul Üniversitesi Edebiyat Fakültesi, 361–442. The new public sultan is also discussed at length in Cengiz Kırlı, 'The struggle over space: coffeehouses in İstanbul 1780–1845', PhD (SUNY, Binghamton), 262–8.

33 Moltke, *Lettres de Maréchal de Moltke sur l'orient*, tr. by Alfred Marchand. 2nd edn. (Paris: Librarie Fischbacker, 1872), 322, 111–12.

34 Population figures for the Empire in the early 1800 period are notoriously unreliable. In the Balkans, contemporaries estimated a ratio of one-third Muslim to two-thirds Christian, but actual figures ranged from 7–9 million Christians in the European territories (David Urquhart, *Turkey and its Resources* (London, 1933), 271–72, where he also supplied a breakdown by ethnic groups, even harder to gauge).

35 Radu R. Florescu, *The Struggle against Russia in the Romanian Principalities: a Problem in Anglo-Turkish Diplomacy, 1821–54* (Iaşi: Center for Romanian Studies, 1997), 152.

36 Afaf Lutfi Marsot, *Egypt in the Reign of Muhammad Ali* (Cambridge: Cambridge University Press, 1984), 211.

37 Judith Mendelsohn Rood, 'Mehmed Ali as mutinous khedive: the roots of rebellion', *International Journal of Turkish Studies* 8 (2002), 115–28, presents the arguments well.

38 For the debate over Mehmed Ali as the putative founder of modern Egypt, see chapter 8, note 4.

39 Moshe Ma'oz, *Ottoman Reform in Syria and Palestine 1840–61*. Oxford: Clarendon Press, 1968, 6–7. Muhammed H. Kutluoğlu, *The Egyptian Question, 1831–1841: The Expansionist Policy of Mehmed Ali Paşa in Syria and Asia Minor and the Reaction of the Sublime Porte* (Istanbul: Eren, 1998), 73–74.

40 L.S. Schilcher, '*Families in Politics: Damascene Factions and Estates of the Eighteenth and Nineteenth Centuries*', 40–43, includes a translated description of the events from a contemporary chronicle.

41 Khalid Fahmy, *All the Pasha's Men: Mehmed Ali, His Army and the Making of Modern Egypt* (Cambridge: Cambridge University Press, 1997), 247–48.

42 Levy, 'Military policy', 452–54; quoted in Roderic H. Davison, *Reform in the Ottoman Empire 1856–76* (New York: Gordian Press, 1972), 31.

43 Fred Lawson, *The Social Origins of the Egyptian Expansionism During the Muhammad Ali Period*. New York: Columbia University Press, 1992, 105, and ch. 5.

44 Kutluoğlu, *The Egyptian Question*, 58–59, 73–74.

45 Fahmy, *All the Pasha's Men*, 62 tells the story of this part of the invasion succinctly, as does Kutluoğlu. Both are well versed in the contemporary accounts. P.M. Holt, *Egypt and the Fertile Crescent 1516–1922* (London: Longman, 1966) remains a standard account.

46 Ma'oz, *Ottoman Reform*, 13.

47 Kutluoğlu, *The Egyptian Question*, 63–67.

48 Kutluoğlu, *The Egyptian Question*, 70.

49 Fahmy, *All the Pasha's Men*, 63.

50 Marshal Auguste Frédéric Marmont, *The Present State of the Turkish Empire*, translated by Frederic Smith. London: John Ollivier, 1839, 275–77.

51 Fahmy, *All the Pasha's Men*, 65; Marmont, *Present State*, 278–79; Kutluoğlu, *The Egyptian Question*, 75–76.

52 Kutluoğlu, *The Egyptian Question*, 77. Fahmy, *All the Pasha's Men*, 67, draws on an Arabic source for his figures, but notes that Marmont provides a much more realistic figure of 50,000 regulars for the Ottoman forces (252). Reşid Mehmed, it will be remembered, was a protegée of Hüsrev Pasha, the fierce old man once again in control of the court in Istanbul. The animosity of İbrahim and Hüsrev, rivals in the Morea, and the supposed intense hatred of Mehmed Ali for Hüsrev, is often used to explain the difficulties between Istanbul and Cairo during the entire decade.

53 The process of pacification and recruitment would be repeated with the Kurds in ten years. See below.

54 Fahmy, *All the Pasha's Men*, 259.

55 Levy, 'Military policy', 597–98. The defter he probably used has been reconsulted by Veysel Şimsek (Kamil Kepeci Askeri Defterleri No. 6799). The figures put total losses between 1826 and 1837 at 55.41% of the total army figure of 161,036. Significantly, 20,117 are listed as deserters, and 21,298 as missing in combat, which accounts for almost a quarter of that percentile. ('Ottoman recruitment and the recruit 1826–53', MA Bilkent 2005, 92.)

56 Fahmy, *All the Pasha's Men*, 161; Marmont, *Present State*, 285–89.

57 Enver Ziya Karal, 'Zarif Paşa'nın Hatıratı, 1816–62', *Belleten* 4 (1940), 443–94. Zarif Pasha is described further below.

58 Marmont was writing at the time of the second crisis in 1839, 289–90. Fahmy includes a translation of Egyptian military historian A. Zaki's description of the battle at Konya on page 163 of *All the Pasha's Men*, and enumerates the number of casualties on 67.

59 Fahmy, *All the Pasha's Men*, 71, elaborates on this particular question, which is much debated. By 1838, Mehmed Ali does begin to refer to a hereditary independence, in a much changed international situation.

60 Kutluoğlu, *The Egyptian Question*, 95–105; Fahmy, *All the Pasha's Men*, 67–68.

61 Roderic H. Davison, 'Britain, the international spectrum, and the Eastern Question 1827–41', *New Perspectives on Turkey* 7 (1992), 15–35. The classic work on this period is Frank Edgar Bailey's *British Policy and the Turkish Reform Movement: A Study in Anglo-Turkish Relations, 1826–53.* Cambridge, MA: Harvard University Press, 1942.

62 Kutluoğlu, *The Egyptian Question*, 106–11.

63 Davison, 'Britain, the international spectrum', 27.

64 Bailey, *British Policy*, 147–49.

65 The reforms of this period are the best studied aspect of the entire reign of Mahmud II and his successors. Their description is readily available in

Bernard Lewis, *The Emergence of Modern Turkey* (London: Oxford University Press, 1961), and Niyazi Berkes, *The Development of Secularism in Turkey* (Montreal: McGill University Press, 1964). See also Ben Fortna, *Imperial Classroom: Islam, the State and Education in the Late Ottoman Empire* (Oxford: OUP, 2002).

66 Lewis, *Emergence*, 87. On the characteristics of the elites that emerged in this period, see especially Carter V. Findley, *Bureaucratic Reform in the Ottoman Empire 1789–22* (Princeton, NJ: Princeton University Press, 1980), 126–87, *passim*.

67 Ahmed Emin, *The Development of Modern Turkey as Measured by its Press* (New York: Columbia University Press, 1914), 30–34. All the histories of reform have relied heavily on both the *Moniteur* and the *Takvim* for the official rhetoric. *Takvim* was eventually published (or translated) into Armenian, Greek, Arabic and Persian.

68 Moltke, *Lettres*, 26–29. The impression is corroborated by Marshal Marmont, who was warmly received by Hüsrev Pasha two years later, 'In the pursuit of his objects he had never put in motion any of the noble springs which best adorn the human heart, and he is said to have been more engaged in forwarding his private interest, than the service of his country or his master.' (Marmont, *Present State*, 55–56.)

69 Enverziya Karal, 'Zarif Paşa'nın hatıratı, 1816–62', Belleten 4 (1940), 443–54.

70 Lewis, *Emergence*, 103, and Levy, 'Military policy', 475–78.

71 Levy, 'Military policy', 479–86.

72 300,000,000 *kuruş* equalled 3 million pounds sterling at the time. Levy, 'Military policy', 490–500. Levy remarks that while enlisted men's salaries did not change, officers' salaries climbed significantly in this period, probably a much-needed incentive.

73 Ahmet Uzun, '1257/1841 tarihli bir belgeye göre Osmanlı devleti'nde mevcut olan askeri birlikler ve bunlara yapılan harcamaların türü ve miktarları', *C.Ü. Sosyal Bilimler Dergisi* 25 (2001), 236–39. My thanks to Veysel Şimşek for drawing this article to my attention.

74 Levy, 'Military policy', 573–89. Musa Çadırcı, 'Redif askeri teşkilatı', in *Yedinci Askerî Tarih Semineri Bildirileri* (Ankara: T.C. Genelkurmay Başkanlığı, 2000), includes copies of the later 1843 reforms to the organisation, and Mübahat S. Kütükoğlu, 'Sultan II. Mahmud devri yedek ordusu Redîf-i Asâkir-i Mansûre', *İ.Ü. Edebiyat Fakültesi, Tarih Enstitüsü Dergisi 12* (1982), 127–57, offers a detailed look at the *redif* and its financing.

75 Kutluoğlu, *The Egyptian Question*, 112–20. Dick Dowes, 'Reorganising violence: traditional recruitment patterns and resistance against conscription

in Ottoman Syria' in J. Zürcher, ed., *Arming the State* (London: I.B. Tauris, 1999), 110–27; see also his *The Ottomans in Syria* (London: I.B. Taurus, 2000), especially 188–210, 121–23; Ma'oz, *Ottoman Reform*, 16–17, 81–86.

76 Martin van Bruinessen, *Agha, Shaikh and State: the Social and Political Structures of Kurdistan* (London: Zed Books, 1992). On the Kurds, see also Ariel Salzmann, *Tocqueville in the Ottoman Empire: Rival Paths to the Modern State* (Leiden: E.J. Brill, 2004), is actually a study of Diyarbakır and its relationship with Istanbul. Hakan Özoğlu, *Kurdish Notables and the Ottoman State: Evolving Identities, Competing Loyalties, and Shifting Boundaries* (Albany, NY: SUNY Press, 2004). Neither book takes up the military situation, or the early nineteenth century, in an effective way. For a thorough survey of Kurdish history, see V. Minorsky, 'Kurds' in *EI2* CD edition.

77 Karal, 'Zarif Paşa'nın hatıratı', 456–7. As with the other quotes from these memoirs, they are both inscrutable and naive. This one is particularly amusing, but its message is quite clear.

78 As explained in Kütükoğlu, 'Redif-i Asâkir-i Mansûre,' 135–6.

79 Moltke, *Lettres*, 205–9.

80 Moltke, *Lettres*, 212 and 264.

81 Kutluoğlu, *The Egyptian Question*, ch. 5.

82 Moltke, *Lettres*, 266–86.

83 Kutluoğlu, *The Egyptian Question*, 138–40; the quotes are from Moltke, *Lettres*, 286–307.

CHAPTER TEN

From Tanzimat to Crimean War

The core of this book has been about the evolution of a military system from the Janissary to conscript army, especially as events unfolded during the reigns of Selim III and Mahmud II. By the latter's death in 1839, the essentials of the new military system were in place, at significant human cost. The regulations, however, remained unevenly applied and largely underfunded. The next period began with resolution of the Mehmed Ali crisis, 1839–41. With the intervention of the five great powers, committed to supporting Sultan Abdülmecid I (1839–61), rather than Mehmed Ali Pasha of Egypt, the Ottomans continued their integration into the European sphere of international relations. By the Anglo-Ottoman Convention of 1838, Britain had essentially nullified the 1833 Russo-Ottoman Treaty of Hünkâr İskelesi, and achieved free-trade access to Ottoman territories, especially in Egypt and the Arab cities of greater Syria. Two further agreements would establish the primacy of the European powers in Ottoman affairs. The first, the Convention for the Pacification of the Levant, or the London Treaty of 15 July 1840, was signed without the agreement of France, but by the other four powers and the Ottomans. If Mehmed Ali were to give up Syria, Adana, Arabia (the Hijaz), and Crete, he and his family could continue to rule Egypt. The second, the Straits Convention of 13 July 1841, signed by all, including France and the Ottomans, followed the military defeat of İbrahim Pasha in Damascus and Beirut. It gave the sultan 'a manifest proof of the respect which they entertain for the inviolability of his sovereign rights, as well as of their sincere desire to see consolidated the repose of this empire.'[1] The maintenance of the territorial Ottoman Empire, later dubbed the 'Sick Man of Europe' by Tsar Nicholas I, had become an international concern, whose recovery was sanctioned by the

European powers. The empire proved an unpredictable patient, however, and manoeuvred its physicians (France, Britain and Russia) into the most foolish and costly war of the entire nineteenth century, the Crimean War 1853–56, acknowledged by many as the first modern world war in terms of scope, immediacy, reporting, public opinion, and global consequences.

Great Britain had embarked on an expansion of colonial and semi-colonial regimes in territories as far flung as Africa, India, the Middle East, and China. Underlying the new, and unequal, treaty relationships with non-western countries was a changing view of international relations. Early theorists of international law, which had been understood to derive from 'natural law', accorded sovereign rights to all nations, regardless of religious or ethnic populations. By the mid-nineteenth century, international law had come to be understood as 'an artifact of Christian civilisation,' and therefore was not fully applicable to Muslim (or other non-Christian) states, who had not yet joined the civilised nations. Civilised in this case meant a state which guaranteed the basic rights of property and person, a political system which could organise self-defence, and which adhered to the rules of international law and diplomacy. International relations theory posited a hierarchy of states, in descending order from 'civilised' to 'barbaric' to 'savage', with the latter two categories remaining outside the realm of international law. This included, of course, the Ottomans, as long as they resisted the introduction of such rights. Foreign representatives used such arguments to justify intervention on behalf of co-religionists in Ottoman territories, and to justify the wide application of the principle of extraterritoriality to very large numbers of Ottoman non-Muslim subjects, a clear intervention in state sovereignty. It is in this international context that the 1839 and 1856 reform proclamations of Ottoman sultans must be understood.

All European powers profited from the informal colonial regime established in the Middle East after 1841, if only to give them respite from the potential conflicts which the Eastern Question had engendered. Britain proved the greatest beneficiary, but diplomats, merchants and missionaries of many nations, including the United States, all streamed to the Middle East. The period between the Treaty of London (1840) and the commencement of Russo-Ottoman hostilities in 1853 was a moment of intense religious proselytising, largely stimulated by the increased presence of missionaries among Ottoman Christian populations, but equally attributable to the imperial posturing of France and Russia, in particular, over the protection of their co-religionists and over access to the sacred places of Jerusalem. It is one of the great ironies of Ottoman history that at precisely the moment when Ottoman rulers aimed at introducing universal citizenship and equality before the law, the European powers seized on the religious *millets* as representing 'national' groups requiring their protection. As Berkes acutely notes, 'The economic interests of the European powers pressed for secularisation while the political-cum-religious interests of the same powers demanded the perpetuation of communal differentiations ranging from the legal and political to the educational field.'[2] Non-Muslim communities were riven by conflicting loyalties, from those who benefited from continuing ties with the dynasty, to those who advocated outright revolt and separation from the empire. For thousands of Muslims, before and after the Crimean War, it was equally a period of immense upheaval as they found their former status undermined by the new laws concerning equality, and their economic foundations crumbling as Anatolia and the Levant became part of the global economy. The seesaw of reform and reaction which characterises 1839–76, the period known as the Tanzimat, is indicative of the long struggle of an Ottoman Muslim dynasty and its bureaucracy to join the 'civilised' nations of Europe on its own terms.

The settlement of the 1839–41 eastern crisis is generally described as the triumph of Palmerstonian diplomacy.[3] The resolution of the Mehmed Ali crisis, and the pacification of Syria and Egypt, will be reviewed here to consider the consequences for Ottoman reconstruction in Syria. The real aim of this chapter is to set out the further changes to the military system from 1841 to 1856, and to describe select episodes of the continued empire-wide resistance to reforms, specifically the conscription regulations that were put in place in 1843 and again in 1846.

The causes of the Crimean War are numerous, and the conflict itself has received considerable attention in a number of new studies and a large exhibition to commemorate the sesquicentennial anniversary of the war. The causes, major confrontations and resolutions will be outlined, but the emphasis will be on an examination of the conflicts on the western and eastern peripheries of the Ottoman northern frontier. For the last time, we will examine the state of the military in Russia, Austria and the Ottoman Empire just prior to the Crimean War, for while the main protagonists of the war on the Danube were the Russians and the Ottomans, the Austrians had a pivotal role to play in the period just before the conflict, much as they had a half-century earlier. The chapter ends with the 12 March 1854 Anglo-Ottoman–French alliance against Russia, prelude to the Crimean War. What concerns us most are the continued Russo-Ottoman disputes over the Principalities, Moldavia and Wallachia, and the Caucasus. The Crimean War is the last time that British and Ottoman officers actually fought on the same side, and the last time British soldiers fought in Europe until 1914. After 1870, with new regulations giving the army a further Prussian cast, the Ottoman army came under the influence of a German military mission which would carry it through World War I.

New sultan, new advisers

Abdülmecid was a boy of sixteen when his father died. In the first two years of his reign, quite apart from the pressing international crisis over Mehmed Ali and İbrahim Pashas, still camped with the Egyptian army on his doorstep, he had to deal with the internecine strife in his court over the question of reform. Hüsrev Pasha, feared ancient enforcer of Mahmud's reign, had one last trick up his sleeve. He is said to have wrenched the seal out of then Grand Vizier Rauf Pasha's hands, and assumed the office for himself on the accession of Abdülmecid, and kept it for almost a year, until June of 1840. The resolution of the Mehmed Ali crisis was likely delayed by this usurpation, as Mehmed Ali refused to consider Ottoman overtures unless Hüsrev were removed from the position. And yet, adding to the enigma of his long life, arch conservative Hüsrev Pasha was Grand Vizier when the Gülhane Edict (known as *Tanzimat Fırmanı* in Turkish, and as the Gülhane Edict, or Rescript in English) was promulgated on 3 November 1839. The text of the proclamation was said to be the work of his greatest rival, Foreign Minister Mustafa Reşid, Anglophile, veteran of diplomatic missions to Paris and London, who represented the reform party of the empire. It inaugurated the period referred to as the Tanzimat, from the decree itself, *Tanzimat-i Hayriye*, or 'Beneficial Reforms'. The period is also characterised by intense rivalries between officials of the grand vizierial offices and the palace, the latter where the military high commanders, the majority married into the royal family, held sway.

Mustafa Reşid is often called the father of the Tanzimat. Six times grand vizier and three times foreign minister, he is certainly one of the main reform figures of the period, along with his protégés Ali and Fuad Pashas. He had served with the army in the entourage of *Serasker* Seyyid Ali Pasha during the Greek rebellion in 1821, and again as army clerk in the Balkans in 1828, but thereafter joined the bureaucracy as a member of the Correspondence Office, which formed part of the Ministry of Foreign Affairs. He was appointed to important posts as part of diplomatic missions to Egypt, Paris and London, and was named head of the Ministry by 1837. Mustafa was instrumental in securing the Anglo-Ottoman

Commercial (Balta Limanı) Convention of 1838, and has generally been credited with obtaining the agreement of Abdülmecid to issuing the formal Gülhane decree declaring the Ottoman reformation. British Ambassador John Ponsonby (1832–41) is given much of the credit for the language of the decree, and many of the narratives of this period make the influence of Reşid and Ponsonby (and later Ambassador Stratford Canning) paramount in all events that unfold.

There are, however, strong arguments for also seeing the influence of both the old guard and new religious advisers in Ottoman ruling circles, voices that had emerged as part of the debate over the dynasty's legitimacy and survival in its struggle with Mehmed Ali. The presence of the Naqshbandi sufi groups in Istanbul in the 1820s has already been noted in relation to the elimination of the Janissary corps. They appear to have dominated court politics for the latter half of the nineteenth century, especially as the imams of the inner circles of sultans Abdülmecid and Abdülaziz. They preached a strict emphasis on the *sunna* and the shar'ia, and the supremacy of the shar'ia in the life of the community and state. Among those whose names were linked to the order were otherwise implacable opponents Hüsrev Pasha and Mustafa Reşid Pasha.

In a directive that Sultan Abdülmecid had sent to his advisers, he declared his dedication to the principles of shar'ia law and its application in all the affairs of state, and asked their advice concerning the problems he should address. His ministers replied that three principles should guide his reign: that the guarantees for life and property and for the preservation of honour and dignity should be extended to Muslim as well as non-Muslim; that taxes should be fixed according to the wealth and means of each subject; and that the burden of military service should be devised according to the size of the population in each province. Among the 38 signatories to this response were Hüsrev Pasha, *Şeyhülislam* Mustafa Asım Efendi and *Serasker* Halil Rifa'at Pasha, as well as Mustafa Reşid Pasha, all veterans of Mahmud II's reforms. They set a restorative agenda for the sultan, which was then read in public and, for the first time, in front of the European diplomatic corps. The text included a paragraph in which the sultan 'pledged to take an oath in the hall of the sacred relics, not to act contrary to its stipulations, and that the senior ulema and state functionaries take a similar oath, an action no Ottoman sultan before Abdülmecid had ever undertaken.'[4]

Mahmud II had begun the process of modernising the image of the dynasty, much in the manner of the Romanovs and Habsburgs, by engaging with the public. We have seen how he reconfigured the image of the sultan by personalising his rule and devising symbolic representations of ancient traditions. He left behind the makings of the modern bureaucracy, and established councils of his advisers. His concept of justice (*adalet*) introduced the idea of equality before the law. Abdülmecid continued in like fashion, with the Gülhane Edict, which, following the preamble above, elaborated on the actual reforms Abdülmecid and his adviers envisioned. The inspiration for its creation is actually less important than what it had to say, which was remarkable. The sultan pledged to eliminate tax farming (*iltizam*); to contravene venality; to eliminate confiscation; to guarantee property rights to individuals; and to institute a public and open court system. These concessions were to be extended to all subjects, regardless of faith. About service in the army, Abdülmecid pledged to reduce service to four or five years, and to continue the establishment of local contingents (*redifs*), in order to prevent the impact on agriculture and industry by depleting manpower in particular areas. The implication was that this too would be universally applied. The Supreme Council of Judicial Ordinances (*Meclis-i Valâ-yi Ahkâm-i Adliye*), established in 1838, became the principal consultative and legislative body. Every aspect of society was to be subjected to scrutiny and reorganisation. While most of the agenda was probably premature and unachievable in 1839, the document did set the parameters for discussion about the nature of late Ottoman society. The edict had a dual character, or more accurately, a split personality. It promoted a universality of citizenship which was grounded in the principle of a religious hierarchy: Muslims and the non-Muslim *millets*. The reformers likely aimed at fostering a new 'Ottomanism' that would rally support for the empire, but in fact, the ambiguities about universal equality alienated many Muslims after the initial fervour.[5] Most of the foreign community in Istanbul was sceptical about the possibility of transformation. For the Ottomans, much of the next two decades was spent in working out what it actually meant. We will return to the actual reforms shortly. But first, the Mehmed Ali crisis had to be resolved, and order restored.

Taming Mehmed Ali

Since the defeat at Nizib, negotiation with Mehmed Ali had continued. Abdülmecid sent a pardon to Mehmed Ali, requesting İbrahim to withdraw to Maraş. France stood behind Egypt and Mehmed Ali, who appeared bent on keeping Syria and was obdurate about wanting Adana as well. In return, he would surrender the Ottoman fleet and Crete. In July 1840,

Mehmed Emin Rauf Pasha replaced Hüsrev as Grand Vizier for his third time in the office. By mid-July, Austria, Prussia, Russia and Britain and the Ottomans signed the Convention of London (15 July 1840). They were to provide assistance in reducing Mehmed Ali's power, use their fleets to cut off his supplies, and protect the straits, and maintain the rule of closing the straits to foreign ships of war. Mehmed Ali was offered Acre for his life, with the hereditary rule of Egypt. He would have to withdraw his forces from Syria, and Egypt was to be restored to the empire.

By August 1840, much of the population of Syria, especially portions of the Druze, and Maronite Christian populations of the mountains, was in arms against the Egyptians, stimulated both by British agents and by İbrahim Pasha's efforts to disarm the population. İbrahim Pasha's second-in-command, Süleyman Pasha, was in Beirut to quell the disorder, with 15,000 Egyptian troops who had landed in Syria to suppress the revolt, with the acquiescence of Druze ally Bashir II. The rest of İbrahim's army was scattered from northern Lebanon to Adana. On 12 August, Sir Charles Napier, on orders from British Admiral Stopford, was anchored before Beirut announcing the Ottoman intention to retake Syria. Mehmed Ali was given an ultimatum which he ignored; by 5 September, the four consuls of the convention signatories in Alexandria declared he had rejected a second offer. Beirut was bombarded by a British fleet of 30 ships, under command of Napier, on 11 September. Napier then landed with a combined force of 5,300 Ottoman troops, who arrived from Cyprus under Mehmed Selim Pasha; 1,500 British troops, and 200 Austrian marines, to secure the territory north of Beirut (Juniya). Some 5,000 Druze and Maronite mountaineers, under Emir Bashir Qāsim, cousin of Emir Bashir II, Beshir (Shihab) II of Mt. Lebanon, joined the Anglo-Ottoman force. On 24 September, four battalions of the Ottoman soldiers with Napier, under command of Mehmed Selim Pasha, and Prussian General Jochmus, drove back İbrahim's advance guard entrenched at Chahaoune. On 26 September, Beirut was again bombarded from the sea as Süleyman Pasha evacuated. Two days later, Napier, with 1,000 mixed British and Turkish troops, confronted and defeated the Egyptian garrison at Sidon, taking some 2,000 prisoners. İbrahim had by then set up headquarters in the formidable fortress of Zahle.

On 8 October, Emir Bashir Qāsim, with his 1,000 mountaineers, attacked Egyptian outposts and dispersed them. By *firman* of the sultan, he was later declared Grand Emir of Lebanon, replacing Bashir II, who had refused to abandon İbrahim. The squeeze of the Egyptian forces by land and by sea continued. On 3 November, Admiral Stopford bombarded the city of Acre for two hours, when the arsenal within the fortress blew up, forcing the surrender of the garrison. The destruction and recapture of Acre, renowned for Cezzar's 1798 holdout against Napoleon and Abdullah's 1831 resistance against İbrahim Pasha, signalled the demise of the Egyptian adventure in Syria. By the end of the month, the Egyptians abandoned Adana, Tarsus, Aleppo, Jaffa and Jerusalem. İbrahim himself took refuge in Damascus with 12,000 troops.

In mid-November, Napier, acting without official approval, secured Mehmed Ali's submission. On 27 November, Mehmed Ali agreed to order İbrahim to evacuate Syria, and restore the Ottoman fleet, if the sultan granted him hereditary rule in Egypt. The sultan rejected the convention and demanded instead an unconditional surrender, which was finally achieved on 11 December, and accepted by Sultan Abdülmecid on 27 December. On 30 December, İbrahim evacuated Damascus, to begin a long and grievous retreat across Gaza to Egypt. The army, stragglers and deserters alike, were harrassed by Ottoman troops under General Jochmus, local Bedouins, Maronite and Druze. İbrahim retreated slowly south, and by 6 January 1841, had lost an estimated 10,000 men to desertion and death. By the time the remains of the Egyptian army arrived home, they had been surviving on mule and donkey flesh, and without water for three days.

In May 1841, the sultan's decree accorded Mehmed Ali hereditary status as Governor of Egypt, but the sultan retained the right of approval of the heir. Mehmed Ali was to pay an annual tribute of one quarter of Egyptian revenues. These latter two points were contested by Mehmed Ali, and he and his progeny subsequently won the right of independent succession, based on the rule of primogeniture, and a much reduced tribute. The Egyptian army was to be maintained at 18,000 men, and the new Viceroy (*Khedive*) had the right to appoint the Colonels (*Miralays*). The senior commanders, Brigadier-Generals (*Mirliva*) and Generals of Divisions (*Ferik*), were to be appointed in Istanbul. The language of Abdülmecid's decree, confirming the hereditary status of Mehmed Ali and his heirs, also reveals how Egypt was to be subordinated within the Ottoman military structure: 'as the land and sea forces of Egypt are designated for the service of my Empire, in time of war it will be permitted to increase the number of soldiers in a way which will be approved by my Empire. As the principle . . . [of] five years' service has been established, it is necessary to implement the same principle in Egypt, so 20,000 of the present Egyptian soldiers will be separated out . . . 18,000 of them will be employed in . . . the internal service in Egypt . . . 2,000 soldiers to be employed here [Istanbul].' While the Egyptians might not stick entirely to the letter of the law, and the degree of independence of the Khedives continued to be contested, both Khedive Abbas Hilmi, 1848–54, and Khedive Said sent contingents of 8,000 troops to the Crimean front.[6]

France, which had supported Mehmed Ali throughout the crisis, reacted angrily at being excluded from the Convention of London in 1840, but failed to support Mehmed Ali in the final confrontation. For a brief moment European war appeared imminent, but Paris was conciliatory by October of the same year, thus averting a diplomatic crisis. On 13 July 1841, the Straits Convention, signed by the five powers and the Ottomans, reiterated the closure of the Dardanelles and Bosphorus straits to foreign warships, confirmed the integrity and independence of the Ottoman empire, and extended the terms of the 1838 Anglo-Ottoman Convention to Egypt. The French rejoined the Concert of Europe, and the Ottomans were brought into the circle. Mehmed Ali's regional empire was shattered, but his fondest wish for his family was achieved, and the dynasty lasted until 1952. Britain achieved an overland route to India from Alexandria to Suez. Egypt's economy was drawn into the world economy as a market for European goods, and a source of grain and cotton. By 1848, both İbrahim and Mehmed Ali, the latter incapable of rule since 1847, were dead. Mehmed Ali's reforms, like those of Mahmud II, foundered on the expense of a reformed army, and bellicose regional ambitions, but laid the foundation for the modern state. The army came to represent the modernising force of Egyptian society, and an effective means of upward mobility. Egypt had been forcibly subdued, and enfolded into the global diplomatic and economic regime.

The Anglo-Russian accord continued, with Tsar Nicholas I visiting England in 1844, to discuss the future of the Ottoman territories with the British Prime Minister and Foreign Secretary. Although no formal agreement was reached, Russian Chancellor Nesselrode later drew up a memorandum that Nicholas I mistakenly took for an Anglo-Russian understanding on the preservation of the empire, part of the misunderstandings that led to the Crimean War. Temporarily defused, the Eastern Question took a backseat to other events in Europe, notably political instability in England and the unrest and revolutions of 1848 in the rest of Europe. The latter brought the problem of the Ottoman future back to the fore, in particular following the struggles in Hungary and the Principalities.

Further military reforms: divisional armies

The respite from the Egyptian challenge gave the new sultan an opportunity to continue his reform agenda, but his chief agent, Foreign Minister Mustafa Reşid, had resigned in March of 1841. Ostensibly his resignation was because of his resistance to Mehmed Ali's demands, but equally because the more conservative forces regained the upper hand in the bureaucracy. Already an effort had been made to prohibit the sale of offices, to regulate the salaries of the state employees, and to reduce the number of internal customs that were levied on a myriad of goods, at least in Istanbul and the provinces of Bursa and Gallipoli, but the speed of the reforms, the unpopularity of many of them, such as the fixed import and export tax rates dictated by the 1838 Anglo-Ottoman Convention, coupled with a bare treasury, forced an inevitable reaction to the ambitious agenda of Reşid and his protégés. He was packed off as Ambassador to Paris for the next three years.

The decade before the Crimean War in Istanbul belongs to Stratford Canning, who returned to Istanbul in 1842, and would serve as British Ambassador, with one visit home, until 1853. His belligerence, self-importance, and religious zeal set a tone for diplomacy in Istanbul that did little to support the reformers, or stimulate the admiration of the city's foreign community. His personal focus was on establishing a Protestant *millet* and church in Jerusalem. (The first Protestant missions had been allowed into Istanbul in 1831). His method was to attack the Ottoman religious establishment and the shar'ia, in the name of religious freedom, and promote the proselytising of Christians (largely Armenians) which was already underway, and was considerable by 1853. Ambassador Canning was personally responsible for a crusade which became a world-wide clamour about Christian martydom by Turks and Muslim fanatics.[7] This had obvious deleterious effects on the reform efforts, especially as Mustafa Reşid served as Canning's ally and chief instrument, and contributed to the poisonous environment of sectarianism which exploded in violence after the Crimean War. It also stimulated much of the fog of misunderstanding around the War itself.

Sultans Abdülmecid and Abdülaziz (1861–76) carried on with the modernisation of the palace and the image of the sultan by making public appearances in Istanbul, travelling in the countryside, enacting ceremonies, and presenting certificates and medals of valour typical of their absolutist neighbours in Moscow and Vienna. They maintained four separate establishments, and their reigns are notable for the increasing size and power of the bureaucracy (the 'Sublime Porte'), the elaboration of the palace service (*mabeyn*), the profligate habits of the dynasty, and financial scandals that created a crisis in 1861. Abdülmecid has the distinction of accepting the first international loans for the empire, when the Ottoman Bank, largely with British capital, was established in 1856. He also built the extravagent Dolmabahçe Palace, completed in 1853.

Meanwhile, military reform was carried on by the old guard who surrounded the young sultan. These included Hüsrev Pasha, Halil Rifat Pasha, (who, it will be remembered, had married into Mahmud II's family, served twice as Grand Vizier, but continued as Grand Admiral for four different terms until 1855), and Damat Mehmed Ali Pasha, also a son-in-law of the royal household, who was Grand Admiral five times, *Serasker* twice (1849–51, 53–54), and Grand Vizier in 1851–53. In 1842, the sultan himself seems to have exerted some control over affairs of state. He was all of nineteen. In August 1842 Rauf Pasha was reappointed Grand Vizier, and Mustafa Nuri Pasha was *Serasker* (1840–43). Mustafa Nuri was a product of Mahmud II's inner circle, the *Mabeyn*, and had been appointed Private Secretary in 1823. Following several short terms as governor in Rumelia, he was appointed *Serasker* in 1840. Hasan Rıza Pasha, his successor, under whose aegis the next, important round of army reforms would be promulgated, was also introduced into Mahmud's inner court at a young age. After his first term as Minister, from 1843–45, he would be reappointed to the post seven more times, until April 1876. Damat Mehmed Ali Pasha (1853–54) and Hasan Rıza Pasha (1854–55) were the *Seraskers* during the Crimean War. Even our old friend Hüsrev Pasha was briefly *Serasker* again during 1846, nine years before his death in 1855.

What is so striking is the closed circle at the top of the military command, and the much-attested parochialism, corruption and ignorance.[8] What we have also seen is the inability of the *Mekteb-i Harbiye* to produce sufficient numbers of qualified officers, a phenomenon which is undoubtedly connected to the grip of the household patronage system on the court. The competent field commanders of the Crimean generation arose in the provinces, not at court. A significant general command, officer training system and a modern Ministry of War are all post-1870 developments. Nonetheless, the Office of the *Serasker* remained extremely influential in the reigns of Abdülmecid and Abdülaziz, especially in the latter, for the reasons suggested, and because of the prominence accorded to defence of remaining Ottoman territories.

New military regulations were instituted in 1843 and 1846. The army was renamed the *Asakir-i Nizamiye-yi Şahane* (or simply *nizamiye*,

Regular Army) in 1841. The law of 6 September 1843 became known as the Rıza regulations, as much of the system survived until after the turn of the century. The Gülhane Edict had noted: 'Although, as we have said, the defence of the country is a matter of importance, and that it is the duty for all the inhabitants to supply soldiers for this purpose, it has become necessary to establish laws to regulate the contingents which each locality, according to the necessity of the moment, must provide, and to reduce to four or five years the period of military service. For to take, without regard to the respective population of different areas, from one more, and from another fewer, men than they can supply, is simultaneously to do an injustice and strike a moral blow to agriculture and industry; in the same way to keep soldiers in the service for the whole of their lives is to reduce them to despair and contribute to the depopulation of the countryside.'[9]

The new law, which was actually an elaboration of the *redif* law of 1834, addressed these problems directly by creating a system of five regional armies (*ordus*). These were located as follows: the first army, the Guards, in Üsküdar (the *Hassa ordusu*); the second in Istanbul (and latterly Ankara, called the *Dersaadet ordusu*); the third army for the European territories, in Manastır (*Rumeli ordusu*); the fourth army in eastern Anatolia, at Harput (*Anadolu ordusu*), and the fifth army to guard the Arab provinces at Damascus and Aleppo (*Şam ordusu*). The Egyptian contingent of 18,000 was to serve as reserve reinforcements for the Damascus army. A sixth army was added in 1848, in Iraq (*Bağdat ordusu*), and was responsible for the Arabian Peninsula. Each of the first five armies was composed of two services: the *nizamiye* active and the *redif* reserve, plus auxiliaries and *başıbozuk* (irregular cavalry), the last to be called up only in times of war. Such irregular cavalry could be Cossacks or Tatars from the Dobruja, or Türkmen and Kurds from eastern Anatolia, as well as a myriad of tribesmen from the Caucasus. Each army commander was to hold the rank of field marshal (*müşir*), appointed by the *Serasker* in Istanbul. Under him the active forces contained two corps, commanded by generals (*ferik*), while the reserves, also divided into two corps, were under the command of a brigadier-general in peace-time. The new organisation allowed for considerable autonomy at the *ordu* level, but centralised control over the divisional command. The *Serasker* continued to embody the functions of both Minister of War and Commander-in-Chief. His office in Istanbul grew during Abdülaziz's reign, and was diversified by a general staff (*Erkân-i Harbiye*) and various departments dealing with the facets of the military, such as artillery, engineering, supply, etc.

A BASHI-BOZOUK.—FROM A DRAWING BY JAMES ROBERTSON, ESQ., OF CONSTANTINOPLE.

PLATE 20 *Sketch of* başıbozuk *by James Robertson (*Illustrated London News, *1854).*

Although there was great variation by location, and by the success or failure of recruitment, the general distribution of soldiers to each *ordu* ran as follows: three regiments each of infantry, and scouts, two of cavalry and one of artillery, with 33 guns. Each infantry regiment (*alay*) was divided into three battalions (*tabur*). These were further divided into companies (*bölük*) and squads (*manga*) of ten men. The novelty lay not in the structure of the corps, which was based on the Hüsrev regulations of 1827–28, but rather in the application of the system across the empire, and to both active and reserve units.

In 1846, a new conscription law (*kur'a nizamnamesi*) was enacted. Conscription was to be regulated by drawing lots. A total of 30,000 recruits, generally aged 18–20, were to be added annually, while a similar number was to be discharged, to maintain an adequate active force. The law included considerable exemptions: members of the scribal and administrative bureaucracy, ulema, kadis and religious students, who had to prove their seriousness by examination in order to be exempt. As one can imagine, the new conscription law had the unintended effect of increasing the numbers attending the medreses.[10]

The estimated strength of the *nizamiye* was 150,000–200,000, that is, at about 30,000 per army. Service in the *nizamiye* was reduced to five years, ultimately two, followed by seven years' service in the reserves. Initially soldiers could be called to their reserve contingent once a month, but that was later stretched to once every two years. The reserve was configured at 120 battalions of 800 men, or approximately 90,000, but with the new seven-year reserve system, it was estimated that a total 400,000-man army could be raised in that fashion. The imposition of the regular conscription system continued to falter, however, so the army likely never reached that size. James Henry Skene, Consul in Aleppo circa 1851, reported that there were 123,000 actives, and 212,000 effective reserves (total 335,000) in the army, but raised the figure to 365,000 by including four detached corps: one on Crete, with three regiments of infantry and one of cavalry, of 11,000 men; another in Tripoli in Libya; a third in Tunis, and a fourth, the veteran artillery brigades which guarded all the fortresses of the empire, amounting to some 9,000 men.[11]

These continued to be Muslim men only, although the reformers made ineffectual efforts to enlist Christians, some of which caused rebellions, to which we will return shortly. Christian or Jewish officers continued to be unacceptable to the enlisted men, and Christian populations understandably preferred a buy-out tax, or other forms of exemptions, to enlistment. In 1855 non-Muslims became eligible for army service, and the old poll tax, the *cizye*, was eliminated. A new exemption tax, *bedel-i askeriye*, quickly replaced it. By the end of the century, the list of possible exemptions was huge, and included Muslims as well as non-Muslims. So to some extent, the problem of weighing too heavily on one part of the population, and on specific regions, continued. Curiously, keeping the non-Muslims out of the army prompted British observers, and Ambassador Canning, to accuse the Ottomans of violating non-Muslim rights by persisting in exempting them from service. During the Crimean War, foreign commanders with the Ottoman forces wondered at the practice when manpower shortages were so acute.

Of the continuities with the *Asakir-i Mansure* army, the continued discrepancy between the salaries of the chief and inferior officers is striking. A *müşir* was paid 100,000 *kuruş* per year, or roughly 11,000 pounds of the time, which included rations. A *ferik* in command of a corps got 50,000, while a colonel (*miralay*) received 3,000 and a lieutenant (*mülazim*) 350 *kuruş*. A private generally earned 20 to 30 *kuruş* and, in theory, was to have access to good rations of meat, bread, rice and vegetables in abundance, with butter and oil for cooking, the equivalent to 60 *kuruş*.

The *Şura-yi Askeriye* continued to serve as the supreme council which oversaw military matters. It was headed by a *müşir* and a *kaziasker*, and it had membership which included civilian as well as military officials. Each army also had its council of war, and could try military offences, but the sentence was passed by the supreme council. The intent was to eliminate abuse among the officer corps as well as among the enlisted men. Corporal punishment was a rarity in late Ottoman armies; imprisonment was preferred. Training and drill were not well implemented, except among the active forces. Some effort was made to post recruits from the same area to a particular *ordu* in order to facilitate the reserves, who worked the land when not called to duty, but training was minimal, especially once they were technically only to report every two years.

One of the striking changes with this reorganisation was the introduction of military preparatory schools at each of the army headquarters, to support the military college Mahmud II had created in Istanbul. Skene says enthusiastically that the enlisted men had learned to read and write. Shaw goes even further, suggesting that the new schools opened the way to secular education, though that might be a bit anachronistic for the 1840s. By the end of the century, a military education and career did offer upper mobility for many Muslims outside the ruling class, as attested by the biographies of the late leadership of the empire, including Mustafa Kemal Atatürk himself.[12]

Two aspects of these regulations deserve comment. One is the reformer's concern with the enlisted man, at least on paper, one characteristic carry-over from the Janissary system, which I have argued is partially attributable to Ottoman-shar'ia idealised notions of justice and protection of the peasant. The second is how local priorities and conditions influenced the design of the system. Although commentators then and now are inclined to see both *Asakir-i Mansure* and *Nizamiye* as derivative from French and Prussian models, and risible, there are clearly aspects of these reforms that reflect the specific realities of an empire under attack on all frontiers. One distinction that was commented on at the time was the extensive reserve system, which apart from Prussia had not been widely implemented in Europe at that point. Austria had in fact abandoned its own *Landwehr* organisation, a people's army founded in the Napoleonic years, once its utility was no longer demonstrable.[13]

Further economic reforms: tax farming

Reform efforts aimed at replacing an indirect tax regime with direct taxation. In 1840 the tax system was standardised by implementation of a ten per cent cultivation tax, fixed head taxes on cattle, and a graduated *cizye* for non-Muslims based on individual incomes and the ability to pay. Salaried civilian tax agents, *muhassıl*, were sent to the districts to assess and collect taxes. The provinces (*eyalets*), divided into *sancaks*, or districts, were subdivided into *kazas* or counties, in an effort to establish control over the countryside and replace the tax farmers, often the governors themselves. Census data was to be collected on all the *kazas*. Furthermore, regional councils representative of the local populations, Muslim and non-Muslim, were established to act as a check on the governors. While the new taxes were imposed and collected with relative ease in urban settings, the new system utterly failed in the countryside. In fact, the first decade of the Tanzimat reform unfolded as a contest between the traditional rural land-holders and the new centralising bureaucracy. It proved impossible to eliminate tax farming, as the net result was the drastic reduction of the revenues available to the sultan. State revenues collapsed so drastically that within the year, the tax-farming system was reinstituted. In order to meet expenditures, the Ottomans issued paper money (*kaime*) for the first time. This was later followed by state loans from wealthy local banking families, opening the way for foreign loans from Britain and other countries as the century wore on. In spite of the success of many other tax-reforms, the tax-farming system proved resilient, was modified again in December 1855, and remained in place for the rest of the empire.

Underfunded and undermanned, the provincial reforms faltered. Sultan Abdülmecid reportedly visited the Sublime Porte in February 1845 and read out loud a proclamation, an unusual intervention in the affairs of state. 'One cannot deny that, in spite of the care brought to the the realization of my desires, none of my projects, with the exception of military reform, have resulted as I promised. And even the military reform lacks a solid base, which is the general prosperity of the country. I am profoundly distressed.' The 22-year-old sultan then ordered his ministers to discuss the problem, but advised them that the solution lay in remedying ignorance, and that schools were needed. One of the results of his personal appeal was the convoking of an assembly of provincial representatives to be held in Istanbul, also an innovation, and the creation of local councils, which had a voice, not always effective, in the pre-Crimean War period, applying directly to Istanbul for available funds for public work projects. Inspection commissions were sent to the provinces; Abdülmecid himself went on such inspections in 1844 and 1846. It is in this period that a start was made on the establishment of public schools, and, as Ergut has argued, on the beginnings of a modern police force.[14] The Crimean War interrupted the evolution of this system, as expenses forced the government to stop paying for the work of the councils and the public projects. In 1852 a new Provincial Regulation Law restored the authority of the governors but retained the local council system, in a further rationalisation of municipal and country-side bureaucracies. In 1848 mixed tribunals of Muslims and non-Muslims attended to interfaith commercial disputes. By 1850 a commercial code became the first western-style law code introduced.

The utilitarian aspects of these reforms masks, of course, what was actually going on at the provincial level. A blizzard of enactments by the sultan and his Supreme Council characterise the period, but good intentions were not generally followed up by enforcement and inspection at village level as was originally anticipated.[15] The solution to unrest was to hand the local administration over to the commanders of the new divisional armies, which helped to stabilise the application of the Tanzimat. This could lead, and often did, to great brutality on the part of the Ottoman commanders. One snapshot of the period has the following individuals in place in the divisional armies circa 1850: 1st Army (*Hassa*): *Müşir* Topçubaşızade Mahmud Pasha; 2nd Army (Istanbul): *Müşir* Damad Mehmed Ali Pasha, *Serasker*; 3rd Army (Manastır): *Müşir* Ömer Lütfi Pasha, soon to assume command of the Ottoman forces in the Crimea, under him *Feriks* Abdi and Avni Pashas; 4th Army (Anatolia, at Harput): *Müşir* Mehmed Reşid (known as Arab, or Darbhor Mehmed Reşid), said to have studied in France, under him *Feriks* Ali Rıza and Selim

Pashas; 5th Army (Arabia, at Damascus and Aleppo): *Müşir* Mehmed Emin (Mühendis) Pasha, who reputedly studied at Cambridge, under him *Feriks* İzzet and Abdülkerim Pashas; 6th Army (Baghdad and Hijaz): *Müşir* Namık Pasha (famed for reorganising the *Hassa* under Mahmud II).[16] By turning to the problem of conscription and direct taxation, intrusions into traditional local systems, we can perhaps gauge the causes of much of the unrest around the Tanzimat reforms in the 1841–56 period.

Provincial resistance: post-1841 Syria

Greater Syria was restored to the Ottomans with İbrahim's withdrawal in 1841. The sultans reclaimed their own territories, which had been badly neglected even before the 1830s, and occupied for a decade by a modernising, and latterly draconian, Egyptian army. İbrahim reportedly taunted an Ottoman general by saying: 'You, with the assistance of the English, have expelled me; you have put arms into the hands of the mountaineers; it cost me nine years and ninety thousand men to disarm them. You will yet invite me back to govern them.'[17] Mahmud II had never been able to establish full control over the territories, and delayed introducing conscription because of the mistrust and unruliness which inevitably followed. The harsh conscription methods of İbrahim Pasha were a more recent memory. Furthermore, the final confrontation with İbrahim had been facilitated in part by arming the countryside. Tribal leaders were loath to surrender their weapons. Entrenched notable families, who over the previous century had acquired significant control over agricultural resources, contested many of the new regulations.

The familiar provincial lines of Aleppo, Damascus and Sidon, absorbing the old province of Tripoli, were re-established. Sidon was the most changed: the new capital was Beirut, and it included most of Lebanon and Palestine. Beirut had gained considerable prominence as a port city, and would continue to do so in this period as the administrative centre of the newly-drawn territories. The province of Damascus was also realigned, and included the al-Biqa valley of Mount Lebanon within its boundaries. Emir Bashir II's rule of Mt Lebanon had ended with İbrahim Pasha's withdrawal from Greater Syria. The Ottomans deposed his successor Beshir III, and ended the reign of the Shihab family. The power vacuum thus created simply perpetuated tensions between the Druze, Muslim and Maronite communities, which was often stimulated by Ottoman governors, who, short of soldiers, set factional tribesmen one against the other. The Ottoman army could not exert control especially over the warrior clans and bedouins of Hawran, Nablus, and the Negev desert, as we have seen in the contest with İbrahim's army. İbrahim had succeeded in a partial disarmament, but the districts reverted to old patterns with the Egyptian departure, and the Ottomans were unable to restore order over the population. Much of Palestine remained that way for the next decade.

Jerusalem (with Nablus and Gaza added) and Mount Lebanon were accorded special status. Jerusalem was governed as a special district under a *mutasarrıf*. Mt Lebanon was divided into Druze and Maronite districts in 1843, after massacres and rebellions alarmed the international community and forced the Ottomans to act. Further negotiations around the special nature of these territories led to another set of regulations in 1845, on the principle that the religious communities could regulate themselves and collect taxes locally, under two *kaymakams*. This was agreed to by all sides in 1846, but the basic tensions remained. Events on the European frontiers of the empire distracted both Ottoman governors and foreign representatives from Mt Lebanon until after the Crimean War.

European diplomatic harmony, which had allowed for the settlement of the Mehmed Ali crisis of 1841, soon settled into the more normal Franco-British disharmony of the Eastern Question. Lebanon, Palestine and Jerusalem were territories which became a special zone of interference by the European concert powers concerned with the brutality inflicted on non-Muslim communities, and who initially supported the re-installation of the Shihab family. The French were known to act as provocateurs among the Maronite Christian communities, and the Ottomans just as often were accused of 'letting loose' the Druze on the Christian communities. The *başıbozuks*, often the substitute for *nizamiye* troops, were just as capable of atrocities as the Maronite and Druze chieftains.

British Ambassador Stratford Canning was newly returned to Istanbul, and was the chief negotiator for the introduction of the *kaymakam* system which gave Lebanon a modicum of peace from 1845 to 1860. The events had the effect of raising the temperature of the international community concerning Ottoman barbarity against non-Muslim subjects, which struck a chord with the zealous evangelism of the Ambassador. Frustrated by the general lack of progress of the Tanzimat, Canning chose this moment to force on Sultan Abdülmecid the abolition of execution for apostasy, one example of his new-found evangelical zeal for protecting Christian (Protestant) rights. He followed that up by establishing a Protestant church in Jerusalem in 1845, and then insisting on the recognition of a Protestant *millet* within the empire. It is not hard to imagine the resentment about Great Britain that must have built up in the Ottoman palace in the years prior to the Crimean War, especially as Canning quickly developed an intimate relationship with young Sultan Abdülmecid.[18]

Rebellion in Aleppo

The Arab provinces continued to be governed by a *vali*, but after 1844 the *serasker* (*müşir*) of the fifth army in Damascus assumed authority over the governors. *Müşir* Namık Pasha served in that capacity from 1844–49. There was a great deal of confusion about military and civilian authority, especially as Istanbul was given to merging and splitting the two centres of power at will. With the reassertion of the powers of the governor in the 1852 provincial regulations, the army assumed a certain independence from civilian affairs, a trend interrupted by the Crimean War.

Hence, the army was the instrument of the Tanzimat in the Arab provinces. To locals, it must not have appeared much different from occupation by Egypt. The quality of administration depended entirely on the quality of the *nizamiye*. Before 1843, insufficient numbers of regular troops (some 15,000–20,000 men) forced the officers to use the *redifs* and local irregulars, such as deserters from Mehmed Ali's army, largely Albanians, and the ubiquitous *başıbozuks*. The 1843 regulations mandated divisional army of 25,000: 19,000 infantry, 4,000 regular cavalry, and 2,000 artillerymen with 72 guns. This was the army of Namık Pasha, who established his headquarters in Aleppo so as to be able to control the *redif* troops drawn from Anatolia and Iraq. He apparently succeeded in dissolving the Albanian bands, and is reported as saying to local chieftains: 'Formerly, we could not compel you to obey us. But now we are strong, and if you are insubordinate, I will . . . throw you into the sea.'[19] After 1845, he moved his headquarters back to Damascus, distributing about half the army across Syria and Palestine and leaving half in Damascus. The army was still only some 20,000, troops who were ill-paid and ill-supplied. Acute financial shortages prevented further pacification of the territories. Before 1849–50, when the sixth army was established in Baghdad under *Müşir* Namık Pasha, the troop presence in Syria represented approximately one-quarter of the manpower that İbrahim had had at his disposal. Neither the fifth nor the sixth army had reserves in this period. The sixth army was split between Baghdad and the Hijaz, with one corps in each.

In 1850 the Ottomans extended conscription to the Arab territories in order to remedy the insufficiency of men, but that prompted rebellion in

Aleppo. An attempt had been made at a census in the years following the 1843 army regulations, with little success. In 1848, a census survey was again attempted, under Governor Mustafa Zarif Pasha, whom we last encountered among the Kurds, but thousands fled to the mountains. In 1850 rioters in Aleppo shouted, 'We shall not give soldiers; we shall not pay the poll tax,' as they had at İbrahim's recruiters. One of the major factors in the 1850 Aleppine revolt concerned an Ottoman attempt to collect tax-farming arrears, to wrench tax-collecting responsibility from Aleppine notable Abdullah, the civil governor and *mütesellim*. The revolt itself was likely instigated by the wealthy notables, but the insurrection grew out of hand and resulted in attacks on Christian quarters. The city was occupied by rebels who demanded the abolition of conscription, the restoration of Abdullah as *mütesellim*, and conversion of the *ferde* (the Egyptian poll tax) to a property tax. Other demands referred to non-Muslim privileges: no church bells should be rung, no crosses in processions and Muslim servants were not to be employed in Christian households. When reinforcements did arrive, the city was restored to order at the cost of 3,000–5,000 lives. Zarif Pasha was immediately dismissed. His successor arrived with more troops, and imposed the Tanzimat reforms more vigorously than Zarif Pasha, who shortly resurfaced as the heavy-handed pacifier of Vidin in 1850–51.

In his memoirs, Zarif Pasha speaks of his experience in Aleppo, as follows:

In 1848, I was appointed as Governor of Aleppo. I was very happy and expressed my profound thanks [to the sultan]. We went by steamship to Rhodes. From there we arrived at İskenderun, disembarked and prepared the animals. We arrived in Aleppo via Antioch. Namık Pasha, shortly to be appointed Müşir *to Baghdad, was there. [His first instructions were to apply the Tanzimat reforms to Urfa, starting with taking a census, nine years after the Gülhane Edict.] First, I sent two battalions. Then the census-takers. At the same time, I went to Birecik with two squadrons of cavalry and 300* başıbozuks. *The census was completed, but the local [financial] records were only partially revealed. I assembled 2,000 of the elders and leaders of Urfa and surrounding villages, and announced the institution of the reforms. [They brought the remainder of the records, which revealed further local expenses. Zarif then set a tax bill for the state, and negotiated exemptions with the elders and tribal leaders.] I spent 140,000* kuruş *on ceremonial robes, watches, and other gifts, and returned to Aleppo . . . The Bedouins started to attack and I stopped them with the* başıbozuk. *I was ordered to collect some 30,000* kese *that İbrahim Pasha had left in Aleppo, and send 14,000* kese *to the [sixth] army in Baghdad. I was then ordered to confiscate any property and goods of anyone in arrears. That was followed by an order for conscription. Arabistan [his word] did not want to give soldiers; as for the rest, they are Aleppo bandits. Abdullah Bey had only 4,000* kese. *[In the midst of trying to impose the new regulations, a riot broke out.] I rode out with a few servants, police and a few guards to the area of the disturbance. Three to five thousand people stood in front of me with weapons in their hands. 'Do not approach or we will shoot,' they said. 'Hey you, it's me!' I shouted. They said, 'Whoever you are, we'll shoot!' I said, 'Oh yeah? Go ahead and shoot!' All those men shot at me.*

He retreated to the front of the inner fort, and called on the guards to close the gates. Then he assembled a company of soldiers, but they told him they did not have any ammunition.

I sent them back. I sent the news to the barracks, but then shots began to be fired from everywhere. I was told they had raided the barracks and were killing the soldiers. I went crazy and wanted to ride to the barracks, but they held me back. One Ömer Agha, an Albanian chief, told me to follow him, and helped me make my way to the barracks.

Arriving at the barracks, he learned that the soldiers had prepared a few artillery pieces, but that there were only 250 men in the place.

The rebels then attacked the Christian districts. He had no choice. Without soldiers, Zarif was forced to summon Abdullah Bey, and with much pleading, begging and promises, he asked him to put a stop to it. It was only then that Zarif learned the reasons for the revolt, which are listed above. By the next morning, the crowd had increased by his estimate to 40,000–50,000. He sent his 250 infantry and two companies of cavalry into the Christian quarters, but they could do nothing. He surrendered the city to Abdullah and the ulema, and retired to the palace for 20 days, having secretly sent for reinforcements. Müşir Emin Pasha sent three battalions of infantry, a regiment of cavalry and about 500 *başıbozuks*. As the *nizamiye* troops approached the city, the inhabitants got wind of it and shut down the market. Ultimately, 2,500 soldiers and ten cannons faced down the revolt. For the most of the next three days, the marketplace was a battleground. Zarif burned down the houses of the insurgents. He praised the regulars who struggled so bravely as they conquered the house of Abdullah. 'Long live our padişah, they cried!' In three days, 8,000 cannonballs were shot but with little effect. The troops behaved well. Zarif claimed to have distributed 1,000 *kese* as encouragement for those who were fighting. He reclaimed some of the goods stolen from the Christian quarters. He also raided the Bedouin camp, and took some 20,000 sheep, camels and oxen. It was over, and business was resumed. The consuls and the Christians came to thank him. He cautioned that they needed 10,000 soldiers to keep order in Aleppo. The consuls protested that should it happen again, they would demand an indemnity. Relief forces arrived, and Zarif Pasha left Aleppo for İskenderun with 150 prisoners, among them Abdullah Bey, who died before they arrived in Istanbul.[20] While Zarif Pasha is no stylist, the story conveys his methods as well as the inadequacy of the forces under his command. He will resurface not just in Vidin but in Erzurum as well, at a crucial stage in the Russo-Ottoman confrontation there during the Crimean War.

Conscription was not fully imposed on Aleppo until 1861; in Damascus conscription was operational after 1860, although urban inhabitants found ways to escape, such as flight or self-mutilation. Here, as elsewhere, it soon became apparent that there were fairly easy ways of being declared exempt from service, the simplest by bribing the officials in charge of the balloting system or by finding substitutes. Sultanic *firmans* reinforced this. Published in both Arabic and Turkish, they confirmed that only one in eleven of men between 20 and 25 would be called up. Invalids, only sons, and students were exempt.[21] After 1853, apart from the Jabal Hawran Druzes, who resisted conscription until 1896, conscription was no longer a contentious issue for local populations but remained deeply unpopular.

Restoring order was inevitably hampered by the reduction in the number of regulars of the fifth army, when troops were redeployed to other parts of the empire to assist in pacification elsewhere. In 1853 17,000 Syrian army troops were transferred to the Crimean battlefront, leaving the Syrian territories largely in the hands of as few as 10,000 *nizamiye*, and the *başıbozuks*. By some reports, no more than a total 4,000 *nizamiye* troops were stationed in Aleppo and Damascus just prior to the massacres of Christians in Damascus in 1860. Only after those massacres provoked an international outcry did the government in Istanbul send an additional 20,000 regular army troops to Damascus.[22]

Similarly, the efforts to impose direct taxation as described above foundered in the Arab provinces. Tax-farming continued throughout the reform period. Initially the Ottomans considered abolishing the individual urban tax, the *ferde*, the unpopular levy imposed by the Egyptians, but it was reinstated at one-third of the Egyptian rate. As we have seen in Aleppo, it too was often the cause of rebellions. After 1852 it was converted to a generalised property tax (*vergi*), more in line with the principles of the Tanzimat.

The local councils, first established by İbrahim and part of the Tanzimat reforms as well, did exert considerable power as an advisory body, with many executive and judicial functions. Unlike those of İbrahim, the new Ottoman *meclis* was less representative of the non-Muslim populations. Local Muslim notables and the ulema consolidated Syrian resources in the hands of the wealthy landowners. After 1852 the councils were returned to the authority of the governors, and their effectiveness was determined by the level of cooperation that could be achieved among the various constituencies. Muslim violence against Christian communities in Aleppo, Nablus, and finally Damascus in 1860, was symptomatic not just of the Ottoman state inability to maintain order and stability, or of the new international trading regime imposed by the 1838 Anglo-Ottoman convention. It was equally emblematic of the turmoil within the old urban order and their militias, and the impact of the Tanzimat reforms on the equilibrium of both urban and rural populations.

Seeds of Russo-Ottoman conflict on the Danube

In Vidin and other towns along the Danube, problems like those of the Arab provinces, and later Bosnia, emerged shortly after the promulgation of the Tanzimat and were linked to developments both in Serbia and the Principalities. The social transformation in north-western Bulgaria was influenced in this early period by its strategic position as the Ottoman European frontier, and by the convulsions occurring across the border after 1848.

As we have seen, the 1826 Russo-Ottoman Convention of Akkirman had promised to fulfil the 1812 Treaty of Bucharest, granting semi-autonomy to Serbia. The Treaty of Adrianople (1829) had once again imposed that condition. In August 1830, Mahmud II confirmed Milos Obrenovic as Hereditary Prince of Serbia, recognising the autonomy of the territory, defining the border, setting out the annual tribute as compensation for abolishing the Ottoman land regime, and regulating the stationing of Ottoman troops in the country. Muslims were prohibited from settling in Serbia. By 1838, a new constitution was declared. In the in-between years, the tension between the Ottoman nominal regiment in Belgrade and the Serbian government often led to strife, and the renegotiating of spatial and 'national' lines. By 1867, the last Ottoman troops were withdrawn from Belgrade.

Thus in north-western Bulgaria, Tanzimat reformers and military men were dealing with the presence of a nascent nation-state in Serbia, and stirrings of constitutionalism and unification in the Principalities of Wallachia and Moldavia. In addition, as we have seen, the 1827 Treaty of London gave the foreign powers carte blanche to intervene on behalf of their religious communities, which they did with regularity, both in the Arab provinces and on this border. However, external pressures influenced the confrontation of old and new Ottoman regimes in distinct ways in this European/Ottoman contact zone. But initially not by much. Rebellions, first in Niş in 1841, and then in Vidin in 1850, focused on the question of property and taxation, as might be expected the Tanzimat reformers had aimed first at abolishing the *timariot* system, in order to reorganise the distribution of land. This resulted in exacerbated conflicts between peasants and large land-holders over rights to land. As elsewhere, the loosening of the bonds of the Ottoman land regime had the net effect of entrenching the power of the local landlords, largely Muslim families. Much of the old system remained in place. The Niş revolt in 1841 was linked to the fundamental conundrum of the old and the new systems: Christians were to be treated as equals, but Muslims benefited most from the land reforms. Governor Sabri Mustafa Pasha began a tax survey at the end of 1840, in order to implement the new reforms. A combination of the new tax regime, and the quartering of troops on families in the area, provoked the revolt, which the Governor put down by calling in some 2,000 Albanian irregulars, whose pacification techniques we have encountered before. Many of the inhabitants fled to Serbia, but peasant appeals called for sultanic intervention, not for liberation, even though the Ottoman governor later blamed Serbian interference. A petition from 1,000 Christians to the Prince of Serbia in 1841 noted: 'People are not revolting against the legitimate government of the Sultan; rather they want that the benevolent terms of the Hatti Sherif of Gülhane be faithfully and exactly carried out.'[23] The Istanbul government was slow to respond, but after the rebellion itself had essentially been pacified, sent commissioners to resettle the claims of returning villagers. Little actually changed to benefit the agricultural sector.

In Vidin, the situation was more complicated. Already a divided city, between the Muslim garrison and the rest of the town, it had become predominantly Muslim as a result of emigration from surrendered lands. A later Ottoman census, which did not include the soldiers of the garrison, estimated some 23,000 inhabitants in Vidin, making it one of the largest cities along the Ottoman Danube. The fortress had been reconstructed according to the Vauban system during the Austrian occupation in the early 1700s, and it continued to be well fortified after it was returned to the Ottomans. Strategically important to the Ottomans, Vidin, as with all the previous Russo-Ottoman wars we have seen, was one of the flash-points preceding the Crimean War.[24]

In 1845, as elsewhere, the government attempted to establish local councils and consultative bodies in the town, but they met with little success. The petitions of those who rebelled in 1850 stressed the abuses of local tax officials, tax farmers and village landlords. As in Niş, initial expectations of the Tanzimat decree were met with disappointment. Anticipating better treatment by the legal system, and an amelioration of the tax burden, the peasants found that abuses continued. Whereas elsewhere in Bulgaria the *timariots* were to be pensioned off by 1842, and some territory did move into individual peasant holdings, it is unclear that the reform was implemented in Vidin to any great degree, again partly because of the defensive position of the town. As in the cases where the *timariot* rights were abolished, the land quickly fell into large Muslim landowners' hands. The other source of peasant anger was the continuation of forced labour, contrary to the promise of the Gülhane Edict.

1848 revolutions and the Ottoman northern frontier

The revolt which broke out in 1850 may have been organised in Serbia and instigated by the Russians, who had troops stationed in Wallachia after 1848, and at one point moved three battalions to Kalafat, across the river from Vidin. This requires a little explanation, and an introduction to Ömer Pasha, Commander of the Third Army, and later *Müşir* of the Ottoman contingent at Sevastopol.

The Treaty of Adrianople between the Russians and the Ottomans had secured virtual autonomy for Wallachia and Moldavia. The Organic Regulations of 1829 set out the principles of self-government for the two territories, but in effect the Principalities became subject to a dual protectorate, the Russians and the Ottomans, with the Russian army in occupation until 1834. Ottoman influence has been reduced to the collection of a fixed tribute and a say in the choice, later election, of the princes. Equally important, a quarantine zone was established along the Danube and the Ottoman monopoly on grain, animals and timber was eliminated. Muslims, as before, were prohibited from residing in the territories. The Organic Regulations, under the administration of Russian Count Pavel D. Kiselev, were issued in July 1831 in Wallachia and January 1832 in Moldavia. Two provinces were established, under princes, with legislative representation by assemblies of the boyars. The move towards a representative government changed very little fundamentally for the peasants. Boyars were made the owners of the land, and still demanded labour services of the peasants on their farms. Serfdom had disappeared from Wallachia and Moldavia in the eighteenth century, but the peasants who remained had little freedom of movement. Still, with the Ottomans prohibited from exacting the agricultural products they had long come to expect from Wallachia and Moldavia, Romanian peasants did benefit from the demand for their agricultural products upstream from Vidin, and their increasing incorporation into the markets and cultural life of Europe.

In 1834, the Ottomans and the Russians signed the Convention of St Petersburg. The Ottomans recognised the new statutes, set the tribute at three million *kuruş*, and Russian troops departed. The two powers appointed new princes, who ruled for the next seven years. Though they were years of prosperity and awakening, it was also a time of much struggle among the landed boyars of the assemblies, princes, and Russian consuls who intervened at will in local affairs. Wallachians and Moldavians technically had the right of appeal to either Russian or Ottoman authorities. By 1840, the Russians were attempting to turn a right into a stipulation that the statutes could not be changed without the approval of the two empires, which threatened the autonomy especially of Wallachia. By 1842, Alexander Ghica, the appointed prince of Wallachia, had been deposed, and George Bibescu was elected by the assembly, but his rule suffered the same interventionism of Russian diplomats and advisers. Moldavia, long drawn more peacefully into the Russian orbit, still chafed at the restrictive regime of the statutes.

Opposition grew as a larger number of boyar children experienced European life, especially in Paris. So when the winds of revolution swept Europe in 1848, two territories were most effected: Hungary and Wallachia. The Hungarian problem in the Habsburg Empire was a long-standing one, and need not be rehearsed here. Chancellor Clemens von Metternich was at the helm, a deeply conservative monarchist and adherent to the historic understanding of the Habsburg empire as a federation of culturally distinct provinces. As we have seen before, the Habsburg involvement in Europe generally meant the neglect of the eastern side of the empire. Preoccupation in this period was with Germany and Italy, and that has been articulated as one of the chief reasons for Habsburg neutrality during the Crimean War. Austro-Hungarian peasants, like their counterparts in the Balkans, had much to complain of regarding their status, condition and lack of voice. In 1848, the call to national liberation and constitutionalism swept from Sicily to Paris, where the monarchy of King Louis Philippe collapsed. The flames spread to Vienna and Budapest in March 1848, when Metternich was dismissed and a new government appointed. The new constitution called for a single centralised regime, single citizenship, and for eighteen-year-old Habsburg Franz Joseph to be crowned as Emperor of Austria, but not as King of Hungary. Further reforms intended the division of the Kingdom of Hungary into new territories, which further stiffened the resolve of the Hungarian revolutionary government. Particularly sensitive for the Hungarians was the removal of Transylvania and its creation as Grand Principality. While most territories of the Habsburg empire had been subdued by the army in mid-1849, Hungary's revolutionary government, led by Lajos Kossuth and backed by its army, was in control of Budapest. Polish revolutionary General Bem, who had defeated the Habsburg forces, held Transylvania.[25]

Meanwhile, revolutionary rebellion spread to Romania. Moldavia remained quiet, except for some petitions by the boyars, but Wallachia became the centre of an international incident. In Bucharest, leaders of the rebellion, many of them boyar sons educated at the Collège de France in Paris, organised a rudimentary republican government, a united Romania, which lasted all of three months. A peasant uprising never materialised, although many left their homes to march with their leaders to Bucharest. Bibescu abdicated, and the Russian consul retired to Jassy. This revolt was less about peasant rights and liberation than about throwing off Russian control and liberalising the administration.[26]

At this point the revolutionaries turned to Istanbul, hoping to prevent a break with one protector while expecting an invasion from the other. Ottoman commissioners were sent to Giurgevo to investigate. The first of them, Süleyman Pasha, was induced to toast the new, illegal government, which prompted Russian protests in Istanbul. A new commissioner, Fuad Efendi, later Grand Vizier Fuad Pasha, was sent at the beginning of September as part of a joint occupation to accompany Russian General Duhanel to put an end to the revolt. Fuad crossed into Wallachia, and camped outside Bucharest with 5,000 soldiers (elsewhere 12,000 or 20,000). The Russians moved 14,000 troops into Moldavia in July, and by October had some 30,000 stationed in Wallachia. (Russia was more concerned with the ongoing Hungarian revolution than the Wallachian one.) With little difficulty, the Ottoman and Russian troops entered

Bucharest in late September, and after a brief encounter with a handful of armed rebels, the revolution was over.[27]

Two inter-related international crises dominated 1849. In January, the Russians demanded a joint protocol with the Ottomans. The military occupation was to extend for seven years, and the two courts were to appoint the *hospodars*, a significant infringement on the autonomy of the Principalities. More significantly for the rest of Europe, the Russians demanded Ottoman acquiescence to their invasion of Transylvania to help Franz Joseph eliminate the rebellion there. The negotiations in Istanbul, led by Ambassador Canning, have been described as a rehearsal for the Crimean War. Throughout the negotiations, Grand Vizier Mustafa Reşid intimated that a show of force by the British fleet would be appreciated. The request was consistently rebuffed by Palmerston, who appreciated the significance of a reconstitution of the conservative imperial order represented by the brief Austro-Russian cooperation over Hungarian rebels. When Habsburg Franz Joseph I requested military assistance in May 1849, the Russians invaded Transylvania with an army of 150,000 and defeated the Hungarians. That was preceded by the signing of the Russo-Ottoman Act of Balta Limanı on 24 April 1849. The Princes of Moldavia were now to be appointed by Russia (effectively) and the Ottomans, rather than elected. Small advisory councils replaced the grander assemblies. The Russian occupation army would remain in the Principalities until 1851.[28] Even more threatening to the tranquillity of European relations was the flood of Hungarian and Polish refugees who entered Ottoman territories, with their leaders Kossuth and Bem among them. Sultan Abdülmecid refused to extradite them, when Russia and Austria demanded the return of their subjects. (Austria evoked the long-standing 1739 Belgrade Treaty stipulations.) Fearful of war, Great Britain reacted swiftly in this instance and a joint French–British squadron headed for the Dardanelles. Meanwhile, Sultan Abdülmecid dispatched Fuad Efendi on a secret mission to Tsar Nicholas I. By November, Nicholas rescinded his extradition demands and the crisis had been averted. Clearly the Eastern Question, exacerbated by revolutionary fervour, and the outcry of nationals and evangelicals alike, was not about to go away.[29]

Provincial resistance: Vidin in 1850

The military commander of the Ottoman troops with Fuad in Wallachia was Ömer Pasha (d. 1871). Born Michael Lattas of Croatia, Ömer Pasha deserted the Austrian army for Bosnia sometime in the mid-1820s. He joined the Ottoman army under Mahmud II after the elimination of the Janissaries in 1826, serving as an infantry captain and instructor in the new military school in Istanbul. An audience with Mahmud II changed the course of his career, when the sultan appointed him as tutor to his son Abdülmecid. After Abdülmecid's accession to the throne, Ömer Pasha was promoted to the rank of *mirliva*. Till that point he had not seen much actual service in the field, but that quickly changed. First he served with the European naval officers during the pacification of Beirut. Next, in 1842–43, he was in Damascus commanding 15,000–20,000 troops aimed at reducing the power of the Druze chieftains in Lebanon. Temperley describes him as 'the ruthless Ömer pasha' with his 'savage Albanians'. He appears to have had some success in reducing the Druze resistance and defeating their leader, before he was reassigned elsewhere. In 1846–47, he was in charge of the pacification of Kurdistan, especially the rebellion of Bedir Khan. At the head of 12,000 troops he moved on Nizib, drew Bedir Khan and his forces in battle, and defeated them. For that triumph, he was promoted to the rank of field marshal (*müşir*) by the sultan.

In 1848, as part of the occupation force in Wallachia, Ömer Pasha was credited with acting in a restrained manner. Of course, as we have seen, there were very few armed rebels in Wallachia but large numbers of Russian troops. The point to be made is that Ömer was an integral part of the reformed military organisation, having become the *Müşir* of the Rumeli third army, which was now called upon to restore order in Vidin and then in Bosnia and Montenegro.[30]

In Vidin, the Ottoman authorities were aware of possible Serbian agitation, and also of the revolutionary fervour which spilled over into Ottoman territories in the shape of Polish and Hungarian refugees. Discontent with the Tanzimat reforms made fertile ground for ideologues of liberation. Serbia, however, does not seem to have been instrumental in the Vidin events, even though Serbian officials served as mediators for Bulgarian petitions concerning their grievances as the revolt wound down. Ali Rıza Pasha was despatched to restore order to Vidin, but local notables had already called in the *başıbozuks* in much the same manner as has been described for Aleppo. Ömer Pasha, on his way to Bosnia, was ordered to negotiate with the rebel camp on the Serbian border. This he appears to have done with some success, persuading the Bulgarians who had fled to Serbian lands to return home by promising them amnesty. Meanwhile, Ali Rıza Pasha, who had arrived in June in Rusçuk, mobilised regular forces to take care of the *başıbozuks* and reported on the numbers killed in the repression of the revolt: 15 Turks and 720 Bulgarians. He issued a warning: 'From now on no objectionable conduct and evil deeds by officials and other persons are permitted.' A delegation of Vidin residents travelled to Istanbul to present their grievances to the sultan, and to request that Ali Rıza be made Governor. They met with the Supreme Council in August. The Council installed Ali Rıza, promised the delegation that a redistribution of property, in accordance with the Tanzimat regulations, would be undertaken in Vidin, but equivocated about the ways in which it might be carried out. When the delegation returned, the word spread that the Ottoman land regime had been abolished.

What had started as a peasant revolt turned into a true test of the viability of the new Ottoman order. In February 1851, the lands of the Muslim lords were ordered to be sold to the peasants, a solution that benefited neither landlord nor peasant. New disturbances brought another inspector from Istanbul, one Şekib Efendi, former ambassador to Vienna and Tanzimat bureaucrat, and the appointment of Zarif Pasha to replace *Vali* Ali Rıza. Zarif is reported as saying he could not understand the selling of property to the reayas, and also as conspiring to have Şekib Efendi removed. He was removed himself in January of 1851, and replaced by Vasıf Pasha, Commander of Belgrade. Both men ended up on the Caucasian battlefront within two years. The example of Vidin is an interesting one. Here, the Tanzimat installation represented a give-and-take between the peasants and the bureaucrats of Istanbul, and a certain amount of restraint is in evidence. Land reforms, however, proved very difficult to impose, and the problem festered for another decade.[31]

Provincial resistance: Bosnia and Montenegro

We left Bosnia in revolt in 1832, when *Kapudan* Hüseyin and Buşatlı Mustafa of Albania challenged the Ottomans. The Ottoman solution to Bosnia was to back rivals to Hüseyin's power, and install Ali Ridvanoğlu as governor of a newly-separate Herzegovina *pashalık*. In 1834, Bosnia was reorganised with six *sancaks*, and smaller districts as elsewhere. In 1835, the Ottomans abolished the hereditary *kapudanlıks* of the Bosnian Muslim nobility, and replaced them with the office of *müsellim*. Many of the former members of the Bosnia militias, *sipahis* and *kapudans* were thus appointed, so little actually changed. Discontent continued in the province sporadically throughout the 1830s and 1840s. After 1848, revolutionary fervour complicated the picture in Bosnia and especially Herzegovina. The chief difference between Bosnia and Vidin was that rebellion was instigated and sustained by Bosnian Muslim chieftains and their militias. The major issue for local populations continued to be tax revenues and the division of land. The largely Christian peasantry protested at the continued dominance and oppression of the Muslim landlords.

Finally, in 1850 *Müsir* Ömer Pasha was appointed as Governor in Bosnia to replace Tahir Pasha. Word of his clemency in Bulgaria preceded him, and the Christian peasants laid down their arms. It was the Muslim landowners who continued to resist. Ömer Pasha had some 8,000 troops and a number of Hungarian and Polish officers who had joined the Ottoman army. He succeeded where others had failed, breaking down the traditional politics of the Bosnian Muslim notables by 1851, and destroying Ali Ridvanoğlu, who had attempted independent rule in Herzegovina. Barbara Jelavich states that suppression probably cost 5,000 lives; others refer to 5,000 refugees in Croatia. Temperley weighs in:

Omer's methods were barbarous, but he 'did not his work negligently' [i.e., he got the job done]. His appeals to the Christians for support were entirely successful. The few Christians taken in arms were liberated, and the mass of them ardently supported him. One by one the castles of their feudal tyrants fell before Omer and his artillery . . . He announced the end of all feudal privileges. He built the first road in Bosnia between Travnik and Sarajevo. He repressed brigandage with a stern hand. He closed the frontier to Croatia and stopped smuggling. He proclaimed the forests to be crown property, and cancelled the very profitable concessions for timber enjoyed by an Austrian sawmill company. He even attempted to construct a line of steamers for the rivers of Bosnia. Omer's methods were as vigorous as they were rough. But it was his vigour which was most hurtful to other powers. For Austria demanded and ultimately obtained his recall, and with his fall ended all hope of permanent reform in Bosnia.[32]

Ömer's stern rule in 1850–52 provoked protests from Austria especially, but he was defended by Great Britain, and therefore emboldened by his own government to take on the Bishopric of Montenegro, which the sultan claimed as part of his sovereignty. In 1852, massive Ottoman forces entered Montenegro and declared war on the new ruler, who had neglected to inform the Ottomans of his intention to secularise the state. The invasion provoked yet another international incident, in which both Austria and Russia gave an ultimatum to the sultan to order Ömer to withdraw. By March 1853, Ömer Pasha was dismissed. Montenegro had become a Principality, with the tenuous threads of connection to the Ottomans broken permanently. The crisis over Montenegro heightened the drifting apart of the Austro-Russian rapprochement, which led to the break by mid-September when Franz Joseph refused to join Russia in the war against the Ottomans, as we shall see.

Meanwhile, as a result of the 1850 revolt, Bosnia and Herzegovina were subdivided into seven *sancaks* under *kaymakams* with military and civil powers. Under Topal Osman Pasha (1861–69), the province made considerable strides in education and infrastructure. A local council was created in 1866. Sarajevo became the official residence of the governor. As we have seen in the Arab provinces, however, it proved impossible to institute significant reforms to the agricultural sector. Attempts were made to respond to the peasant demands. In 1859, they were guaranteed certain rights to use of the land, but large estates continued to be run by their Muslim land-owners as tax-farms, leading to further discontent among the peasantry and further rebellions in the 1870s that convulsed large parts of the Balkans and provoked further European intervention.

As regards conscription, Bosnia did not contribute much manpower to the Crimean front. Only afterwards did the idea of a Bosnian militia (*mustahfız*) finally settle the problem of Bosnian Muslim service in the *nizamiye*. In 1862, at the suggestion of Grand Vizier Fuad Pasha, a military commission had met secretly to discuss the creation of a home guard, which would solve the major objection of the Muslim population to enlistment in the Ottoman army. They did not want to fight outside their province, and they wished to wear their own uniforms. In 1864–65, a census was taken in Bosnia–Herzegovina, and volunteers filled six companies and 800 officers. The aim was to have 4,800 men under permanent arms, who would serve three years as regulars, which was to be followed by nine years in the *redif* or seven years in the *mustahfız*, but in the home province. The first two regiments, commanded by local officers, paraded before Abdülaziz (1861–76) in Istanbul in 1865. Besides the regulars, Bosnia eventually had eighteen *redif* battalions. By 1874, the Bosnian regular and reserve troops equalled 30,000 men.[33] The agreements sound much like the Bosnian *ocaklık* system of the eighteenth century. The new organisation represents one more example of the way in which the Ottoman reformers were forced to compromise by local conditions and traditions. Conscription everywhere remained a combination of volunteerism, balloting, and violent impressment.

Austrian and Russian armies revisited circa 1850

That is how things stood in Ottoman territories as the imperial squabbles and restless populations threatened to engulf all of Europe in a general war. Having assessed the rather dismal record of the Ottoman reforms, we should pause a moment to consider the state of the military organisation of Austria and Russia as they stood just prior to the Crimean War. The brutal clash of traditional and liberalising ideas exemplified in the Tanzimat were a caricature of similar conflicts in autocratic settings like Vienna and Moscow. After the fall of Napoleon in 1815, Chancellor Metternich saw to the maintenance of stabillity and tranquillity in Habsburg territories by a combination of diplomacy and armed force. Military command was subordinated to dynastic concerns. The wars with France had left both Austria and Russia impoverished, and insufficient and inefficient taxation often mandated the reduction of military expenditures. On paper, the diverse territories of Austria could mobilise 400,000 soldiers, but training, command and supply all lagged behind Prussia and Russia in the period just prior to 1850. The military command, the *Hofkriegsrat*, was composed of a mixture of civilian and military men, and was ruled over by the *Staatsrat*. The somewhat cumbersome command structure and the General Staff, in contrast to Prussia, for example, remained much as they were in the eighteenth century. Metternich preferred the supremacy of civilian over military authorities. The military high command remained multi-national, aristocratic, and resistant to change, which set the tone for the officer class and discouraged innovation or assumption of responsibility.

The people's army, the *Landwehr*, was disbanded by 1831, and selective and unequal conscription was restored, based on the system inaugurated by Joseph II. For example, in Hungary service was for life, while elsewhere it was fourteen years. After 1845, a uniform eight years was fixed throughout the empire. Conscription was by lots, as in Ottoman territories, but often the 'lots' drew society's misfits. Austria was second only to Russia in the poor levels of pay and rations, and the liberal application of punishments, such as flogging, caning, and the gauntlet was routine.[34]

There were two exceptions to the general system: one, the Tyrolean militias, and the other, the military border system. In an era of poor financing, the latter especially had an important role to play in our context. In 1848, 16,000 Tyrolean marksmen were mobilised when Garibaldi threatened the Austrian borders. The military border system, four General Commands – Croatian, Slavonian–Symrian, Banat and Transylvania – could be counted on to raise 100,000 troops when necessary, as in 1831 when 12 battalions were sent to Italy.

In 1848 the Austrian army had, on paper: 315,000 infantry; 49,000 cavalry, 26,000 artillerymen, 5,400 men in technical services, and a wagon train of 4,000, organised across twelve different command centres. Underfunding and desertion meant that actual strength was likely closer to 250,000 men.[35] Armament and equipment, with the exception of infantry arms, saw little change in this period. Cavalrymen still largely relied on the sabre, although some carried carbines. First-line units were equipped with percussion rifles starting in 1835. Proficiency remained high in the technical branches. Elsewhere, however, training was highly deficient. Ceremonial exercises, rather than combat drill, were the norm, although the *Grenzer* on the Military Border, as always, were better prepared.

Austria maintained cordial relations with Prussia and Russia throughout the period, largely due to the general paranoia about the spread of French revolutionary ideas. The emperor's interests in the German Federation remained, and Austro-Prussian cooperation as a bulwark against France on the Rhine characterised the period. As we have seen in Hungary and Transylvania, Austrian and Russian monarchs did not hesitate to call on one another for military support. Italy was the focus of Austrian concern, though even there, the army was reduced in size from 104,500 men in 1831 to 34,000 by 1846, stationed in the best-armed fortresses of the empire. The Italian force was under command of General Radetzky, who created a climate of training and morale which made it the superior part of the Habsburg forces. By and large, the remaining defences of the monarchy had become obsolete.[36]

Austria's chief difficulty lay in the nationalities question, a problem made abundantly clear in 1848 and thereafter. Most of the diverse ethnicities sought their own 'national' armies, so the solution remained to station national corps outside their own territories where they could also be used as a police force when necessary. Poorly funded, undermanned, much scattered, with pockets of expertise on the frontiers, the Austrian army maintained a sense of itself as guardian of the empire, and so it proved when the revolutions broke out in 1848.

Hungary, however, was another matter. In 1848–49, Hungary made a sustained bid for independence which was predicated on establishing a national army. In July 1848, Kossuth persuaded the Hungarian parliament to finance the *Honvéd*, a 200,000 strong Hungarian army which would be Magyar in character, language and command. In response to the rebellions in Italy and elsewhere, the Austrian high command mounted a sustained military counter-revolution which eventuated in the restoration of the monarchy in Vienna in October of the same year. Franz Joseph was crowned emperor, but not, as we have seen, King of Hungary, in December 1848. By April 1849 Franz Joseph proceeded to take over supreme command of the army, a decision which influenced the direction of the Austrian military for the following decade. The Hungarian army, which had given up Buda in early January 1849, proved more resilient in other parts of the country, prompting the Hungarian parliament to depose the Habsburg dynasty in April and put Kossuth in charge. Some 60,000 troops under Field Marshal Haynau faced the 80,000-man army of the Hungarian rebels. That prompted Franz Joseph to call on Nicholas I for troops, as previously described. The Russians were commanded by Paskievich, a veteran of the Caucasus Russo-Turkish war of 1828–29. The final confrontation took place near Temeşvar, when the Hungarians were surrounded by the Russian and Austrians.[37] The Hungarian revolution was finished by October. A brief reign of brutal retribution by Vienna followed, with some 490 officers court-martialled, of whom 231 were condemned to death, although most eventually had their sentences commuted. A total of 13 Hungarian officers, leaders of the revolution, were executed, along with 114 civilians. A further 1,765 were imprisoned. The restoration of a centralised Habsburg monarchy came at a large cost to the future viability of a multi-cultural empire, although one of the immediate results was the strengthening of the imperial army, which in 1849 reputedly stood at 648,000 men, with 1,200 mobile field guns. The collaboration with Russia also caused considerable unease among Austrian leaders, as the presence of Russian troops among Slavs of the Military Corridor, for example, stimulated Orthodox *Grenzer* to join the Russian regiments in saying mass. It did not bode well for the coming Crimean conflict.

For the next decade, the emperor himself ran the military, as head of a newly centralised supreme command, with four armies headquartered in Vienna, Verona, Buda and Lemberg. The emperor intervened in the direction of operations in all aspects of the military. His adviser was Field Marshal Grünne, who was patronised by Archduchess Sophie, the emperor's mother, and was the former chamberlain of Franz Joseph. The reorganisation reflects Franz Joseph's passage from constitutional monarch to absolute ruler, especially after the office of Prime Minister was abolished in 1852. Army expenses exploded over the next decade, at the same time as army effectiveness declined. Before 1848, 50,000,000 florins annually was the normal outlay for the military; by 1854 this rose to 198,219,783 florins, some 42 per cent of the budget. Pay remained abysmally low, even though there were plenty of candidates. The officer corps was more than fifty per cent foreign-born by 1859, with deleterious effects on camaraderie and loyalty. In July 1852, a univeral conscription law was passed, with eight years in full service and two in reserve. Annual recruits probably never exceeded 80,000, and few were kept on for the full term. Pay, rations, and rates of corporal punishment did not improve significantly as we pass through the Crimean War period. New weapons were introduced, such as the percussion rifle, but training and tactical support did not follow. As elsewhere, the artillery received considerable attention, its numbers doubling in the decade 1849–59. The 1850s saw the establishment of the gendarmerie service, which provided police and security in cities and rural settings. Gendarmes were better paid and housed than the regular army.

It was a time of preoccupation with internal security, and the nationalities were governed with considerable severity. Foreign policy was poorly coordinated in the absence of a prime minister, although Franz Joseph had ambitions to restore Austria to European greatness. Confrontations with Prussia over German territories intensified, though they remain outside our story. French power threatened central Europe when Louis Napoleon became Napoleon III in 1851, and Austria hastened to sign a military convention with Prussia against the possibility of French hostility in 1852. Relations with Russia continued to deteriorate over geopolitical conflicts in the Balkans, as previously described, and their further resolutions lie outside the chronology of this book.

During the Montenegro crisis, the Austrians had mobilised 50,000 *Grenzer* at the same time as they issued the ultimatum which forced Ömer Pasha to desist. Austrian generals advised their civilian counterparts that Serbia, Albania and Bosnia should be occupied, to prevent the Russians from doing so. Then Russia reoccupied Moldavia and Wallachia in July 1853, precipitating the Crimean War. The Austrian field army was deployed in Transylvania and along the Danube. The third army in Hungary, and also the fourth in Galicia, were mobilised. In April 1854 Austria and Prussia completed a military alliance, with Prussia promising to support Austria, should Russia not withdraw from the Principalities. By mid-August 1854 Nicholas I had removed the troops from the Principalities, following ten months of Russo-Ottoman confrontation on the Danube, but Franz Joseph continued to mobilise further. By mid-October he was calling for a full mobilisation of the army. Warned of the Russian colossus by his military advisers, he desisted by November, joined the French and British and their Italian allies in December 1854, and declared Austrian neutrality. In the event, Austria did not send troops to the Crimea.

The Austro-Prussian–Russian interlude presaged later conflicts, and Habsburg reticence here likely extended the life of the dynasty for another half century.[38] There are striking parallels here with the Ottoman experience, not in terms of constitutionalism but in the struggles to maintain the traditional, autocratic order in a revolutionary age, and the crippling financial debt which the effort entailed. Vienna could not simultaneously be federative, centralised, and nationalist.

Concerning the Russian 'colossus', Nicholas I's experience in the 1828–29 Russo-Turkish War had left him with an understanding of the limits of expansion to the south, especially in the matter of logistics. He cooperated with Britain based on his understanding of the Eastern Question throughout the decade of the Mehmed Ali crisis, and joined the Concert of Europe in signing the 1840 Treaty of London, but he also undertook to reorganise the military. From 1832 to 1836, War Minister Chernyshev transfered control over the military from the palace general staff to the Ministry of War, a structural move which clarified the lines of command and eliminated dual structures. This was a significant reform which lasted until the end of the Romanovs, and even into the Soviet era. As with both Austria and Turkey, military expenditures outran even peacetime budgets, but the rationalisation of ministerial staff also began in 1832. One of the aims was the elimination of corruption and improvement of administrative efficiency.

A more serious problem of the period was the question of manpower, which was tied to the continued conscription of serfs. The Ottomans created a reserve system in 1834, which became one of the branches of the divisional armies after 1843. The Habsburgs eliminated a *Landwehr* people's army, but had 100,000 *Grenzer* at their disposal. The Romanovs maintained an army three times the size of any army in Europe because they did not have a reserve system. Conscription had dipped so deeply into rural resources as to provoke resistance which could not be controlled by the dispersed army. Landowners were unwilling to part with able men, so recruits were the expendable of rural communities. The excessive 25 years of active service had to be sustained to keep up the numbers in the army. The era under discussion was rife with peasants' revolts, with origins as much economic as military.

Multiple frontiers, as with the Habsburgs and the Ottomans, made for strategies which scattered massive numbers of forces and raised military costs to an intolerable degree. Nicholas I experimented with furloughs, a type of reserve for soldiers who had served 20 of their 25 years, who were presumed to be trained, and could be recalled in emergencies. In 1849 such troops made up those who entered Transylvania, but their numbers were insufficient to the demand. When the Crimean War broke out, Russian military strategists were faced with an unenviable paradox: the army had grown so large that it threatened to bankrupt the state, but it remained too small and ill-equipped to defend the vast territories, and take on Istanbul, as the 1828–29 campaign had demonstrated. The Crimean War demonstrated it again, as nearly two million men were mobilised in the course of the conflict. While the combined allied force amounted to 300,000, Russia's stood at 31,954 officers and 1,742,343 men in 1856. With irregulars, that number might have exceeded 2,500,000. The problems were legion, but largely due to the vast terrain, poor leadership, inadequate supply, and 'strategic immobility'. The colossal failure of the Russian military resulted in the greatest of the empire's defeats in the nineteenth century, and a complete overhaul of the system from 1861 forward, when the Emancipation Edict eliminated the system of serfdom.[39]

Origins of the Crimean War

Revolutionary fervour did not convulse Russia in 1848 as it did elsewhere, but the stationing of Russian troops in Moldavia and Wallachia, as we have seen, did awaken ghosts of previous such provocations in Vienna and Istanbul. I have argued that the introduction of Tanzimat reforms, combined with revolutionary fervour among ethnic communities made for a volatile stew on the Ottoman Danubian frontier. Others have asserted that diplomatic blundering, especially of Russian envoy Menshikov, sent with ultimatums to Istanbul, and of British Ambassador Stratford Canning, were responsible for the curious conflict that unfolded after 1853. To the extent that the British and (less so) the French diplomatic communities misread the Ottoman environment of 1853–54, which turned hostile and bellicose, they should be held accountable. For they shortly found themselves in an alliance with the Ottomans and mobilising large armies for the Crimean battlefront, which would bring together the world's largest sea power with the world's largest land power.

The focus of the diplomatic wrangling in Istanbul was Jerusalem, where a dispute was under way after 1850 over the primacy of Catholic or Orthodox communities over the holy places. The Russian and French diplomats, respectively, argued their supposed rights to protection of the Ottoman Christian communities, by reference to the language of the treaties of Küçük Kaynarca (1774), for the Orthodox; the Treaty of Belgrade (1739) was the referent for the Catholics. The religious question obscured what most interpreted as the thrust by Napoleon III (installed as Emperor on 2 December 1852) to destabilise the European concert, and reassert some influence over the Levant. By the end of December 1852, the sultan issued a *firman* giving the Latins (Catholics) control of the Church of the Nativity in Bethlehem in preference to the Orthodox Greeks. This was the reversal of a previous *firman* which had promised the Orthodox (and Russia) the same privileges (November 1851). Foreign Minister Fuad Pasha had argued a French alliance would follow, and may have been impressed by French threats about invading Syria.

This of course was a direct insult to the Russians, who had argued their right to protect the Orthodox Greek community since 1774. The Orthodox community was in fact substantially larger than the small Catholic community of the empire, and the 1851 decision much more in keeping with Russo-Ottoman diplomacy of previous eras. Tsar Nicholas anticipated British compliance with his view that the Ottomans were on the verge of collapse, and that Russia and Britain had a gentlemen's understanding about the Eastern Question. That may, in fact, have accounted for initial British reticence over the crisis as it escalated. Tsar Nicholas I, having survived a number of years of passivity and even cooperation with Sultan Abdülmecid, miscalculated about the possible impact of the Menshikov mission on Istanbul politics.

On 28 February 1853, Prince Menshikov, accompanied by a large number of naval and military officers, presented the sultan with several notes from the Tsar concerning the question of the Orthodox subjects in Ottoman territories, especially in the Holy Lands. The most demanding was that an international agreement be drawn up to give Russia what amounted to sovereignty over all the orthodox subjects of the sultan. Menshikov bullied his way around the Ottoman bureaucracy. He engineered the resignation of Foreign Minister Fuad Pasha, assumed to be pro-French, by simply refusing to meet with him. The sultan simply replaced Fuad Pasha with pro-Austrian Sadık Rıfat Pasha, another of the Tanzimat reformers. The grand vizier of the period, Mehmed Ali Pasha, was informed by Menshikov that he planned a visit to Hüsrev Pasha, ancient and out of office but still valuable as an informant, and ostensibly pro-Russian. Menshikov is also implicated in stirring up the Greeks over a Greek–Ottoman border incident, and in fostering the downfall of the Serbian premier, advocate of the South Slav nationalism described earlier.[40]

Mehmed Ali Pasha responded to the Russian demand on two levels: Rıfat was told to continue negotiating, while he approached the British and French about naval aid. He also began secret communications with the military about Ottoman preparedness. The French responded immediately, and by April, the French fleet was anchored near Athens. The British hesitated, misinterpreting the seriousness of the situation. Negotiations between Menshikov and Rıfat continued, but the Ottomans simply refused to offer more than guarantees of such privileges by sultanic order. On 13 May, Mehmed Ali Pasha's government was dismissed by the sultan, probably at the instigation of Menshikov's staff and in collusion with Mustafa Reşid, his replacement. Reşid represented the peace party, and was presumed to be more amenable to Russian demands, while Mehmed Ali was secretly preparing for war. Mehmed Ali's views were shared in the streets of Istanbul. Mehmed Ali was moved to the *Seraskerlik*, and proved to be just as influential in the direction of discussions on the Supreme Council. Muslims of the empire had become more vociferous about the restoration of Ottoman prestige, after decades of humiliating treatment by Christian powers, and this sentiment, shared by Mehmed Ali and others on the council, would ultimately affect the decisions made there.

In a final gesture to save a diplomatic break with Russia, Reşid called a general assembly of 48 members of government, including both Hüsrev and Fuad Pashas, to deliberate on the possibility of war. The greatest impediment was the lack of the guarantees of great power support. A diluted response to the Menshikov ultimatum was crafted, which offered a *firman* rather than a treaty, reasserting the existing privileges of the Orthodox Church and a return to the status quo over the Holy Places. This was flatly rejected by 18 May, and the Russian mission left with the entire Istanbul embassy by 21 May. Tsar Nicholas subsequently threatened to occupy the Principalities, and, if that did not force a capitulation to the ultimatum, to blockade Istanbul. Menshikov blamed Stratford Canning for the failure of the mission.

Meanwhile, an Ottoman bank was established specifically to organise state finances in the case of war. A £450,000 pound loan was negotiated in London. By 18 May, the entire empire was put on naval and military alert, including the reserves. Orders were issued by the sultan that all measures should be taken to protect the Christians of the empire from Muslim abuse during the crisis. On 2 July 1853 50,000 Russian troops, under Prince M.D. Gorchakov, crossed the frontier into the Principalities, and by the middle of July had occupied Bucharest. The Ottomans stalled for time. Some 10,000 Egyptian troops, three battleships and six other vessels joined the Ottoman fleet by August. Adolphus Slade, then an Ottoman naval adviser in Istanbul, may have influenced *Serasker* Mehmed Ali's thinking that the Ottomans were strong enough to defend Istanbul. The city's defences continued to be upgraded, with thirteen forts manned with more than 300 guns.[41] Such warlike preparations had the Russians contemplating a joint land–sea operation from Varna, which was rejected in preference for the initial occupation of Wallachia and Moldavia in the hopes of preventing a wider conflict.

British Ambassador Canning, who had returned from a short absence in London just as Menshikov arrived, had begun to distance himself from the reformers and from the Ottoman reform agenda altogether. In 1852, a financial scandal involving the sultan's closest advisers, including Mustafa Reşid, discouraged the Ambassador. His attitude was hardly unique, as it was an era when British attitudes hardened against the Ottomans in general. Prime Minister Aberdeen was known for his hatred of Muslims and Ottomans, confiding to his Lord Admiral: 'I should as soon think of preferring the Koran to the Bible, as of comparing the Christianity and civilization of Russia to the fanaticism and immorality of the Turks.'[42] British and French discord, and inflexible British diplomatic attitudes, meant the two powers were initially powerless to influence the drift to war. Still, British policy aimed at maintaining the Ottoman status quo, while the Russians were already envisioning its demise and partition among the great powers. Such fundamental misunderstanding certainly contributed to the escalation of hostilities.

By mid-June, both British and French fleets had anchored just outside the Dardanelles, part of the reason why Nicholas abandoned the proposed double-pronged land attack on Varna and the naval attack on Istanbul. The Russian occupation plan depended on continued good relations with Austria, which, as we have seen, failed to materialise after mid-1853. In July, with the Russians in occupation in the Principalities, the Austrians proposed a four-power diplomatic conference in Vienna to solve the crisis. The discussions were to produce the Vienna Note, which became the central document of controversy in the next stage of the crisis. In Istanbul, Mehmed Ali, alarmed at Reşid's passive reaction to the Russian occupation, attempted to replace his rival with Ali Pasha. Stratford Canning demanded the return of Reşid, and entangled the British even further in the looming war by interfering in local politics to support Canning's protégé. Grand Vizier Mustafa Naili and Foreign Minister Reşid Pasha were reinstated on 9 July, but were reduced in the eyes of the public to the status of British lackeys.

Their solution to the crisis was to call another general assembly to discuss the Russian occupation. The ultimatum of 23 July, as it is called, reiterated Ottoman sovereignty in relation to its own subjects, and promised to grant the same rights to the Orthodox Greeks which other *millets* already had, rejecting Tsar Nicholas's demands once again. The ultimatum sounds remarkably like that of Mahmud II on another occasion (1828), and was considered by the diplomatic community to be the last concession the Ottomans would make. Should the Russians accept the condition, the Note would go with a special Ottoman envoy to St Petersburg, and the Russian troops could then be withdrawn from the Principalities.

Meanwhile, negotiators in Vienna had crafted a similiar Note by the end of July, but without consulting Istanbul. Their version had significant variations, which favoured the Russian view of its right to intervene, drawn from the language of the Küçük Kaynarca Treaty, and insisted that any changes to such language be approved by both France and Russia. While the Ottoman Ultimatum of 1853 preserved the fiction of Ottoman independence from foreign intervention, the Vienna Note did not. The Vienna conference gave preference to the great power version; the Austrian representative is said to have repressed the Ottoman version on orders from Franz Joseph, hardly the behaviour to gain the confidence of the Istanbul government.

The story now takes on aspects of an Oriental tale, with a blizzard of textual variations and attempts to intercede between Russia and the Ottomans. The diplomatic community had miscalculated (or simply dismissed) the strength of Ottoman resistance to the question of Russian intervention in what Ottoman negotiators insisted was a matter of sovereignty. 'There was very general irritation with the Ottomans who, according to *The Times*, were in themselves of no more account "than the red Indians of Yucatan," but the very circumstance that they were perceived as the cause of the crisis reflects the fears routinely evoked by any new instalment of the Eastern Question.'[43] On 18 August the Ottomans rejected the Vienna Note and offered another one of their own, which was also rejected by Nicholas. A diplomatic impasse had been reached.

Ottoman fears of Russian intentions concerning Serbia and Greece, as well as the tense atmosphere of the northern border areas that has been described, were paramount concerns among Ottoman bureaucrats.

Austria maintained observation troops in Transylvania, which must also have been seen as menacing to Ottoman security. Much discussion centred in council on the Caucasus frontier as a place where the Ottomans could make a stand against Russia. By early September, the 'war party', likely stimulated by Mehmed Ali Pasha and his entourage, had begun to dominate street politics in Istanbul. Mustafa Reşid pressed for a four-power guarantee of the Ottoman Empire, which was resisted by both France and Britain. News spread that Nicholas and Franz Joseph intended a meeting at Olmütz to discuss differences. This was interpreted as a war council. Petitions calling for holy war were submitted to the sultan, from prominent members of the ulema, in the midst of disturbances in Istanbul in the second week of September. Reşid pressed France and Britain to move their ships into the Bosphorus. The demonstrators in Istanbul further petitioned the sultan, who urged his religious officials to argue for the virtues of the diplomatic avenues.

While it appeared to many that the sultan's throne was in danger, as during the accession of Mahmud II, wiser observers saw the hand of Mehmed Ali Pasha and the small war party at work. Foreign observers, however, gauged that an influx of refugees from the Balkans, largely Christian, and war-enthusiastic volunteers from all over the empire might have numbered a quarter of a million.[44] This was a volatile situation for any government. Sultan Abdülmecid convened a large general council on 26 September. They were as yet without news of the Olmütz meeting on 22 September, when Nicholas I made further concessions which might have made the Vienna Note more palatable to the Istanbul government. An extensive discussion ensued, but neither side had proper information to make a considered decision about the Ottoman ability to defeat the Russians. Mustafa Reşid was unable to persuade the council to wait for guarantees from Britain and France. They concluded that accepting the current Russian version of the Vienna Note was tantamount to 'drinking poison and simply dying'. According to Temperley, as the debates proceeded, 'the naked face of fanaticism peered forth.'[45]

On 29 September, Abdülmecid accepted a recommendation from the council to declare war, and 40,000 reservists were ordered mobilised. The war manifesto avoided the call to holy war, and spoke of the protection of the sultan's subjects. 'This declaration of war, which concerns Russia, is required by the need to work for the defence and interest of the Exalted State. Even more care and attention than previously will be taken to defend and protect the Christian subjects of the Exalted State. It will be of the utmost importance and a primary responsibility of government to evade anything which would invite the hostility of other states by reason of bad treatment of them.'[46]

As the hostilities escalated, the *Takvim-i Vekayi* recorded speeches by leaders of the various Christian *millets* attesting their loyalty and commitment to the conflict.[47] Ömer Pasha, stationed at Şumnu, gave Russia two weeks to withdraw from the Principalities on 6 October. Instead, Nicholas I and his military advisers made plans to take the war to the Caucasus, aiming at Kars. On 22 October, Istanbul instructed Ömer Pasha to begin military operations against Russia in the Principalities immediately.[48]

The British, who apparently truly failed to appreciate the strength of Ottoman resistance to their interference in internal religious matters, likely because of Canning's obsession with them, found themselves virtual guarantors of Ottoman survival without having acquired any concessions. For the rest of the year 1853, further attempts were made to bring the sides to the negotiating table while the British and French fleets hovered in the Bosphorus. British public opinion grew 'fanatic' as anti-Russian sentiments had escalated, stimulated by old Ottoman hands such as pamphleteer David Urquhart. Describing Anglo-Russian relations, and excoriating British administrators, he wrote: 'You [Britain] discover that she [Russia] is illogical – in rhetoric which has convulsed an Empire; you denounce her to be insane – in marching with your aid on the Bosphorus; you send your men-of-war to the Dardanelles – she seizes the Sound; she grasps the Danube – you send – no, you do *not* send, a note; she usurps the vastest plain of Europe, and you *do* send an apology. When she has overrun a province you convoke a conclave . . . And all this is peace; two campaigns and a dozen fortresses are offered up for the sake of the tranquillity of Europe; and two Empires [Austria and Ottoman] are stripped of their arms in the interest of their defence.'[49] In the days that followed, war hysteria put pressure on the British government to act.

On 22 October, as Ömer Pasha and his army began to cross the Danube, Stratford Canning summoned the British fleet from the Dardanelles to Istanbul. Hostilities had already begun in the Caucasus. For more than a month, Mustafa Reşid urged the French and British fleets to move into the Black Sea, while diplomatic efforts in Vienna were revived with little result. Mustafa Reşid argued for an enlargement of the London Treaty of 1841, which would guarantee the freedom of the straits as before, but establish buffer zones between Russia and the empire, such as the long-standing Serbia, Moldavia and Wallachia in the west, and Circassia and Dagistan in the east, the latter two having been surrendered to the Russians in 1829.[50] Confidence was high. The Ottoman army won the first contests of the war, at Vidin and Oltenitsa on the Danube, and St Nicholas in the Caucasus, and raised hopes in Istanbul. In late October the Ottoman navy made an attempt to maintain a token force in the Black Sea, as news of similar patrols by the Russian navy reached Istanbul. The British issued an ultimatum to Russia not to venture out of Sevastopol to attack the Ottoman navy, yet another provocation to war. By mid-November, the remainder of the British and French fleets in the Dardanelles had moved to the Bosphorus, and the entire Ottoman navy had set sail and was anchored at Sinop. On 30 November, six Russian vessels and two frigates (716 guns) overwhelmed and defeated the Ottoman force of seven frigates and four light ships (472 guns) in two hours. Some 4,000 are said to have perished in the bombardment and fires which followed. As the news of that disaster sank in, word arrived in Istanbul of a Russian victory at Başgedikler on the road to Kars.[51] The European allies had reached the point of no return: either they supported the Ottomans, or they watched an uncertain dissolution of the empire following Russian certain victory.

Taking advantage of a moment of uncertainty in Istanbul, Stratford Canning is said to have pressed Mustafa Reşid to commit to international negotiations for peace in order to engage the joint Franco-British expedition into the Black Sea. Hints of such a capitulation set off riots in Istanbul, likely stimulated by *Serasker* Mehmed Ali. A squadron of French and English ships made a show of force in Istanbul, the sultan and his advisers broke the back of the resistance, and a grand council agreed to continue negotiations. But the damage to British–Ottoman relations was severe. Then news arrived from Europe that a joint Franco-Ottoman force was to sail into the Black Sea. With the new year, on 3 January 1854, three squadrons of the joint command left Istanbul for the east, some packed with Ottoman reinforcements for the eastern battlefront. On 12 January the French and British Ambassadors informed the tsar's government that Russian ships who left port would be treated as hostile, another deliberate provocation. Even then the allies dallied, likely because of weather and the need to prepare. On 12 March the Ottomans, French and British signed an alliance. On 27 and 28 March, Britain and France declared war on Russia. Palmerston argued at that time that British aims ought to include 'the restoration of Poland as of 1772; the union of Finland with the kingdoms of Sweden and Norway instead of with Russia; Austria's retention of the Danubian Principalities, with the surrender of Lombardy and Venice; Turkey's enlargement through the acquisition of the Crimea and Georgia, and the enlargement of Austria and Prussia in a

Germany free from the domination of the tsar.'[52] The Crimean adventure had begun.

Notes

1 M.S. Anderson, ed. *The Great Powers and the Near East*, 1774–1923 (London: Edward Arnold, 1970), 52.

2 Niyazi Berkes, *The Development of Secularism in Turkey* (Montreal: McGill, 1964), 147. The discussion of international relations draws on Richard S. Horowitz, 'International law and state transformation in China, Siam and the Ottoman Empire during the nineteenth century', *Journal of World History* 15 (2005), 445–86. His is the 'artifact of Christian civilization' (453).

3 Frank E. Bailey, *British Policy and the Turkish Reform Movement: A Study in Anglo-Turkish Relations, 1826–53* (Cambridge, MA: Harvard University, 1942); Harold Temperley, *England and the Near East: the Crimea*, (London: Archon Books, 1964; reprint of 1936 edn) are the two classic sources. Muhammad H. Kutluoğlu, *The Egyptian Question, 1831–41: The Expansionist Policy of Mehmed Ali Paşa in Syria and Asia Minor and the Reaction of the Sublime Porte* (Istanbul: Eren, 1998), and Roderic H. Davison, 'Britain, the international spectrum, and the Eastern Question 1827–41', *New Perspectives on Turkey* 7 (1992), 15–35, have both been used here.

4 Butrus Abu Manneh, 'The Islamic roots of the Gülhane Rescript', in his *Studies on Islam and the Ottoman Empire in the 19th Century* (Istanbul: ISIS, 2001), 89.

5 This story is well told in Niyazi Berkes, *The Development of Secularism in Turkey* (Montreal: McGill University Press, 1964). Selim Deringil has explored the evolving image of the sultan in his *The Well-Protected Domains, Ideology and the Legitimation of Power in the Ottoman Empire, 1876–1909* (London: I.B. Tauris, 1998).

6 The full *firman* is reproduced in Kutluoğlu, 195–97, from which much of this description is taken. The readily accessible account is by Harold Temperley, *England and the Near East: The Crimea* (London: Archon Books, 1964), and more recently, Caesar E. Farah, *The Politics of Interventionism in Ottoman Lebanon* (London: Center for Arab Studies & I.B. Taurus, 2000), chapter 2.

7 Berkes, *Development of Secularism*, 148–51.

8 Veli Şirin, *Asâkir-i Mansure Ordusu ve Seraskerlik* (Ankarai TATV Yayınları, 2002), has a list, with short biographies of all the *seraskers* from 1826 to 1908. As a point of interest, the military high command of the major players at Crimea are all criticised for their ineptitude, and age, with the exception of

the French, who had 'democratised' their officer training schools in the age of Napoleon and after.

9 M.S. Anderson, ed. *The Great Powers and the Near East, 1774–1923*, 61.

10 The copy of the law can be found in *Kanunname-i Askeri Defterleri* no. 7, Istanbul (for further discussion, see Veysel Şimşek, 'Ottoman military recruitment and the recruit 1826–53', MA Bilkent, 2005, and Musa Çadırcı, 'Osmanlı İmparatorluğu'da askere almada kura usulüne geçilmesi "1846 Tarihli Askerlik Kanunu" ', *Askeri Tarih Bülteni* 18 (1985), 59–75).

11 Necati Tacan, 'Tanzimat ve ordu', *Tanzimat I* (Istanbul, 1940), 129–37; James Henry Skene, *The Three Eras of Ottoman History: A Political Essay of the Late Reforms of Turkey* (London, 1851), 62–63, who seems to be the source for many of the other contemporaries commenting on Ottoman military strength in the case of war with Russia. This piece was likely written for Stratford Canning. It is hard to evaluate these figures which likely suffer from the inflation disease, as almost every record of this era draws on the *Takvim-i Vekayi*, the official, and undoubtedly rosy, picture of the new organisation.

12 S.J. Shaw and Ezel Shaw, *History of the Ottoman Empire and Modern Turkey* (Cambridge: Cambridge, University Press, 1976–77), vol. 2, 86. Skene, *The Three Eras*, 67.

13 Gunter E. Rothenberg, *The Army of Franz Joseph* (Purdue: Indiana University Press, 1976). Of course, the Military Border system supplied the Austrians with over 100,000 troops when need be (13–15).

14 E. Engelhardt, *La Turquie et le Tanzimat, ou Histoire des Réformes dans l'empire Ottoman* (Paris: A. Cotillon et Cie., 1882), 75–77. Ferdan Ergut, 'State and social control: the police in the late Ottoman Turkey and the early republican Turkey, 1839–1939', PhD, New School for Social Research, 1999, 99. He notes that in 1845, the word *polis* was actually used as a synonym for *zaptiye*, but the latter prevailed in all later documents. Initially, the *zaptiye* were to be created from the *timariots* abolished at the same time.

15 The provincial administrative reforms are fully described in Shaw, *History of the Ottoman Empire*, vol. 2, 88–95.

16 Skene, *Three Eras*, 77–80.

17 Maʿoz, *Ottoman Reform in Syria and Palestine 1840–61* (Oxford: Clarendon Press, 1968); also Farah, *Politics of Interventionism*, chapter 3, 85.

18 Stanley Lane-Poole, *The Life of the Right Honourable Stratford Canning*, 2 vols (London: Longmans, 1888), vol. 2, 96–97.

19 Maʿoz, *Ottoman Reform*, 78.

20 Enver Ziya Karal, 'Zarif Paşa'nın hatıratı 1816–62', *Belleten* 4 (1940), 467–70.

21 Maʿoz, *Ottoman Reform*, 82.

22 Maʿoz, *Ottoman Reform*, chs 4 and 5; Dick Dowes, 'Reorganizing violence: traditional recruitment patterns and resistance against conscription in Ottoman Syria', in *Arming the State*, ed. Erik J. Zürcher, 121–25 (London: I.B. Tauris, 1999).

23 Mark Pinson, 'Ottoman Bulgaria in the first Tanzimat period – the revolts in Nish (1841) and Vidin (1850)', *Middle Eastern Studies* (1975), 109. Evidence of Serbian interference is much disputed. Pinson has made extensive use of the well-known study on these revolts by Halil İnalcık, *Tanzimat ve Bulgar Meselesi* (Ankara: Türk Tarih Kurumu, 1943).

24 Svetlana Ivanova, 'Widin' *EI2*, vol. 9 (2001), 205–08.

25 Jelavich, *History of the Balkans*, chapter 6.

26 Radel Florescu, *The Struggle Against Russia in the Romanian Principalities: a Problem in Anglo-Turkish Diplomacy 1821–54* (Iasi: Centre for Romamain Studies, 1997), chapter 8.

27 There were some 200 casualties in the Bucharest rebellion. The Ottomans were later blamed for the use of excessive force (Florescu, *The Struggle*, 220–21).

28 Barbara Jelavich, *The History of the Balkans: Eighteenth and Nineteenth Centuries* (Cambridge: Cambridge University Press, 1983), chapter 6, and Florescu, *The Struggle*, 234–42.

29 Temperley, *England and the Near East*, 260–68. He cites 3,600 Hungarian and 800 Polish refugees, most of whom returned to Austria by 1851. The leaders either joined the Ottoman army, or emigrated to the United States.

30 James J. Reid, *Crisis of the Ottoman Empire: Prelude to Collapse, 1839–78* (Stuttgart: Franz Steiner, 2000), 466–79.

31 Pinson, 'Ottoman Bulgaria', 123–33.

32 Temperley, *England and the Near East*, 212–28.

33 This description is drawn from 'Bosnia–Herzegovina', *EI2*, CD edition, and Odile Moreau, 'Bosnian resistance', 130–35. Halil İnalcık and Donald Quataert, eds, *An Economic and Social History of the Ottoman Empire 1300–1914* (Cambridge: Cambridge University Press, 1994), part IV, for a general overview of the reforms; Barbara Jelavich, *History of the Balkans: Eighteenth and Nineteenth Centuries*, vol. 1 (Cambridge: Cambridge University Press, 1983), 348–60.

34 Rothenberg, *The Army*, chapter 2.

35 Rothenberg, *The Army*, 14–15.

36 Rothenberg, *The Army*, 17–18.

37 Rothenberg, *The Army*, chapter 3.

38 Rothenberg, *The Army*, chapters 4–5.

39 Kagan, 'Russia's small wars', in Frederick W. Kagan and Robin Higham, eds., *The Military History of Tsarist Russia* (New York: Palgrave, 2002), 131–35; Robert F. Baumann, 'The Russian army 1853–81', in Kagan, *The Military History*, 136–50.

40 Ann Pottinger Saab, *The Origins of the Crimean Alliance* (Charlottesville, VA: University Press of Virginia, 1977), 31–32, still the best book on the diplomatic origins of the war. A more recent summary is David M. Goldfrank, *The Origins of the Crimean War* (London: Longman, 1994).

41 Saab, *Origins*, 34. Also Clive Ponting, *The Crimean War* (London: Chatto and Windus, 2004), 11.

42 Ponting, *Crimean War*, 11.

43 It was actually referred to as the romance of 'Mille et une notes'. (Lane-Poole, *Stratford Canning*, vol. 2, 317). There were perhaps as many as eleven versions in circulation. Allan Cunningham, 'The preliminaries of the Crimean War', *Eastern Questions in the Nineteenth Century: Collected Essays*, vol. 2, ed. Edward Ingram (London: Frank Cass, 1993), 196–97. Austrian culpability is noted on 209.

44 Cunningham, 'The preliminaries', 203.

45 Temperley, *England and the Near East*, 360.

46 Ali Fuat Türkgeldi, *Mesâil-i Mühimme-i Siyâsiyye*, ed. Bekir Sıtkı Baykal, 4 vols (Ankava: Türk Tarih Kurumu, 1960), vol. 1, 321.

47 The Russian response of 28 October spoke of the 'holy and just cause that has found at all times ardent defenders among our pious ancestors'. Saab, *Origins*, 102–03; Ali Fuat Türkgeldi, *Mesâil*, vol. 1, 320–21.

48 The events are described with great clarity in Saab, *Origins*, chapter IV. She has made full use of Turkish documents and historians of the Crimean events. A protocol of the events at that meeting is reproduced in Türkgeldi, *Mesâil*, vol. 1, 315–20.

49 Quoted in Anderson, *The Great Powers*, 77.

50 Saab, *Origins*, 97.

51 Saab, *Origins*, 114–17.

52 John Shelton Curtiss, *Russia's Crimean War* (Durham, NC: Duke University Press, 1979), 236–37.

CHAPTER ELEVEN

Completing the northern arc

Of all the wars of the pre-modern Middle East, the Crimean War of 1853–56 has probably attracted the most attention, not only at the time, but in every generation of historians since. As the first major conflict documented on a daily basis by telegraph, photograph and newspaper, it brought a new age of immediacy to global conflicts. It need not consume a great deal of our time here. As a means of drawing this story of the Ottoman military and social transformation to a close, however, this chapter will focus on the Ottoman–Russian confrontations in Vidin and Silistre in 1853–54 and then in the Caucasus, ending with the siege of Kars in 1855. The siege of Sevastopol is somewhat irrelevant to the survey of this war, as beyond some early skirmishes, the Ottomans were largely inactive during the fall and winter, and only tangentially active in the spring of 1855. The stories of the Franco-British alliance, including the failure of logistics and supply systems, and the incompetence in military command on the Russian and British side in particular, have been repeatedly told. Indeed, the tales of sufferings of all the troops, and much revisionist history concerning the myths of the battles in the Crimea, have exposed even more starkly the folly of the enterprise. With the Treaty of Paris in 1856, the Ottomans had to accept the collapse of their military system, reassert their commitment to further social reforms, and accept the further erosion of economic independence. The siege of Sevastopol underlined the desperate shape of the Russian land forces and the vulnerability of their navy, and forced decades of radical reorganisation of Russian society. For neutral Austria, it delayed the moment of military reckoning and isolated the Habsburgs from allies east and west.

To focus on Kars is to complete the promised description of the eastern end of the northern arc of Ottoman defence. Kars and Erzurum were immensely important to the Russian–Ottoman confrontations from 1800 to the collapse of both empires by 1918. The siege of Kars is particularly illustrative of the failure of the regular Ottoman army in that region, and of the irreconcilable cultural barriers of British and Ottoman military styles, once British commanders were placed in charge of the campaign. There are myriads of British first-hand accounts of this war, most of which exude contempt for the Ottoman high command. Small wonder, then, that Ottoman military reformers looked elsewhere for inspiration after the Crimean War, and found ready advisers among the Germans, especially after 1870. Our narrative ends with 1869, when a new set of reforms signalled a further effort to construct an Ottoman army based on universal conscription.

On the Danube once again: Vidin and Silistre in 1853–54

On 23 October 1853, the Ottoman army crossed the Danube at Vidin and occupied Kalafat. On 27–28 October, Ottoman troops on the Transcaucasus border had crossed at St Nicholas, north of Batum on the Black Sea, under Ahmed Pasha, and captured the garrison. The Russo-Ottoman war of 1853–54 began with these small confrontations, which were hailed as great victories in Istanbul, but soon escalated into the contest for the control of the Danubian fortresses which had characterised all previous campaigns in Bulgaria.

Ömer Pasha began the war with 18,000–20,000 regulars under his command at Şumnu. His strategy was to defend the lower passes of the Danube River Basin, but also to cross the Danube to harass the Russian troops which had moved into Wallachia in July 1853, and to prevent their communications with Serbia. Hence, the first crossing to Kalafat passed with considerable ease. It was not until January 1854 that Russian reinforcements could be brought to bear on the north-western frontier of Bulgaria. By then, the Ottoman forces were arrayed in three corps: 34,000 effectives at Karasu, to defend the Danube shores from Hirsova to Tulcea; the centre, at Sistova, under Mustafa Pasha, also with some 35,000 troops, and a left wing, under Vidin Governor Sami Pasha, with 60,000 men, to guard Tutrekan, Niğbolu and Vidin. Of those latter troops, 16,000 were stationed at Kalafat, under Ahmed Pasha. Ömer Pasha, Commander-in-Chief, remained at Şumnu, with reserves in Sofia. The initial embarrassment at Kalafat forced the Russian commanders to ask for reinforcements, so that by the end of 1853, Gorchakov commanded

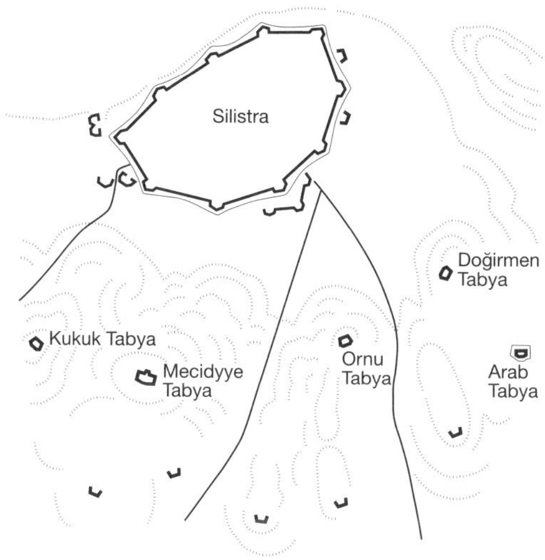

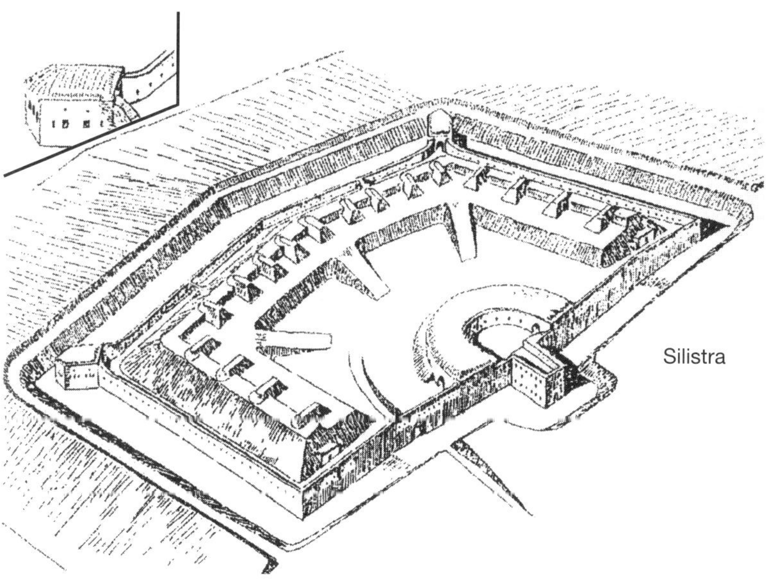

Top: Silistra was a Turkish frontier fortress on the Danube. Its development in the nineteenth century proceeded as follows: the enceinte of the town, 1811; the detached forts of Sultan Abdülmecid I, 1852/53; and the outer line of redoubts, 1877.

Bottom: The Mecdidiyye Fort was a typical example of a polygonal fort of mid-nineteenth-century type, with caponiers instead of bastions, a Carnot wall instead of a stone-revetted escarp wall, and "bomb-proof" semicircular barracks. This fort was up to date with the best contemporaneous European forts.

PLATE 21 *Sketch of Silistre fortress system and the Mecidiye* tabya *still standing (Kiel, Machiel, 'Urban development in Bulgaria', from the* International Journal of Turkish Studies, *4, 1989, 79–159*).

some 160,000 troops, of which 58,000 were placed under command of General Anrep, in Bucharest, with a reserve of 8,000 in Craiova.[1]

Once Kalafat was secured, Ömer Pasha moved troops to Tutrekan; under command of Selim Pasha, they crossed the Danube on 30 October 1853, occupied the island of the quarantine station, and set up batteries below Oltenitsa with 3,000 men. This had been the intended Russian crossing point, along with Silistre. The Russians advanced on 2 November with 9,000 troops, but as Ömer Pasha had fortified the right bank of the Danube, they were beaten back three more times with considerable casualties. Ömer Pasha's despatch to Istanbul was reproduced in the *Illustrated London News* on 3 December 1853. British officers General Lord Worsley, Captains Bathurst and Herbert Wilson, and Lieutenant Buckley, with Ömer Pasha at the battle, commended both infantry and artillery for steadfastness and accuracy.[2]

In early January, Kalafat was threatened by the arrival of Russian reinforcements. A report in the *Illustrated London News* described the subterranean barracks, modelled on a Wallachian peasant home, which the Ottoman troops inhabited in the miserable winter. A large room was dug out of the earth, sloping down to its greatest depth of about seven feet. A roof of poles was erected over the open hole, which was then covered with mats and a layer of earth. Commander Ahmed Pasha invited his visitor to a night inspection of the troops in these barracks. 'Standing in double rank within this subterranean room were about 200 men all under arms. In front of them were soldiers who held aloft long candles. The dimness of the light, the smoke of the candles, and the hoarse call of the drill-sergeants, together with the clank of the sabre of the commanding officer, gave the scene the aspect of a meeting of military conspirators. It must be owned, however, that the scene was calculated to give a high idea of the cleanliness and discipline of the Turkish soldiers, who went through that exercise with great precision and spirit.'

Ahmed Pasha and İsmail Pasha (Hungarian General Kmety), later hero at Kars, led ten battalions of those Ottoman troops, and two batteries of six guns, to confront the Russian army at Citate on 7 January 1854. They were accompanied by four regiments of cavalry and 500–600 *başıbozuks*. The Russians had 3,000 troops quartered in the village, part Russian soldiers and part Wallachian levies. During the first engagement, the Ottomans routed the Russians. They then confronted a Russian relief force of three regiments, with 600 cavalry and nine guns, and scattered them after a few hours of fierce street fighting. At this point the Ottomans retreated, and failed to follow up on the victory, while the Russians remained just long enough to bury the dead, which one estimate placed at 6,000, as against 1,000 for the Ottomans. Casualties to disease were 20–25 per cent of the effectives in the Russian army, contributing to an estimated 35,000 dead and wounded by the end of 1853.[3]

On 13 February 1854 the Russians attacked the Ottomans entrenched at Giurgevo and pushed them back to the south of the Danube. On 15 March the Ottomans repulsed the Russians at Tutrekan, across from Oltenitsa. The Russians, however, proved successful in occupying Tulcea Maçin, İsakcı, and Babadağı on the right bank of the lower Danube, and by early April they stood in much the same position as they had in 1828–29. The army was held from further southern advance largely because General Paskevich, veteran of the 1828–29 campaigns, took control, and proved more cautious in his command of the Danubian front. By the end of March, France and Britain had declared war on the Tsar. French troops numbering 30,000, the bulk of them from Algeria, left for the east in March, while 27,000 British troops shipped out soon after, both contingents disembarking first at Gallipoli. The whole British force eventually camped at Üsküdar in Istanbul, at the Selimiye Barracks which later served as the famous military hospital of Florence Nightingale.

The Ottomans and their allies were momentarily distracted from the Danube by sporadic Greek revolts, which spread from Epirus to Macedonia, usually involving small bands of 500 to 1,000 Greek irredentists, encouraged by King Otto I and pro-Russian elements of the population. These small bands engaged in guerrilla tactics thoughout 1853, but became a serious threat in confrontations with Ottoman regulars and irregulars, mostly Albanians, from January to May 1854. The French and British found themselves in the peculiar position of supporting the Ottomans against Greek partisans, first by using their ships to transport Ottoman troops to the places of insurrection, then by threatening economic sanctions, blockading Piraeus with warships, and landing troops to reinforce the threats. King Otto of Greece was forced to submit, but the allied occupation would continue until after the Crimean War, when the financial affairs of Greece were submitted to an international commission. An estimated 40,000 inhabitants of Piraeus, Athens and environs succumbed to the cholera which accompanied the French troops.[4]

Meanwhile, the siege at Silistre dragged on. On the advice of General Paskevich, Russian forces had been withdrawn from Wallachia and concentrated on the lower fortresses of the Danube, with Şumnu and Silistre the main targets for the establishment of bases for future deployment. General Paskevich himself took charge of the siege of Silistre, which began on 14 April 1854. Following three weeks of skirmishing, the Russians had surrounded the south-eastern side of the city, facing an outer defence of the main fortress named Arab Tabya (Arab Redoubt). It will be remembered that both Mahmud II and Abdülmecid I paid particular attention to the defences of Silistre, which included an outer ring of forts, one of which, the Mecidiye, still stands. As the town was not completely surrounded, the garrison had access to food and supplies, much as in the 1828–29 siege.

Two British officers, Captain James A. Butler and Lieutenant Charles Nasmyth, were part of the officer staff at the siege, under Musa Pasha, in whose command there were an estimated 12,000 to 18,000 troops. Prussian Colonel Grach, an expert on entrenchments, was also with the Ottoman officers. The Russians initially had 30,000 troops, which grew to 50,000. By May, the siege was reduced to a daily Russian bombardment in what became a monotonous and lethal routine, as the new explosive shells being used for the first time on these battlefields caused considerable damage. The Ottoman commanders resisted making frontal assaults on the Russian positions, as Russian sappers moved the trenches forward in zigzag fashion. A major assault by the Russians at the end of May was repulsed.

Ömer Pasha conferred with newly-arrived allied French Commander St Arnaud and British Commander Raglan at Varna on 18 May 1854. The discussion centred on a general allied assault on the Russians. Finally, Ömer Pasha sent a relief force from Şumnu to Silistre in mid-June. Diversionary attacks were ordered upon Giurgevo and Oltenitsa as well. Ömer Pasha's widely dispersed army had prevented him from sending the troops earlier, and only as the allied armies moved into Bulgaria could local troops be relieved to come to the defence of Silistre. In mid-June the relief force entered Silistre, where the Arab Tabya had fallen to the Russians, and a general assault on the main fortress was set for 21 June. By that time General Paskevich had resigned and returned to Warsaw. Within another 48 hours, the Russians lifted the siege and withdrew across the Danube. Lieutenant Nasmyth later reported: 'The Turkish army may well talk with pride. Their opponents had an army on the right bank of the Danube, which at one time amounted to 60,000 men. They had sixty guns in position and threw upwards of 50,000 shot and shell, besides an incalculable quantity of small-arm ammunition. They constructed more than three miles of approaches, and sprang six mines: yet during forty days not one inch of ground was gained, and they abandoned the siege, leaving the petty field work, against which their principal efforts had been directed, a shapeless mass from the effects of their mines and

PLATE 23 *The Mecidiye* tabya *at Silistre (author's photograph).*

batteries but still in the possession of the original defenders.'[5] Of course, the withdrawal of the Russians from the Principalities was also driven by an awareness of the immediate threat of an Austro-Prussian alliance, and an ultimatum from Austria sent on 3 June, which Franz Joseph emphasised by ordering the mobilisation of troops in Hungary and Galicia. By 14 June, the Austrians signed a Convention with the Ottomans for a joint occupation of the Principalities. By mid-August Russian troops had withdrawn into their own territory, and the Austrians occupied Bucharest with the Ottomans.

Ottoman forces under Behram Pasha (General Cannon) followed the Russians as they retreated across Moldavia. A fierce battle occurred at Giurgevo on 7 July 1854, but in general the withdrawal occurred in an orderly fashion, although Austrian and Ottoman contingents replacing the retreating Russians squared off in Bucharest in early September. The Ottomans reached İbrail first by the third week of September, but the Austrians entered Jassy ahead of them in the first week of October. Bucharest was divided into Ottoman and Austrian zones of influence. For the remainder of their stay, the Austrian presence was resented. They had to declare martial law to maintain order, and face the disapproval of Britain and France concerning their intentions regarding the long-term status of Wallachia and Moldavia.[6] Peace was within the grasp of all concerned, but the wartime build-up began to exert its own logic. By August 1854, the allied commands required some major demonstration of force to justify the expense and human loss.

By that time, the British, French and Ottoman troops numbered some 130,000 at Varna. For over six weeks, they remained in bivouac at the pestilential harbour, as cholera raged among their troops. In a month, 5,000 of the French and 850 of the British soldiers had succumbed. On 21 July 1854, French Commander St Arnaud had decided to send a new unit, called the *Spahis d'Orient*, under command of Algerian General Yusuf, to Dobruja, along with the 1st Division of French regulars under General Espinasse, to chase remaining Russians from the area. The 2nd and 3rd Divisions were to follow a day or two later. The *Spahis* were a unit made up of 4,000 *başıbozuks* and regular cavalry, recruited from among the Ottomans, to be the advance guard of the French troops. These included Albanians, Kurds, Türkmen and Arabs. The expedition wandered the marshes of Dobruja for almost a month, with 150 dying daily as time went on. The *başıbozuks* deserted, raided villages, and died of cholera. In small skirmishes with the Russians, mostly Cossack irregulars, they simply vanished, or demanded compensation for the heads of dead and wounded they brought to General Yusuf. Close to 3,000 died of cholera in that time. Such was the French effort to instruct the Ottomans how to deal with the *başıbozuks*.

The more famous exploits of General Beatson's irregular cavalry of the same source, the British *başıbozuks*, represent a nadir in the romance between European soldiers-of-fortune and their native armies. Beatson operated from mid-1853 to October 1854, recruiting his *başıbozuks* from the Balkans to the Caucasus, with little success. His experience in India made him over-confident of making a fine body of troopers out of undisciplined tribesmen who were tolerated, *faute-de-mieux*, as Ottoman irregulars, who were ever-present at the edge of the Tanzimat order, and routinely escaped justifiable punishment by commanding officers. Severe, exemplary discipline by British officers had little effect. Frustrated European officers grew to detest the *başıbozuks*, and Beatson was finally forced to abandon the project. Stratford Canning exemplifies the attitudes of British officials, rooted still in the romance of the 'nomad': 'even the irregulars submitted with good will to the command of Christian officers and to a degree of discipline which they had not previously undergone. On returning to their respective provinces they expressed so much satisfaction with the good treatment they had experienced in our service that when the Indian Mutiny broke out it would have been easy to raise an auxiliary force from among the population of their creed . . . The Bashi-Bozuks were never wanting in military courage. The crimes, by which they had earned a bad name, were necessary consequences of ill-usage on the part of their commanders, who cheated them of their pay and let them loose for subsistence upon the countries which they passed through or occupied.' [7]

Crimea and Sevastopol

Seeking some success in the time that remained of the campaign year, Napoleon III and the British Minister of War, the Duke of Newcastle, decided on the attack on Sevastopol. By late July 1854, the assembled allied officers were ordered to proceed to the Crimea, much to the dismay of seasoned soldiers, who were without detailed maps, charts, or knowledge of the strength of the enemy. Politicians in London were already planning on the disposition of the territories after the fall of Sevastopol: 'Palmerston thought the Crimea ought to go back to the Ottomans, probably with Georgia, Circassia, the Sea of Azov and the mouths of the Danube. "An adverse critic might say catch and kill your bear before you determine what you will do with his skin, but I think our bear is as good as taken." '[8]

Embarkation from Varna began on 28 August, but loading the ships was a difficulty from the first, and most of the British ships did not sail until 7 September 1854. A total of 3,000 horses were loaded on ships, but 5,000 were left to starve to death. Much of the hospital equipment had to be abandoned, and as the ships delayed in harbour, the bodies of the cholera dead were thrown overboard and floated in the sea around the ships. Among those who were transported to the Crimea were 7,000 Ottoman troops in six battleships, under the command of Ömer Pasha. In the end, the British used 52 sailing ships, 27 steam transports and some of their battleships, carrying 22,000 infantry, 1,000 cavalry, 3,000 engineers and 60 field guns. The French transported 25,000 infantry, 2,800 engineers and 68 field guns, in 200 small vessels in addition to their battleships. It has been reckoned 'the largest amphibious operation ever attempted until that moment.'[9]

In the fall of 1854, European allies confronted one another on the battlefield for the first time in 40 years, in strange, unpredictable terrain. The French are credited with reliable leadership and supply systems, but Napoleon III covertly commanded the French army in the early months of 1855, leading to confusion in the chain of command. Lack of coordination with the navy, which included six Ottoman battleships, and disagreement among the officers in charge of land operations, characterised much of the early days of the campaign. Most historians now agree that the entire enterprise suffered the lack of strategic definitions of war aims, and exposed the limitations of poorly centralised and badly managed auxiliary systems, such as transport and medical care. The Crimea was terra incognita in 1853–54, and the international army brought to bear was drawn from national as well as colonial armies. Misinformation and botched communications abounded, features exposed by the instancy of reporting. It bears repeating that this phase of the war both closed the Napoleonic Age and also opened the imperial surge to the total war of the First World War. It was closely followed by the Indian Mutiny of 1857 and the American Civil War 1861–65, both of which had global consequences.

The participation of the Ottomans at Crimea was minimal, first by deliberate policy of the allies, and then by a poor showing at the Battle of Balaklava, on 25 October, when the Ottoman infantry battalions, as well as Tunisians in four other such positions, were stationed on redoubts in a critical place in front of the allied army. After an hour of attack by the Russians, and seeing no allied relief in sight, 500 of them broke and ran as the Russian army bore down on them, abandoning the guns that Cardigan was ordered to recover with his infamous Light Brigade. Ottoman losses stood at 170; the Light Brigade lost 113.[10]

It was the last such opportunity for the Ottomans to distinguish themselves, as they were branded as cowards and left to peripheral assignments. Thereafter, the Ottoman troops largely sat out the siege of Sevastopol, from late November to early 1855, freezing, starving and dying of various afflictions just as did their French and British allies over the exceptionally brutal winter of 1854–55. The troops were turned into labour battalions, or left to die. The European allies had agreed to feed the Ottoman troops, but could only offer them biscuit, rum and salt pork, leaving the Muslims only the biscuit to eat. 'Turkish bodies lay in grisly rows until famished comrades could gather the strength to bury them under a few inches of dirt. Each night, wild dogs dug up the bodies and ate their fill. The Turkish surgeon in charge, who had been trained in London and spoke very good English, showed [correspondent] Woods the hospital: "The deadly fetid air which issued from this charnel-house made me involuntarily shrink back from the door with loathing." The building had previously been used for the Russian prisoners, many of whom had died of cholera, the "hospital" was given to the Turks, but despite "ankle-deep" blood and faeces, it had never been cleaned. Hundreds of starving Turks now lay in this filth without bed, blankets, or even clothing.' A six-gun Ottoman frigate had been turned into a 300-bed hospital ship, with four surgeons and other medical personnel. When it arrived in the Crimea, it stood at anchor for ten days waiting for a berth in any of the harbours of the allies, but returned empty in the end, having been refused a place to load the sick and dying.[11] Of course, Ottoman officials were just as culpable for the lack of supplies as were the allies, but doubtless prejudice against the 'cowardly' Turkish soldiers and sailors also contributed to the miseries of the exceptionally harsh winter. Two important impediments to improving the lot of all were the size of the harbours just described, and the impassability of the roads after their effacement during the rainy, icy winter.

In February 1855, the situation changed. Raglan, reduced to perhaps 10,000 effectives in his army, entertained Stratford Canning's offer to procure an agreement with the Ottomans for an additional 20,000 troops, what became known as the Turkish Contingent, to be paid for and commanded by the British. In February, an Anglo-Ottoman convention was signed to that effect, but the troops, who assembled at Şumnu, did not arrive in the seat of war until later in the summer. Ömer Pasha's existing Ottoman force defended Eupatoria (Kozlov), a port north of the Alma River, and defeated a Russian attempt to take the place in February 1855. Nicholas I had ordered the attempt in anticipation of the arrival of the Turkish Contingent, but the Ottoman victory reportedly was a final blow to his spirits. He died on 2 March 1855, and was succeeded by his son,

Alexander II (1855–81), who chose to continue the war, rejecting the peace proposals of the allies in Vienna. Ömer Pasha moved 30,000 of his troops, many of them new recruits from Sudan and Syria, to Sevastopol, where they stood idle until 9 April 1855, when he led twelve battalions of infantry in a reconnaissance before Sevastopol. They were made to cover the same route as the Light Brigade of the previous fall, and followed by many observers. *Times* correspondent Russell noted their drab, dark blue uniforms, gray woollen socks and sheepskin sandals, but observed that they marched well and that their steel was highly polished. The infantry marched up an incline and scattered a few Russian officers and Cossacks, and then were shipped back to their northern outpost at Eupatoria.[12] Perhaps the parade was in recognition of the new convention and a rehabilitation of the 'Turkish cowards'. At any rate, the Ottomans participated no further in the final months of the siege of Sevastopol.

The Caucasus and the Ottomans pre-1853

By mid-1855, Ömer Pasha was pressing the allies to be allowed to take his troops to the Caucasus, where General Muraviev had begun the approach to Kars. The entire episode suffered from the vexing problem of command and communication. British Commissioner Colonel (later General) W.F. Williams, in command of Kars Ottoman garrison during the siege, Ömer Pasha stuck in the Crimea with some 20,000 troops, and Ambassador Stratford Canning in Istanbul operated with three separate and obstinate agendas. Canning was fielding angry communiqués from Williams, and much criticism from London, while Ömer Pasha lobbied to have his troops sent to the Caucasus.

Much confidence was placed in the new Turkish Contingent by Raglan and his staff, as the Contingent was to be under the direction of British officers. They were assembling on the banks of the Danube, while General Williams in Kars demanded, and got, the dismissal and court martial of many of the Ottoman officials in the province of Erzurum, including Governor Zarif Mustafa. It was the first time a foreign, non-Muslim officer, was accepted into the Ottoman command structure on his own terms, and it must have rankled deeply.

Serasker Mehmed Ali was disgraced and exiled in February 1854, to be replaced by Hasan Rıza Pasha, who was a Mehmed Ali protégé and was himself replaced by Mehmed Rüşdü in June 1855. Zarif Mustafa was Rıza Pasha's protégé, likely accounting for the appointment to Erzurum after his dismissal from Vidin. Mehmed Rüşdü, while part of the palace clique around Abdülmecid, appears to have been more compliant concerning the British high-handedness during the Crimean War. Ömer Pasha, a product of the court environment, commanded the respect of all the factions in Istanbul.

There was little apparent reason for keeping Ömer and his men in the Crimea, except likely as reserves, and because of the natural prejudices against the 'cowardly' Turkish soldier. Stratford Canning alludes to intrigues in Istanbul around the disgraced War Minister Mehmed Ali Pasha, which may or may not have involved Ömer Pasha, but likely the focus on Sevastopol, the hopes for the new Turkish Contingent, and matters of transport and manpower played a greater role in the allied commanders' delay in releasing the troops to be sent to relieve Kars.[13] On 12 July 1855, Ömer Pasha requested permission to go with his 25,000-man force from Eupatoria to cause a diversion in the Caucasus, and come to the aid of Kars. He was clearly opposed to using the newly formed Turkish Contingent, as the men were not yet acquainted with their officers, who did not speak their languages, and it was as yet too small. He argued that he was more knowledgeable, and therefore could be more persuasive among the inhabitants. If his existing force were to threaten Russian communications with Georgia, he argued, they would be forced to abandon the campaign. On 23 July, Ömer Pasha was in Istanbul, conferring with Mehmed Rüşdü and the sultan. The plan was to take the 25,000 men, add another 10,000 from Batum, and after landing near Poti, move to isolate the Russian troops in the south Caucasus. The British commanders thought the relief force should land at Trabzon. The dispute about aims in the Caucasus continued over the next month, with the British insisting on reinforcing Erzurum and surrounding areas from Trabzon, while Ömer Pasha argued more broadly for defending the Anatolian frontier and the Black Sea port line. He persuaded the council in Istanbul, although Sultan Abdülmecid expressed willingness to keep the discussion with the allies open. Ömer Pasha's plan was vetoed in London, and Ömer Pasha did not leave the Crimea until 29 September 1855. By that time, General Muraviev had launched an all-out attack on the fortress at Kars.[14]

Before continuing the story of the siege of Kars, however, we need to review the previous confrontations in the area. For half a century, the Russian and Ottoman armies had confronted one another in the Caucasus, home of myriad mountainous tribal peoples who sometimes sided with the Russians, sometimes with the Ottomans. Until 1812 the area was of little strategic significance to the Ottomans. Much like the mountains of western Greece and Albania, the mountains of the Caucasus not only served as a natural barrier but remained virtually impenetrable, and extremely difficult for military manoeuvres. For the Russians '[a] century and a half ago, Moscow was twenty times farther removed from the Caucasus than it is from Washington, D.C., in our time.'[15] Much of eastern Anatolia was governed indirectly, and virtually autonomous until the early days of the Tanzimat reforms as we have seen. The chief city of the area, Erzurum, with a large fortress, was the seat of a governor consistently from the earliest days of the empire, and served as a second centre of empire after Istanbul. Erzurum lay on the great Anatolian trunk road from Sivas via Erzincan. It was one of the first centres targeted for the earliest round of military reforms after 1826, but half of the forces mobilised for any campaign into the area were certain to be irregulars: from Georgians (Imeretians, Mingrelians and Gurias, Acars, Lazis, Abkhazians (Abazas)), Circassians (Çerkes), Ossetians, and Dagistanis (Chechens, Andis and Avars). Armenians and Kurds, Türkmen and Azeris were also native to the area. Erzurum was a city probably numbering some 50,000 in 1853.

Trabzon and Batum, on the Black Sea, and Kars were the other strong fortresses of the area; Ardahan, Bayburt and Beyazid less so. Kars gave access to Tiflis, which made it strategically important to the Transcaucasus. Mountain passes were tracks; transportation largely by horse, camel and donkey. Even more than in the Balkans, the Caucasus frontier remained untamed. Successive Russo-Ottoman wars had given the Russians access to significant Caucasus ports, from Anapa near the Azov Sea as far south as Poti just above Batum, as well as control of Georgia and much of Dagistan.

Until the late eighteenth century, this was territory which was disputed by the Persian Safavids and their successors and the Ottomans. As early as the 1768–74 war, however, Russian armies had penetrated as far as Kutaisi, capital of Imeretia, and laid siege to Poti. The treaty of 1774, much contested by the Ottomans, gave Russia control over Kabarda (Georgia) and Ossetia; by 1783, the Georgian Kartlis signed a treaty with the Romanovs, and Derbent in Dagistan was briefly occupied. The Georgian military road through the Deryal Pass was fortified. During the 1787–92 confrontations, the Ottomans made several attempts at recovering some of the north Caucasus, with heavy fighting around Anapa, near the Kerch Straits. Field Marshal Rumiantsev established a series of forts and military communities called the Azov–Mozdok line up the Kuban River from Kerch. During the same war, Caniklizade Battal Pasha led a great army up the Kuban but was defeated by an inferior force of Russians and Cossacks.

In the 1806–12 war, the Ottomans still held the seaports of Poti, Anakliya and Sokhum, and claimed nominal hegemony over the Circassians and the Abkhazians. The new Qajar Turco-Persian dynasty made a bid to capture Tiflis and surrounding areas in 1795, which resulted in the consolidated Russian control of the eastern Transcaucasus, and extension of control up to the Aras River ın Karabağ. Georgia had been officially annexed to Russia as of 1801. The Treaty of Bucharest (1812), negotiated so tenaciously by Mahmud II and Galib Pasha, resulted in the Ottomans retaining only Poti and Anapa, as well as the frontier fortresses of Akhaltzikhe (Ahiska) and Atskhur. Circassia and the mountainous areas of Dagistan continued to be contested and mutinous territory. At this point, various British commissions began to interest themselves in the Persians, the Russians, and the Muslim territories of the south Caucasus, owing to British Foreign Office concerns about Central Asia and access to India. British diplomacy prevented an Ottoman–Persian war in 1821, when a particularly ambitious Persian prince, Abbas Mirza, occupied Kars and Beyazid. In 1827, Muslim revolts in Karabağ prompted the Russians to besiege Erivan, and they easily defeated the rebels. The Perso-Russian Treaty of Türkmençay of 1828 consolidated Russian control of Dagistan and removed Qajar Persia from the military equation.[16]

The strategic line of ports and fortresses from Trabzon to Anapa, with the interior fortresses described, was the strong Ottoman position in 1828, but the fortresses were very poorly equipped with men and supplies. Garrisons in Batum and Poti numbered 2,000, with 5,000 at Anapa. At Akhaltzikhe, there were 4,000 *Asakir-i Mansure* troops, with 6,000 irregulars, while 30,000 regulars and 10,000 irregulars were concentrated at Erzurum and Kars. They faced a Russian army with decades of victories in the region. General Paskevich had 60,000 troops under his command, includıng 17 regiments of Cossacks, as well as 36 battalions of regular infantry, and several squadrons of cavalry. Ottoman Anapa surrendered early in the 1828 campaign, as without a navy, sunk at Navarino in 1827, Ottoman land forces could not withstand both the Russian naval and military attack. The fall of Anapa freed the Russians to direct their attack on Kars. On 14 June 1828, Paskevich established his army south of Kars, to counter any relief force from Erzurum. In fact, Köse Mehmed Pasha sent such a force to Kars, which was intended to encounter the Russians on the Erzurum road. By 23 June, the Russians occupied Kars. Paskevich then decided to attack the fortress of Akhaltzikhe from the east instead, and the forces under Köse Mehmed moved via Ardahan to counter the Russians from the south. The Ottomans had 35,000 troops, but two-thirds of them were irregulars. The Ottoman camp did not seize the advantage over the outnumbered troops, so Paskevich attacked first, on 7 August, and scattered the Ottoman forces. The *Serasker* himself retired into the fortress with 5,000 regulars. Some 6,000 were killed. Fleeing Lazis and Kurds left behind great piles of supplies and animals. Akhaltzikhe, which put up a fierce struggle, was bombed on 13 and 14 August, and capitulated. The other fortresses fell with minimal resistance. 'In the campaign of 1828, which only lasted five months, three pashaliks had been conquered, namely Kars, Akhiska, and Bayazeed: there had been captured three fortresses, three castles, 313 cannon, 195 standards, 11 horsetails, and 8,000 prisoners.' The Russians had lost around 5,000 by the end of the summer, about half of them to disease and epidemics. Monteith adds, 'the magazines taken from the enemy reduced the expenses to 5,000,000 of rubles, which was little more than the troops would have cost in time of peace.'[17] The Ottomans, as we have seen on other occasions, subsidised their enemy.

In the 1829 campaign year, Mahmud II ordered a new mobilisation under Salih Pasha, formerly Superintendent of the Gümüşhane mines, and his deputy, Hakkı Pasha of Sivas. Only 10,000 effectives had survived the previous year's campaigns. The aim was to raise 60,000 troops around Erzurum, but as so often, conscription failures meant they fell back on the irregular Lazis and Acars for manpower. The Kurds proved mutinous, anticipating the series of revolts in the 1830s and 1840s against both İbrahim Pasha and the Ottoman Tanzimat armies. The Ottomans, however, reorganised an army of 50,000, 30,000 of them regular troops, based between Erzurum and Hasankale. Ahmed Bey of Hulo, an Acar, raised a force of 12,000 Lazis and Acars to besiege Akhaltzikhe, while the Ottomans proceeded to Ardahan. General Muraviev advanced to the relief of Ardahan, and the Ottomans were again defeated, with 8,000 dispersed into the mountains. General Paskevich had assembled 12,000 regulars, 12 regiments of Cossacks and some 6,000 irregulars in the area of Kars. Between Kars and Erzurum was Soğanlı Dağ, difficult mountain terrain. The pass traversed the village of Sarıkamış, the site of a disastrous engagement of the First World War for the Turks. The main Ottoman camp in 1829 was in Zivin, on the south-west side of the pass. Paskevich split his forces into three columns, marched them over different passes, and began the offensive on 13 June 1829. By 19 June, the entire Ottoman force at Zivin had fled down the Erzurum road, leaving artillery and supplies behind. Monteith suggests one way in which Paskevich operated:

Two Turkish prisoners, Mamisch Agha and Bekir Agha, both natives of Erzeroum, undertook to carry the Russian proclamation to the inhabitants of their native city. They were accordingly set at liberty by the Russians, and arrived at Erzeroum in the evening; but finding the lower classes had been induced by the Seraskier to insist on the city being defended, they proceeded to the residence of the Aghan Agha, or civil governor of the town, who called a meeting of the chief people and those who were in favour of a capitulation. The proclamation of Prince Paskievich was then read to them, in which he exhorted the people to put no faith in the promises of the Seraskier, who, he asserted, was on the point of abandoning the city and leaving the inhabitants to suffer like those of Kars and Akhiska; he also added that the Seraskier, by his delays and bad management, had already caused the loss of two armies and the abandonment of the strong pass of Soganlook. As usual in such cases, the rabble were of the same opinion as the last speaker, and the empty tents of the troops who had returned to their homes [deserted] were struck. The Seraskier now appeared really inclined to abandon the town, but was prevented from so doing by the mob, who vehemently reproached him with having destroyed his army, and then wishing to abandon the city.[18]

By 25 June both Hasankale and Erzurum had capitulated. It was a brilliant campaign for Paskevich, which continued to be the military example for the Russians over the next half-century.

In terms of numbers of troops, however, Paskevich was over-extended at that point. While tolerated, even welcomed in Armenian and Kurdish territories, he remained vulnerable to guerrilla tactics all over the territories tenuously under Russian control, but he understood that control of the port of Trabzon was required to secure the territory they had rather easily acquired. On 6 July 1829, he sent two battalions and a detachment of Cossacks and Muslim irregulars to occupy Bayburt, on the road to Trabzon. Confronted with a group of Lazi tribesmen to the north of Bayburt, the Russians were turned back, losing 300 men. Paskevich then marched with his 6,000 men to take Bayburt, and defeated the new *Serasker* Osman Pasha. Paskevich apprehended his forces' vulnerability, especially so in the Acar and Laz strongholds around Trabzon and Batum, and as winter approached, began to withdraw his troops back into Georgia. He was accompanied by large numbers of Armenians, who would make Erivan, also acquired during this war, their new home. By capturing Akhaltzikhe, the Russians had a new defensive outpost, which would cripple future Ottoman campaigns around Kars.

As we know, Erzurum, Kars and Bayburt were all returned to the Ottomans, as well as Ardahan, in the 1829 Treaty of Adrianople. The frontier was marked at St Nicholas, north of Batum, on the coast. Anapa and Poti went to the Russians, and the Ottomans renounced suzerainty over the Çerkes. Paskevich's name joined the ranks of Rumiantsev, Suvorov, and Kutuzov, as commanders of the great triumphs over the Turks. Two things Paskevich and the Russian commanders of the Caucasus understood: that control of the Black Sea was of primary importance for control of Anatolia, and that guerrilla action could easily disable dispersed regular forces. The Ottomans must have learned the first lesson as well: in 1853–54, 1877–78 and 1914 conflicts with the Russians, they controlled the sea, the first time with allies; the second time alone; the third time, in 1914, with the Goeben and Breslau battleships. Lesson Two they never seem to have mastered, or rather, they mastered too well.[19]

The siege of Kars: the British and the barbarians

At the beginning of the war, in fall of 1853, the Ottomans had reinforced Trabzon, Erzurum and Batum, and manned the garrisons, in theory from the Harput headquarters of the the Fourth Army, which existed mostly on paper. In fact, Erzurum remained the pivotal fortress for Transcaucasian defence and supply. In October 1853, Ottoman detachments of 15,000–20,000 regulars were distributed across Erzurum, Kars, Ardahan and Batum. The Russians, meanwhile, faced the distractions not just of the Danube, but also of Imam Shamil and his Murids, the Muslim rebels of long standing in the Caucasus, who could raise 10,000 men and attack the Russian garrisons, as they did in August of 1853. Defeated, Shamil retreated, but the incursion did reveal Russian vulnerability. Prince Vorontsov commanded some 23,000 men, two divisions, ten regiments of Cossacks, and Georgians, Imeretians and Gurian irregulars with which to defend the Transcaucasus frontier. The fortresses were not in good condition, but the Russians concentrated on field forces, well apprised of Ottoman ineffectiveness in manoeuvring in open battle.

As we have seen, the first action of the war belonged to Ahmed Pasha at the frontier of St Nicholas. The Russians withdrew from Poti and surrounding areas, and the Ottomans failed to follow up. At the same time, Ali Pasha attacked Akhaltzikhe, from a high point at Suflis, with 18,000 men. A frontal attack by Commander Andronikov scattered Ali Pasha's army, and he lost 4,000 men, to 450 Russians. At the beginning of November 1853, Abdi Pasha, with 30,000 men and 40 guns, marched on

Aleksandropol from Kars. The Ottoman forces, with gun emplacements on the heights, won the first round but failed to follow up, and reinforcements allowed the Russians to counter the attack in order to withdraw, with 1,000 dead. It was the single opportunity for the Ottoman forces to break the Russian strength in the early days of the campaign, before reinforcements arrived. Cautious Abdi Pasha withdrew to Kars instead, and left Ahmed Pasha in a strong position on the Aleksandropol–Kars road, at Başgedikler, with 48 guns. He had 36,000 men, 20,000 of them regular soldiers, and the rest the usual assortment of irregulars. While their position was strong, the engagement by the Ottomans on 1 December, with superior numbers, resulted in a rout, with the remainder of the force retreating to Kars. Half the artillery and supply trains were lost, and 6,000 men were killed or wounded. The Russians lost fewer, but among them a significant number of senior officers. The shelling of the Ottoman navy at Sinop on 30 November 1853, plus the defeat at Başgedikler, left the Russians masters of the Caucasus and Black Sea ports over the winter, until the allies moved ships into the Black Sea in early 1854.[20]

Over that same winter, the state of the troops in Kars deteriorated even further because of the rivalries and corruption of pashas Ahmed and Abdi. In February 1854, an Ottoman investigatory mission, headed by Hayreddin Pasha, was sent to Kars to investigate the abuses of command under the two pashas. We know of this and subsequent events largely because every step of the summer of 1854 is detailed by the various British officers who were allies of the Ottomans for the season. Hayreddin ascertained that Ahmed Pasha had been filing fraudulent muster rolls, numbering 30,000 troops, and received munitions and supplies for that number which he sold to various contractors. He had arrangements with local businesses, such as the bakers, to supply bread to the troops. Skimming the funds delivered for bread, the bakers were delivering sub-standard loaves to the troops. Both Hayreddin and General Williams, who arrived later in the year, found the troops in a desperate state, poorly clothed, freezing and starving, largely the result of the venality of the commanders and local businessmen. Hayreddin is reported as humiliating Ahmed Pasha in front of his officers and troops, and sending both Abdi and Ahmed to Istanbul for trial. (Ahmed Pasha was ultimately tried and executed for such malfeasance in 1860.)

Hayreddin Pasha ordered the bakers to improve the quality of the bread. On the third day, when the loaves had not improved, Hayreddin forced the contractor to eat the 'filthy, black coarse crumb' of five large loaves. Similar efforts were undertaken in Erzurum.[21] When General

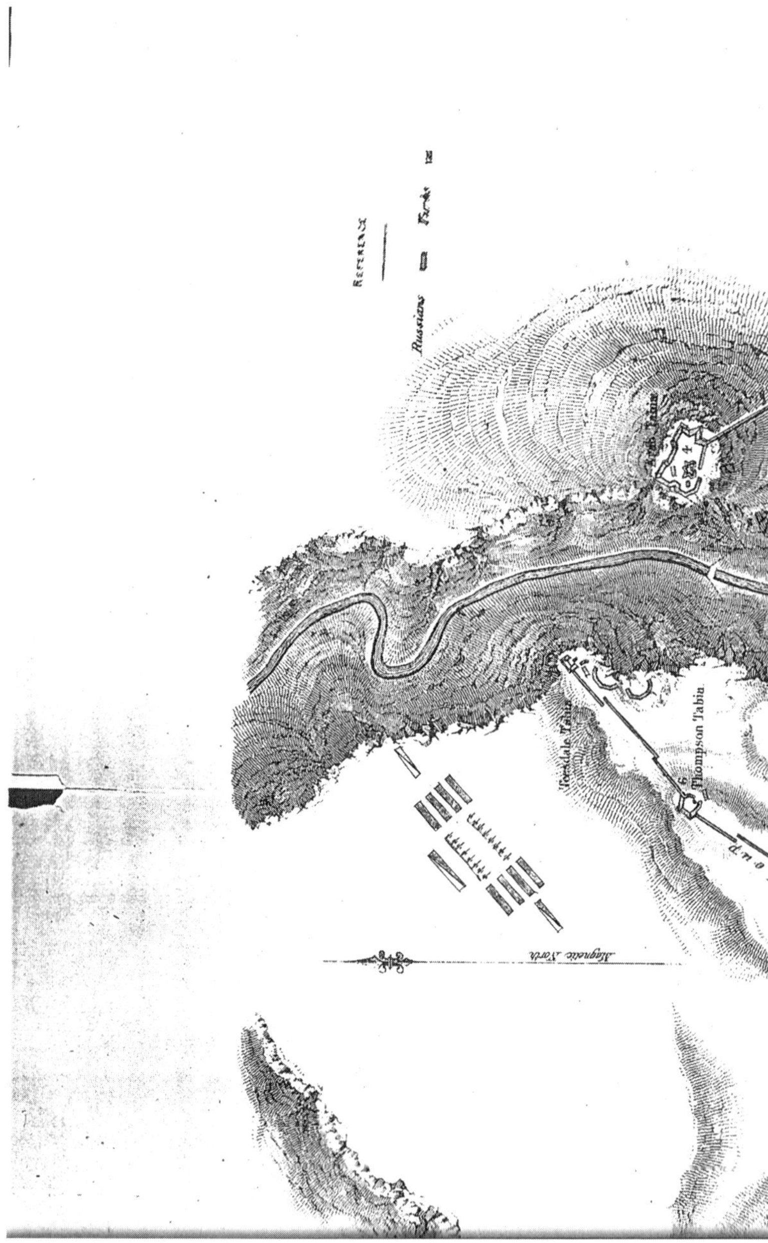
Reference
Russians
Turks
Teesdale Tabia
Thompson Tabia
Magnetic North

PLATE 23 *Map of fortress of Kars, 1854–55 (Colonel Atwell Lake,* Narrative of the Defence of Kars, Historical and Military. *London: Richard Bentley,* 1857).

Williams and his entourage arrived in Erzurum in May, they found substantially little had changed. Williams would spend much of the summer haranguing Istanbul, and pursuing the dismissal of Zarif Mustafa, Governor of Erzurum, who was appointed as *Müşir* of the Army of Anatolia in spring of 1854. As Governor, Zarif had this to say about mobilising men and supplies the summer before. 'It was decided to send a division of soldiers to each of Kars, Ardıhan, Beyazid and Batum. The order to bring and collect the reserves [*redif*] came, and my orders to gather irregular soldiers [*başıbozuk*] were sent out. . . . A report of the provisions needed by the army was requested, and only five thousand *himar* [mule load] of wheat was demanded, the commanders saying they would manage the rest themselves. Upon this, the staff officers were invited to a two-night dinner in which it was stressed that they should estimate what they needed, as it was winter, nothing else could be found. They insisted that the five thousand *himar* would suffice, and they would do the rest by contract. I set out to deliver the requested amount, but understood it would be insufficient. I ordered great quantities from the surrounding towns as a precaution, and warned Çıldır, Kars and Beyazid to do the same, and acquire oil and supplies (*zahire*) there. I was ordered by the Minister of War to assist Abdi Pasha in completing the supplies for the army. As it was my duty to the state, I worked very hard in this matter. The artillery was insufficient and the carriages and wheels old and defective, so the repair was undertaken, for which Salih Pasha must also be given credit. In short, I worked so hard, that even Ahmed Pasha regarded me as a rival.'[22]

The summer of 1854 proved no different in terms of the Ottoman ability to regroup and take the war to the Russians. The allies chose to concentrate on the Crimea, rather than the Caucasus, as we have seen, which meant another round of Ottoman–Russian fighting in and around Kars before the arrival of Ömer Pasha's troops. Surprisingly, given the description of the state of the army, new contingents poured into the area. Selim Pasha's troops, stationed between Batum and Ozurgeti, grew to 40,000, half of whom were regulars. *Müşir* Zarif Mustafa Pasha had 40,000 *nizamiye* and 20,000 irregulars at Kars, with 20,000 *redifs* in Erzurum. The Russians concentrated their forces at Kutaisi (9,000 men), Akhaltzikhe (9,000 men), Aleksandropol (20,000 men) and Erivan (2,000 men). In the first confrontation, in early June 1854, Commander Andronikov forced Selim Pasha and his army back over the frontier, the Çoluk River, to Batum. During June, the army at Kars proceeded with leisure to its advance post at Hacıveli near Aleksandropol. General Bebutov had 20,000 troops (13,000 infantry, 3,000 regular cavalry, and

4,000 mounted irregulars) at Aleksandropol, but was outnumbered by those of Zarif Pasha at Hacıveli. Under Zarif's command were Generals İsmail Pasha (Hungarian Kmety) and Hurşid Pasha (British–Hungarian Guyon), veterans of the 1848 revolt in Hungary in Ottoman service. The Russians set up camp on a prominence called Karayal Hill and remained there until early August, in sight of the Ottoman encampment at Hacıveli for over three weeks. The battle plan belonged to Guyon, who aimed at splitting the superior Ottoman forces into three discrete details, which worked to their disadvantage in the end. On the evening of 4 August 1854, a group of *başıbozuks* and crack and regular Ottoman riflemen attacked Karayal Hill at dawn, and took the hill, giving them momentary advantage. The battle of Kurudere, as it came to be called, was a fierce one, 'made up of three separate consecutive actions, in each of which the Turks were defeated. The Turkish *nizams*, particularly the Syrian battalions, fought with great courage; the fire of the new Turkish rifle battalions was excellent; the new regular cavalry units attacked with boldness and a spirit of sacrifice; and the guns were served competently but with a certain lack of mobility. It was clear to the Russians that the newly organised Turkish army was not yet capable of a war of manoeuvre but might be a serious enemy behind fortifications.' Most of the army retired to Kars in order, saving a majority of the artillery. The Russians made no further move on Kars. Zarif Pasha lost 8,000 killed and wounded, and an additional 2,000 were made prisoners. Perhaps 10,000 of the *başibozuk* returned home (largely Kurds); Bebutov lost 15 per cent (3,000 men) of his force.[23]

By early September 1854, Colonel Fenwick Williams arrived in Erzurum, head of a British Commission, under orders from Raglan and British Foreign Minister Clarendon. He was accompanied by Captain C.C. Teesdale, his *aide-de-camp*, who had been given orders to survey the state of the Ottoman army in the Caucasus. Shortly thereafter, Williams moved to Kars, where he was joined in March of 1855 by three more officers: Colonel Atwell Lake, author of *Narrative of the Defence of Kars*, the best known of the memoirs; Captain Henry Langhorne Thompson; and Captain Humphrey Sandwich, an army surgeon, who served as Williams's military adviser, and left a journal of his observations. Sandwich eventually headed up a team of some fifty surgeons and pharmacists, both foreign and Ottomans. Colonel Lake, who was assigned the responsibility of rebuilding/building the defences of the fort, described it as an irregular polygon, made of blocks of stone in cement with a double *enceinte* of walls, and four towers. The citadel itself sat above the town, with a covered passage leading from the citadel to the river. The outer walls of the fortress measured 2,600 yards. In 1855 it lacked a fuller defence system, without outpost on the hills surrounding the town, or a system of redoubts.[24]

Müşir Zarif Pasha had been dismissed in disgrace after the most recent defeat. He was replaced temporarily by Şükrü Pasha in Kars. Williams's first report to *Serasker* Ali Rıza Pasha condemned the entire Ottoman general staff for widespread corruption and venality, which included false muster rolls, skimming of accounts, and collusion with local suppliers. As a result, perfectly good soldiers were dying from starvation and a lack of proper clothing. A quick roll call of the actual troops in the fortress revealed that some 10,000 troops registered in the muster rolls were unaccounted for. Allen and Muratoff add that in May 1854 Turkish effectives on the Caucasus frontier had numbered some 140,000; by the following year, that had fallen to 70,000 men. An estimated 20,000 men had deserted, but perhaps as many as 30,000 men had perished because of disease and wretched conditions.[25]

Williams then began a campaign to punish the culprits and demand the resupply of the frontier with men and equipment, and continued his condemnation of the entire military organisation of the *Serasker*. He was particularly acute about the problem of old school officers, raised from the ranks, and the new military school graduates. 'Fourteen of these young men, after completing their studies at the Galata Serai, were sent to this army; they found themselves exposed to every description of insult and degradation; not one of them received a paid appointment in the état-major, and several have, in consequence, disappeared altogether from this army. [I]n short, the officers at present in command, as well as those in subordinate posts, will always endeavor to keep the young cadets out of employ in order that their own promotion may secure for them those illicit sources of peculation on which they at present fatten, at the expense of the unfed and badly-clothed soldiers.'[26] Such an attack must have gone over extremely well in the War Ministry in Istanbul; nonetheless, Williams was given the rank of *ferik* in the Ottoman army at the end of 1854. Vasıf Pasha, known for compliance rather than military experience, was appointed *Müşir* early in 1855, and the two officers appear to have worked well together during the subsequent siege. *Serasker* Ali Rıza and Ömer Pasha resisted the demands for trials of the miscreants sent to Istanbul, but a few were ultimately punished. Effective command lay in Williams's hand. Lacking supplies, and especially horses, he chose not to reorganise a field army, but to fortify Kars as the main defensive barrier to the Russian route to Anatolia. In January 1855, for example, an inspection of the horses within the garrison had led to 700 of them designated for slaughter, as they were unfit for service. 'Cavalry outpost duty thenceforward became a dead letter.'[27] The decision was also based on the condition of much of the Kars army, ill with scurvy and malnutrition, but the strategy did mean holding the bulk of the Ottoman forces immobile in the fortress.

In May 1855, the Kars garrison amounted to sixteen battalions of *nizamiyes*, three battalions of Istanbul riflemen, eight battalions of *redifs*, one of cavalry, and eighty field pieces. The total number was 15,000 men. Additionally, there were 9,000 Lazi and Kurd irregulars. Some 5,000 reserves were in Erzurum, Selim Pasha commanded 15,000 in the area around Batum, and Veli Pasha commanded 14,000 troops, half of them irregulars, in the region of Eleşkirt, on the Beyazid to Erzurum route. The Russians, under General Muraviev, regrouped their forces around Akhaltzikhe (16,000 infantry, 7,000 cavalry) and Aleksandropol (16,000 infantry, 7,000 cavalry), the latter under command of Muraviev himself. Reserves of 16,000 were stationed at Marani, east of Poti. Neither side was particularly flush with manpower, but by contrast with the previous campaign season the Russians had the advantage, and sufficient forces to press the operation against Kars and Erzurum.[28]

Lake and Sandwich spent the early summer months building defences and preparing the hospital in Kars, the latter a 2,000-bed facility which provided care superior to that anywhere else during the Crimean War. *Başıbozuks* poured into the fortress. Lake had considerable help from the citizens of Kars. On 16 June, the Russians made the first exploratory attack on the fieldworks, to be repulsed with heavy losses. They faced the best troops the reformed Ottoman army had to offer, as acknowledged by Williams himself. Muraviev, fully apprised of the supply difficulties of Kars, operated with a strategy that understood that the fortress would starve by November if not substantially relieved. He remained cautious about approaching Erzurum, ever expecting the relief to come from Trabzon via Erzurum. In fact he had the opportunity to do so, as Veli Pasha fell back even from Hasankale when confronted with Russian advance troops and essentially left the road to Erzurum open, but Muraviev preferred to concentrate his forces on Kars, just as Sevastopol fell and Ömer Pasha was released to take his troops to Mingrelia (Georgia). Throughout August, Kars was gradually invested by the Russians. Williams reported that 8,000 Russian cavalry had effectively cut off their access to supplies. The garrison went on half rations. By 20 September, Muraviev had word that Ömer Pasha had made a first landing with 8,000 troops at Batum, with the intention of making a diversionary attack on Kutaisi and perhaps Tiflis.

This prompted Muraviev to revise his strategy, and make an all-out assault on the newly fortified Kars. At dawn of 29 September 1855, hoping for a surprise attack, the Russian forces encountered stiff resistance. '[T]he Turks then opened a converging fire upon the head of the column, and though many of the front rank fell, the rest moved up from the rear, pressing forward those in advance, but they could not withstand the deadly fire of the elite of the old soldiers, who deployed, and protected by a breastwork did not fire a ball in vain. This horrid carnage continued until the Russians, stopped by a mound of dead bodies, and dislocated by repeated charges of grape, were brought to a stand-still. The Turks leaping over the breastwork, and led on by the gallant Kméty, finished with the bayonet the utter rout of their assailants. This column left eight hundred and fifty corpses upon a space not exceeding an acre in area. In this attack nearly every Russian superior officer fell.' By eleven a.m., after seven hours of fighting, they were forced to withdraw. Muraviev had lost 60 per cent of his infantry force, some 8,000 killed and wounded. Those defending Kars, killed and wounded, numbered 1,500 men. Pursuit of the retreating Russians was impossible, as the remaining horses had died or been killed for food.[29] In spite of the victory, Muraviev continued the siege of Kars, which was eventually starved into submission. Repeated entreaties to Veli Pasha in Erzurum were ignored. Selim Pasha, known to be in Erzurum with a small relief force, failed to march to Kars. On 6 November, when 100 a day, including citizens, were dying of famine, Captain Teesdale approached the Russians with the flag of truce, and two days later, the garrison capitulated; 24,000 prisoners, including 4,000 sick and wounded, surrendered finally on 28 November, 1855. Muraviev is justifiably commended for having treated the captured Kars garrison and its commanders well. The siege and its collapse generated great interest in the public, so much so that the defenders were commemorated in numerous paintings created after their release by the Russians, one of which, entitled *General Williams and His Staff Leaving Kars*, painted by Barker, is now in the National Army Museum in London.[30] No such honour awaited the Ottoman survivors of Kars.

Meanwhile the relief force under Ömer Pasha, numbering some 30,000–35,000 men in the end, was crippled by the time of year, the lack of horses, and Ömer's confidence in the assistance of local Çerkes and Abaza irregulars. He was not prepared for an attack until the beginning of October 1855, when he moved his forces to the right bank of the Ingur River. In a confrontation with the troops of General Bagration, on 5 November, Colonel Simmons, one of numerous British and other foreign officers with Ömer Pasha, defeated the Russian force. The gallantry of the Turkish soldiers, the steadiness of the rifles, and the bravery of the charge was noted by observers with the troops.[31] Ömer Pasha dallied on the left bank of the Ingur River, however, allowing the Caucasian winter to embrace the endeavour. Neither Russian nor Ottoman commander appears to have been particularly effective as November passed, and 'the Turkish army was submerged in Mingrelian mud.' The Russian commander burned all remaining supplies for his troops at Marani, which made an attack on the mired Ottomans impossible. An approach to the relief of Kars and Erzurum never happened; over the winter, the Ottomans regrouped in Redut-Kale, and by February 1856, were transported to Batum. Ömer Pasha was back in Istanbul, as the war drew to a close. Within a year, he was appointed Governor of Baghdad.[32]

Goldfrank has described the Crimean War as 'a classic case of diplomacy interrupted by warfare,' largely because the Vienna Conference continued to offer various solutions to the impasse between the Russians and the allies. Protracted negotiations among the allies, as well as the lack of a definitive victory, prevented all sides from overcoming the inertia. In the end, the Sevastopol humiliation and the Kars victory (and, it must be added, the difficulties in the Baltic provinces, once Sweden joined the alliance, untouched in this narrative) sufficiently convinced the Russians to capitulate. The Treaty of Paris, signed on 30 March 1856, left six powers in a position to intervene in Ottoman sovereignty over its Christian subjects. The autonomy of Moldavia and Wallachia was guaranteed under Ottoman suzerainty. The Ottomans did recover control of the mouth of the Danube, the single territory the Russians lost, and the fortress of Kars. The treaty of 1841 on the subject of the Dardanelles was renewed; Russia and the Ottomans agreed separately on the neutralisation of the Black Sea, and the number of vessels allowed by either side was severely restricted.[33]

Debate about winners and losers continues. Most see the Treaty of Paris as an ongoing territorial rearrangement of central Europe, which would make itself felt next in the Franco-Prussian War of 1870–77, and play out even further in the 1877–78 Russo-Ottoman War in Bulgaria. Russian humiliation guaranteed a return to many of the same unsolved problems, but first, the imperial army had to undergo several decades of radical reform. Austria's belligerent negotiations alienated many of her allies, and we have already observed the problems with the Habsburg military system. Keeping the Crimean War out of the Balkans may have 'cushioned' the Austrians, but 'it also led to their own almost unbearable responsibilities for the Balkan ethnic mess.'[34] While the France of

Napoleon III appears to have been the big winner in the Crimea, the next decade would prove otherwise. The human cost: 450,000 Russians, 80,000–95,000 French, 20,000–25,000 British, 2,000 Piedmontese, and possibly 200,000–400,000 Ottomans, was the greatest loss of life between 1815 and 1914.[35]

Reform and the Concert of Europe

In retrospect, the Crimean War put one more nail in the coffin of the three imperial settings that have been the subject of this study. Undoubtedly, 1848 revolutionary fever, and the presence of two armies on the Danube and the Caucasus, had the effect of stimulating further Ottoman unrest and resistance to reforms. Similarly, while Ottoman success on the Danube, and the return of territory, gave them a respite from warfare, Ottoman indebtedness crippled the war recovery and guaranteed the continued semi-colonialism of external financing. Furthermore, the cost of continued autonomy, however subscribed, was a reiteration of promises to reform. The Treaty of Paris was not signed until the Ottomans decreed their commitment to reform in the *Hatt-i Humayun*, promulgated on 18 February 1856. It was only then that they were assumed to join the Concert of Europe.

Sultan Abdülmecid confirmed the equality of his subjects before the law, irrespective of class or religion. All privileges granted to non-Muslims were to be continued. Non-Muslim communities were to be given the opportunity to reaffirm those privileges. Freedom of religion was proclaimed; freedom of access to government offices and the military was declared; mixed tribunals for Muslim–Christian disputes, codification of penal and commercial laws, and direct taxation instead of tax farming all were reaffirmed in this reiteration of intentions. Over the next few decades, the promises made could not all be kept. In effect, the new document was an absolute guarantee of continued resistance, prompting sustained revolts in Damascus, Armenian and Kurdish borderlands, Bosnia, as well as in Albania. Refugees poured into Ottoman territories, from the Crimea and the Caucasus, most settling in Anatolia in the decade after the Crimean War. In spite of that, Istanbul continued to press for reform, such as provincial reorganisation, regular inspection of municipalities and the countryside, and the sorting out of the rights of the *millets*.[36]

The post-Crimean War period is known as the era dominated by the bureaucracy headed by Ali and Fuad Pashas, both originally protégés of Reşid Pasha, who held the offices of Grand Vizier and Foreign Minister

Mustafa for over ten years. Both have left small political testimonies as to their understanding of reform. Fuad wrote 'that the aim of the Ottoman administration should be the absolute equality and fusion of all races. Separatisms based on religious differences should be stifled. To achieve effective quality it will be necessary to institute a new system of justice, and a new system of public instruction.' Neither felt that the aims at reform contravened Islamic law. Sultan Abdülaziz (1861–76), whose throne speech confirmed the continuation of reforms, also stressed, as had his predecessor, the primacy of shar'ia law, as the Ottomans reconstrued themselves as Muslim constitutionalists.[37] Not all politicians signed on wholeheartedly, but the reality of survival pressed them into service. Hence, it is possible to see a commitment in Istanbul to the promised equality, which resulted in giving considerable power to local councils and appointees. Such minor improvements occurred against the increasing clamour and resistance of the Muslim populations, whose economic and cultural prominence in the empire continued to slide. The deterioration of relations between Muslim and non-Muslim communities across the empire evinced itself in major riots in Lebanon, Crete and Bosnia–Herzegovina, but outbreaks occurred in many other parts of the empire, such as among Anatolian Armenians and Kurds in eastern Anatolia, with increasing frequency and ferocity after 1856.

A turn to the Prussian – 1869 conscription reforms

It seems a good place to draw this narrative to a close, for 1856 represents a turning point for the armies of Europe, as well as for the Ottomans. The model for military reform in the decades after 1860 was Prussia. Apart from the Napoleonic period, viewed as exceptional, European armies remained generally small until the 1860s, when conscription of the population became the standard method of mobilising national armies. From 1866 to 1870, the Prussian army expanded from 300,000 to 1,200,000 men. France aimed at 1,000,000, but stood at 500,000 in 1870. The organisation of these huge numbers was predicated on the Prussian 'short/active' and 'long/reserve' system. In 1874, of a population of 83,000,000, 496,000 men in peace, and 1,600,000 war-time troops were estimated as available for the Russian military. For Austria, with a population of 36,000,000, it was presumed that 300,000 men in peace and 1,000,000 troops in wartime could be raised. Financial considerations had a large role to play in the actual numbers. In the period between 1874 and 1896, Russian and German expenditures for their armies increased by 79 and 75 per cent respectively. The Russians maintained a large army with difficulty, perhaps 750,000 in 1874 with a further reserve of 250,000.[38]

The Ottoman population ranged from 30,000,000 to 40,000,000 in this period, and we have already seen the experimentation with the *redif* reserve system in place after 1843. Evidence for the Crimean War indicates that a maximum of 250,000–300,000 men was raised by the Ottomans. The irregulars continued to be a source of manpower regularly tapped, but the impact of the *başıbozuk* behaviour on local societies tends to exaggerate their actual number. Perhaps not more than 50,000 were utilised in the Crimean War, and likely less than half of that were mobilised in the later decades of the nineteenth century. *Başıbozuks* became a significant liability in the Balkans and Anatolia under Abdülhamid II (1876–1909), who created the *Hamidiye* Kurdish cavalry regiments in the 1890s as a partial remedy to the disorders of the Balkans and elsewhere, setting the stage for a more lethal confrontation between the Kurdish and Armenian communities in eastern Turkey.

In 1869, under Hüseyin Avni Pasha, the Ottomans introduced another round of military reforms, which were overtly modelled on the Prussians. Initially the reorganisation was intended to increase the capacity of the Ottoman army for external defence. It was based on the short/long principle described above, a ratio of four years in the regular army and sixteen years in various levels of reserves. Service in the *nizamiye* was designated as four years, with an additional two years in the active reserve (*ihtiyat*); three years in the first *redif*, and three years in the second *redif*; eight years in the general reserve, or militia. (*mustahız*). The army estimated that such a system would produce 702,000 men: 150,000 *nizam*, 60,000 *İhtiyat;* 96,000 men each in *redif* one and two, and 300,000 *Mustahfız*. An annual intake of 37,500 was expected, chosen by lot from the 21–24 age group. While some have said that the Ottomans raised 750,000 in the war against Russia in 1877–78, that does not square with the 250,000 troops remaining at the end of the war, unless one is willing to accept one-half million dead and wounded.[39]

The Ottomans were no different than the other powers in Europe in following the Prussian model, except that they were continually hampered by the lack of finances, and a shrinking tax base, preventing them from investing properly in the reformed system. In 1869, the Ottoman army budget was 4,700,000 pounds sterling, of which 3,600,000 pounds were intended for the *nizamiye* and its overhead alone. The second *redif* and the *Mustahfız* remained paper forces.[40] The dynasty, under the new Sultan Abdülaziz, who initially pledged to curb palace spending, continued to view state funds as a personal patrimony, in much the same fashion as his predecessor, and the levels of corruption and cronyism continued in the bureaucracy as well as the royal household. In 1861, the financial crisis had come to a head, when the Ottomans were barely able to cover the interest payments they had accrued on loans from Europe made during the Crimean War and after, and could raise no further loans. By 1863–64, with further aid from France and Britain, the Ottomans created the Imperial Ottoman Bank, backed by European financiers, but the fundamental problem remained.[41]

In short, after the Crimean War, while imperial aspirations may have aimed higher, the fact was that the Ottomans slipped quickly behind their major rivals. Beyond the obvious budgetary problems, significant, major differences lay in the functionality of the army, and in conscription. For the Ottomans it proved impossible to separate internal from external enforcement, even though the *zaptiye*, or police force, was a product of the post-Crimean reforms. Most European monarchies succeeded to varying degrees in removing the army from such internal disciplinary activities, but the late Ottomans never managed to rid themselves of the need for regular and irregular forces for security purposes. It was not so much a technological gap, which military historians generally emphasise, but an internal security gap and a manpower gap that hindered the evolution of the Ottoman military in the last decades of the nineteenth century.

The impact of conscription fell heaviest on the Muslims of the remaining Ottoman territories, and even more intensely on the population of Turkic Anatolia, which grew by perhaps more than 2,000,000 refugees in the latter half of the nineteenth century. The Christian population and at least one quarter of the Muslim population was exempt from service, leaving some estimated 12,000,000 Muslims available for conscription. Attempts such as that in Bosnia, to create special forces of ethnic groups, were repeatedly made in Albania and the tribal areas in Syria and Iraq, but with little success. Revolt inevitably followed. The exemption tax (*bedel*) could produce more revenue than the previous poll tax, which had been abolished in 1846, a large incentive for *not* conscripting Christians.[42] Some proposals were made by the reformers, among them Ömer Pasha, to enrol Armenians and Bulgarians, but not Greeks or Christian Bosnians, so the proposal went nowhere. Not until 1909, under the Young Turks, is it possible to speak of a universal conscription law, and that was evaded by many. The additional obstacle was the resistance of the Muslim recruits to serve under Christian officers, a problem that persisted into the First World War.[43]

Notes

1 Sir James Porter, *Turkey: Its History and Progress, from the Journals and Correspondence of Sir James Porter, Continued to the Present Time, by his Grandson* Sir George Larpent (London: Hurst and Blackett, 1854) vol. 2, 429–38, a summary taken from the *Morning Chronicle* so the military figures cannot be presumed to be definitive.

2 *Illustrated London News*, 3 December 1853, 479, included here to give some sense of the immediacy of the war and the initial war enthusiasm in Britain.

3 *Illustrated London News*, 28 January 1854, 73–74 [Constantin Guys]; James J. Reid, *Crisis of the Ottoman Empire: Prelude to Collapse, 1839–78* (Stuttgart: Kranz Steiner, 2000), 244. Reid's is the most consistently useful for the (Ottoman) military context. John S. Curtiss, *Russia's Crimean War* (Durham, NC: Duke University Press, 1979) has also been consulted for this discussion. Of any number of general surveys, R.L.V. ffrench Blake's *The Crimean War* has been used here (London: Sphere Books, 1973); On Greek irredentism of the period, see John Koliopoulos, 'Brigandage and insurgency in the Greek domains of the Ottoman Empire, 1853–1908', in Dmitri Gondicas and Charles Issawi, eds., *Ottoman Greeks in the Age of Nationalism* (Princeton, NJ: Darwin Press, 1999), 143–60.

4 Clive Ponting, *The Crimean War* (London: Chatto and Windus, 2004), 61.

5 Quoted in Reid, *Crisis*, 260–61.

6 Ponting, *The Crimean War*, 67–68.

7 Ponting, *The Crimean War*, 74; Reid, *Crisis*, 264–66, 274–77. Reid is very good on the problem of such irregular troops in the Ottoman provinces.

8 Quoted in Ponting, *The Crimean War*, 84.

9 Ponting, *The Crimean War*, 85.

10 Robert B. Edgerton, *Death or Glory: The Legacy of the Crimean War* (Cambridge, MA: Westview Press, 2000), 168–71. Of the 14,000 Egyptian reservists sent to the Danube, half died, surviving a 28-day voyage from Egypt on dry biscuit and water. Of the 10,000 Tunisians in the Crimea, 7,000 died. Edgerton seems determined to smash every 'myth' there ever was about the Crimean War.

11 Quoted in Edgerton, *Death or Glory*, 170.

12 Reid, *Crisis*, 268–69. Edgerton, *Death or Glory*, 172–73.

13 Stanley Lane-Poole, *The Life of the Right Honourable Stratford Canning* (London: Longman, Green, 1888), vol. 2, 408–9.

14 Tim Coates, *The Siege of Kars, 1855. Defence and Capitulation Reported by General Williams* (London: The Stationery Office, 2000), 41–46.

15 Dmitrii I. Oleinikov, 'The Caucasus factor in Russian military reform', in *Reforming the Tsar's Army: Military Innovation in Imperial Russia from Peter the Great to the Revolution*, David Schimmelpenninck van der Oye and Bruce Menning, eds (Cambridge: Cambridge University Press, 2004), 205–31.

16 W.E.D. Allen and Paul Muratoff, *Caucasian Battlefields: A History of the Wars on the Turco-Caucasian Border 1828–1921* (Cambridge: Cambridge University Press, 1953), chapter 1.

17 William Monteith, *Kars and Erzeroum: With the Campaigns of Prince Paskiewitch in 1828 and 1829* (London, 1856), 221–22.

18 Monteith, *Kars and Erzeroum*, 268–69.

19 I have followed Allen and Muratoff, *Caucasian Battlefields*, chapter 2, here.

20 Allen and Muratoff, *Caucasian Battlefields*, chapter 4.

21 Reid, *Crisis*, 279, quoted Charles Duncan, who accompanied Hayreddin Pasha.

22 Enver Ziya Karal, 'Zarif Paşa'nın hatıratı, 1816–62', *Belleten* 4 (1970) 474–76. This memoir was written after the fact, and ends after his dismissal from the front. He appears to be shifting the blame for the corruption onto the staff officers.

23 Allen and Muratoff, *Caucasian Battlefields*, 76–80.

24 Colonel Atwell Lake, *Narrative of the Defence of Kars, Historical and Military*. London: Richard Bentley, 1857.

25 Allen and Muratoff, *Caucasian Battlefields*, 81.

26 Lake, *Narrative*, 62.

27 Lake, *Narrative*, 155.

28 Allen and Muratoff, *Caucasian Battlefields*, 84.

29 Allen and Muratoff, *Caucasian Battlefields*, 89–95, and Lake, *Narrative*, 200–01.

30 Peter Harrington, 'The defence of Kars: paintings by William Simpson and Thomas Jones Barker', *Journal of the Society for Army Historical Research* 227 (1991), 26. Williams and his staff were exonerated, while Stratford Canning was blamed for the British failures to relieve the garrison.

31 Edgerton, *Death or Glory*, 182.

32 Allen and Muratoff, *Caucasian Battlefields*, 100–01.

33 David N. Goldfrank, *Origins of the Crimean War* (London: Longman, 1994), 289–93.

34 Saab, *Origins*, 161.

35 Goldfrank, *Origins*, 289.

36 Reproduced in Anderson, *Great Powers*, 63–65. The best rendering of the later reforms is Roderic Davison, *Reform in the Ottoman Empire 1856–76* (New York: Gordian Press, 1973).

37 Davison, *Reform*, 87–110.

38 Yapp, M.E., 'The modernization of Middle Eastern armies in the nineteenth century', in V.J. Parry and M.E. Yapp, eds, *War, Technology and Society in the Middle East* (London: Oxford University Press, 1975), 338–42.

39 Yapp, 'Modernization', 348–49.

40 Yapp, 'Modernization', 350–54. The reforms are reproduced in H. Zboiński, *Armée ottomane: son organisation actuelle telle qu'elle résulte de l'exécution de la loi de 1869* (Paris: Librairie Militaire, 1877).

41 Christopher Clay, *Gold for the Sultan: Western Bankers and Ottoman Finance 1856–81: A contribution to Ottoman and International Financial History* (London: I.B. Taurus, 2000) provides the details.

42 On the refugees, Kemal Karpat, who has written extensively on both the Crimean period and later immigration, states that approximately five to seven million people immigrated into remaining Ottoman territories between 1860 and 1914. See 'The *hijra* from Russia and the Balkans: the process of self-definition in the late Ottoman State', in *Muslim Travelers: Pilgrimage, Migration, and the Religious Imagination*', Dale F. Eickelman and James Piscatori, eds (Berkeley, CA: UC Press, 1990, 131–52). The first Ottoman immigration commission was established in January 1860 (Halaçoğlu, Yusuf. *XVIII. Yüzyılda Osmanlı İmparatorluğu'nun İskân Siyaseti ve Aşiretlerin Yerleştirilmesi* 2.baskı (Ankara: Türk Tarih Kurumu, 1991), 42; Ufuk Gülsoy, Osmanlı *Gayrımüslimlerinin Askerlik Serüveni* (Istanbul: Simburg, 2000). In his study of the problem of non-Muslims in the army, Gülsoy reports a roundup in Trabzon in 1855, Christians and Muslims alike, but most escaped (55–56). The *bedel-i askeriye* tax averaged 30–50 *kuruş* per individual form 1856–70 (computed on a communal basis). A young Muslim man could only buy out completely with 15,000 *kuruş* (*bedel-i nakdi*).

43 Yapp, 'Modernisation', 350–54. For the post-1870 period, see Handan Nezir Akmeşe, *The Birth of Modern Turkey: the Ottoman Military and the March to World War I* (London: I.B. Taurus).

Conclusion

Historians point to a bifurcation in the intellectual classes of the empire in the last decades of the nineteenth century, between non-Muslims, whose affiliations with the French and British created a wealthy middle class, determined to separate from the empire, and Muslims, for whom the army came to represent the passage to wealth and power. As a template it has considerable merit, but I do not think it represents the further divisions within the Muslims themselves, a voice until very recently absent in our narratives. Arguably, within Muslim communities there was also a bifurcation of the imagination of what a reformed Ottoman state might resemble, between a governing elite and a society increasingly defensive and conservative.[1]

In 1859, for example, Istanbul was treated to a revolt called the Kuleli Vakası, which was probably triggered by some officers who survived the Caucasus battlefields, combined with dissatisfied theological students stirred up by a local preacher. Described as an inchoate rebellion, organised by a secret society, it likely aimed at the removal of Sultan Abdülmecid, and may well have been anti-reform in origin, stirring up considerable support in Istanbul. Discontent was fuelled by the pay arrears of the demobilised army, neglect of veterans, and corruption in the dynastic household, as much as by the new wave of reforms. The conspiracy was betrayed to the officials, the ring-leaders rounded up and exiled. The few that were condemned to death had their sentences commuted. Among the conspirators was General Hüseyin Daim, veteran of the siege of Kars.[2] Other such conspiracies dot the decades to follow.

In terms of military leadership, while the Ministry of War and the general staff remained little changed in the immediate post-Crimean period, the effect of the military schools began to be felt. Hüseyin Avni Pasha, War Minister four separate times and architect of the 1869 reforms, for example, was a product of the military colleges, as were many of the War Ministers who followed him. As late as 1877, however, only

15–20 per cent of Ottoman officers had been trained in the state-run military schools. They were known as the *mektebli*, while the majority, often functionally illiterate, were *alaylı*, that is, they rose through the ranks. The need for educated politicians and reformers in Istanbul meant that many of the graduates of the military schools stayed on as teachers, or moved into civilian posts, rather than into the ranks of command. The result was that the War College (part of it now the Military Musuem in Istanbul) became a school for the Ottoman elite, the aim of ambitious young Muslim men from all over the empire. Originally modelled on the French system of the 1840s, the curriculum focused on technology and theory, but little practical application.

In the 1880s, after the British occupied Egypt, Abdülhamid II invited the Germans to send a Military Advisory Mission to reorganise and re-equip the Ottoman army. Then head of the Mission General Kaehler called on Major Colmar von der Goltz to inspect the curriculum of the War College. Goltz recommended the replacement of a semi-engineering curriculum with the German officer training system, but as with many of Goltz's other recommendations, it was ignored. Officials of the academy resisted the changes, saying that advanced mathematics and physics had to be maintained in order to train professionals who could handle both civilian and military duties. In fact a reform-minded military general staff posed a significant threat to the sultan himself, so Abdülhamid was complicit in paying no more than lip service to the curriculum recommendations of the mission.

After 1885, Goltz was put in charge of the mission, and effected significant changes to the entire Ottoman military system, including replacing British and French contractors with German suppliers of military arms and equipment. He also trained the generation of young officers who would take the Ottomans into the First World War, the likes of Enver Pasha and Mustafa Kemal Atatürk.[3]

The story of the last days of the empire belongs to others, but a word is in order about the evolution of Ottomanism, the prevailing ideology of the ruling class, which was one of many 'identities' contested in the empire in the chaotic decades after 1856. In an era of fading imperial and monarchical traditions, the Ottomans identified themselves with their neighbours, and sought recognition based on the manipulation of the symbols of rule and reciprocal diplomacy. 'The Ottoman sultan, the Meiji emperor, the Russian tsar, the Habsburg emperor . . . [a]ll invested in a recharged state mythology, which they sought to inculcate through mass education. All invested to varying degrees in the inevitable technical trappings of modernity: railways, the telegraph, factories, censuses, passports, steamships, world fairs, clock towers and art-deco palaces. All looked to each other to see how their peers were playing the role of "civilized monarchy." '[4] But beneath such posturing of the sultan and household, the Ottoman reformers understood in differing ways the need for the introduction of the 'rule of law' in the empire, a process which did not begin with the Gülhane promulgation of 1839, but rather on the battlefields of 1768–74, when defeat and insolvency required explanation. It has been the habit of historians to locate reform in 'secular' texts, to hunt for the political debates among bureaucrats, to deny to the Muslim texts arguing for change any authority, but as more and more of them become available to us, it is possible to trace an Ottoman Muslim road to the modern. The reasons why that road is located in debates about the state of the military, and appear to be largely defensive, even paranoid, should by now be obvious.

From a military point of view, while this book began with 1699 and the Treaty of Karlowitz, the Treaty of 1739 is even more significant, as the triumph of the return of Belgrade masked the true state of affairs and delayed a real assessment of both military and financial systems. That assessment began only in 1774, was accelerated by the public's outrage at the loss of the Crimea in 1783, and finally concluded with the very public elimination of the Janissaries in 1826. To justify the radical reordering of society, an evolution of the authority and legitimacy of Ottoman rule ('Ottomanism') can be traced to the reign of Mahmud II, who embodied the modern Ottoman monarch and exemplified the new Muslim autocrat, insisting on reciprocity in international relations and equality among his subjects. That he accomplished much, with great ferocity, cannot be denied. Under Mahmud, the northern arc was refortified, and discussions about citizenship and obligations became part of public discourse, especially as a modern, if heavily censored, press first made its appearance. Called the 'infidel sultan', Mahmud II nonetheless made use of the mosque and its rhetoric for reconfiguring the imagined space of the remaining Ottoman territories, and for sorting the loyal and disloyal among those who were within its constricted borders. It took most of the rest of the nineteenth century for Ottoman intellectuals to make sense of the merging of Ottoman–Islamic political traditions and European thought. Ottoman Tanzimat statesmen had access to Vienna, London, Paris, and Berlin, where a similar and vigorous debate over alternative systems of law was under way. This was reflected in the debates which began among Ottoman intellectuals of the 1840s over the Ottoman place in the pantheon of civilizations, and in the assertion of a civilising mission over their rebellious nomads and tribes in the last years of the century: Druze, Arab Bedouin, Kurd, Albanian mountaineers, and latterly Yemenites. In particular, the conquest of Yemen was explicitly argued on the model of British India.

According to one recent analysis, the Tanzimat ideology aimed at 'a balanced, but nevertheless complete, enforcement of the state's law and prerogatives – be they religiously justified or not'. Public order and just rule remained the bedrock of Ottoman Islamism throughout its existence. The obsession with rebellion (*fitne*), to which we have often alluded, and the emphasis on justice (*adalet*) is better meant as the restoration of the equilibrium, where order and stability equalled prosperity. The state's function then was to ensure that restoration, and a return to the security and tranquillity of the flock (*reaya*). Studying bureaucratic language from the nineteenth century, Reinkowski has discerned the creation of a new 'circle of equity', which assumed that part of the state's role in restoring order was to punish and educate (*te'dib ve terbiye*) Ottoman subjects (*tebaa*, a neutral term) who had caused the disruption, by creating a new order through discipline and civilization (*inzibat ve medeniyet*).[5] One is reminded of the novel disciplinary code that accompanied the Hüsrev military reforms of 1827.

In fact, the imagined stability was never fully achieved, primarily because of financial difficulties. Threatened by a new military order in European armies, the Ottomans were forced to transform a land-based *timariot* system, which included many Janissaries of the countryside by 1800, into a new, costly standing army. This also meant the necessity of transforming an intricate system of indirect taxation, which they had allowed to fall under the purview of provincial *ayan*. That process continued for a hundred years, and lay at the base of a majority of the revolts of the post-Janissary empire.

Costs increased as technology, especially mobile field artillery and arms and ammunition, became essential to winning wars. The evidence indicates a continued Ottoman investment in the renewal of technology, and the imitation and adoption of new arms and equipment in rapid fashion, but an inability to sustain the manufacturing of such items, again largely due to lack of finances or the misuse of what was available. Concentrating then on maintaining and manning the frontiers of conflict, and the large fortresses we have described, to the west and north, the post-Janissary army neglected the southern and eastern tiers of empire and continued to rely on local tribal affiliations, and natural warrior inclinations of some populations to serve as the frontier forces and later police. By contrast with their Habsburg and Romanov neighbours, after 1700 there was very little effort to settle such groups as frontiersmen, who could and did serve as a reserve for the Habsburg and a first line of defence for the Russians. The preference, as with the Principalities and the Crimea, was for a contractual relationship between the Ottomans and their clients, which for a long time benefited both parties, but it encouraged, in fact often engendered, the continuation of volunteer warrior societies. When the time came to discipline and 'civilize' the nomadic and semi-nomadic ethnicities of the peripheries, both for the purposes of conscription and the imposition of the Tanzimat regime, the newly-organised army had neither the manpower nor the resources to do more than restore order temporarily. Imperial aspirations and arrogance at the centre always outran the resources and capabilities of a system crippled by poor financing and a patrimonial household. Indebtedness and disorder contributed to a growing siege mentality, and resulted in an atmosphere of desperation, urgency, conspiracy and anger at the dynasty itself.

Still, the picture is very much complicated by the arrival of the seaborne powers of France and Great Britain. 'For historians of the Eastern Question, the centerpiece of the struggle has been the problem posed by the Sick Man of Europe, or, to view the situation from the other side, the impediments to modernization through consensus that were exacerbated by the self-serving policies of meddling foreign powers',[6] but the age of transformation chronicled by this book was much more complicated than foreign interference with sovereign rights. Perceptions of the strategic importance of Eurasia to the British in India turned the Eastern Question into the Great Game, as the remainder of the world was drawn into the international trading system.

After 1841, British consuls were to be found in every major port city of the Arab southern tier of the empire where they had not been before. Foreign powers, in effect, established an informal colonial rule over a large part of the Ottoman territories, establishing an international market regime that continues to the present. In 1882 the British occupied Egypt, and turned it into a colony, arguing about the temporary nature of their stay for the next 50 years. The beneficiaries of the new colonial relationship were the non-Muslim communities. The British found themselves contesting territory with a rival colonial power, the Ottomans themselves, as in Yemen at the turn of the century.

That contest for the Ottomans centred on loyalty to an ideal, Ottomanism, and played itself out as a religious quarrel in part, with

Muslim attacking non-Muslim and vice-versa. The clamour for autonomy and self-determination continued, as it would in the Habsburg and Romanov realms, as radical individualism, and a new world order spread east across Europe to Eurasia. Missionaries of all faiths became instruments of colonial power as well, and stoked the fires of difference with the compliance of the foreign representatives in Istanbul. We have tended to populate this story with Muslim fanatics and barbaric Turks because of the final years of collapse, but in 1918, Ottoman, Habsburg and Romanov houses all fell, with unprecedented violence, and with consequences still unfolding in the successor states of the Middle East and the Caucasus.

Notes

1 Fatma Müge Göçek, *Rise of Bourgeoisie, Demise of Empire: Ottoman Westernization and Social Change* (Oxford: Oxford University Press, 1995); Reşat Kasaba, 'A time and a place for the non-state: social change in the Ottoman Empire during the long nineteenth century', in Joel S. Migdal, Atul Kohli and Vivienne Shuh, eds, *State Power and Social Forces: Domination and Transformation in the Third World* (Cambridge: Cambridge University Press, 1994), 207–30.

2 Roderic H. Davison, *Reform in the Ottoman Empire 1856–76* (New York: Gordion Press, 1972), 100–1; James J.; Reid, *Crisis of the Ottoman Empire: Prelude to Collapse, 1839–78* (Stuttgart: Franz Steiner, 2000), 303–4.

3 Mesut Uyar and A. Kadir Varoğlu, 'In search of modernity and rationality: the case of [the] Turkish Military Academy', unpublished paper. Glen W. Swanson, 'War, technology and society in the Ottoman Empire from the reign of Abdülhamid II to 1913: Mahmud Şevket and the German military mission', in V.J. Parry and M.E. Yapp (eds), *War, Technology and Society in the Middle East* (New York, Oxford University Press, 1975), 366–85; C. Max Kortepeter, 'Ottoman military reform during the late Tanzimat (1885–95): The Prussian General von der Goltz and the Ottoman Army', in *V. Milletlerarası Türkiye Sosyal ve İktisat Tarihi Kongresi, Tebliğler* (Ankara: Türk Tarih Kurumu, 1900), 174–86. Debates about educated officers and a general staff resonate with the Russian experience after the Crimean War. See Oleg Airapetov, 'Miliutin contra Moltke: Russia's refusal to adopt a Prussian-style General Staff', in *Reforming the Tsar's Army: Military Innovation in Imperial Russia from Peter the Great to the Revolution*, D. Schimmelpenninck van der Oye and Bruce Menning, eds (Cambridge: CUP, 2004), 292–303.

4 S. Deringil, *The Well-Protected Domains* (London: I.B. Tauris, 1998), 171.

5 Birgit Schaebler, 'Civilizing others: global modernity and the local boundaries (French/German, Ottoman and Arab) of savagery', in Birgit Schaebler and Leif Stenberg, eds, *Globalization and the Muslim World: Culture, Religion and Modernity* (Syracuse: Syracuse University Press, 2004), 18; Deringil, *Well Protected Domains*, chapter 2; and Maurus Reinkowski, 'The state's security and the subjects' prosperity: notions of order in Ottoman bureaucratic correspondence (19th century)', in Hakan T. Karateke and Maurus Reinkowski, eds *Legitimising the Order: The Ottoman Rhetoric of State Power* (Leiden; Boston: Brill, 2005).

6 Ann Pottinger Saab, *Origins of the Crimean Alliance* (Charlottesville, VA: University Press of Virginia, 1977), 161.

Guide to further reading

The following discussion of recent works largely in English should serve as an introduction to the extended time frame and complicated story this book covers. The 'Works Cited' bibliography contains the complete list of titles utilised in this study, which is naturally informed by the significant work of our Turkish and European colleagues.

On the Ottomans

General works on Ottoman history must include *Osman's Dream* by Caroline Finkel (London: John Murray, 2005; New York: Basic Books, 2006), the latest version of the Ottoman story. Over 600 pages, the book has both political and aesthetic appeal, as it reaches for a holistic (political, social and material culture) view of the empire. As a general study, it is more accessible than that of Colin Imber, who writes entirely from the Ottoman chronicles (*The Ottoman Empire, 1300–1650: The Structure of Power* (Palgrave Macmilllan, 2002), and eschews context and analysis. Suraiya Faroqhi's *The Ottoman Empire and the World Around It* (London: I.B. Tauris, 2004), is a select social history of the empire, set in its geographical context, with special emphasis on diplomacy, dynastic life, peasants and trade.

Dan Goffman and Donald Quataert have each written a study of the Ottomans for the Cambridge series *New Approaches to European History*. Goffman's *The Ottoman Empire and Early Modern Europe* (2002) places the pre-1700 Ottomans squarely among the monarchs of Europe. Quataert's *The Ottoman Empire 1700–1922* (2nd edn 2005) offers glimpses into late Ottoman social history. On Ottoman perceptions of sovereignty, an especially interesting collection covering the entire span of the empire is that edited by Hakan T. Karateke and Maurus Reinkoski, *Legitimizing the Order: The Ottoman Rhetoric of State Power* (Leiden: E.J. Brill, 2005). On the economy, while numerous studies exist in

Turkish, Şevket Pamuk's *The Monetary History of the Ottoman Empire* (Cambridge: Cambridge University Press, 2000) supplies us with concrete evidence of what he calls the great debasement of the currency under Mahmud II (1808–39), so central to the explanations for the halting recovery of the military after 1826.

One of the major questions still surrounding the Ottomans is why they survived. This has been approached in a number of ways by recent work: Gabriel Piterberg takes up the transformation of the state in the seventeenth century in *An Ottoman Tragedy: History and Historiography at Play* (Berkeley: University of California Press, 2003). He joins Karen Barkey's *Bandits and Bureaucrats: The Ottoman Route to State Centralization* (Ithaca: Cornell University Press, 1994) in the study of political relations of the state and its provincial subjects. Both books help us to imagine the construction of the Ottoman official narrative at the centre, as it was challenged by provincial warlords and rebels in the repeated political crises of the middle period. Ariel Salzmann prefers an economic analysis for the same era, constructing a theory of the decentralisation of power and the privatisation of property through life-lease tax farms (*Tocqueville in the Ottoman Empire: Rival Paths to the Modern State* (Boston: E.J. Brill, 2004). Her use of Tocqueville provokes a consideration of the Ottoman ancien régime, thus drawing the Ottomans into the larger circle of pre-modern societies, with the French. Her close study is of the city of Diyarbakır in present-day. All these works have been influenced by Rifaʿat ʿAli Abou El-Haj, *Formation of the Modern State: The Ottoman Empire, Sixteenth to Eighteenth Centuries* (2nd edn. Syracuse: Syracuse University Press, 2005), who has argued for normalising the empire, and studying its social evolution as one would other parts of the world. As this book goes to press, the first volume of a projected four volumes of *The Cambridge History of Turkey (vol. 3: The Later Ottoman Empire, 1603–89)*, edited by Suraiya Faroqhi, has been published (Cambridge University Press, 2006), which has a number of articles representing the latest work on the Ottoman middle era.

Two histories of Istanbul make a good introduction to the court and dynasty itself: Phillip Mansel, *Constantinople: City of the World's Desire, 1453–1924* (New York: St Martin's Press, 1995), and John Freely, *Istanbul: The Imperial City* (New York: Penguin, 1998), which is actually more about the Byzantine than the Ottoman city. Two further studies of the period of Süleyman the Magnificent round out our understanding of the nature of Ottoman legitimacy and power: Metin Kunt and Christine Woodhead, eds, *Süleyman the Magnificent and his*

Age (London: Longman, 1995), and Colin Imber, *Ebu's-su Su'ud: The Islamic Legal Tradition* (Stanford: Stanford University Press, 1997), a study of the *Şeyhulislam* of the sixteenth century.

Local provincial and urban history of the transitional period of the empire is one of the most vibrant fields in Ottoman history at the moment. Molly Greene, *A Shared World: Christians and Muslims in the Early Modern Mediterranean* (Princeton, NJ: Princeton University Press, 2000), analyses Crete in the context of Venetian–Ottoman contest in the seventeenth century; Dina Rizk Khoury, *State and Provincial Society in the Ottoman Empire: Mosul, 1540–1834* (Cambridge: Cambridge University Press, 1997), examines the give-and-take of centre and periphery in Mosul; Rossitsa Gradeva has just published a collection of her articles on Bulgaria under the Ottomans: *Rumeli Under the Ottomans, Fifteenth–Eighteenth Centuries: Institutions and Communities* (Istanbul: Isis, 2004). The proceedings of the most recent of the *Halcyon Days* symposia: 'Provincial elites in the Ottoman Empire,' edited by Antonios Anastasopoulos (Rethymno: Institute for Mediterranean Studies, 2005), include a collection of articles on local elites' exercise of power throughout the empire. Bruce Masters, *Christians and Jews in the Ottoman Arab World: The Roots of Sectarianism* (Cambridge: Cambridge University Press, 2001), addresses the much-vexed debate over minorities and tolerance, in his work on the Catholic communities of Syria. In that vein, the recent *Minorities in the Ottoman Empire*, edited by Molly Greene, includes an interesting article on eighteenth-century Zagora Greece, and the emergence of the *kocabaşı* class there (Socrates D. Petmezas, 'Christian communities in eighteenth- and early nineteenth-century Ottoman Greece: their fiscal function,' 71–126. Princeton, NJ: Markus Wiener, 2005).

On the truly vexed question of immigration and population, a new collection of articles on Bosnia includes a remarkable record of an exchange among workshop attendees, in response to a talk by Kemal Karpat (among others): 'The migration of the Bosnian Muslims to the Ottoman States, 1878–1914: an account based on Turkish sources', *International Journal of Turkish Studies* 10 (2004), 121–40.

On the Ottoman neighbourhood

The core of this book deals with the defence of the northern frontiers of the Ottomans from Belgrad to Kars and the period from 1700 to 1870. For an engaging view of the history of the Black Sea, Charles King's *The*

Black Sea (Oxford: Oxford University Press, 2004) is very informative. So too is John Stoye's *Marsigli's Europe 1680–1730* (New Haven, CT: Yale University Press, 1993), which narrates the life and Habsburg service of Luigi Ferdinando Marsigli, one of the first to map large parts of the lower Danubian basin. I do not pretend to cover either the Habsburgs or the Romanovs in any detail in this volume, but do make broad comparisons of their difficulties in raising and modernising large armies from diverse populations. A general introduction to the European context is Jeremy Black's *The Rise of the European Powers 1679–1793* (London: Arnold, 1990). Paul W. Schroeder, *The Transformation of European Politics 1763–48* (Oxford: Oxford University Press, 1994) continues the story. Equally pertinent is Hamish M. Scott, *The Emergence of the Eastern Powers, 1756–75* (Cambridge: Cambridge University Press, 2001), which pinpoints the emergence of Russia (and the Eastern Question) prior to the Ochakov crisis at the end of the century. The standard remains M.S. Anderson's *The Eastern Question 1774–1923: A Study in International Relations* (New York: St. Martin's, 1966), reworked but substantially unchanged by A.L. MacFie (London: Longman, 1996).

I have found Charles Ingrao's, *The Habsburg Monarchy 1618–1815* (Cambridge: Cambridge University Press, 1994; 2nd edn. 2000) and J. Bérenger's *A History of the Habsburg Empire 1700–1918*, [vol. 2] translated by C.A. Simpson (NY and London: Longman, 1997), of particular use. Michael Hochedlinger's *Austria's Wars of Emergence: War, State and Society in the Habsburg Monarchy 1683–1797* (London: Longman, 2003) is a thorough study of the Habsburg imperial moment through the prism of its military. Paul Dukes, *The Making of Russian Absolutism 1613–1801* (2nd edn, London: Longman, 1990) gives an introductory overview to the Russia of Peter the Great and Catherine. Tim Chapman, *Imperial Russia, 1801–1905* (London: Routledge, 2001), carries the story forward. So too does John P. LeDonne, *The Russian Empire and the World, 1700–1917: The Geopolitics of Expansion and Containment* (New York: Oxford University Press, 1997). Dominic Lieven, *Empire: The Russian Empire and its Rivals* (New Haven, CT: Yale University Press, 2000), has little time for the Ottomans, as he is more concerned with the distinction between the maritime versus land empires. Still, he lays out the common problems of pre-modern imperial settings. Jane Burbank and David Ransel, eds, *Imperial Russia: New Histories for the Empire* (Bloomington, IN: Indiana University Press, 1998), offer new readings on the eighteenth and nineteenth centuries of the Romanovs.

Michael Khodarkovsky, *Russia's Steppe Frontier: The Making of a Colonial Empire, 1500–1800* (Bloomington, IN: Indiana University Press, 2004), takes up Russia's policies regardings its ethnic and religious minorities. Similarly, *Ottoman Borderlands: Issues, Personalities and Political Changes* (Madison: University of Wisconsin Press, 2003, also published as vol. 9 of the *International Journal of Turkish Studies*), edited by Kemal Karpat and Robert Zens, brings together a number of new studies and interpretations of Ottoman frontiers. It includes a 'typology' of Ottoman borderland by Kemal Karpat. Fikret Adanır and Suraiya Faroqhi have edited a volume of historiographic essays on the Balkans, which offers good insights into post-Soviet approaches to the Ottoman period of Balkan history (*The Ottomans and the Balkans*. Leiden: E.J. Brill, 2002). Fred Anscombe has put together a collection of essays, *The Ottoman Balkans, 1750–1830* (Princeton: Markus Wiener, 2006) which includes my 'Whose territory and whose peasants? Ottoman boundaries on the Danube in the 1760s', describing an Ottoman attempt to regulate cross-Danubian peasant traffic in the new multi-lateral diplomatic environment with the Habsburgs. Of interest in that regard is Viorel Panaite's *The Ottoman Law of War and Peace: The Ottoman Empire and Tribute Payers*, on the long client relationship between Ottoman sultans and Wallachian and Moldavian princes (Boulder, CO: East European Monographs, 2000).

The Crimea and Caucasus borders are less well studied, at least in English, for the period of this book. King's book above touches on the Crimea in passing, while Brian Glyn Williams, *The Crimean Tatars: the Diaspora Experience and the Forging of a Nation* (Leiden, E.J. Brill, 2001), though more concerned with the return of the Tatars to the Ukraine after 1989, has very thorough introductory chapters on the Crimean Khanate and its dissolution in the eighteenth century. While there is a whole new generation of Russian/Ottoman scholars in the making, much of the work has concentrated on late Ottoman Caucasus and Balkan politics, such as Moshe Gammer, 'Şamil and the Muslim powers: the Ottomans, the Qajars and Muhammad 'Ali of Egypt'. in *Caucasia Between the Ottoman Empire and Iran, 1555–1914* (edited by Raoul Motika, Michael Ursinus, (Wiesbaden: Dr Ludwig Reichet Urlag, 2000) 11–20), represents the kind of work emerging on the Caucasus, as well as Firouzeh Mostashari, *On the Religious Frontier: Tsarist Russia and Islam in the Caucasus* (London: I.B. Tauris, 2005). An intriguing pair of translations of texts on Shamil, who challenged the tsar's government in the Caucasus for close to thirty years, brings to life that darkest corner of the Ottoman world (*Russian–Muslim Confrontation in the Caucasus: Alternative Visions of the Conflict between Imam Shamil and the Russians*, 1830–59, London: Routledge, 2004).

War and society in the neighbourhood

Of recent years, research into the evolution of military institutions in the three empires of this book has produced a handful of studies which address the nature of state power and the transformation of armies in much the same fashion as here. In addition to Hochedlinger, mentioned above, Robert I. Frost's *The Northern Wars: War, State and Society in Northeastern Europe 1558–1721* (London: Longman, 2000) sets the stage for the expansion of Russia described in this volume, and has one of the best descriptions of the military transformation of the 1600s and 1700s as it affected Sweden, Russia, Poland and Lithuania. While there is little to replace Gunther E. Rothenberg's *The Army of Francis Joseph* (Purdue: Indiana University Press, 1976), Ingrao is very succinct and clear about the Napoleonic period. Matthew Z. Mayer's two articles in the *International History Review* remind us of the precariousness of the Habsburg geopolitical ambitions. ('The price for Austria's security: Part I: Joseph II, the Russian Alliance, and the Ottoman War, 1787–89', *International History Review* 26 (2004), 257–99, and 'part II: Leopold II, the Prussian threat, and the peace of Sistove, 1790–71', *IHR* 26 (2004), 473–514.)

Russia is better served since the publication of John Keep's *Soldiers of the Tsar: Army and Society in Russia*, 1462–1874 (Oxford: Clarendon, 1985). William Fuller's *Strategy and Power in Russia, 1600–1914* (New York: Free Press, 1992) is a very useful survey of the entire sweep of Russia policy. Brian L. Davies, 'Russian military power 1453–1815', in *European Warfare, 1453–1815*, edited by Jeremy Black, 145–79 (New York: St. Martin's, 1999) has much to offer the novice reader. Frederick W. Kagan and Robin Higham have gathered together a truly remarkable set of articles by Bruce Menning, Robert F. Bauman and Kagan himself, among others (*The Military History of Tsarist Russia*. New York: Palgrave, 2002). Kagan has also published *The Military Reforms of Nicholas I: The Origins of the Modern Russian Army* (New York: St Martin's Press, 1999). *Reforming the Tsar's Army: Military Innovation in Imperial Russia from Peter the Great to the Revolution*, edited by David Schimmelpenninck van der Oye and Bruce Menning (Cambridge: Cambridge University Press, 2004) caps off two decades of work on the

Russian transition after 1700, which has been particularly useful for my comparisons with the Ottoman side.

War and society in the Ottoman Empire

The first half of the empire, especially the Hungarian period (through 1699), has been better served than the post-1700 era. Gábor Ágoston has an article in Black's *European Warfare*, 'Ottoman warfare in Europe, 1453–1812', 118–44, which alongside Rhoads Murphey's idiosyncratic *Ottoman Warfare 1500–1700* (London: UCL Press, 1999), and my 'Ottoman war and warfare, 1453–1812', in Jeremy Black, ed., *War in the Early Modern World, 1450–1812* (London: UCL Press, 1999), 147–75, represents most of what is currently available to an English-speaking audience. Godfrey Goodwin published a book called *The Janissaries* (London: Saqi Books, 1994 which is a collection of oriental tales of the excesses of the corps rather than a real history; see my review of it and many of the titles mentioned here: 'Ottoman military matters', *Journal of Early Modern History* 6 (2002), 52–62). Caroline Finkel's *The Administation of Warfare: Ottoman Military Campaigns in Hungary 1593–1606* (Vienna, 1988), is a stand-out as a monograph in English on a single sustained Austro-Ottoman war. Géza Dávid and Pál Fodor have edited two volumes on Hungarian–Ottoman military affairs: *Hungarian–Ottoman Military and Diplomatic Relations in the Age of Süleyman the Magnificent* (Budapest: Loránd Eötvös, 1994), and *Ottomans, Hungarians and Habsburgs in Central Europe: The Military Confines in the Era of Ottoman Conquest* (Leiden: E.J. Brill, 2000), both marvellous examples of military micro-history, if somewhat out of scope for this volume. Victor Ostapchuk's long-awaited *Warfare and Diplomacy Across Sea and Steppe: The Ottoman Black Sea Frontier in the Seventeenth Century* (forthcoming in Harvard's Middle East Monographs), will carry on that tradition with an intimate and detailed look at the Crimean Tatars, the Cossacks and their relations with the Ottomans. Michael Hickok's *Ottoman Military Administration in Eighteenth Century Bosnia* (Leiden: E.J. Brill, 1997), looks at the military administration of that most loyal of provinces in terms of military service. Rossitsa Gradeva's 'War and peace along the Danube: Vidin at the end of the seventeenth century', in *The Ottomans and the Sea*, Kate Fleet, ed. *Oriente Moderno* n.s. XX (LXXXI) no. 1 (2001), 149–75, gave us a tantalising look at the possibility of utilising court records to study the Janissaries, an approach barely under way in Ottoman studies. Nothing has appeared to replace W.E.D.

Allen and Paul Muratoff, *Caucasian Battlefields: a History of the Wars on the Turco-Caucasian Border 1828–1921* (Cambridge: Cambridge University Press, 1953). For a useful look at the military historiography of the past half century in Turkish, see Kahraman Şakul, 'Osmanlı Askerî Tarihi Üzerine Bir Literatür Değerlendirmesi', *Türkiye Araştırmaları Literatür Dergisi* 1 (2003), 529–71.

Of recent, more specialised studies, Colin Imber has just published *The Crusade of Varna, 1443–45* (Aldershot: Ashgate, 2006), which includes lots of new translations by him. Ostapchuk and Finkel have teamed up to publish 'The archeology and construction history of the Black Sea fortress of Özü', an article which considerably overturns much of the previous understanding of Ottoman military architecture and administrative capabilities in an outpost of empire (*Muqarnas* 2005, 150–88). My own 'Manning a Black Sea garrison in the eighteenth century: Ochakov and concepts of mutiny and rebellion in the Ottoman context', *International Journal of Turkish Studies* 8 (2002), 63–72, adds speculations about defending the fortress they have so ably described. In the same vein, Fariba Zarinebaf, John Bennet and Jack L. Davis have collaborated on *A Historical and Economic Geography of Ottoman Greece: The Southwestern Morea in the 18th Century* (Athens: American School of Classical Studies at Athens, 2005), which includes remarkable photographs, plans and a documentary record of the Morea before and after the Venetian moment, 1685–1715.

Meanwhile, Ágoston has published *Guns for the Sultan: Military Power and the Weapons Industry in the Ottoman Empire* (Cambridge: Cambridge University Press, 2005), which stops around 1800, but thoroughly dispels any notion of Ottoman total dependence on Europe, or as the empire of the 'giant cannons' so pervasive in the European literature on military technology. Istanbul was an ideal location for what Ágoston calls the technological dialogue, and relocates the Ottoman problems, as I do, with financing, and the loss of control over resources for the production of war material, as far more important than the technology deficit.

Jane Hathaway has edited two volumes of essays on Ottoman rebellion and mutiny, one of which was volume 8 of *IJTS* mentioned above; the other, *Rebellion, Repression, Reinvention: Mutiny in Comparative Perspective*, was published by Praeger in 2001. Palmira Brummett's 'Classifying Ottoman mutiny: the act and vision of rebellion', published in the *Turkish Studies Association Bulletin* 22 (1998), 91–107, gave the Ottoman community the idea for the conference.

On the navy, which is largely untouched in this study, the work of Palmira Brummett is also a good place to start: *Ottoman Seapower and Levantine Diplomacy in the Age of Discovery* (Albany, NY: SUNY, 1994). Salih Özbaran has worked on the Ottoman and Portuguese confrontation in the Red Sea and Indian Ocean: *The Ottoman Response to European Expansion* (Istanbul: Isis, 1994). A recent edited volume, *The Kapudan Pasha, His Office and His Domain* (by Elizabeth Zachariadou, Rethymnon, 2002), includes the generously illustrated 'Selimian Times: a Reforming Grand Admiral, Anxieties of Repossession, Changing Rites of Power', 7–49, by Tülay Artan and Halil Berktay, about the era of the reforms of Selim III. Bernd Langensiepen, and Ahmet Güleryüz, *The Ottoman Steam Navy, 1828–1923*, edited and translated by James Cooper, London: Conway Maritime Press, 1995, is representative of recent work on the post-1800 period. Daniel Panzac has just published *Barbery Corsairs: The End of a Legend 1800–20* (Leiden: E.J. Brill, 2005). Panzac is, of course, the most prolific author on the pre-modern Ottoman navy.

A recent edited volume, *The Kapudan Pasha, His Office and His Domain* (by Elizabeth Zachariadou, Rethymnon, 2002), includes a remarkable article, generously illustrated, by Tülay Artan and Halil Berktay, 'Selimian times: a reforming Grand Admiral, anxieties of repossession, changing rites of power', 7–49, about the era of the reforms of Selim III (1789–1807). Bernd Langensiepen and Ahmet Güleryüz *The Ottoman Steam Navy, 1828–1923*, edited and translated by James Cooper (London: Conway Maritime Press, 1995), is representative of recent work on the post 1800 period.

For the era of transformation, most of what is available, such as the work of Khaled Fahmy, or Hakan Erdem, Avigdor Levy and Zürcher's edited *Arming the State*, has been extensively used in this study. Of my own work, 'Locating the Ottomans among early modern empires', *Journal of Early Modern History* 3 (1999), 21–39, serves as an introduction to the seventeenth and eighteenth centuries, along with 'The one-eyed fighting the blind: mobilization, supply and command in the Russo-Turkish War of 1768–74'. *International History Review* 15 (1993), 221–38, which first addressed many of the themes that concern us here. Fourteen of my articles are reproduced in a collection of fourteen essays: Virginia H. Aksan, *Ottomans and Europeans: Contacts and Conflicts* (Istanbul: Isis Press, 2004). New work by Amira K. Bennison on Morocco offers a glimpse into Maghrebian military transformation in a colonial context, even as the contacts with Istanbul remained. ('The "New

Order" and Islamic Order: the introduction of the Niẓāmī Army in the Western Maghrib and its legitimation, 1830–73', *International Journal of Middle East Studies* 36 (2004), 591–612).

On the specific reforms, a recent German thesis covers much of the same ground that I do here: Tobias Heinzelmann, *Heiliger Kampf oder Landesverteidigung? Die Diskussion um die Einführung der allgemeinen Militärpflicht im Osmanischen Reich 1826–56* (Frankfurt: Peter Lang, 2004). Kemal Beydilli has been assiduous in publishing with his students heretofore inaccessible manuscripts from the eighteenth century, and discusses them in 'Küçük Kaynarca'dan Tanzimât'a Islahat Düşünceleri', *İlmi Araştırmalar* 8 (1999), 25–64. My own *An Ottoman Statesman in War and Peace: Ahmed Resmi Efendi 1700–83* (Leiden: Brill, 1995), is a useful place to start for the reformers of the eighteenth century. Stanford Shaw's *Between Old and New: The Ottoman Empire Under Sultan Selim III, 1789–1808* (Cambridge, MA: Harvard University Press, 1971), remains the standard account.

For the transition from empire to nation-state, and the role of the military, two recent studies have appeared: Ayşe Gül Altınay, *The Myth of the Military-Nation: Militarism, Gender, and Education in Turkey* (London: Palgrave, 2004), and Handan Nezir-Akmeşe, *The Birth of Modern Turkey: The Ottoman Military and the March to WWI* (London: I.B. Tauris, 2005). For one crossover study, see Eugene Rogan's *Frontiers of the State in the Late Ottoman Empire: Transjordan, 1850–1921*, Cambridge: Cambridge University Press, 2002. On the question of the incorporation of Arab tribal leaders into the late Ottoman army, see Alisan Akpınar and Eugene L. Rogan, *Aşiret, Mektep, Devlet. Osmanlı Devleti'nde Aşiret Mektebi*, Istanbul: Aram 2001.

I would also single out Edward J. Erickson's *Ordered to Die: A History of the Ottoman Army in the First World War* (Westport, CT: Greenwood Press, 2001) reviewed by Erik Jan Zürcher, 'Scholarship on the Ottoman Empire in World War I', *Middle Eastern Studies Association Bulletin* 39 (2005), 183–92).

Appendix

[Excerpts from Travels in Turkey, Asia Minor, Syria and Egypt by William Whittman, Surgeon with the Anglo-Ottoman forces as they crossed the desert from Gaza to Alexandria in 1801.]

On the 28th, at ten in the morning, the army began its march, the baggage having been sent off at a very early hour. Previously to our quitting the ground, further dispatches were received from Sir Ralph Abercrombie and Admiral Lord Keith, the contents of which the Vizier read and communicated at our first halt. We were then six miles distant from Gaza, and had been obliged to cross a river, formed by the late very abundant falls of rain, the waters of which were so high, and the current so strong and impetuous, that the passage was effected by the army with great difficulty, several of the camels, with the baggage, tents, &c. having had a very narrow escape from destruction. The dispatches contained the glorious news of the success of the British troops in Egypt since their landing, detailing the capture of Aboukir, as well as of the three actions which had terminated so favourably to the British arms, and in which the French had lost four thousand men, comprehending that of three of their generals. On our reaching, at five in the afternoon, our new ground of encampment at Kahnyounes, the Vizier fired a royal salute of twenty-one guns, to celebrate the above events; and in the evening, at sun-set, we heard distinctly a discharge of eleven guns, which appeared also to have been fired as a salute by the advanced troops.

Our day's march had been extremely pleasant; and we now occupied a dry and elevated ground, supplying us with excellent water. Kahnyounes is a small village, situated in a plain on the border of the desert. From the rising ground on which the mission was encamped to the eastward, it exhibited a very pleasing appearance, containing not only a handsome mosque, but a ruinous castle, which added greatly to the picturesque effect. The Turks had taken up their position close to the village, the inhabitants of which are exclusively Arabs. The fine plains through which we had passed this day, on our route thither, afforded excellent pasturage for cattle, and contained a greater proportion of grasses and clover than I had seen in any other part of the country.

At day-break of the 29th the signal was made to march; and by eight o'clock the whole of the army was in motion, with the most favourable weather for its progress, and a fine refreshing breeze from the westward, to add to the gaiety which the glad tidings we had received had diffused over every countenance. Immediately on our quitting the village we entered on the desert, in which I observed a shrub, resembling our furze bush, shooting up at intervals from beneath the sand. After a march of about two hours, we reached the boundary which separates Asia from Africa.

At this place, and in the vicinity of a well, which promised us a supply of good water, we made a halt, and afterwards rode between two columns of Egyptian granite, erected there, we were told, to mark distinctly the limits which define each of these quarters of the globe. Several portions of the same material lay scattered on the ground, apparently connected in past ages with buildings erected on the spot.

Our late march was not so dreary and unpleasant as we had reason to apprehend; but we were told that the desert, in proportion as we should penetrate still further, would assume a more dismal and solitary aspect. At one o'clock we arrived at a place called by the Arabs Zaca, or Sheick Sahwych, distant about sixteen miles from Kahnyounes, where, for the first time, we pitched our tents in a desert. We could not complain of the quality of the water we met with, notwithstanding it was blended with a considerable portion of sand. We saw the holes, which were still open, said to be dug by the French for their corn magazines. The desert exhibited an appearance sufficiently barren; but we contrived to pick up a few shrubs for the purpose of cooking. Barley was still so scarce in the camp, that it sold at two piastres and a half, nearly four shillings English the feed.

We left Zaca at seven in the morning of the 30th, in the midst of a thick fog, which was extremely disagreeable, and the more so as our tents and baggage had been sent forward very early, insomuch that we could neither shift ourselves, nor procure any shelter on our halts. Our small party, hitherto unprovided with horses, suffered greatly from the intense heat, as well as from the fatigues necessarily attendant on a march over the heavy sands. The Vizier was not an unconcerned spectator of the sufferings to which the troops were thus exposed, and promised to furnish them, on the subsequent marches, either with horse or camels. After a march of nearly sixteen miles, we arrived at El-Arish at one o'clock. In approaching the encampment before that place, the Grand Vizier was met by Taher Pacha, at the head of a large party of the troops, who had been marched out for that purpose. We pitched our tents on a barren sand, at the distance of about half a mile from the sea-shore, to the north of the fortress. Several vessels were lying at anchor, and their crews employed in landing the cargoes of provisions, barley, &c. At El-Arish the barley was sold at forty paras the measure. The Turkish ships of war, lately arrived from Aboukir, were also lying off the port.

The troops encamped at El-Arish were healthy, having had no appearance of plague among them for the last fortnight. Shortly after our arrival, I visited Taher Pacha, with whom I took coffee, and who sent to our camp ready dressed provisions, which were highly acceptable. We were supplied with excellent water, furnished by the late heavy rains which had fallen on the mountains. The prospect of the country around us, the surface of which was almost entirely sand, with here and there a few small shrubs growing in tufts, was wretched and dreary. The village of El-Arish, near to the ruins of which of fort was still standing, had been destroyed by the French, who had, when they captured the place, cut down nearly the whole of the date-trees by which it was ornamented. It was situated at the distance of two miles and a half from the sea. Our camp was supplied with oranges, lemons, raisins, dried figs, and other fruits, brought by the merchant-vessels from Cyprus, and the price of which was extravagantly high. With the exception of onions, no vegetables were to be obtained.

The Vizier took measures, on the 31st, for the speedy departure of the army from the ground it then occupied. According to the arrangements which had been made at Gaza the troops were in future to march in three divisions.

In consequence of demands recently made by the Arnauts, who had had a meeting to debate on their grievances, respecting the future supply to be made to them of water, biscuit, and barley, the Grand Vizier adopted the following regulations for the troops: In the first place, that no other tents or baggage, except such as should be found indispensably necessary, should be carried on the march. Secondly, that all such tents and baggage as should be useless at the moment, should follow the army. And, lastly, that the camels should be employed for the express purpose of carrying the water, barley, and biscuit, together with such tents, and such a proportion of baggage as might be allowed to accompany the army. – I dreaded, I must confess, the consequences of these regulations to our people, who would thus be left without tents to shelter them from the effects of a hot and parching sun, and who were not, like the Turkish troops, inured to the climate, and accustomed to the privations to which the latter had been habituated from an early age. Under these circumstance, which gave rise to bitter and distressing reflections, I lamented the nature of the service in which they were engaged, and the melancholy situation in which I saw them plunged.

Two thousand five hundred troops arrived in camp on the evening of the above day. They consisted principally of the Arabs, Avarees, who inhabit the eastern desert, the western parts of which are occupied by the Mograbians.

It was finally settled on the 1st of April, that a certain portion of the baggage and tents should be carried with the army; and that the remainder should be left in the charge of a trusty and responsible person, to follow its movements with all convenient dispatch.

Taher Pacha, accompanied by Captain Leake, marched on the 2d towards Salahieh, with three thousand men, and three pieces of artillery. Summons to the garrisons of Tineh and Salahieh were delivered to Captain Leake by Colonel Holloway, under the authority of the Grand Vizier. Mahomed Pacha arrived in the encampment from Gaza, with three thousand men, and four pieces of artillery.

Much firing among the Turkish troops was heard in the camp on the 3d. It was occasioned by a violent dispute between two companies of Janissaries, the 37th and 65th, who, in the division of a quantity of barley which they had found and appropriated to themselves, had come to hostilities. In this conflict several of them were killed, and many others wounded. It happened that an Arnaut, who was passing by during the affray, received a slight wound. This man made an immediate representation to his corps, that it was the intention of the Janissaries to fall on and butcher the Arnauts without distinction. The effect of this mis-statement was, that the latter had recourse to their arms, and were proceeding to the most alarming measures, which were, however, fortunately prevented by the strenuous interference of the principal Turkish officers in the camp.

During the whole of the morning it blew a violent gale, which raised in the air tremendous clouds of sand, extremely harassing to the eyes. At one o'clock there was a hail storm, accompanied by thunder and lightning. In the afternoon the violence of the wind was not abated; but it was attended by heavy showers, which impeded the elevation of the sands: the impulsion they received was, however, so great, that large hillocks were suddenly formed in different parts. The vessels were driven by the gale from El-Arish, so as to subject us to much eventual distress.

In the evening I rode to the fort, a square building provided with four towers, one at each of the angles. The French had begun two bastions, which they had not time to finish; and to these Major Fletcher, of the royal engineers, made some additions. Originally, the fort stood in the centre of the village, which was now a heap of ruins.

The high winds, and the drifting of the sands, continued to annoy us greatly on the 4th. On the following day there was not merely a scarcity, but an absolute want, of barley in the camp, insomuch that the horses and other animals, deprived of pasturage since our arrival at El-Arish, were left without food. As the blowing weather rendered the return of the vessels driven out to sea impracticable, the Grand Vizier sent to Gaza, about fifty miles distant, for a small supply of barley.

Mahomed Pacha, with eight thousand men, and five pieces of artillery, marched on the same day. He was accompanied by Captain Lacey, of the royal engineers, who was invested, previously to his departure, with a pelice.

It was officially announced by the Reis Effendi, in the evening, that the Vizier would march forward on the 9th. His Excellency sent twelve horses for the dismounted men belonging to the mission. I was sorry to learn from him that four cases of plague had occurred on the preceding day in the camp.

An individual died of plague on the 6th, within fifty yards of our tents. The indifference of the Turks to this disease was truly surprising. Ibrahim Bey was positively encamped on the burial ground of El-Arish, where the bodies of several thousands of persons, who had fallen victims to that disease during the course of the last six weeks, were interred. His own tent covered a part of the graves!

This being the fourth day our unfortunate horses had passed without food, we made strong representation in their behalf, in consequence of which a small proportion of damaged biscuit-dust and decayed rice was issued to them. To such an extremity had these animals been reduced by hunger, that they had eaten their halters, together with the dung, and whatever lay within their reach. Several of them, as well as many of the camels, had perished. Our own situation, reduced as we were to bad biscuit and water, was almost as deplorable.

Riley, the person who had been sent with dispatches, returned on the morning of the 7th. He had exchanged his dispatches with an English officer commanding a schooner; but the vessel on board which he had embarked on his return, having been shipwrecked, those he was entrusted to bring back were unfortunately lost. He reported, that the Capitan Pacha had reached Aboukir with ten thousand men; but that Damietta was still in the possession of the French.

We were exposed on the 9th to a true kampsin. The heat and closeness of the air were so extremely oppressive, as to induce an extraordinary languor and faintness. The atmosphere was hazy throughout; and the wind blew on the body as if it had passed through the medium of a heated oven. During the continuance of this morbid state of the weather, it was from the south-east, south, and south-west.

In the evening several vessels appeared off the coast, but they could not come to anchor, on account of a smart and unfavourable breeze from the south-west. In the mean time the Vizier received a small supply of barley from Gaza. The party by which it was escorted, had, on its return, been attacked by a band of Bedouin Arabs, who had taken several of the mules, and killed one of the people.

Five of the above vessels anchored on the 10th, and landed eight hundred Arnauts from Constantinople, together with a supply of corn, which was highly acceptable.

Intelligence reached us on the 12th, that Salahiech had been taken possession of by Taher Pacha, and by the troops who had quitted the encampment a few days before. It appeared that on the approach of the Turkish forces, the French, about five hundred strong, had made a precipitate retreat towards Cairo. Previously to their departure, they had destroyed the works in the interior of the place, and set fire to the buildings.

A French deserter, who came into camp, and surrendered himself to the Vizier, reported, that he had left Damietta four days before, having travelled alone, and on foot, from Tineh, and remained during three days without water. He represented the French at Damietta as being in such critical circumstances, that they would rejoice at the approach of the Turkish army, to furnish them with a pretext for relinquishing the place.

A salute was fired in the camp on the occasion of the capture of Salahieh, an account of which was transmitted over land to the commander in chief of the British forces.

During the night we lost the greater part of the camels, together with their drivers, who had fled to the mountains. This unfortunate event was likely to subject the army to great risks, and very serious difficulties, since our speedy departure from El-Arish was thus rendered impracticable. Detachments of cavalry were on the 13th, sent off in their pursuit. Thus, with the above loss, the occasional deprivation of provisions, the plague, and the frequent intestine quarrels among the Turkish soldiery, we were surrounded, in our forlorn situation in the desert, by a train of threatening evils, among which may be enumerated, pestilence, and famine, and battle, and murder, and sudden death.

Several vessels anchored in the road. It was reported that a part of those which had been driven off by the late gales, had been wrecked on the adjacent coast.

A considerable number of vessels laden with corn came to anchor on the 14th. To enable us to pass the desert with greater ease, it was determined to send the spare tents and heavy baggage by water to Tineh. For this purpose eleven tents, and such of the baggage as was not indispensably necessary to the mission, were embarked on board a Turkish vessel bound to that place, together with gunner Foster, and the whole of the civil artificers, whose services would not be necessary in the interim. Five tents only were retained for the party. The Vizier and principal Turkish officers attended on the beach, to superintend the embarkation of the artillery and stores destined for Tineh, as well as to prevent any confusion which might result from the landing of the barley. After such an interval of distress as had been experienced in the camp, it was natural to presume that the troops would manifest great impatience on the landing of the supplies.

The camels which had been carried off by their drivers were still missing. Two hundred, however, of these animals were brought in on the 15th by a sheick, who, after having received a handsome present, set off with his camels and people in the night . . .

An English schooner, commanded by Lieutenant Milne, arrived on the 17th, having on board Major Missett, with dispatches. We learned through this channel that Major McKerras, of the engineers, had been killed, and Major Fletcher made prisoner, in a reconnoitring party on the water, previously to the landing of the British army. A French fleet with seven thousand troops was expected at Alexandria from Toulon. L'Africaine, one of the frigates belonging to this fleet, had been captured by the English frigate the Phœbe, Captain Barlow. The English, under the command of Colonel Spencer, and the Turks, with the Capitan Pacha, were before Rosetta, which was expected to fall very speedily.

An overland dispatch to the two British commanders in chief was sent off by an Arab; and an order transmitted by the Vizier to Mahomed Pacha, to send a body of Turkish cavalry to act with the British army.

At six in the morning of the 19th the trumpet was sounded to prepare for the march. In consequence of the scarcity of camels, the Vizier was to move forward with a part only of the Turkish army. For the same reason the party belonging to the mission to march with the Vizier's detachment, was confined to Colonel Holloway, Major Hope, Mr. Whiteman, secretary, Mr. Pink, draughtsman, myself, a serjeant, and five men. The others were to follow with the remainder of the Turkish army, as soon as beasts of burden could be procured. Strict orders were given to keep a watchful eye over the camel-drivers, in crossing the desert, as there was a possibility of their deserting with their camels – a disaster which would expose the whole of the army to the risk of perishing on its passage . . .

The Vizier quitted the ground of encampment at noon; and at three in the afternoon we followed. Prior to our departure, Major Missett embarked for Aboukir, with answers from the Vizier and Colonel Holloway to the dispatches with which he was charged.

After a very agreeable march of two hours along the sea-shore, we arrived at the wells of Messoudieh, the place where the traveller takes his supply of water, before he proceeds to cross the remaining part of the desert. Neither habitation nor inhabitant was to be seen; nor was there so much as the appearance of brush-wood to diversify the surface of the arid sands. It was with difficulty that a few dried vegetables could be raked together to heat a little coffee. By digging wells in the sands to the depth of two or three feet, water was, however, readily procured: and notwithstanding we were so near to the sea-shore, it was good and sweet. As the route from this station to Catieh occupies three days, in the course of which no drinkable water can be procured, it was necessary to make an appropriate provision of this necessary article; and for this purpose we were furnished with the skins of goats properly prepared. The camels and other animals were laden with biscuit, barley, and water, the three essentials for the crossing of the desert.

The Reis Effendi, who came up after us with the remainder of the Turkish army, informed us that when the troops commanded by Taher Pacha arrived at Salahieh, they pursued the enemy, several of whom they killed in their slight. Among the pursuers, there were several Moors, who were well mounted, and very expert in the use of their fire-arms, which they discharge with great effect when on full speed. Their chief was unfortunately killed in the attack.

At midnight our baggage was sent off under an escort; and at six in the morning of the 20th, we proceeded on our route. At the expiration of two hours and a half we halted for three quarters of an hour; and at half past one reached our new ground of encampment, Barrahcat, where we immediately pitched our tents. This spot had been evidently the bed of a salt lake in past ages, large quantities of marine shells lying scattered on its surface, together with much saline matter. We took the precaution to dig wells; but the water was so salt and brackish as to be unfit for the cattle. On subjecting it to an analysis, I found it to contain a large proportion of common salt. The progress we had made on this day's march was about fifteen miles; and in pursuing our route we were gratified by the view of the surprising visual deception, which the French term *mirage*, and which has been described with great ingenuity by Monge, of the French National Institute. It is peculiar to the desert, and presents the distant appearance of water with such an air of truth and reality, that the shadows of the camels who were in advance, seemed to be reflected on the apparent watery surface. To give a more extraordinary effect to the illusion, the bodies of these animals appeared to be narrowed and elongated upwards, so as to give them the appearance of trunks of trees, the branches of which had been lopped off. The most elevated of the distant sand hills represented light clouds; while the smaller ones appeared like ships under a press of sail in the midst of beautiful lakes. This phenomenon was more particularly apparent on the levels, which were in some parts covered with a saline substance, finely crystallized, and very shining and brilliant. A portion of this saline matter I collected for future experiments.

We left our ground on the 21st, at between three and four in the morning; and, as it was not yet day-light, were preceded by guides carrying torches. We lost our way, notwithstanding, and were detained for nearly two hours. During this day's march we made two halts; and at length, at half past four in the afternoon, reached our ground of encampment at Bir-el-habt, after a very laborious and fatiguing march of thirty miles. Great numbers of skeletons of camels, horses, and other animals were scattered on the route. The earlier part of our march was tolerably agreeable, with the exception of our having lost our way; but during the last nine miles we had to pass over very heavy sands, and an uneven ground. The heat of the weather was at the same time very oppressive . . .

We marched at six in the morning of the 22d, and at noon reached Theah, distant from Catieh about a mile and a half. The route thither was very heavy and irregular, insomuch that to perform a distance of fifteen miles we were under the necessity of making two halts. In the vicinity of our encampment there were several fine date-trees. The wells in general contained good water; but in one of them it was black and offensive, like that which is procured at Harrowgate. The Turks asserted that the French had thrown into it a quantity of gunpowder.

On our route great numbers of carcases were scattered. We were met by a messenger from Mahomed Pacha, who transmitted the information that Tineh had been taken possession of by a detachment of Turkish cavalry. In abandoning the above place, the French had left behind two guns, and a quantity of barley, and other stores . . . a great mortality, supposed to be occasioned by the plague, prevailed among the Turks. It was more probably, however, owing to constant fatigues, and the want of a good nourishing diet.

The early part of the morning had been clouded over, which rendered the heat less oppressive; but at ten o'clock, when we were within two hours march of our destination, the heat suddenly became so extremely intense, that, to use a vulgar, but expressive phrase, we were literally broiled by the sun as we trod the scorching sands. The appearance of the desert was here, however, much improved: we met with several plants and flowers on our way, together with a few date-trees; and in approaching Theah, found a greater proportion of green shrubs than we had encountered in any other part of the desert. In several of the hollows there was a sufficient degree of moisture to promote vegetation; and in these spots there is no doubt but that good water might be procured with little trouble. The numbers and varieties of the birds which were flying about them, were a strong evidence of the truth of this observation.

It appeared that a village had once stood on the site of our encampment, fragments of red bricks, &c. being scattered over the surface of the sands. In the Arabic language also, Theah implies a village; but we could not discover the vestiges of any buildings. The wells were situated beneath a groupe of fine date-trees. It was the intention of the Vizier to halt at this place for two days, to give time for the artillery and stragglers to come up, as well as to allow some repose to his army, which had had to perform a harassing march of nearly seventy miles, in the space of four days, over what was certainly to be considered as the worst part of the desert . . .

On the 25th we rose at half past two in the morning, and at three o'clock the baggage moved off the ground. We were, however, detained until near eight o'clock by some arrangements it was necessary to make for sending off a party of camels to Tineh; and the interval which elapsed hung very heavily on our hands. At ten o'clock we halted for nearly an hour, and reached our ground of encampment at Bir-denedar at two o'clock, after having performed a march of eighteen miles. There were several date-trees on the spot; and the red bricks which lay scattered on the grounds indicated that it had been formerly covered by dwellings. The water which we found at this place was salt, bitter, and of the worst quality. The weather was extremely hot during our march, which, being over a rough and uneven ground, covered with deep and heavy sands, was performed with infinite fatigue. There was, however, a greater abundance of shrubs than we had met with on our preceding marches. Half an hour after we had quitted our late encampment, we passed over a very rough piece of ground, which, being covered with large quantities of saline matter, appeared to have been a salt work. Several pits in which the salt water had been left to evaporate, had been apparently dug out for that purpose, and still contained a quantity of pure and white salt. Many of the date-trees in the vicinity of Catieh lay on the ground, having been cut down by the French for various purposes on their retreat from Syria.

In the course of the day's march the Vizier sent forward a party of Dehlis to prepare a causeway over a river, which had formerly been provided with a capacious stone bridge. To impede the progress of the Ottoman army, the enemy had thrown down this bridge; but the impediment was soon removed by the Dehlis, who effected their purpose in the course of a few hours. The river had apparently been formed by the Nile in its periodical inundation . . .

We rose at two in the morning of the 26th, and set out at five. After a march of two hours and a half, we met with large pieces of water, which had been in several places formed into lakes and rivulets by the overflowing of the Nile. It was brackish, and far from being of a good quality. In this part of the desert there was a great quantity of green brush-wood. The ground was more level, with a less proportion of sand, and the travelling by no means disagreeable. At eight o'clock we halted for half an hour; and at ten o'clock passed the river over which the causeway had been thrown. At eleven o'clock we reached the ground destined for our encampment at Kantara, and found there plenty of good water, together with shrubs and grasses for the cattle. As an abundance of pigeons and ducks were flying, I took my gun and shot several of the former, which were highly acceptable to our mess, we being reduced to the spare diet of bread, coffee, and a little rice.

In consequence of the advices which were received of the capture of Rosetta by the combined British and Turkish forces, the Vizier fired a *feu de joye* and royal salute.

We rose at half past two on the 27th, and at five o'clock were on foot. We halted twice, about an hour each time, during our march, and arrived at Salahieh at noon. The latter part of the route was very sandy and laborious. On our approaching Salahieh the quantity of shrubs gradually diminished, and at length totally disappeared, insomuch that we could not procure sticks for fuel to boil our coffee. The aspect of the country which immediately surrounded us was dreary, consisting entirely of an extensive desert plain, or level; but to the westward and northward of Salahieh lay immense woods of date-trees, which extended for several miles. We were encamped to the south of the sort, at the distance of nearly a mile.

On the approach of the Grand Vizier to Salahieh, his Highness halted under a grove of date-trees, to arrange the manner in which he should make his public entry, as well as the form of his encampment. The Turkish army afterwards marched in the following order: – First, a line of cavalry, small parties of horsemen riding up and down in front of the line, and firing while on full speed. Next another line of Arnauts, with the led horses of his Highness, and the priests, or imaums, singing hymns. Next followed Colonel Holloway, Major Hope, &c. the Turkish officers of state in succession, and his Highness the Vizier, with his bands of music, and attendants. And lastly, a body of cavalry, Dehlis, closed the rear.

Thus had we surmounted a troublesome, fatiguing, and hazardous march across the desert (a distance of about one hundred and fifty miles from Kahnyounes to Salahieh), but not without the loss of many animals, and several men. The lamentable scenes which occasionally presented themselves were truly distressing to a feeling mind.

Glossary of places

Place	Variant
Abkhazia	Republic in Caucasus
Aboukir	Abu Qir, Egypt; site of land and sea confrontations between Anglo-Ottoman and French forces, 1798–1801
Acre	Akka, Israel
Adakale	Ada Kale; Danube Island submerged by Iron Gate Dam, also Orsova Island
Adana	town and province in southern Turkey
Ağrıboz	Negreponte, Euboia, Greece
Akhaltzikhe	Ahiska; town in Armenia
Akkirman	Akkerman, Bilhorod-Dnistrovskiy, Ukraine
Aleppo	Halab, Syria
Aleksandropol	now Gyumri in Georgia
Alma River	Crimea, Ukraine
Anapa	Black Sea port; entrance to sea of Azov
Anatolia	Anadolu; Asian Turkey
Antakya	Antioch
Ardahan	town in northern Turkey
Astrakhan	on Volga River
Avlonya	Vlora, Albania
Aynalıkavak	on the Golden Horn, Istanbul
Azov	on Don River, 3 km from Sea of Azov
Babadağı	Babadag, south of Danube in Romania
Baban	a principality (Sulaimaniya) in northern Iraq
Bahçesaray	Bakhchisaray, former court of Crimean Khanate
Banjalucka	Banja Luka, in Bosnia and Herzegovina
Bayburt	fortress town and province in eastern Turkey
Beyazid	northeastern Turkey; Doğubeyazit
Bazarcık	Pazardzik, Bulgaria
Bender	Tighina, Moldova
Bessarabia	Ottoman territory ceded to Russia in 1812; Moldova
Bitlis	town and province in eastern Turkey
Brasov	Braşov, Romania
Bujak	Budjak, Budzhak, Bucak (Tr.), southern Bessarabia; now Ukraine,
Buda	Budapest, Budin, Hungary
Bug River	Buh, Boh, Ukraine
Burgas	on Black Sea, Bulgaria
Butrinto	in Albania
Canik	Ottoman province, Trabzon region

Place	Variant
Çankırı	Kangra, Gangra, northeast of Ankara, Turkey
Chios	Island in Aegean Sea, Sakız (Tr.)
Çerkes	Circassian
Çeşme	site of Ottoman naval disaster; bay (present town) near Izmir, 1770
Craiova	Kraiova, Crajova, Krayova, Romania
Dacia	Ancient name Romania and Moldova
Dagistan	Dagestan, present Republic in Russian Federation
Damietta	Dimyat, Dumyat, Egypt
Davut Paşa	suburb of European Istanbul, once the Janissary assembly area for European campaigns in Rumeli
Delvine	Delvinë, Albania
Derbent	town in Dagistan, Russian Federation
Diriya	Dariyya, Riyadh, Saudi Arabia
Diyarbakır	city and province in SE Turkey
Dobruja	Dobruje; territory at mouth of Danube, surrendered to Russians in 1829
Drava	Drave River
El Arish	Al-Arish, Egypt
Elbasan	in Albania
Eleşkirt	in Ağrı Province, eastern Turkey
Erivan	Yerevan, capital of Armenia
Esztergom	Gran, Estergom (Hungary)
Eupatoria	port in Crimea; Kozlov, Gozlov
Falça	Falçı; Larga, Moldavia
Farshut	northern Egypt
Fethülislam	Ottoman border fortress, upper Danube River
Filibe	Plovdiv, Bulgaria
Foksani	Focşani, on Milcov River, Romania
Galata Mevlevi hanesi	Sufi Monastery in Istanbul, Turkey
Galați	Galatz, on Danube River
Gelibolu	Gallipoli on Dardanelles Peninsula
Gradiska	north of Banja Luka, Bosnia and Herzogovina
Grocka	Kroska, Groka, suburb of modern Belgrade
Guirgevo	Guirgiu, Yergöğü (Tr.), Yerköy, Iurgiu, Romania
Gümüşhane	town in northeastern Turkey
Guria	Region in Georgia
Hakkari	town and province in southeastern Turkey
Harput	fortress in southeastern Turkey
Hasankale	fortress; now town Pasinler in province of Erzurum
Hasköy	section of Istanbul
Hawran	Hawran plateau, Syria & Lebanon
Hezargrad	Razgrad, Bulgaria
Hirsova	Danube town, Romania
Hotin	Khotin, Khotyn, Ukraine
Hudaybiya	battle in early Islamic history
Hüdavendigar	now Province (Bursa), Turkey
İannina	Ioannina, Yanya (Tr.), Greece, seat of Ali Pasha 1786–1821
İbrail	Brăila Romania
Iflak; Eflak (Tr.)	Wallachia
İnebahtı	Lepanto, Castel, Naupactus, Greece

Place	Variant
Imereti	province in Georgia, Imertiz, Imertia
Iron Gate	Demirkapı (Tr.), Gorge on Danube River; site of present dam
İsakçı	Isaccea, Romania
İslimiye	Slivno, Croatia
İsmail	Izmail, Ismail, Ukraine
İşkodra	now Shkodër or Shkodra in Albania
İzmit	capital of Kocaeli Province, northwestern Turkey
İzvornik	Zvornik, Bosnia and Herzogovinia
Nablus	West Bank, Palestine
Jassy	Yasi, Iaşi, Romania
Kabarda	Kabardino-Balkaria, North Caucasus
Kaffa	Kefe, Caffa, Feodosiya or Theodosia in Crimea
Kalafat	former fortress on the Danube River (Romania)
Kamenice	Kameniçe, Kamieniecz, Poland
Kandia	Candia, Heraklion Iraklion, Crete
Kanizsa	Nagykanizsa, Hungary
Karaferye	Veroia, Veria, Greece
Karahisar	Afyonkarahisar, Turkey
Karinabad	now Karnabat, Bulgaria
Kartal	Kagul, Kahul, Moldova; site of major Russo-Ottoman battle, 1770
Kerch	Kerç, Crimea
Kilburun	Kilburn, Ukraine
Kilya	Kilia, Kiliya, Ukraine
Köstence	Constanţa, Romania
Kuban	Kuban River region flanked by Don River basin and Caucasus annexed by Russia in 1783
Kurdufan	Sudan
Kütahya	City in western Turkey; site of major Ottoman defeat 1833
Kutaisi	town in Imereti, Georgia
Little Wallachia	Küçük Eflak, (Tr.) Oltenia
Lom	town on Danube River, Bulgaria
Maçin	Măcin, Romania
Maros river	Mureş, Hungary and Romania
Mingrelia	province in Georgia, also Minrelia
Misivri	Missuri
Missolonghi	Mesolonghi, Greece, Gulf of Patras, Greece
Modon	Peloponnesus, Greece
Moldavia	Boğdan (Tr.) Romania and Moldova
Monastir	Bitola, Macedonia
Morea	Peloponnesus, Greece
Muş	town in eastern Turkey
Nauplion	Napoli, Nafplion, Nauphlia Greece
Navarino	Anvarin, Peloponnesus, Greece
Niğbolu	Nicopol, Nikopolis, Nikopol, on Danube River, Bulgaria
Niş	city in Serbia
Nizib	in province of Gaziantep; site of major Ottoman defeat 1839
Ochakov	Özü, Ochakiv, Ukraine
Odessa	Russian city in Bessarabia, present-day Moldova
Of	town and province on Black Sea, Turkey
Okhri	Ohrid, Ochrid, Macedonia
Olt River	River in Romania

Place	Variant
Oltenia	Little Wallachia; also Küçük Eflak (Tr.)
Oltenitsa	Oltenitza, town on Danube, Romania
Osijek	Ösek (Tr.), Croatia
Ossetia	region in North Caucasus Mts.
Ozurgeti	town in western Georgia
Pecs	Peçy (Tr.), Hungary
Perekop	Or Kapı; isthmus connecting Crimea to mainland Ukraine
Plovdiv	Filibe (Tr.), Bulgaria
Poti	port on Black Sea, Georgia
Pravadi	Pravadia
Prut	Pruth, River and site of major Russo-Ottoman battle in 1711
Qusayr	Qosseīr, Cosser, Egypt
Rakka	Raqqa, al-Raqqah, Iraq
Rawandiz	Iraq
Rosetta	now Rashīd, port in Egypt
Rumeli	formerly Ottoman Europe; Rumelia
Rumeli Kavağı	town on upper Bosphorus, Istanbul
Rusçuk	Russe, Ruse, on Danube, Bulgaria
Salonika	Selanik (Tr.); Thessaloniki, Greece
Sarıkamıs	town in northeastern Turkey; site of Ottoman defeat in WWI
Sarıhan	Manisa Province, Turkey
Sevastopol	Port on Black Sea; also Sebastopol, Akyar, Ukraine
Selimiye	Barracks in Istanbul, Asian side, built by Selim III
Semendre	Smederevo, Serbia
Shorofi	Soroki, Soroca, Moldova
Sibiu	Transylvania, on Cibin River in Romania
Sidon	Sayda, Lebanon
Siirt	town in SE Turkey
Silistre	Silistri, Silistra, fortress town on Danube, Bulgaria
Sinop	Black Sea fortress and port in Turkey
Sistova	Svishtov, northern Bulgaria; also Zistova
Sıvas	Town and province in central Turkey
Siverek	town in southeastern Turkey
Sizeboli	Sozopol, Bulgaria
Slatina	city on River Olt in Romania
Sofya	Sofia, Bulgaria
Sokhum	Suhum, Sukhumi, capital of Abkhazia
St. Gotthard	Szentgotthárd, Hungary
Şumnu	Shumla; Shumen, Bulgaria
Taganrog	Russia's first naval base, on Sea of Azov 1698
Taman	Peninsula in Sea of Azov and town (Temryuk)
Temeşvar	Timoşoara, Romania
Tiflis	Tbilisi, capital of Georgia
Tirana	capital of Albania
Tokat	town in central Turkey
Tophane	arsenal in Istanbul
Trabzon	city and province on Black Sea, eastern Turkey
Travnik	city in Bosnia and Herzogovina
Tulcea	Tulçu, Tolçu, Horatepe (Tr.), Romania
Turnu	Turnu Măgurele, Romania
Urfa	Şanlıurfa, Turkey

Place	Variant
Uyvar	Nové Zániky in Slovakia
Üsküdar	Asian side of Istanbul; assembly point for Janissary campaigns to points east
Üsküb	Skopje, Macedonia
Varna	Black Sea port, Bulgaria
Vidin	fortress town, on the Danube,Bulgaria
Viranşehir	town in Şanlıurfa Province, Turkey
Visegrad	Višegrad, on Drina River in Bosnia and Herzogovina
Verbas River	in Bosnia and Herzogovina
Wallachia	Iflak; also Eflak, Romania
Zahle	Zahlah, town in Lebanon
Zemun	Zemun, Bulgaria; formerly Semlin

Glossary of terms

Term	Definition
acemioğlan	novice soldier
akçe	silver coin; *asper*
akıncı	light cavalry raider
alay	regiment
alemdar, bayraktar	standard, or flag-bearer
altın	gold coin
arabacı	waggoneer
askeri	tax-exempt military class; military
avariz-i divaniye	extraordinary taxes
avarizhane	village tax households
ayan (s. & pl.)	provincial magnate, notable, warlord; Ottoman provincial officials
azab	infantryman; local infantry
bailo (baili pl.)	Venetian consul in Istanbul
başıbozuks	irregulars; mercenaries
bastinado	falaka (Tr.); punishing a miscreant by beating the soles of the feet
bedel-i askeriye	military exemption tax for non-Muslim men
bedel-i nakdi	military exemption, buy-out for Muslim men
beldar	trench digger
berat	licence of privilege
beşe	term for Janissary, or Janissary associate, like *yoldaş*
beşlü	elite cavalry corps
binbaşı	major
bölük	company
bostancı	palace guard (Janissaries)
boyar	landowner; notable, ruler in Wallachia, Moldavia
cebeci	armorer
cizye	non-Muslim poll tax
Concert of Europe	The great powers of Europe (Prussia, Britain, France, Austria; latterly, Russia) who managed international relations by diplomacy rather than aggression, 1815–56
corvée	forced labor
dakik	flour
dar-al-Islam	abode of Islam
dayı	Also *dey*, protector or enforcer, used for Janissary commanders acting as rulers, here in Belgrade

Term	Definition
defterdar	chief accountant
delis	irregulars, mercenaries; identified with Kurds
derbend başbuğu	warden of the pass
derebey	provincial notable; warlord; also *ayan*
devşirme	Janissary levy of Balkan Christians
divan	council of state, or ruler
dizdar	local fortress commander
enderun	sultan's private apartments in palace
esame	technically, muster roll; but also Janissary pay and ration ticket
esnaf	merchants; artisans; guilds
Eşkinci	name briefly of reorganized infantry just prior to 1826
eşkıya	bandit; brigand
farisan	cavalrymen
fellahin	Egyptian peasants
ferde	head tax imposed on Syria by İbrahim Pasha
ferik	general (military rank)
fetva, fatwa	religious ruling
firman	ruling; edict of the sultan
fitne	rebellion
gazi, ghazi	mercenary; raider; warrior for the faith
gönüllü	volunteer; aspiring Janissary
grenzer	border troops of Habsburg military corridor, Croatians and Serbians
haidamak	peasant
hammal	porter; manloads [as measurement]
haydut	*hayduk, haiduk*: mercenary, marauder, brigand, thief, highwayman, freedom fighter
hinta	wheat
hospodar	local official, but also ruler, esp. Principalities (Rumanian); also *voyvoda*
hüccet	court affidavit
humbaracı	bombardier; mortar corpsman
iltizam	tax-farming, awarded by auction
imdad-i seferiye, imdadiye	extraordinary [then annual] tax granted to provincial governors for campaign expenses
İrad-ı Cedid	New Revenue (Treasury) – under Selim III
kadi	judge [also kazi, kadı, qadi]
kadiasker	chief military judge; *kaziasker*
kantar	unit of measure = 56.449 kilograms
kanun, kanunnames	sultanic ordinances, law, collections of laws
kapu halkı, kapılı, kapı halkı	household entourage, provincial governor's forces
kapu kulu, kapıkulu	Janissary
kapudan pasha	also *kapudan-ı derya*; grand admiral; *kapudan* also used as commander, captain, or local notable in Balkans
kapıcıbaşı	chief gatekeeper of the palace
kapısız levendat	unattached or demobilized irregulars
kaymakam	substitute official, for example, in absence of grand vizier from Istanbul; also governer of *kaza* under Mahmud II; see also *kethüda*

Term	Definition
kaza	administrative district under *kadi* – under *kaymakam* with Mahmud II
kese	bag, purse, usually of 500 *kuruş* – accounting notation
kethüda	*kahya*, deputy, substitute, second-in-command to grand vizier, steward
khedive	viceroy of Egypt 1841–1952
kile	unit of weight = 25.659 kilograms
kleft, klepht	bandit, mercenary, freedom fighter, south Balkans
knezes	Serbian notables
kul	slave; Janissary, or officials in sultan's service
kuruş	unit of money, silver coin equal to 120 *akçes*; piaster
lağımcı	sapper
layiha	report; essay
levend, levent, levendat (pl.)	mercenary militia, *sarıca*, *sekban*, irregular; also *deli*, *başıbozuk*
malikane	lifetime tax farms bid on at annual auction
mamluk/Mamluk	slave soldier, largely from Caucasus; latterly style of rule through households in Cairo and Baghdad
manga	squad
martolos	local troops; cavalry, Christian
maslaha	the common good
meclis	council; assembly
mektupçu	secretary of a diplomatic delegation; part of *reis* staff
menzil	way-station, bivouac
meşveret	council, normally grand vizier presiding
mevacibli	salaried soldiers
Mevlevi	sufi order
millet	religious community
miralay	colonel
miri	belonging to the state land; also used for fixed price (state regulated)
miri levendat (pl.)	state-funded mercenaries; militias
mirliva	brigadier general
mubayaacı	state purchaser/commissary, regulated prices (*mubayaa*)
mukataa	revenue source
mülazim	lieutenant
müsadere	confiscation (of disgraced official's property and wealth)
müşir	field marshal
müstahfiz	local fortress guard
mutasarrıf	governor (sometimes deputy governor) of district (*sancak*); tax collector
mütesellim	tax collector; deputy governor; local official
nan-i aziz	special bread for Janissaries
Naqshbandi	sufi order
nasihatname	advice manual
nazır	supervisor
nefer-i' am	call-to-arms of general population
Nizâm-ı Cedid	New Order – Selim III

Term	Definition
nizamiye	*Asakir-i Nizamiye-yi Şahane,* or *nizamiye,* the Ottoman army after 1843
ocak	regiment; the Janissary Corps
ocaklık	group of tax farms supporting Janissaries
okka	unit of weight = 1.282945 kilograms
ordu	army
ordu akçesi	cash contribution for campaigns by guilds
orducu	campaign supply system (Janissary)
orta	battalion; later *tabur*
palanka	Small fort; outposts on frontier; often wooden
pandours	Serbian militia
pasha	governor, ministers of state, honorific
peksimed	biscuit, hardtack
Philike Hetairia	Greek liberation secret society
pişkeş	annual tribute (here, from *hospodar* of Principalities)
Pomaks	Muslim Bulgarians
rayic	market price
reaya	peasants, tax category
redif	reserve soldier; later *ihtiyat, yedek*
reis	chief, head, of chancery; latterly became foreign affairs minister
Reisülküttab	Chief of the Scribes
riyal	Egyptian coin
Rum taifesi	here, Orthodox Christians, Greeks
salyane	annual tax
sancak	district – below vilayet
sarıca	see *levend*
sekban	see *levend*; also briefly used as the name the reorganized troops of Alemdar Pasha in 1808
Sened-i İttifak	Deed of Agreement 1808
Serasker	commander-in-chief; minister of war; also regional commanders during campaigns
serdengeçti	Janissary vanguard; shock troops
serhad	frontier; border
serhad kulu	frontier troops
serseri	vagrant
silahdar	sword-bearer
sipahi	fief-based cavalry
sipahiyan	cavalrymen
Sublime Porte (Bab-ı Ali)	office of the grand vizier, and seat of Ottoman government
Sunna	way, teachings of the Prophet Muhammad
süratçi	rapid-fire artilleryman
talimli asker	trained soldier, Selim III's troops
telhis	summary report to sultan by grand vizier
tercüman (dragoman)	Translator, interpreter
tevzi	extraordinary taxes, village assessment
şayka	Danube river boats
şeyhülislam	grand mufti
tabur	battalion; earlier *orta*
tekke	sufi, dervish lodge

Term	Definition
timar	land grant, fief
timariot, timarlı	fief-based cavalry
topçu; topçubaşı	artilleryman; artillery corps commander
topraklı	countryside forces, timariot
ulufeciyan-i süvari	salaried cavalrymen
vakf, vakıf, waqf	charitable endownment
vali	governor of province (*vilayet*)
vekalet	representative of the sultan
verst	Russian unit of distance = 1.0668 kilometers
vilayet	province (*eyalet*)
voynuk	local corps
voyvoda	local official, but also see *hospodar*
yamak	new recruit, irregular, Janissary in waiting
yerli, yerli kulu, yerliye	local auxiliary services, local Janissaries
yoldaş	Janissary fellow traveler, comrade in arms
yüzbaşı	captain
zimmi; dhimmi	protected non-Muslims; 'people of the book'
zorba	bully

Bibliography

Documents

Başbakanlık Arşivleri (Prime Mınister's Archives BA)

CA (Cevdet Askeri) 11773 on impure grain shipments

CA 14462, 1772, from Rusçuk garrison, supply shortages

CA 15612, March 1773, rapid fire artillery corps ordered to front

CA 20344, July 1773, rapid fire artillery corps rations

CA 22902, 1772, Ochakov defterdar report

CA 29732, May 1773, supplies for rapid fıre artıllery

CA 30301, Nov. 1772 order for gunpowder

CA 32630, 1772, on Ochakov defterdar

D.BŞM (Başmuhasebe) 3913, 2–3, on grain shipments

D.BŞM 4274, an account of salaries, ration and clothing allowances of the resident guards in border fortresses, beginning date 1185 (1771)

D.BŞM 3949, on supplies for 1768–74 war, esp. camels.

D.BŞM 5665, 1787, on supplies

D.BŞM.BNE (Bina Emini) 15970 September 1768–April 1769 instructions on mobilizing for war

D.MKF (Mevkufat) 30413, 1769/70, on biscuit and other provisions

HH (Hatt-ı Humayun) 20896, Halet Efendi to Mahmud II, 1225 (1810)

HH 21173-D, 15 April 1831, on Albanian and revolt

KK (Kamil Kepeci) 2929, 5–6, on grain trade problems in 1769, and Ahmed Resmi as *mubayaacı*

KK 6799 Askeri defter, 1826–37 statistics on *Asakir-i Mansure* army

MM (Maliyeden Müdevver) 4844 on rapıd fıre artıllery

Bulgaria State Library, Sofia S166 (court records)

Manuscripts

Anon. 'Eflak ve Bogdan'. Topkapı Palace Library Ms H 445 [1760].

Ahmed Resmi. 'Layiha'. Istanbul University Ms TY 419, undated [1771].

Canikli Ali Efendi. 'Nasayih al-Muluk'. Süleymaniye Esad Efendi Ms 1855, 1782.

[Hasan Kurdi] *Anonymous History of Prut*. Russian National Library, St. Petersburg, Fund 933, TNS A Collection 155, folios 24–47, listed as anonymous, but author is Hasan Kurdi (folio

Mustafa 'Ibretnuma-yi Devlet.' Millet Kütüphanesi: Ali Amiri Efendi Ms 484, 1182/1768.

Sadullah Enverî, 'Tarih'. Istanbul University Library Ms T 5994, 1780.

Encyclopedia of Islam articles

''Ali Pasha Tepedelenli'. H. Bowen. *Encyclopedia of Islam*, *EI2*, CD edn.

'Bosnia–Herzegovina'. *EI2*, CD edn.

'Dustūr'. Bernard Lewis. *EI2*, CD edn.

'Hâlet Efendi'. M. Şihâbeddin Tekindağ, *İslam Ansiklopedisi*, vol. 5, 123–25.

'Husayn Pasha'. Münir Aktepe. *EI2*, vol. 3 (1971), 627–28.

'Izzet Molla'. Fahir İz. *EI2*, vol. 4 (1978) 293–96.

'Ibn 'Abd al-Wahhab'. H. Laoust. *EI2*, CD edn.

'Istanbul', Halil İnalcık, *EI2*, CD edn.

'Küčük Kaynardja'. Heywood, C.J. *EI2*, CD edn.

'Khosrew Pasha'. *EI2*, vol. 5 (1979), 35–36.

'Kurds'. V. Minorsky. *EI2*, CD edn.

'Mehmed Sa'īd Ghālib Pasha'. E. Kuran. *EI2*, CD edn.

'Mora'. N.A. Bées. In *EI2*, CD edn, 236–41, 1993.

'Muhammad 'Ali Pasha'. E. Toledano. *EI2*, CD edn.

'Mustafa Pasha, Alemdar'. İ. Uzunçarşılı. *İslam Ansiklopedisi*, vol. 7, 10–27.

'Mustafā Pasha Bushatlı (1797–1860)'. A. Popovic. *EI2*, CD edn.

'Selim III'. V.H. Aksan. *EI2*, CD edn.

'Sirb'. M.O.H. Ursinus. *EI2*, CD edn.

'Timar'. J. Deny, *EI1*, vol. 8, 767–66; as edited by Halil İnalcik. *EI2*, CD edn.

'Widin'. Svetlana Ivanova. *Encyclopedia of Islam*, 2nd edn. vol. 9 (2001), 205–8.

Published works

Abou El-Haj, Rifa'at A. *Formation of the Modern State: The Ottoman Empire, Sixteenth to Eighteenth Century*. 2nd edn. Syracuse, NY: Syracuse University Press, 2005.

Abou El-Haj, Rifa'at A. 'Ottoman attitudes towards peace making: the Karlowitz case'. *Der Islam* 51 (1974): 131–37.

Abou El-Haj, Rifa'at A. *The 1703 Rebellion and the Structure of Ottoman Politics*. Leiden: Brill, 1984.

Abu-Manneh, Butrus. 'Islamic roots of the Gülhane Rescript'. In Abu-Manneh, B., *Studies on Islam and the Ottoman Empire in the 19th Century* (1826–76). Istanbul: ISIS, 2001: 73–97.

Abu-Manneh, Butrus. 'The Naqshbandi-Mujaddidi and the Bektashi Orders in 1826'. In Abu-Manneh, B., *Studies on Islam*, 59–71.

Adanir, Fikret and Suraiya Faroqhi. *The Ottomans and the Balkans*. Leiden: E.J. Brill, 2002.

Ágoston, Gábor. 'Gunpowder for the Sultan's Army: new sources on the supply of gunpowder to the Ottoman Army in the Hungarian campaigns of the sixteenth and seventeenth centuries'. *Turcica* 25 (1993): 75–96.

Ágoston, Gábor. *Guns for the Sultan: Military Power and the Weapons Industry in the Ottoman Empire*. Cambridge: Cambridge University Press, 2005.

Ágoston, Gábor. 'Ottoman artillery and European military technology in the fifteenth and seventeenth centuries'. *Acta Orientalia Academiae Scientiarum Hungaricae* 47 (1994): 15–48.

Ágoston, Gábor. 'Ottoman warfare in Europe, 1453–1826'. In Jeremy Black, ed., *European Warfare 1453–1815*. Basingstoke: Macmillan, 1999: 118–44.

Airapetov, O. 'Miliutin contra Moltke'. In David van Schimmelpennick der Oye and Bruce Menning, eds., *Reforming the Tsar's Army: Military Innovation in Imperial Russia from Peter the Great to the Revolution*. Cambridge University Press, 2004.

Aksan, Virginia H. 'Enlightening the Ottomans: Tott and Mustafa III'. In *Proceedings of the International Congress on Learning and Education in the Ottoman World*, Istanbul: IRCICA, 1999: 163–74.

Aksan, Virginia H. 'Feeding the Ottoman troops on the Danube 1768–74'. *War and Society* 13 (1995): 1–14.

Aksan, Virginia H. 'Is there a Turk in the *Turkish Spy*?' *Eighteenth Century Fiction* 6 (1994): 201–14.

Aksan, Virginia H. 'Locating the Ottomans among early modern empires', *Journal of Early Modern History* 3 (1999): 1–32.

Aksan, Virginia H. 'Manning a Black Sea garrison in the 18th century: Ochakov and concepts of mutiny and rebellion in the Ottoman context', *International Journal of Turkish Studies* 8 (2002): 63–72.

Aksan, Virginia H. 'Mutiny and the eighteenth century Ottoman Army'. *Turkish Studies Association Bulletin* 22 (1998): 116–25.

Aksan, Virginia H. 'The one-eyed fighting the blind: mobilization, supply and command in the Russo-Turkish War of 1768–74'. *International History Review* 15 (1993): 221–38.

Aksan, Virginia H. 'Ottoman military matters': review article, *Journal of Early Modern History* 6 (2002): 52–62.

Aksan, Virginia H. 'Ottoman military recruitment strategies in the late eighteenth century'. In Erik J. Zürcher, ed., *Arming the State: Military Conscription in the Middle East and Central Asia 1775–1925*. London: I.B. Tauris/Palgrave Macmillan, 1999: 21–39.

Aksan, Virginia H. 'Ottoman political writing, 1768–1808'. *International Journal of Middle East Studies* 25 (1993): 53–69.

Aksan, Virginia H. 'Ottoman sources of information on Europe in the eighteenth century', *Archivum Ottomanicum* 11 1886 [1988]: 5–16.

Aksan, Virginia H. *An Ottoman Statesman in War and Peace: Ahmed Resmi Efendi, 1700–83*. Leiden: Brill, 1995.

Aksan, Virginia H. 'Ottoman war and warfare, 1453–1812', in Jeremy Black, ed., *War in the Early Modern World, 1450–1812*. London: University College London Press, 1999: 147–75.

Aksan, Virginia H. *Ottomans and Europeans: Contacts and Conflicts*. Istanbul, Isis Press, 2004.

Aksan, Virginia H. 'Whatever happened to the Janissaries?' *War in History* 5 (1998): 23–36.

Aksan, Virginia H. 'Whose territory and whose peasants? Ottoman boundaries on the Danube in the 1760s', in F.F. Anscombe, ed., *The Ottoman Balkans, 1750–1830*. Princeton, NJ: Markus Wiener, 2006: 61–86.

Akpinar, Alisan and Eugene L. Rogan, *Aşiret, Mektep, Devlet. Osmanlı Devleti'nde Aşiret Mektebi*. Istanbul: Aram, 2007.

Aktepe, M. Münir. 'Ahmed III devrinde şark seferine iştirâk edecek ordu esnafı hakkında vesikalar'. *Tarih Dergisi* VII, no. 10 (1954): 17–30.

Alexander, John. *Autocratic Politics in a National Crisis: The Imperial Russian Government and Pughachev's Revolt, 1773–75*. Bloomington IN: Indiana University Press, 1969.

Alexandrescu-Dersca, M. 'Contribution a l'étude de l'approvisionnement en blé de Constantinople au XVIIIe siècle'. *Studia et Acta Orientalia* (1957): 13–37.

Allen, W.E.D., and Paul Muratoff, *Caucasian Battlefields: a History of the Wars on the Turco-Caucasian Border 1828–1921*. Cambridge: Cambridge University Press, 1953.

Altinay, Ayşe Gül *The Myth of the Military-Nation: Militarism, Gender, and Education in Turkey*. London: Palgrave, 2004.

Anastasopoulos, Antonios. 'Imperial institutions and local communities: Ottoman Karaferye 1758–74'. PhD, Cambridge, 1999.

Anastasopoulos, Antonios, ed. 'Provincial elites in the Ottoman Empire', *Crete*. Rethymno: Institute for Mediterranean Studies, 2005.

Anderson, M.S. *The Eastern Question, 1774–1923*. New York: St Martin's Press, 1966.

Anderson, M.S. *Europe in the Eighteenth Century, 1713–83*. London, 1961; 4th edn, London: Longman, 2000.

Anderson, M.S., 'Great Britain and the Russian fleet, 1769–70', *Slavonic and East European Review* 31 (1952–3): 148–63.

Anderson, M.S. ed. *The Great Powers and the Near East, 1774–1923*. London: Arnold, 1970.

André, Raymond. *Le Caire des Janissaires: l'apogée de la ville ottomane sous 'Abd al-Rahman Katkhud*. Paris: CNRS Editions, 1995.

Andrić, Ivo. *The Bridge on the Drina*. Chicago: University of Chicago Press, 1977.

Anscombe, Frederick F., ed. *The Ottoman Balkans 1750–1830*. Princeton: Markus Wiener, 2006.

Anscombe, Frederick F. 'Reform, resistance and Islam in the Ottoman Empire'. Unpublished paper.

Arıkan, V. Sema. 'Nizâm-ı cedit'in kaynaklarından Ebubekir Ratib Efendi'nin 'Büyük Lâyiha'sı'. PhD, Istanbul University, 1996.

Artan, Tülay and Halil Berktay, 'Selimian times: a reforming Grand Admiral, anxieties of re-possession, changing rites of power'. In Elizabeth Zachariadou, ed., *The Kapudan Pasha, his Office and his Domain*. Rethymnon: Institute for Mediterranean Studies, 2002: 7–49.

An Authentic Narrative of the Russian Expedition Against the Turks by Sea and Land. London, 1772.

Aynural, Salih. *İstanbul Değirmenleri ve Fırınları: Zahire Ticareti (1740–1840)* Istanbul: Tarih Vakfı, 2001.

Bağış, A.I. *Britain and the Struggle for the Integrity of the Ottoman Empire: Sir Robert Ainslie's Embassy to Istanbul, 1776–94*. Istanbul: ISIS, 1984.

Bailey, Frank E., *British Policy and the Turkish Reform Movement: A Study in Anglo-Turkish Relations, 1826–53*. Cambridge, MA: Harvard University Press, 1942.

Bareilles, Bertrand. *Un Turc à Paris 1806–11: Relation de Voyage et de Mission de Mouhib Efendi, Ambassadeur Extraordinaire du Sultan Selim III (d'après un manuscrit autographe)*. Paris: Bossard, 1920.

Barker, Thomas H., *Double Eagle and Crescent: Vienna's Second Turkish Siege and its Historical Setting*. Albany: SUNY Press, 1967.

Barkey, Karen. *Bandits and Bureaucrats: The Ottoman Route to State Centralization*. Ithaca: Cornell University Press, 1994.

Barkey, Karen and Mark van Hagen, eds. *After Empire: Multiethnic Societies and Nation-building*. Boulder, CO: Westview Press, 1997.

Bartlett, R. and Hartley, J.M. *Russia in the Age of the Enlightenment*. Basingstoke: Macmillan, 1990.

Baumann, Robert F. 'The Russian army, 1853–81'. In *The Military History of Tsarist Russia*, ed. F.W. Kagan and R. Higham. New York: Palgrave/St Martin's, 2002: 137–50.

Bayly, C.A. 'The British military-fiscal state and indigenous resistance: India 1750–1820'. In *An Imperial State at War: Britain from 1689 to 1815*, Lawrence Stone, ed. London: Routledge, 1994: 322–54.

Bayly, C.A. 'Knowing the country: empire and information in India'. *Modern Asian Studies* 27, no. 1 (1993): 3–43.

Bayly, C.A. 'The first age of global imperialism, c. 1760–1830'. *Journal of Imperial and Commonwealth History* 26 (1998): 28–47.

Bennison, Amira K. 'The "New Order" and Islamic order: the introduction of the *Niẓāmī* Army in the western Maghrib and its legitimation, 1830–73', *International Journal of Middle East Studies* 36 (2004): 591–612.

Bérenger, Jean. *A History of the Habsburg Empire 1700–1918*. [vol. 2] Translated by C.A. Simpson. White Plains, NY and London, Longman, 1997.

Berkes, Niyazi. *The Development of Secularism in Turkey*. Montreal: McGill University Press, 1964.

Beydilli, Kemal. *Büyük Friedrich ve Osmanlılar*. Istanbul: İ.Ü. Edebiyat Fakültesi, 1985.

Beydilli, Kemal. 'İlk mühendislerimizden Seyyid Mustafa ve Nizâm-ı Cedîde dair risalesi', *Tarih Dergisi* 13 (1983–87): 387–479.

Beydilli, Kemal. 'Küçük Kaynarca'dan Tanzimât'a islâhât düşünceleri'. *İlm Araştırmalar* 8 (1999): 25–64.

Beydilli, Kemal. *1790 Osmanlı-Prusya İttifâkı*. Istanbul, 1984.

Beydilli, Kemal. 'Sekbanbaşı Risalesi'nin müellifi hakkında', *Türk Kültürü İncelemeleri Dergisi* 12 (2005): 221–24.

Beydilli, Kemal. *Türk Bilim ve Matbaacılık Tarihinde Mühendishâne ve Kütüphânesi (1776–1826)*. Istanbul: Eren, 1995.

Birn, Raymond. *Crisis, Absolutism, Revolution: Europe 1648–1789*. 2nd edn. Fort Worth, TX: Broadview Press, 1992.

Black, Antony. *The History of Islamic Political Thought from the Prophet to the Present*. Edinburgh: Edinburgh University Press, 2001.

Black, Jeremy, ed., *European Warfare 1453–1815*. Basingstoke: Macmillan, 1999.

Black, Jeremy. *European Warfare 1660–1815*. London: UCL Press, 1994 New Haven, CT: Yale University Press, 1994.

Black, Jeremy. *The Rise of the European Powers 1679–1793*. London: Arnold, 1990.

Black, Jeremy. *War in the Early Modern World 1450–1812*. London: UCL Press, 1999.

Blaquière, Edward. *Letters from Greece*. London, 1828.

Bonaparte, Napoleon. *Napoleon in Egypt: Al-Jabarti's Chronicle on the French Occupation of Egypt*. Translated by Shmuel Moreh. Princeton, NJ: Markus Wiener, 1993 [1798].

Boppe, A. 'La France et le "militaire turc" au XVIIIe siècle', *Feuilles d'histoire* (1912): 386–402, 490–501.

Brøndsted, Peter Oluf. *Interviews with Ali Pacha of Joanina in the Autumn of 1812*. Athens: Danish Institute at Athens, 1999.

Brower, Daniel R., and Edward J. Lazzerini, eds. *Russia's Orient: Imperial Borderlands and People, 1700–1917*. Bloomington, IN: Indiana University Press, 1997.

Brown, L. Carl. *International Politics and the Middle East: Old Rules, Dangerous Game*. Princeton, NJ: Princeton University Press, 1984.

Brue, Benjamin. *Journal de la Campagne que le Grand Vesir Ali Pacha a Faite en 1715 pour la Conquête de la Morée*. Reprint of the 1870 Paris (Ernest Thorin) edn. Athens, 1986.

Bruinessen, Martin van. *Agha, Shaikh and State: The Social and Political Structures of Kurdistan*. London: Zed Books, 1992.

Brummett, Palmira. 'Classifying Ottoman mutiny: the act and vision of rebellion'. *Turkish Studies Association Bulletin* 22 (1998): 91–107.

Brummett, Palmira. *Ottoman Seapower and Levantine Diplomacy in the Age of Discovery*. Albany, NY: SUNY Press, 1994.

Burbank, Jane, and David Ransel, eds. *Imperial Russia: New Histories for the Empire*. Bloomington, IN: Indiana University Press, 1998.

Çadırcı, Musa. 'Ankara sancağında Nizâm-ı Cedîd Ortasının teşkili ve Nizâm-ı Cedîd Askeri Kanunnamesi', *Belleten* 35 (1972): 1–13.

Cadırcı, Musa. 'Osmanlı İmparatorluğu'da askere almada kura usulüne geçilmesi "1846 Tarihli Askerlik Kanunu" ', *Askeri Tarih Bülteni* 18 (1985): 59–75.

Çadırcı, Musa. 'Redif askeri teşkilatı', in *Yedinci Askerî Tarih Semineri Bildirileri*. Ankara: T.C. Genelkurmay Başkanlığı, 2000.

Cantemir, Demetrius. *The History of the Growth and Decay of the Othman Empire*. London, 1756.

Cassels, L. *The Struggle for the Ottoman Empire, 1717–40*. London: John Murray, 1966.

Çataltepe, Sipahi. *19. Yüzyıl Başlarında Avrupa Dengesi ve Nizâm-ı Cedit Ordusu*. Istanbul: Göçebe Yayınları, 1997.

Ceauşescu, Illie. 'Military aspects of the national struggle of the Rumanian principalities in the eighteenth century'. In G.E. Rothenberg, B. Király and P.F. Sugar, eds., *East Central European Society and War in the Pre-Revolutionary Eighteenth Century*. New York: Columbia University Press, 1982: 371–85.

Cevdet, Ahmed. *Tarih*. 1st edn, 12 vols. Istanbul, 1858.

Cezar, Mustafa. *Osmanlı Tarihinde Levendler*. Istanbul: Güzel Sanatlar Akademisi, 1965.

Cezar, Yılmaz. *Osmanlı Maliyesinde Bunalım ve Değişim Dönemi (XVIII.yy dan Tanzimat'a Mali Tarihi*. Istanbul: Alan Yayıncılık, 1986.

Chandler, David. *The Art of Warfare in the Age of Marlborough*. London: Batsford, 1976.

Chapman, Tim. *Imperial Russia, 1801–1905*. London: Routledge, 2001.

Chesney, F.R. *Russo-Turkish Campaigns of 1828 and 1829 with a View of the Present State of Affairs in the East*. New York: Redfield, 1854.

Christensen, Stephen T. 'The heathen order of battle'. In Stephen T. Christensen, ed. *Violence and the Absolutist State: Studies in European and Ottoman History*. Copenhagen: Copenhagen University, Humanistiske Forskningcenter, 1990: 75–198.

Çınar, Ali Osman. 'Sevânihü'l-Levâyih'ı ve değerlendirmesi'. MA (Marmara Üniversitesi), 1992.

Çızakça, Murat, 'The Kapudan Pasha and the shipowners (18th and 19th centuries)', in *The Kapudan Pasha: His Office and His Domains*,

Elizabeth Zachariadou, ed. Rethymnon: Institute for Mediterranean Studies, 2002: 203–13.

Clay, Christopher. *Gold for the Sultan: Western Banks and Ottoman France 1856–81. A Contribution to Ottoman and International Financial History*. London: I.B. Taurus, 2000.

Clogg, Richard. *A Concise History of Greece (Cambridge Concise Histories)*. Cambridge: Cambridge University Press, 1992.

Cohen, Amnon. 'The army in Palestine in the eighteenth century – sources of its weakness and strength', *Bulletin of the School of Oriental and African Studies* 34 (1971): 36–55.

Collins, James B. *Fiscal Limits of Absolutism*. Berkeley: University of California Press, 1988.

Comstock, John L. *History of the Greek Revolution Compiled from Official Documents of the Greek Government*. New York: William W. Reed, 1828.

Constantiniu, Florin. 'Tradition and innovation in the eighteenth-century military structures of the Rumanian lands'. In G.E. Rothenberg, B. Király and P.F. Sugar, eds., *East Central European Society and War in the Pre-Revolutionary Eighteenth Century*, New York: Columbia University Press, 1982: 387–99.

Creasy, Edward S. *History of the Ottoman Turks*. London: R. Bentley and Son, 1877.

Crecelius, Daniel. *The Roots of Modern Egypt: A Study of the Regimes of 'Ali Bey al-Kabir and Muhammad Bey Abu al-Dhahab, 1760–75*. Minneapolis: Bibliotheca Islamica, 1981.

Cunningham, Allan. *Anglo-Ottoman Encounters in the Age of Revolution: Collected Essays, vol. 1*, Edward Ingram, ed. London: Frank Cass, 1993.

Cunningham, Allan. *Eastern Questions in the Nineteenth Century: Collected Essays, vol. 2*, Edward Ingram, ed. London: Frank Cass, 1993.

Cunningham, Allan. 'The Ochakov debate'. In A. Cunningham, *Anglo-Ottoman Encounters*, 1–31.

Cunningham, Allan. 'The preliminaries of the Crimean War'. In A. Cunningham, *Eastern Question*, 130–224.

Cunningham, Allan. 'Robert Liston at Constantinople'. In A. Cunningham, *Anglo-Ottoman Encounters*, 51–102.

Cunningham, Allan. 'Stratford Canning and the Treaty of Bucharest'. In A. Cunningham, *Anglo-Ottoman Encounters*, 144–87.

Cunningham, Allan. 'Stratford Canning, Mahmud II, and Greece'. In A. Cunningham, *Anglo-Ottoman Encounters*, 276–323.

Curtiss, John S. *Russia's Crimean War*. Durham, NC: Duke University Press, 1979.

Cvetkova, Bistra. 'The Bulgarian Haiduk movement in the 15th–18th centuries'. In *East Central European Society and War in the Pre-Revolutionary Eighteenth Century*, 301–37.

Daly, M.W., ed. *Cambridge History of Egypt. Vol. 2: Modern Egypt from 1517 to the End of the Twentieth Century*. Cambridge: Cambridge University Press 1989.

Daniel, Norman. *Islam, Europe, and Empire*. Edinburgh: Edinburgh University Press, 1966.

Darling, Linda T. *Revenue-Raising and Legitimacy: Tax collection and Finance Administration in the Ottoman Empire, 1550–1660*. Leiden: Brill, 1996.

Dávid, Géza and Pál Fodor, eds., *Hungarian–Ottoman Military and Diplomatic Relations in the Age of Süleyman the Magnificent*. Budapest: Loránd Eötvös, 1994.

Dávid, Géza and Pál Fodor, eds., *Ottomans, Hungarians and Habsburgs in Central Europe: Military Confines in the Era of Ottoman Conquest*. Leiden: Brill, 2000.

Davies, Brian L. 'Russian military power 1453–1815'. In *European Warfare, 1453–1815*, 145–79.

Davison, Roderic H. 'Britain, the international spectrum, and the Eastern Question 1827–41', *New Perspectives on Turkey* 7 (1992): 15–35.

Davison, Roderic H. *Reform in the Ottoman Empire 1856–76*. New York: Gordian Press, 1972.

de Grunwald, Constantin. 'Ambassade russe à Constantinople au XVIIIe siècle'. *Miroir de l'histoire* 7 (1956): 491–99.

Deringil, Selim. *The Well-Protected Domains: Ideology and the Legitimation of Power in the Ottoman Empire, 1876–1909*. London: I.B. Tauris, 1998.

D'Ohsson, Mouradgea, *Tableau Général de l'Empire Othoman*. Paris, 1788–1823.

Dowes, Dick. *The Ottomans in Syria: a History of Justice and Oppression*. London: I.B. Tauris, 2000.

Dowes, Dick. 'Reorganizing violence: traditional recruitment patterns and resistance against conscription in Ottoman Syria', in *Arming the State*, edited by Erik J. Zürcher, 110–27. London: I.B. Taurus, 1999.

Downing, Brian M. *The Military Evolution and Political Change: Origins of Democracy and Autocracy in Early Modern Europe*. Princeton: Princeton University Press, 1992.

Duffy, Christopher. *Russia's Military Way to the West 1700–1800*. London: Routledge and Kegan Paul, 1981.

Dukes, Paul. *The Making of Russian Absolutism, 1613–1801*, 2nd edn. London: Longman, 1990.

Dykstra, Darrell. 'The French occupation of Egypt'. In M.W. Daly, ed., *Cambridge History of Egypt*, 113–38.

Edgerton, Robert B. *Death or Glory: The Legacy of the Crimean War*. Cambridge, MA: Westview Press, 2000.

Emin, Ahmed, *The Development of Modern Turkey as Measured by its Press*. New York: Columbia University Press, 1914.

Engelhardt, Edouard. *La Turquie et le Tanzimat, ou Histoire des Réformes dans l'Empire Ottoman*. Paris: A. Cotillon et Cie., 1882.

Erbakan, Cevat. *Osmanlı-Rus ve Avusturya Savaşları*. Istanbul: Askeri Matbaa, 1938.

Erdem, Hakan. ' "Do not think of them as agricultural labourers": Ottoman responses to the Greek war of independence'. In Thalia G. Dragonas and Faruk Birtek, eds., *Citizenship and the Wation-State in Greece and Turkey*, London: Routledge, 2005, 67–84.

Erdem, Hakan. 'Recruitment of "Victorious Soldiers of Muhammad" in the Arab Provinces, 1826–28'. In Israel Gershoni, Hakan Erdem, Ursula Woköck, eds., *Histories of the Modern Middle East: New Directions*. Boulder, CO: Lynne Riener, 2002, 189–204.

Erdem, Hakan. 'The wise old man, propagandist and ideologist: Koca Sekbanbaşı [sic] on the Janissaries, 1807', in K. Virtanen, ed.,

Individual, Ideologies and Society. Tampere: Peace Research Institute, 2000.

Ergut, Ferdan, 'State and social control: the police in the late Ottoman Turkey and the early Republican Turkey, 1839–1939', PhD, New York: New School for Social Research, 1999.

Erickson, Edward J. *Ordered to Die: A History of the Ottoman Empire. A History of the Ottoman Army in the First World War*, Westport, CT: Greenwood Press, 2001.

Esad, Mehmed, Efendi. *Précis Historique de la Destruction du Corps des Janissaires par le Sultan Mahmud, en 1826*. Traduit du turc par A.P. Caussin de Perceval. Paris, 1833.

Esad, Mehmed, Efendi. *Üss-ü Zafer*. Istanbul, 1827.

Esad, Mehmed, Efendi. *Vak'a-Nuvis Es'ad Efendi tarihi*. Istanbul: Osmanlı Araştırmaları Vakfi, 2000.

Eton, William. *A Survey of the Turkish Empire*. London, 1798.

Evans, Guy. 'George Stavrakoğlu and the rule of rüşvet', in K. Fleet, ed., 'The Ottoman Empire in the Eighteenth Century', *Oriente Moderno* 18 (79)n.s.1 (1999): 147–60.

Fahmy, Khalid. *All the Pasha's Men: Mehmed Ali, His Army and the Making of Modern Egypt*. Cambridge: Cambridge University Press, 1997.

Fahmy, Khalid. 'The era of Muhammad 'Ali Pasha, 1805–48'. In M.W. Daly, ed., *Cambridge History of Egypt*. Cambridge: Cambridge University Press, 1988: 139–79.

Fahmy, Khalid. 'The nation and its deserters: conscription in Mehmed Ali's Egypt'. In *Arming the State*, 59–77.

Farah, Caesar E. *The Politics of Interventionism in Ottoman Lebanon*. London: Centre for Arab Studies and I.B. Taurus, 2000.

Faroqhi, Suraiya. *The Ottoman Empire and the World Around It*. London: I.B. Tauris, 2004.

ffrench Blake, R.L.V. *The Crimean War*. London: Sphere, 1973.

Findley, Carter V. *Bureaucratic Reform in the Ottoman Empire 1789–1922*. Princeton, NJ: PUP, 1980.

Findley, Carter V. 'Ebu Bekir Ratib Efendi's Vienna embassy narrative: discovering Austria or propagandizing for reform in Istanbul?'

Wiener Zeitschrift für die Geschichte des Morganlandes 85 (1995): 40–80.

Finkel, Caroline. *The Administration of Warfare: Ottoman Military Campaigns in Hungary 1593–1606*. Vienna: VWGÖ, 1988.

Finkel, Caroline. *Osman's Dream*. London: John Murray, 2005; New York: Basic Books, 2006.

Finkel, Caroline. 'The provisioning of the Ottoman army during the campaigns of 1593–1606'. In *Habsburgisch–osmanische Beziehungen, CIEPO Colloque*. Vienna, 1983: 107–23.

Finlay, George, *History of the Greek Revolution and the Reign of King Otho*. Oxford: Clarendon Press, 1877, reprint, London: Zeno, 1971.

Fisher, Alan W. *A Precarious Balance: Conflict, Trade and Diplomacy on the Russian–Ottoman Frontier*. Istanbul: Isis, 1999.

Fisher, Alan W. *The Russian Annexation of the Crimea, 1772–83*. Cambridge: Cambridge University Press, 1970.

Fleming, K.E. *The Muslim Bonaparte: Diplomacy and Orientalism in Ali Pasha's Greece*. Princeton, NJ: Princeton University Press, 1999.

Florescu, Radu. *The Struggle against Russia in the Romanian Principalities: A Problem in Anglo-Turkish Diplomacy, 1821–54*. Iaşi: Centre for Romanian Studies, 1997.

Florovsky, A.V. 'Russo-Austrian conflicts in the early 18th century'. *Slavonic and East European Review* 47, no. 108 (1969): 94–114.

Fortna, Ben, *Imperial Classroom: Islam, the State and Education in the Late Ottoman Empire*. Oxford: Oxford University Press, 2002.

Frankland, Charles Colville. *Travels To and From Constantinople in the Years 1827 and 1829*. London: 1829.

Freely, John. *Istanbul: The Imperial City*. New York: Penguin, 1998.

Freeze, Gregory L. 'Institutionalizing piety: the Church and popular religion, 1750–1850'. In Jane Burbank and David L. Ransel, eds., *Imperial Russia: New Histories for the Empire*. Bloomington; IN: Indiana University Press, 1998.

Frost, Robert F. 'Napoleon's diplomats in the Danubian principalities: revolutionaries, opportunists, and spies'. In *Proceedings: Consortium on Revolutionary Europe* 20 (1990): 853–60.

Frost, Robert F. *The Northern Wars: War, State and Society in Northeastern Europe 1558–1721*. London: Longman, 2000.

Fuller, William G. *Strategy and Power in Russia, 1600–1914*. New York: Free Press, 1992.

Gammer, Moshe. 'Şamil and the Muslim powers: the Ottomans, the Qagjars and Muhammad 'Ali of Egypt'. In Raoul Motika, Michael Ursinus, eds., *Caucasia Between the Ottoman Empire and Iran, 1555–1914*. Wiesbaden: Dr Ludwig Reichert Verlag, 2000, 11–20.

Genç, Mehmet. 'L'économie ottomane et la guerre au XVIIIe siècle'. *Turcica* 27 (1995): 177–96.

Gershoni, I., H. Erdem and U. Wökock, eds. *Histories of the Modern Middle East: New Directions*. Boulder, CO: Lynne Riener, 2002.

Ghorbal, Shafik. 'The missions of Ali Effendi in Paris and of Sedki Effendi in London, 1797–1811'. *Bulletin of the Faculty of Arts, University of Egypt* 1 (1933): 114–29.

Göçek, Fatma Müge. *Rise of Bourgeoisie, Demise of Empire: Ottoman Westernization and Social Change*. Oxford: Oxford University Press, 1995.

Goffman, Dan. *The Ottoman Empire and Early Modern Europe*. Cambridge: Cambridge University Press, 2002.

Goldfrank, David M. *The Origins of the Crimean War*. London: Longman, 1994.

Goodwin, Godfrey. *The Janissaries*. London: Saqi Books, 1994.

Goodwin, Jason. *Lords of the Golden Horn: A History of the Ottoman Empire*. New York: Holt, 1998.

Gosu, Armand. 'The third anti-Napoleonic coalition and the Sublime Porte'. *International Journal of Turkish Studies* 9 (2003): 199–237.

Gouranis, Basil G. 'Social cleavages and national awakening in Ottoman Macedonia'. *East European Quarterly* 29 (1996): 409–26.

Gradeva, Rossitsa. 'Osman Pazvantoğlu of Vidin: between old and new'. In *The Ottoman Balkans*, Frederick F. Anscombe, ed. Princeton: Markus Wiener, 2006, 115–61.

Gradeva, Rossitsa. *Rumeli Under the Ottomans 15th–18th Centuries: Institutions and Communities*. Istanbul: Isis, 2004.

Gradeva, Rossitsa. 'Shipping along the lower course of the Danube (end of the 17th century)'. In Elizabeth Zachariadou, ed. *The Kapudan Pasha, His Office and His Domain*. Rethymnon, 2002: 301–24.

Gradeva, Rossitsa. 'War and peace along the Danube: Vidin at the end of the seventeenth century'. In *The Ottomans and the Sea*, Kate Fleet, ed. *Oriente Moderno* n.s. XX (LXXXI) no. 1 (2001): 149–75.

Grant, Jonathan. 'Rethinking the Ottoman "decline": military technology diffusion in the Ottoman Empire, fifteenth to eighteenth centuries'. *Journal of World History* 10 (1999): 179–201.

Greene, Molly, ed. *Minorities in the Ottoman Empire*. Princeton, NJ: Markus Wiener, 2005.

Greene, Molly. *A Shared World: Christians and Muslims in the Early Modern Mediterranean*. Princeton, NJ: Princeton University Press, 2000.

Greenwood, Anthony. 'Istanbul's meat provisioning: a study of the Celepkeşan system'. PhD, Chicago, 1988.

Gresh, Alain. 'Malevolent fantasy of Islam'. *Le Monde Diplomatique* August 2005, 16.

Groot, Alexander H. de. 'The Ottoman Mediterranean since Lepanto (October 7th 1571). Naval warfare during the seventeenth and eighteenth centuries'. *Anatolica* 20 (1994): 269–93.

Güran, Tevfik. 'The State role in the grain supply of Istanbul: the Grain Administration, 1793–1839'. *International Journal of Turkish Studies* 3 (1984). 27–41.

Hagen, Gottfried. 'The Prophet Muhammad as an exemplar in war: Ottoman views on the eve of World War 1'. *New Perspectives on Turkey* 22 (2000): 145–72.

Halaçoğlu, Yusuf. *XVIII. Yüzyılda Osmanlı İmparatorluğ'nun İskân Siyaseti ve Aşiretlerin Yerleştirilmesi* 2.baskı. Ankara: Türk Tarih Kurumu, 1991.

Hammer-Purgstall, Joseph von. *Histoire de l'Empire Ottomane Depuis son Origine Jusqu'à nos Jours*. J.J. Hellert, tr. 18 vols. Paris, 1835–43.

Harrington, Peter. 'The defence of Kars: paintings by William Simpson and Thomas Jones Barker'. *Journal of the Society for Army Historical Research* 227 (1991): 22–28.

Hastings, Adrian. *The Construction of Nationhood: Ethnicity, Religion and Nationalism*. Cambridge: Cambridge University Press, 1997.

Hathaway, Jane. *The Politics of Households in Ottoman Egypt: The Rise of the Qazdağlis*. Cambridge: Cambridge University Press, 1997.

Hathaway, Jane. *Rebellion, Repression, Reinvention: Mutiny in Comparative Perspective*. New York: Praeger, 2001.

Hegyi, Klara. 'The Ottoman military force in Hungary'. In *Hungarian–Ottoman Military and Diplomatic Relations in the Age of Süleyman the Great*, Géza Dávid and Pál Fodor, eds. Budapest: Loránd Eötvös University, 1994: 131–48.

Heinzelmann, Tobias. *Heiliger Kampf oder Landesverteidigung? Die Diskussion um die Einführung der allgemainen Militärpflicht im Osmanischen Reich 1826–56*. Frankfurt: Peter Lang, 2004.

Hellie, R.F. 'The Petrine Army: continuity, change and impact'. *Canadian–American Slavic Studies* 8 (1974): 237–53.

Hertz, A.Z. 'Ada Kale: the key to the Danube'. *Archivum Ottomanicum* 3 (1971): 170–84.

Hertz, A.Z. 'Ottoman conquest of Ada Kale 1738'. *Archivum Ottomanicum* 6 (1980): 151–209.

Hickok, Michael. *Ottoman Military Administration in Eighteenth Century Bosnia*. Leiden: Brill, 1997.

Hitzel, Frédéric. 'Défense de la place turque d'Oczakow par un officer du génie français (1787)'. In *İkinci Tarih Boyunca Karadeniz Kongresi Bildirileri*, Samsun, Turkey: 1990: 639–55.

Hitzel, Frédéric. ''Osmân Ağa, captif ottoman dans l'Empire des Habsbourg à la fin du XVIIe siècle', *Turcica* 33 (2001): 191–212.

Hitzel, Frédéric. *Prisonnier des Infidèles: Un Soldat Ottoman dans l'Empire des Habsbourgs*. Arles: Sindbad, 1998.

Hochedlinger, Michael. *Austria's Wars of Emergence: War, State and Society in the Habsburg Monarchy 1683–1797*. London: Longman, 2003.

Holt, P.M. *Egypt and the Fertile Crescent 1516–1922*. London: Longman, 1966.

Holt, P.M. *A Modern History of the Sudan*. London: Weidenfeld and Nicholson, 1967.

Horowitz, Richard S. 'International law and state transformation in China, Siam and the Ottoman Empire during the nineteenth century', *Journal of World History* 15 (2005): 445–86.

Hughes, B.P. *Open Fire: Artillery Tactics from Marlborough to Wellington*. Chichester: A. Bird, 1983.

Hurewitz, J.C. 'The Europeanization of Ottoman diplomacy: the conversion from unilateralism to reciprocity in the nineteenth century'. *Belleten* 25 (1961): 455–66.

Hurewitz, J.C., ed. *The Middle East and North Africa in World Politics: A Documentary Record*. 2 vols. New Haven: Yale University Press, 1975.

Imber, Colin. *Ebu's-su Su'ud: The Islamic Legal Tradition*. Stanford: Stanford University Press, 1997.

Imber, Colin. *The Ottoman Empire, 1300–1650: The Structure of Power*. London: Palgrave Macmilllan, 2002.

İnalcık, Halil. 'Military and fiscal transformation in the Ottoman Empire 1600–1700'. 6 (1980): 283–337.

İnalcık, Halil. *The Ottoman Empire*. London: Orion, 1973.

İnalcık, Halil. *Tanzimat ve Bulgar Meselesi*. Ankara: Türk Tarih Kurumu, 1943

İnalcık, Halil and Donald Quataert, eds. *An Economic and Social History of the Ottoman Empire 1300–1914*. Cambridge: Cambridge University Press, 1994.

Ingrao, Charles W. *The Habsburg Monarchy 1618–1815*. Cambridge: Cambridge University Press, 1994, 2nd edn, 2000.

Ingrao, Charles W. 'Habsburg strategy and geopolitics during the eighteenth century'. In G.E. Rothenberg, B. Kiràly and P.F. Sugar, eds. *East Central European Society and War in the Pre-revolutionary Eighteenth Century*. New York: Columbia University Press, 1982: 49–66.

Ismail, F. 'The making of the Treaty of Bucharest, 1811–12'. *Middle Eastern Studies* 15, no. 2 (1979): 163–92.

Israel, Fred L., ed. *Major Peace Treaties of Modern History*, vol. 2. New York: Chelsea House, 1967.

Ivanova, Svetlana, 'Sketches from the life of a Kapudan Pasha on the Danube'. In E. Zachariadou, ed. *The Kapudan Pasha, His Office and His Domain*. Rethymnon: Institute for Mediterranean Studies, 2002: 325–45.

Jelavich, Barbara. *History of the Balkans: Eighteenth and Nineteenth Centuries*. Vol. 1. Cambridge: Cambridge University Press, 1983.

Jennings, R.C. 'Firearms, bandits and gun-control: some evidence on Ottoman policy towards firearms in the possession of the Reaye, from judicial records of Kayseri, 1600–27'. *Archivum Ottomanicum* 6 (1980): 339–58.

Jersild, Austin Lee. 'From savagery to citizenship: Caucasian mountaineers and Muslims in the Russian Empire'. In Daniel R. Brower and Edward J. Lazzerini, eds. *Russia's Orient: Imperial Borderlands and People, 1700–1917*. Bloomington, IN, 1997: 101–14.

Jewsbury, George F. 'Comte de Langeron and his role in Russia's armies against the Porte and France, 1805–14'. *The Consortium on Revolutionary Europe 1750–1850. Proceedings*, vol. 1. Athens, GA: Consortium on Revolutionary Europe, 1980: 222–30.

Jewsbury, George F. *The Russian Annexation of Bessarabia: 1774–1828: A Study of Imperial Expansion*. Boulder, CO: Westview Press, 1976.

Jones, Robert. 'Opposition to war and expansion in late eighteenth century Russia', *Jahrbücher für Geschichte Osteuropas* 32 (1984): 34–51.

Juchereau de Saint-Denys, A. de. *Révolutions de Constantinople en 1807 et 1808: Précédées d'Observations Générales sur l'État Actuel de l'Empire Ottoman*. 2 vol. Paris: Brissot-Thivars, 1819.

Kaçar, Mustafa. 'Osmanlı İmparatorluğu'nda askeri teknik eğitimde modernleşme çalışmaları ve Mühendishanelerin kuruluşu (1808'e kadar)'. In *Osmanlı Bilimi Araştırmaları* (1998): 68–137.

Kafadar, Cemal. *Between Two Worlds*. Berkeley, CA: University of California Press, 1995.

Kagan, Frederick W. *The Military Reforms of Nicholas I: The Origins of the Modern Russian Army*. New York: St Martin's Press, 1999.

Kagan, Frederick W. 'Russia's small wars, 1805–61'. In W. Kagan and Rebin Higham, eds. *The Military History of Tsarist Russia*. New York: Palgrave/St Martin's, 2002: 122–36.

Kagan, Frederick W., and Robin Higham, eds. *The Military History of Tsarist Russia*. New York: Palgrave, 2002.

Kaiser, Thomas. 'The evil empire? The debate on Turkish despotism in eighteenth-century French political culture'. *Journal of Modern History* 72 (2000): 6–34.

Kàldy-Nagy, Gulda. 'The first centuries of the Ottoman military organization'. *Acta Orientalia Academiae Scientiarum Hungaricae* 31 (1977): 147–83.

Karal, Enver Ziya. 'Nizâm-ı Cedîd'e dâir lâyihalar', in *Tarih Vesikaları* 1 (1941/2), 414–25, and 2 (1942/3), 104–11, 342–51, 424–32.

Karal, Enver Ziya. 'Osmanlı Tarihine Dair Vesikalar'. *Belleten* 4 (1940): 175–89.

Karal, Enver Ziya. 'Zarif Paşa'nın Hatıratı, 1816–62', *Belleten* 4 (1940): 443–94.

Karateke, Hakan, and Maurus Reonkowski, eds. *Legitimizing the Order: The Ottoman Rhetoric of State Power*. Leiden: Brill, 2005.

Karpat, Kemal H. 'The *hijra* from Russia and the Balkans: the process of self-definition in the late Ottoman state'. In Dale F. Eickelman and James Piscatori, eds. *Muslim Travelers: Pilgrimage, Migration, and the Religious Imagination*'. Berkeley, CA: University of California Press, 1990: 131–52.

Karpat, Kemal H. 'The migration of the Bosnian Muslims to the Ottoman States, 1878–1914: an account based on Turkish sources', *International Journal of Turkish Studies* 10 (2004): 121–40.

Karpat, Kemal H. and Robert Zens, eds. *Ottoman Borderlands: Issues, Personalities and Political Changes*. Madison: University of Wisconsin Press, 2003. (Also published as vol. 9 of the *International Journal of Turkish Studies*.)

Kasaba, Reşat. 'A time and a place for the non-state: social change in the Ottoman Empire during the long nineteenth century', in Joel S. Migdal, Atul Kohli and Vivienne Shuh, eds. *State Power and Social Forces: Domination and Transformation in the Third World*. Cambridge: Cambridge University Press, 1994, 207–30.

Keep, John L.H. 'Feeding the troops: Russian Army supply policies during the Seven Years War'. *Canadian Slavonic Papers* 29 (1987): 24–44.

Keep, John L.H. *Soldiers of the Tsar: Army and Society in Russia, 1462–1874*. Oxford: Clarendon, 1985.

Kefeli, Agnès. 'Constructing an Islamic identity: the case of Elyshevo village in the nineteenth century'. In *Russia's Orient: Imperial Borderlands and People, 1700–1917*. 271–91.

Keralio, Louis Felix. *Histoire de la Guerre Entre la Russie et la Turquie Particulièrement de la Campaigne de 1769*. 2 vols. St Petersburg, 1773.

Kesbi, Mustafa. *Ibretnümâ-y-Devlet (Tahlil ve Tenkitli Metin)*. Ahmet Öğreten, ed. Ankara, 2002.

Kesbi, Mustafa. *Sbornik Nniekotorykh Vazhnykh Izviestii I Ofitsial'nykh Dokumentov Kasatel'no Turtsii, Rossi I Kryma. Devlet-i 'Aliye iyle Rusya Devleti ve Kirim Hakkinda Baze Malumat Muhimme ve Tahrirat Resmiye-yi havi Mecmu'a dir*. By V.D. Smirnov, ed. (St Petersburg, 1881)

Khodarkovsky, Michael. 'Not by word alone: missionary policies and religious conversion in early modern Russia'. *Comparative Studies in Society and History* 38 (1996): 267–93.

Khodarkovsky, Michael. *Russia's Steppe Frontier: The Making of a Colonial Empire, 1500–1800*. Bloomington, IN: Indiana University Press, 2004.

Khoury, Dina Rizk. *State and Provincial Society in the Ottoman Empire, 1540–1834*. Cambridge: Cambridge University Press, 1997.

King, Charles. *The Black Sea*. Oxford: Oxford University Press, 2004.

Kırlı, Cengiz, 'The struggle over space: the coffeehouses of Istanbul, 1700–1845', PhD, State University of New York, Binghamton, 2000.

Kohut, Zenon. *Russian Centralism and Ukranian Autonomy*. Cambridge, MA: Harvard University Press, 1988.

Kolipoulous, John. 'Brigandage and insurgency in the Greek domains of the Ottoman Empire, 1853–1908'. In Dmitri Gondicas and Charles Issawi, eds. *Ottoman Greeks in the Age of Nationalism*. Princeton, NJ: Darwin Press, 1999: 143–60.

Kortepeter, C. Max. 'Ottoman military reform during the late Tanzimat (1885–95): the Prussian General von der Goltz and the Ottoman Army', In *V. Milletlerarası Türkiye Sosyal ve İktisat Tarihi Kongresi, Tebliğler*. Ankara: Türk Tarih Kurumu, 1900, 174–86.

Küçük, Hülya. *The Role of the Bektāshīs in Turkey's National Struggle*. Leiden: Brill, 2002.

Küçükerman, Önder. *Türk Giyim Sanayii Tarihindeki Ünlü Fabrika 'Feshane' Defterdar Fabrikası*. Ankara: Sümerbank, 1988.

Kunt, Metin, and Christine Woodhead, eds, *Süleyman the Magnificent and his Age*. London: Longman, 1995.

Kuran, Ercüment. *Avrupa'da Osmanlı Elçiliklerinin Kuruluşu ve İlk Elçilerin Siyasi Faâliyetleri 1793–1821*. Ankara: Türk Kulturu Araştırma Enstitüsü, 1960.

Kurat, Akdes Nimet. *Prut Seferi ve Barışı 1123* (1711). Ankara: Türk Tarih Kumuru, 1951.

Kutluoğlu, Muhammad H. *The Egyptian Question, 1831–41: the Expansionist Policy of Mehmed Ali Paşa in Syria and Asia Minor and the Reaction of the Sublime Porte*. Istanbul: Eren, 1998.

Kütükoğlu, Mübahat S. 'Sultan II. Mahmud devri yedek ordusu Redîf-i Asâkir-i Mansûre', İ.Ü. *Edebiyat Fakültesi, Tarih Enstitüsü Dergisi* 12 (1982): 127–57.

Lake, Colonel Atwell, *Narrative of the Defence of Kars, Historical and Military*. London: Richard Bentley, 1857.

Lane-Poole, Stanley. *The Life of the Right Honourable Stratford Canning*. 2 vols. London: Longmans, Green, 1888.

Langensiepen, Bernd and Ahmet Güleryüz. *The Ottoman Steam Navy, 1828–1923*. Ed. and tr. by James Cooper. London: Conway Maritime Press, 1995.

Laoust, H. 'Ibn 'Abd al-Wahhab'. In *Encyclopedia of Islam*. Leiden: Brill, 1992.

Laurens, H. *Les Origines Intellectuelles de l'Éxpedition d'Egypte: L'orientalisme Islamisant en France (1698–1798)*. Istanbul: Isis, 1987.

Laugier, Marc Antione. *The History of The Negotiations for the Peace Concluded at Belgrade September 18, 1839. Between the Emperor,*

Russia and the Ottoman Porte, by the Mediation and under the Guarantee of France, Shewing the Grounds of the Present War between the Russians and the Turks. Tr. from the French of M. L'Abbé Laugier. London: J. Muray, 1770.

Lawson, Fred H. *The Social Origins of Egyptian Expansionism During the Muhammad 'Ali Period*. New York: Columbia University Press, 1992.

LeDonne, John P. *The Russian Empire and the World, 1700–1917: The Geopolitics of Expansion and Containment*. New York: Oxford University Press, 1997.

Levy, Avigdor. 'The contribution of Zaporozhian Cossacks to Ottoman military reform: documents and notes'. *Harvard Ukrainian Studies* 6 (1982): 372–413.

Levy, Avigdor. 'Formalization of Cossack service under Ottoman rule'. In *East Central European Society and War in the Pre-revolutionary Eighteenth Century*. 491–505.

Levy, Avigdor. 'The military policy of Sultan Mahmud II, 1808–39'. PhD, Harvard, 1968.

Levy, Avigdor. 'The officer corps in Sultan Mahmud II's new army, 1826–39'. *International Journal of Middle East Studies* 2 (1971): 21–39.

Levy, Avigdor. 'Ottoman attitudes to the rise of Balkan nationalism'. In B.K. Király, ed. *War and Society in East Central Europe*. New York: Columbia University Press, 1979: 325–45.

Lewis, Bernard. *The Emergence of Modern Turkey*. London: Oxford University Press, 1961.

Lewis, Bernard. *The Muslim Discovery of Europe*. New York: Norton, 1982.

Lewis, Bernard. *What Went Wrong? Western Impact and Middle Eastern Response*. New York: Oxford University Press, 2001.

Lieven, Dominic. 'Dilemmas of empire 1850–1918: power, territory, identity'. *Journal of Contemporary History* 34 (1999): 163–200.

Lieven, Dominic. *Empire: The Russian Empire and its Rivals*. New Haven, CT: Yale University Press, 2000.

Lynn, John A. *Giant of the Grand Siècle: The French Army 1610–1715*. Cambridge: Cambridge University Press, 1997.

McConachy, Bruce. 'The roots of artillery doctrine: Napoleonic artillery tactics reconsidered'. *The Journal of Military History* 65 (2001): 617–40.

Macfarlane, Charles. *Turkey and its Destiny: The Result of Journeys Made in 1847 and 1848 to Examine the State of that Country.* Philadelphia: Lea and Blanchard, 1850.

Macfarlane, Charles. *Constantinople in 1828: A Residence of Sixteen Months in the Turkish Capital and Provinces, with an Appendix to the Autumn of 1829*, 2nd edn. 2 vols. London: Saunders and Otley, 1829.

Macfie, A.L. *The Eastern Question 1774–1923*. Revised edn. London: Longman, 1996.

McNeill, William H. *Europe's Steppe Frontier, 1500–1800*. Chicago: University of Chicago Press, 1966.

McNeill, William H. *Keeping Together in Time*. Cambridge, MA: Harvard University Press, 1995.

McNeill, William H. *The Pursuit of Power: Technology, Armed Force and Society since AD 1000*. Chicago: University of Chicago Press, 1982.

McNeill, William H. *Venice: The Hinge of Europe, 1081–1797*. Chicago: University of Chicago Press, 1974.

Madariaga, Isabel de. *Russia in the age of Catherine the Great*. New Haven: Yale University Press, 1981.

Makdisi, Ussama. *The Culture of Sectarianism: Community, History and Violence in Nineteenth Century Ottoman Lebanon*. Berkeley: University of California Press, 2000.

Mansel, Philip. *Constantinople: City of the World's Desire, 1453–1924*. New York: St Martin's Press, 1995.

Maʿoz, Moshe, *Ottoman Reform in Syria and Palestine 1840–1961*. Oxford: Clarendon Press, 1968.

Mardin, Şerif. 'The just and the unjust'. *Daedalus* 120 (1991): 113–29.

Marmont, Auguste Frédéric, Marshal. *The Present State of the Turkish Empire*. Translated by Frederic Smith. London: John Ollivier, 1839.

Marsigli, Comte de. *L'Etat Militaire de l'Empire Ottoman*. 2 vols. La Haye, 1732.

Marsot, Afaf Lutfi al-Sayyid. *Egypt in the Reign of Muhammad Ali.* Cambridge: Cambridge University Press, 1984.

Masters, Bruce. *Christians and Muslims in the Ottoman Arab World: The Roots of Sectarianism.* Cambridge: Cambridge University Press, 2001.

Mayer, Matthew Z. 'Joseph II and the campaign of 1788 against the Ottoman Turks'. MA, McGill University, 1997.

Mayer, Matthew Z. 'The price for Austria's security: part I: Joseph II, the Russian Alliance, and the Ottoman War, 1787–89', *International History Review* 26 (2004): 257–99.

Mayer, Matthew Z. 'The price for Austria's security: part II: Leopold II, the Prussian threat, and the peace of Sistove, 1790–71', *International History Review* 26 (2004): 473–514.

Meeker, Michael. *A Nation of Empire: The Ottoman Legacy of Turkish Modernity.* Berkeley, CA: University of California Press, 2002.

Menning, Bruce W. 'G.A. Potemkin and A.I. Chernyshev: two dimensions of reform and Russia's military frontier', in *Reforming the Tsar's Army the Great to the Revolution*, eds. David Schimmelpenninck van der 2004: 273–91.

Menning, Bruce W. 'The Imperial Russian Army, 1725–96'. In *The Military History of Tsarist Russia*, eds. Frederick W. Kagan and Robin Higham. New York: Palgrave/St Martin's, 2002: 47–75.

Menning, Bruce W. 'Paul I and Catherine II's military legacy, 1762–1801'. In *The Military History of Tsarist Russia*: 77–105.

Menning, Bruce W. 'Russian military innovation in the second half of the eighteenth century'. *War and Society* 2 (1984): 23–41.

Mikhailovsky-Danilevsky, Alexander. *Russo-Turkish War of 1806–12.* Translated by Alexander Mikaberidze. vol. 1–2. West Chester, OH: The Nafziger Collection, 2002.

Moltke, Helmuth von. *Essays, Speeches and Memoirs.* New York: Harper and Brothers, 1893.

Moltke, Helmuth von. *Lettres de Maréchal de Moltke sur l'Orient*, traduites par Alfred Marchand. 2e éd. Paris: Librarie Fischbacher, 1872.

Moltke, Helmuth von. *The Russians in Bulgaria and Rumelia in 1828 and 1829, During the Campaigns of the Danube, the Sieges of*

Brailow, Varna, Silistria, Shumla, and the Passage of the Balkans. Translated by Marshal Diebitch. London: John Murray, 1854.

Montefiore, Simon Sebag. *Prince of Princes: The Life of Potemkin.* London: Weidenfeld and Nicolson, 2000.

Monteith, William. *Kars and Erzeroum: With the Campaigns of Prince Paskiewitch in 1828 and 1829*. London: Longman, Brown, Green, 1856.

Moreau, Odile. 'Bosnian resistance against conscription in the nineteenth century'. In *Arming the State*: 129–37.

Morier, John Philip. *Memoir of a Campaign with the Ottoman Army in Egypt*. London, 1801.

Mostashari, Firouzeh. *On the Religious Frontier: Tsarist Russia and Islam in the Caucasus*. London: I.B. Tauris, 2005.

Murphey, Rhoads. 'Continuity and discontinuity in Ottoman administrative practice during the late seventeenth century'. *Poetics Today* 14 (1993): 425.

Murphey, Rhoads. 'Ottoman resurgence in the seventeenth-century Mediterranean'. *Mediterranean Historical Review* 8 (1993): 186–200.

Murphey, Rhoads. *Ottoman Warfare, 1500–1700*. London: University College Press, 1999.

Murphey, Rhoads. 'The functioning of the Ottoman army under Murad IV (1632–39/1032–49): key to the understanding of the relationship between center and periphery in seventeenth century Turkey. (PhD thesis, Chicago, 1979).

Müteferrika, İbrahim. *Traité de Tactique* [Usul ul-Hikem fi Nizam ül-Ümem]. Translated by K.I.S. Revicsky. Vienna, 1769.

Müteferrika, Ibrahim. *Usul ül-Hikem fi Nizam ül-Ümen*. Istanbul, 1731.

Naff, Thomas, 'The Ottoman Empire and the European states system'. In Hedley Bull and Adam Watson, eds. *The Expansion of International Society*. Oxford, UK: Oxford University Press, 1984: 143–69.

Nagata, Yuzo, *Muhsinzade Mehmed Paşa ve Ayanlık Müessesesi*. Tokyo: Institute for the Study of Languages and Cultures of Asia and Africa, 1976.

Napoleon in Egypt: Al-Jabarti's Chronicle on the French Occupation of Egypt, 1798, tr. by Shmuel Moreh. Princeton: Princeton University Press, 1993.

Nezir-Akmeşe, Handan. *The Birth of Modern Turkey: The Ottoman Military and the March to WWI*. London: I.B. Tauris, 2005.

Nieuwenhuis, Tom. *Politics and Society in early Modern Iraq: Mamlūk Pashas, Tribal Shayhs and Local Rule Between 1802 and 1831*. The Hague: Martinus Nijhoft, 1982.

Oğuzoğlu, Yusuf. 'Osmanlı Şehirlerindeki Askerlerin Ekonomik Durumuna İlişkin Bazı Bilgiler'. In *Birinci Askeri Tarihi Semineri Bildirileri*. Ankara: Genelkurmay Basımevi, 1983.

Oleinikov, Dmitrii. 'The Caucasus factor in Russian military reform', in D. Schimmelpenninck van der Oye and B. Menning, eds., *Reforming the Tsar's Army: Military Innovation in Imperial Russia from Peter the Great to the Revolution*. Cambridge: Cambridge University Press, 2004, 205–31.

Olson, Robert W. *The Siege of Mosul and Ottoman–Persian Relations 1718–43*. Vol. 124, *Uralic and Altaic Series*. Bloomington, IN: Indiana University Press, 1975.

Orhonlu, Cengiz. 'The geography of Wallachia written by a Turkish politician'. *Revue des Études Sud-est Europeénnes* 13 (1975): 447–52.

Osman of Temeşvar, *Autobiographie der Dolmetschers Osman Ağa aus Temeswar*. Cambridge: Gibb Memorial Trust, 1980.

Osman of Temeşvar, *Kendi kalemiyle Temeşvarlı Osman Ağa*, Harun Tolasa, ed. Konya: Selçuk Universitesi, 1986.

Ostapchuk, Victor and Caroline Finkel, 'The archeology and construction history of the Black Sea fortress of Özü', *Muqarnas* 2005: 150–88.

Özcan, Abdülkadir. 'II. Mahmud'un memleket gezileri' in *Prof. Dr. Bekir Kütükoğlu'na Armaganı*. İstanbul: İstanbul Üniversitesi Edebiyat Fakültesi: 361–442.

Özcan, Abdülkadir, ed., *Eyyubî Efendi Kanûnnâmesi*. Istanbul: Eren, 1994.

Özcan, Besim. 'Tatarcık Abdullah Efendi ve ıslahatlarla ilgili lâyihası'. *Türk Kültürü Araştırmaları* 25 (1987): 55–64.

Özkaya, Yücel. 'Canikli Ali Paşa'. *Belleten* 144 (1972): 483–525.

Özoğlu, Hakan. *Kurdish Notables and the Ottoman State: Evolving Identities, Competing Loyalties, and Shifting Boundaries*. Albany: SUNY Press, 2004.

Pacey, Arnold. *Technology in World Civilization*. Cambridge, MA: MIT Press, 1990.

Pamuk, Şevket. *A Monetary History of the Ottoman Empire*. Cambridge: Cambridge University Press, 2000.

Pamuk, Şevket. 'Prices in the Ottoman Empire, 1469–1914'. *International Journal of Middle East Studies* 36 (2004): 451–68.

Panaite, Viorel, *The Ottoman Law of War and Peace: The Ottoman Empire and Tribute Payers*. Boulder, CO: East European Monographs, 2000.

Panaite, Viorel, 'The *Re'ayas* of tributary-protected principalities: the 16th through to the 18th centuries'. *International Journal of Turkish Studies* 9 (2003): 79–104.

Papoulia, Vasiliki. 'The impact of *Devshirme* on Greek society'. In *East Central European Society and War in the Pre-Revolutionary Eighteenth Century*.

Parry, V.J. and Yapp, M.E., eds. *War, Technology and Society in the Middle East*. London: Oxford University Press, 1975.

Parvev, Ivan. *Habsburgs and Ottomans between Vienna and Belgrade (1683–1739)*. Boulder, CO: East European Monographs, 1995.

Peirce, Leslie. *The Imperial Harem*. New York: Oxford University Press, 1993.

Perjes, Geza. 'Army provisioning, logistics and strategy ın the second half of the seventeenth century'. *Acta Historica* 16 (1970): 1–51.

Petmezas, Socrates D. 'Christian communities in eighteenth- and early nineteenth-century Ottoman Greece: their fiscal function'. In Molly Greene, ed. *Minorities in the Ottoman Empire*: Princeton, NJ: Markus Wiener, 2005: 71–126.

Petrosjan, Irina E. 'The *Madba-i kanun-i yeniçeri ocaği tarihi* on the system of *devşirme*'. In György Kara, ed. *Between the Danube and the Causcasus*. Budapest: Anademiai Kiadó, 1987: 217–27.

Philipp, Thomas. *Acre: the Rise and Fall of a Palestinian City, 1730–1931*. New York, NY: Columbia University Press, 2001.

Pingaud, L. *Choiseul-Gouffier: La France en Orient sous Louis XVI*. Paris: Picard, 1887.

Pinson, Mark. 'Ottoman Bulgaria in the first Tanzimat period – the revolts in Nish (1841) and Vidin (1850)', *Middle Eastern Studies* 1975, 11: 103–46.

Piterberg, Gabriel. *An Ottoman Tragedy: History and Historiography at Play*. Berkeley: University California Press, 2003.

Poe, Marshall. 'The consequences of the military revolution in comparative perspective'. *Comparative Studies in Society and History* 38 (1996): 603–18.

Pollock, Sean. 'Empire by invitation: Russian empire-building in the reign of Catherine II'. PhD, Harvard, 2006.

Ponting, Clive. *The Crimean War*. London: Chatto and Windus, 2004.

Porter, Sir James. *Turkey. Its History and Progress, from the Journals and Correspondence of Sir James Porter, Continued to the Present Time, by his Grandson* Sir George Larpent. 2 vols. London: Hurst and Blackett, 1854.

Prevelakis, E., and K. Kalliataki Mertikopoulou, eds. *Epirus, Ali Pasha and the Greek Revolution: Consular Reports of William Meyer from Preveza*. Athens: Academy of Athens, 1996.

Puryear, Vernon J. *France and the Levant: From Bourbon Restoration to the Peace of Kutiah*. Berkeley, CA: University of California Press, 1941, reprint New York: Archon Books, 1968.

Quataert, Donald. *The Ottoman Empire 1700–1922*. Cambridge: Cambridge University Press, 2000.

Radushev, Evgenii. 'Les dépenses locales dans l'Empire Ottoman au XVIIIe siècle'. *Etudes Balkaniques* 16, no. 3 (1980): 74–94.

Radushev, Evgenii. 'Ottoman border periphery (*serhad*) in the Nikopol Vilayet, first half of the sixteenth century'. *Etudes Balkaniques* (1995): 141–160.

Raif, Mahmud, Efendi. *Osmanlı İmparatoluğu'nda Yeni Nizamların Cedveli*. Arslan Terzioğlu and Hüsrev Hatemi. Istanbul, [1798].

Raif, Mahmud, Efendi. *Osmanlı İmparatorluğu 'nda Yeni Nizamların Cedveli*. Tr. And ed. by Arslan Terzioğlu and Hüsrev Hatemi. Istanbul: Türkiye Turing ve Otomobil Kurumu, n.d., includes French original, *Tableau des nouveaux reglemens de l'Empire ottoman*. Constantinople, 1798.

Raif, Mahmud, Efendi. *Mahmud Râif Efendi ve Nizâm-I Cedid'e Dair Eseri*. Kemal Beydilli and İlhan Şahin, eds. Ankara, Türk Tarih Kurumu, 2001, includes a much better French facsimile and translation.

Raymond, André, *Le Caire des Janissaires: l'Apogée de la Ville Ottomane sous 'Abd al-Rahman Katkhud*. Paris: CNRS Editions, 1995.

Raymond, André. 'Soldiers in trade: the case of Ottoman Cairo', *British Journal of Middle Eastern Studies* 18 (1991): 16–37.

Reed, Howard Alexander. 'The destruction of the Janissaries by Mahmud II in June, 1826'. PhD, Princeton, 1951.

Reid, James J. *Crisis of the Ottoman Empire: Prelude to Collapse, 1839–78*. Stuttgart: Franz Steiner, 2000.

Reinkowski, Maurus. 'The state's security and the subject's prosperity', in H.T. Karateke and M. Reinowski, eds., *Legitimizing the Order: The Ottoman Rhetoric of State Power*. Leiden: Brill, 2005.

Rogers, Clifford J. *The Military Revolution Debate: Readings on the Military Transformation of Early Modern Europe*. Boulder, CO: Westview, 1995.

Roider, Karl A. *Austria's Eastern Question 1700–90*. Princeton: Princeton University Press, 1982.

Roider, Karl A. 'Futile peacemaking: Austria and the Congress of Nemirov (1737)'. *Austrian History Yearbook* 12/13 (1976/77): 95–107.

Roider, Karl A. *The Reluctant Ally: Austria's Policy in the Austro-Turkish War, 1737–39*. Baton Rouge, LA: Louisiana State University, 1972.

Rolt, Richard, ed. *Memoirs of the Life of the Late Right Honorable John Lindesay, Earl of Craufurd*. London, 1753.

Rood, Judith Mendelsohn. 'Mehmed Ali as mutinous khedive: the roots of rebellion', *International Journal of Turkish Studies* 8 (2002): 115–28.

Rothenberg, Gunther E. *The Army of Francis Joseph*. Purdue, IN: Indiana University Press, 1976.

Rothenberg, Gunther E. 'The Habsburg military border system: some reconsiderations'. In Béla K. Király and Gunther E. Rothenberg, eds.

War and Society in East Central Europe, vol. 1: Special Topics and Generalizations on the 18th and 19th centuries. New York: Columbia University Press, 1979: 361–92.

Rothenberg, Gunther E. *The Military Border in Croatia, 1740–1881*. Chicago: University of Chicago Press, 1966.

Rothenberg, Gunther E. *Napoleon's Great Adversary: Archduke Charles and the Austrian Army, 1792–1814*. New York: Sarpedon, reprint 1995.

Rothenberg, Gunther E., Béla K. Király and Peter F. Sugar, eds., *East Central European Society and War in the Pre-Revolutionary Eighteenth Century*. New York: Columbia University Press, 1979.

Roudometof, Victor. 'From rum *Millet* to Greek Nation: enlightenment, secularization, and national identity in Ottoman Balkan society, 1453–1821', *Journal of Modern Greek Studies* 16 (1998): 11–48.

Rumiantsev, P.A., ed. *Sbornik dokumentov* [Collected documents] 2 vols. Moscow, 1953. [SIR10].

Russian–Muslim Confrontation in the Caucasus: Alternative Visions of the Conflict between Imam Shamil and the Russians, 1830–59, Thomas Sanders, Ernest Tucker and Gary Hamburg, eds. London: Routledge, 2004.

Russkie polkovodtsy: Dokumenty i materialy, vol. 2: Generalissimus Suvorov, ed. N. Korobko (Leningrad, 1947).

Saab, Ann Pottinger *The Origins of the Crimean Alliance*. Charlottesville, VA: University Press of Virginia, 1977.

Sadat, Deena R. 'Rumeli ayanlari: the eighteenth century'. *Journal of Modern History* 44 (1972): 346–63.

Safrastjan, Ruben. 'Ottomanism in Turkey in the epoch of reforms in XIX c: ideology and policy I'. *Etudes Balkaniques* 24, no. 4 (1988): 72–86.

Sahillioğlu, Halil. 'The Income and the expenditure of the Ottoman Treasury Between 1683 and 1740'. In *Studies on Ottoman Economic and Social History*. Istanbul: Organisation of the Islamic Conference (IRCICA), 1999, 65–82.

Şahin, Canay. 'The economic power of Anatolian *ayan* in the late 18th century: the case of the Caniklizâdes', *International Journal of Turkish Studies* 11 (2005): 28–47.

Sajdi, Dana. 'Peripheral visions: the worlds and worldviews of commoner chronicles in the 18th century Ottoman Levant'. PhD, Columbia, 2002.

Şakul, Kahraman. 'Osmanlı Askerî Tarihi Üzerine Bir Literatur Değerlendirmesi', *Türkiye Araştırmaşları Literatur Dergisi* 1 (2003): 529–71.

Salim, Necati. *Prut [1711]: Türk Ordusunun Eski Seferlerinden bir İmha Muharebesi*. Istanbul, 1931.

Salzmann, Ariel. *Tocqueville in the Ottoman Empire: Rival Paths to the Modern State*. Leiden: Brill, 2004.

Salzmann, Ariel. 'Toward a comparative history of the Ottoman state, 1450–1850'. In *Proceedings of the XIIth Congress of CIEPO*, 1996: 351–66.

Sarı Mehmed, Defterdar. *Ottoman Statecraft [Nasaih ül-Vüzera ve ül-Ümera]*. by W.L. Wright, ed. Princeton, NJ: Princeton University Press, 1935.

Sarkisyanz, Emanuel. 'Russian imperialism reconsidered'. In *Russian Imperialism from Ivan the Great to the Revolution*, edited by T. Hunczak, 45–81. New Brunswick, NJ: Rutgers University Press, 1974; also Lanham, MD: University Press of America, 2000.

Schaebler, Birgit. 'Civilizing others: global modernity and the local boundaries (French/German, Ottoman and Arab) of savagery'. In Birgit Schaebler and Leif Stenberg, eds. *Globalization and the Muslim World: Culture, Religion and Modernity*. Syracuse: Syracuse University Press, 2004.

Schilcher, Linda Schatkowski. *Families in Politics: Damascene Factions and Estates of the 18th and 19th Centuries*. Stuttgart: Franz Steiner, 1985.

Schimmelpenninck van der Oye, David, and Bruce Menning, eds. *Reforming the Tsar's Army: Military Innovation in Imperial Russia from Peter the Great to the Revolution*. Cambridge: Cambridge University Press, 2004.

Schmidt, Jan. 'The adventure of an Ottoman horseman: the autobiography of Kabudlı Vasfī Efendi, 1800–25', in *The Joys of Philology: Studies in Ottoman Literature, History and Orientalism (1500–1923)*. Istanbul: Isis Press, 2002, vol. 1: 165–286.

Schroeder, Paul W. *The Transformation of European Politics 1763–48.* Oxford: Oxford University Press, 1994.

Scott, H.M. *The Emergence of the Eastern Powers, 1756–75.* Cambridge: Cambridge University Press, 2001.

Scott, H.M. 'Russia as a European power'. In R. Bartlett and J.M. Hartley, eds. *Russia in the Age of the Enlightenment: Essays for Isabel de Madariaga.* Basingstoke: Macmillan, 1990: 7–39.

Shaw, S.J. *Between Old and New: The Ottoman Empire under Sultan Selim III, 1789–1807.* Cambridge, MA: Harvard University Press, 1971.

Shaw, S.J. 'The established Ottoman army corps Under Selim III', *Der Islam* 40 (1965), 142–84.

Shaw, S.J. 'The origins of Ottoman military reform: the Nizâm-i Cedid of Sultan Selim III', *Journal of Modern History* 37 (1965), 291–306.

Shaw, S.J. and Ezel Shaw. *History of the Ottoman Empire and Modern Turkey*, 2 vols. Cambridge: Cambridge University Press, 1976–77.

Shay, Mary L. *The Ottoman Empire from 1720 to 1734 as Revealed in the Dispatches of the Venetian Baili.* Urbana, IL: University of Illinois Press, 1944.

The Siege of Kars, 1855. Defence and Capitulation Reported by General Williams. London: The Stationery Office, 2000.

Şimşek, Veysel. 'Ottoman military recruitment and the recruit: 1826–53'. MA, Bilkent 2005.

Şirin, Veli. *Asâkir-i Mansure Ordusu ve Seraskerlik.* Anhara: Tarih ve Tabiat Vakfi Yayınları, 2002.

Skene, James H. *The Three Eras of Ottoman History: A Political Essay of the Late Reforms of Turkey* (London, 1851).

Slade, Adolphus. *Record of Travels in Turkey, Greece, etc.* New edn. London: Saunders and Otley, 1854.

St. Rémy, S. de *Memorial de l'Artillerie*, 2 vols. Paris, 1693; Amsterdam, 1702.

Stein, Mark. 'Seventeenth Century Ottoman Fortresses and Garrisons on the Habsburg Frontier', PhD Thesis, University of Chicago, 2001.

Stevens, Carol Belkin. *Soldiers on the Steppe: Army Reform and Social Change in Early Modern Russia*. Dekalb: Northern Illinois University Press, 1995.

Stoianovich, T. 'The conquering Balkan Orthodox merchant'. *Journal of Economic History* 20 (1960): 234–313.

Stoye, John. *Marsigli's Europe, 1680–1730: The Life and Times of Luigi Ferdinando Marsigli, Soldier and Virtuoso*. New Haven: Yale University Press, 1993.

Stoye, John. *The Siege of Vienna*. New York: Holt, Rinehart and Winston, 1964.

Sugar, Peter. 'A near perfect military society: the Ottoman Empire'. In *War: A Historical, Political and Social Study*, ed. L.L. Farrar. Santa Barbara, CA: University of California Press, 1978.

Sugar, Peter. *Southeastern Europe under Ottoman Rule 1354–1804, A History of East Central Europe, 5*. Seattle, WA: University of Washington, 1977.

Süleyman Penah Efendi, 'Mora İhtilâli veya Penah Ef. Mecmuası', ed. Aziz Berker. *Tarih Vesikaları* 2 (in four parts) (1941–43).

Sumner, B.H. *Peter the Great and the Ottoman Empire*. Hamden, CT: Archon Books, 1965.

Swanson, Glen W. 'War, technology and society in the Ottoman Empire from the reign of Abülhamid II to 1913: Mahmud Şevket and the German military mission' In *War, Technology and Society*: 366–85.

Tabakoğlu, Ahmet. *Gerileme Dönemine Girerken Osmanlı Maliyesi*. Istanbul: Dergâh, 1985.

Tacan, Necati. 'Tanzimat ve Ordu', In *Tanzimat I*, 129–37. Istanbul: Maarif Matbaası, 1940: 129–37.

Tatarcıkzade Abdullah Efendi, [Layiha] *Tarih-i Osmani Encümeni Mecmuası*, no. 41: 257–84; no. 42: 321–46, no. 43: 15–34 (1916/17).

Temperley, Harold. *England and the Near East: The Crimea*. New York: Longman, 1936; North Haven, CT: Archon Books, 1964.

Thornton, Thomas. *The Present State of Turkey*. London, 1807.

Tilly, Charles. *Coercion, Capital and European States, AD 990–1990*. Oxford: Blackwell, 1990.

Toledano, Ehud. 'The emergence of Ottoman-local elites (1700–1900): a framework for research'. In Ilan Pappé and Moshe Ma'oz, eds. *Middle Eastern Politics and Ideas: A History from Within*. London and New York: Tauris, 1997: 145–62.

Tott, Baron Francis de. *Memoirs of Baron de Tott*, 2 vols. London, 1785; New York: Arno Press, 1973 [1785].

Turgay, A.U. 'An aspect of Ottoman–Russian commercial rivalry: confiscation of Ottoman merchant ships in the Black Sea'. In Heath W. Lowry and Ralph S. Hattox, eds., *Proceedings of the IIIrd Congress on the Social and Economic History of Turkey*: 59–65, Istanbul, 1990: 59–65.

Türk Silahlı Kuvvetleri Tarihi Osmanlı Devri vol. 2, pt. 4: *Ikinci Viyana Kuşatmasından Nizâm-ı Cedidin Teşkiline Kadar Olan Devre*. Ankara: 1982.

Türkgeldi, Ali Fuat. *Mesâil-i Mühimme-i Siyâsiyye*. Bekir Sıtkı Baykal, ed. 2nd edn, 4 vols. Ankara: Türk Tarih Kurumu, 1987.

Unat, Faik Reşat. *Osmanlı Sefirleri ve Sefaretnameleri*. Ankara: Türk Tarih Kurumu, 1968.

Ungermann, Richard. *Der Russisch–türkische Krieg 1768–74*. Vienna: W. Braumüller, 1906.

Urquhart, David. *The Military Strength of Turkey*. London: Diplomatic Review Office, 1868.

Urquhart, David. *Turkey and its Resources*. London: Saunders and Otley, 1833.

Uyar, Mesut, and A. Kadir Varoğlu, 'In search of modernity and rationality: the case of [the] Turkish military academy', unpublished paper.

Uzun, Ahmet. '1257/1841 tarihılı bir belgeye göre Osmanli devleti'nde mevcut olan askeri birlikler ve bunlara yapılan harcamaların türü ve miktarları', *Cumhuriyet Üniversitesi Sosyal Bilimler Dergisi* 25 (2001): 235–45.

Uzunçarşılı, İ. Hakkı. *Osmanlı Devleti Teşkilatından Kapukulu Ocakları*, 2 vols. Ankara: Türk Tarih Kurumu, 1944.

Vasıf, Ahmed, *Précis Historique de la Guerre des Turcs Contre les Russes*. Paris, 1872.

Vernadsky, George et al., eds. A *Source Book for Russian History from Early Times to 1917*. New Haven: Yale University Press, 1972.

Volney, C.F.C. *Voyage en Egypte et en Syrie pendant les années 1783, 1784, 1785*. Paris, 1787.

Voltaire, François Marie. *Russia under Peter the Great*, translated M.F.O. Jenkins. Rutherford, NJ: Fairleigh Dickinson University Press. London: Associated University Presses, 1983.

Walsh, Thomas. *Journal of the Late Campaign in Egypt, Including Descriptions of that Country, and of Gibraltar, Minorca, Malta, Marmorice, and Macri: With an Appendix Containing Official Papers and Documents*. London, 1803.

Whitcombe, Thomas Douglas. *Campaign of the Falieri and Piraeus in the year 1827, or, Journal of a Volunteer, Being the Personal Account of Captain Thomas Douglas Whitcombe; Edited with an Introduction, Notes, and Appendices by C.W.J. Eliot, Gennadeion Monographs 5*. Princeton, NJ: American School of Classical Studies at Athens, 1992.

Wilkinson, William. *An Account of the Principalities of Wallachia and Moldavia*. Reprint New York: Arno Press, 1970 [1820].

Williams, Brian Glyn. *The Crimean Tatars: the Diaspora Experience and the Forging of a Nation*. Leiden: Brill, 2001.

Wittman, William. *Travels in Turkey, Asia Minor, Syria and Egypt*. London, 1803; reprint New York: Arno Press, 1971.

Wolff, Larry. *Inventing Eastern Europe: The Map of Civilization on the Mind of the Enlightenment*. Stanford, CA: Stanford University Press, 1994.

Yalçinkaya, Mehmet Alaaddin. 'Mahmud Raif Efendi as the Chief Secretary of Yusug [sic] Agah Efendi, the first permanent Ottoman–Turkish ambassador to London (1793–97)'. *Ankara Üniversitesi Osmanlı Tarihi Araştırma ve Uygulama Merkezi Dergisi* (1994): 385–434.

Yapp, M.E. 'The modernization of Middle Eastern armies in the nineteenth century': 330–66.

Yates, William Holt. *The Modern History and Condition of Egypt: Its Climate, Diseases and Capabilities with an Account of the Proceedings of Mohammed Ali Pascha from 1801 to 1843*. London: Smith Elder and Co., 1843.

Yerasimos, Stéphane. *Deux ottomans à Paris sous le Directoire et l'Empire: relations d'ambassade*. [Paris]: Sindbad, 1998.

Zachariadou, Elizabeth, ed. *The Kapudan Pasha, His Office and His Domain*. Rethymnon, Crete: Institute for Mediterranean Studies, 2002.

Zarinebaf, Fariba, John Bennet and Jack L. Davis, *A Historical and Economic Geography of Ottoman Greece: The Southwestern Morea in the 18th Century*. Athens: American School of Classical Studies at Athens, 2005.

Zboinski, H. *Armée Ottomane: Son Organisation Actuelle Telle qu'elle Résulte de l'Exécution de la Loi de 1869*. Paris: Librairie Militaire de J. Dumaine, 1877.

Ze'evi, Dror. '*Kul* and getting cooler: the dissolution of elite collective identity and the formation of official nationalism in the Ottoman Empire'. *Mediterranean Historical Review* 11 (1996): 177–95.

Zeidner, Robert. 'The army of the Tanzimat: the evolution of a new Ottoman martial tradition under the stimuli of European diplomatic and military pressures'. MA, American University, 1961.

Zens, Robert. 'Pasvanoğlu Osman Pasa and the Paşalik of Belgrade, 1791–1807'. *International Journal of Turkish Studies* 8 (2002): 89–104.

Zürcher, Erik J., ed. *Arming the State: Military Conscription in the Middle East and Central Asia, 1775–1925*. London: I.B. Tauris, 1999.

Zürcher, Erik J. 'The Ottoman conscription system in theory and practice, 1844–1918'. In *Arming the State*.

Zürcher, Erik J. 'Scholarship on the Ottoman Empire in World War I', *Middle Eastern Studies Association Bulletin* 39 (2005): 183–92.

Index

The index includes Chapters 1 to 11; their associated footnotes; and 'By Way of Conclusion'. Filing order is word-by-word and numbers are filed as if spelled out. References to footnotes are indicated by the page number, followed by '*n*' and the footnote number, for example: 'Abd al-Wahhab 337*n*7.

Abaza (Abkhazian) Mehmed 144
Abbas Hilmi, Khedive 407
Abbas Mirza, Prince 463
Abd al-Rahman al-Jabarti 240
Abdi Pasha 152–54
  and Crimean War 467
  rivalries and corruption 467–70
Abdülaziz 323, 403, 431, 479
  reform agenda 408–10
Abdülcelilzade Pasha 286
Abdülhamid I 160, 202, 220, 235, 247
  and Janissaries 317
Abdülhamid II 85, 332
  military reforms 478
Abdullah (Aleppo) 419
Abdullah Bey 420–21
Abdullah Pasha (Arnavut) 156
Abdullah Pasha (al-'Azm) 232, 367
  campaign against 371–72
  and Siege of Acre 370, 371
Abdullah Ramiz Efendi 248
Abdullah (Wahhabi leader) 309
Abdülmecid I 323, 394*n*23, 399, 427, 461, 483
  accession 402
  and apostasy 417–18
  and Crimean War 442–43
  and Gülhane Edict 402–3
  and Menshikov 437–38
  protection of Christians 417–18
  reform agenda 402–3, 408, 476–77
  and Sublime Porte 415
Abdurrahman Pasha 196, 257*n*65, 260–64, 286
Abdürrezzak 155
Abercromby, General 236
Aberdeen, Prime Minister 440
Abkhazia 234
  and Russia 281
Abkhazians 33, 463
Aboukir 236
absolutism 47–48
Acre 63, 230, 239–40, 405
  Albanian troops 255*n*39
  and Napoleon 232
  siege 367–71
  surrender 406
Acre-Sidon-Damascus triangle: disorder 364–65
Acropolis 294
Adair, Robert 279
Adakale (New Orsova) 111, 113–15
  see also Orsova (Old Orsova)
Adana 221, 370–71
  Egyptians abandon 406
  unrest 392
Aden 385
Adrianople, Treaty of (1829) 194, 283, 306, 357, 424
  negotiations 354–56
  terms 361–63, 466

Adrianople, Treaty of (1829) (*continued*)
Adriatic Islands 289
advice manuals *(nasihatnames/ islahatnames)* 186–92
Afsharids 85
Ágoston 158
Ağriboz (Negreponte) 100, 296
Ahmed Agha 270, 271
Ahmed Cevdet Pasha 184, 207*n*3, 222, 245–46, 252–3*n*10, 258*n*77, 269–70, 289, 301*n*22; *n*24
  on cavalry mutiny 182–84
Ahmed Efendi 249
Ahmed Fevzi Pasha 378–79
  and Ottoman fleet 365
Ahmed II 52
Ahmed III 37, 52, 98
  and Prut campaign 91
  and Venice 99
  westernisation of court 72
Ahmed Bey of Hulo 464
Ahmed Pasha:
  and Crimean War 450, 452, 466, 467
  rivalries and corruption 467–70
Ahmed Pasha (Laz Ahmed) 273–80
  and Treaty of Bucharest 277–78, 281
Ahmed Resmi Efendi 55, 90, 145–48, 158, 191
  embassy to Prussia 121*n*19, 139, 172*n*20, 225, 253*n*20
  Hulâsat ül-İ'tibar 188–89
  Layiha 188–89
  military reform 187–89
  1768–74 War 174*n*34, 176*n*59
Ahmed Vasıf Efendi 152, 154, 158
  and need for reform 181–84
  author of *Koca Sekbanbaşi Risalesi* 192, 207*n*2
  1768–74 War 176*n*59
Ainslie, Robert 159–61, 226
Akdeş Nimet Kurat 122*n*34
Akhaltzikhe 464
  and Crimean War 466, 471–73
  and Treaty of Bucharest 463
akinci (raiding parties) 63
Akkirman 32, 68, 158
  Russians and 166
Akkirman, Convention of (1826) 283, 297, 355, 422
Aktiar, Gulf of 159–60
Akyol 271
Albania 22, 60, 288
  and Ali Pasha of Iannina 287
  anti-reform movement 366
  autonomy 392
  and Crimea 106
  ethnic forces 480
  Muslims 359
  nationalism 394*n*23
  rebellions 486
  resistance to Mahmud 282–83
  revolt 357–59, 364
Albanian troops 150, 153, 189–90, 221, 232, 237–41, 251, 355
  at Acre 255*n*39
  1828–29 War 345, 349–51
  and Mehmed Ali Pasha 310
  and reform 352
  at Şumnu 271
Aleksandropol: and Crimean War 467, 471–73
Alemdar Incident 262–64
Alemdar Mustafa Pash (Mustafa Bayraktar) 204, 223, 245–50, 252, 264, 270, 272, 315
  military reform 262–64
  religious toleration 360
Aleppo 223, 370
  attacks on Christians 419–22
  Egyptians abandon 406
  and İbrahim Pasha 384
  post-1841 416
  rebellions 253*n*10, 255*n*39
  reform of fiefs 336
  and Tanzimat 418–22
Alexander I, Tsar 136, 243, 265–66, 269, 271, 273, 281, 290
  death 297
  and Slobozia armistice 247–48
  and Ypsilanti 290
Alexander II, Tsar: and Crimean War 460
Alexander the Great 91
Alexandria 364
  and Britain 243, 307
  and France 228, 229, 306
  and Napoleon 225

Algeria: French occupation 392
Algiers 364
Ali Agha 148
Ali Aziz Efendi 226
Ali Bey Agha 230
Ali Bey el-Kabir 215, 230
  pacification of Egypt 234–35
Ali Pasha (grand vizier) 377, 477
    See also Mehmed Ali Pasha
  and Crimean War 467
  and Gülhane/Tanzimat 377
  and İbrahim Pasha 371
Ali Pasha, Diyarbakir Governor 96
Ali Pasha of Iannina 56, 84, 215–16, 229, 237, 255*n*38, 261, 273, 285, 303*n*53, 355
  and Albanian national narrative 287
  challenge to Selim III 240
  death 293
  and foreign diplomacy 287
  and Greek national narrative 287
  loyalty to 303*n*51
  neutralisation 288–89
  rise 221
  and Greek Revolt 292–94
  sale of leases 380
  source of power 288
Ali Rıza Pasha: at Vidin 427–73
Alo Pasha (Kürd) 221
Amasya 216
Amasya, Treaty of (1555) 31
American Civil War (1861–65) 458
Amiens, Treaty of (1802) 236
Anakliya 463
Anapa 166, 462–63, 466
Anatolia 33, 63, 244, 259, 261
  anti-reform movement 366
  campaign routes 68–69
  cavalrymen 182, 232
  centralised control 287
  frontier 461, 462
  guerrilla tactics 465, 466
  and İbrahim Pasha 373
  and Mehmed Ali 237
  military judges 53
  Muslims and 60
  and *Nizâm-ı Cedid* 197–98
  nomadisation 85
  provincial elites 215–16
  and Prut 95
  reform of fiefs 336
  sipahis 55
Andric, Ivo: *Bridge on the Drina, The* 283–84
Andronikov, Commander: and Crimean War 466, 471
Anglo-French conflicts 87
  global 119
Anglo-Ottoman Convention (1838) 388, 408
  extension of terms 407
Anglo-Ottoman Convention (1855) 459–61
Anglo-Ottoman cooperation 236
  defensive Alliance (1799) 230
  and Napoleon 310
  rout of Mehmed Ali 404–6
Anglo-Ottoman-French alliance 401
Anglo-Ottoman-Russian relations:
  coalition 237
  naval confrontations 181
Anglo-Russian relations 443
Ankara:
  anti-reform movement 366
  and *Redif* 383
Antakya 371, 384
Arab provinces 62–63, 261, 309
  anti-reform movement 366
  conscription 419
  ethnic/religious diversity 286
  informal colonialism 388–89
  and Mahmud II 364–65
  recruits 144
  unrest 392
  and Wahhabis 308–9, 312
Arab Tabya (Arab Redoubt): and Crimean War 454
arabacıs (waggoners) 144
Arabian Nights 227, 255*n*33
Arad 66–67
Aranjuez Pact (1752) 86
Aras River 463
Ardahan 464, 466
al-Arish, Convention of 236
Armenians 33, 136, 462
  borderland rebellions 477
  and Erivan 341n53, 466
Army and Navy Engineering Schools 203

Army Finance Secretariat 380
Arnavut Abdullah Pasha 156
Asakir-i Mansure *see also* *Nizâm-ı Cedid*; *Nizamiye*; Ottoman military
  Anatolian recruits 359–60
  artillery 315–16, 332–3, 339*n*25, 356
  assessment of 356–61
  cavalry 330–32
  compensation 332, 397*n*72
  conscription 344, 380–82
  and Cossacks 331–32
  discipline 357
  1828–29 War 344, 352
  financing 325–26, 341*n*53, 380–82
  formation 326–28
  Hassa (palace guard) 330
  industrial support 341*n*60
  and Janissaries 321
  Moltke's description 333–34
  name changed 410
  night attacks 350, 356
  and *Nizâm-ı Cedid* 328–29
  officers 326–27
  organisation of corps 328–29
  prisoners-of-war 357
  in the provinces 330
  provisioning 357
  recruits/recruitment 344, 357–60, 392*n*4
  regulations 326, 328–29, 329–36
  at Şumnu 349, 352
  uniforms 335*plate*
Astrakhan 93–94
Aswan 312
Atatürk, Mustafa Kemal 392, 413, 484
Ataullah Efendi *(Seyhülislam)* 246–47, 248–49
Athens 293, 303*n*58
  European response to fall 297–98
  siege 295, 316
Auerstädt 242
Augsburg, Peace of (1555) 31
Augustus II, King of Poland 93
Augustus III 139
Auspicious Occasion 306, 321, 333 *see also* Janissaries
Austerlitz, Battle of (1805) 242
Austria 18, 24, 83, 166, 442 *see also* Habsburg Empire
  agrarian reform 137
  and Balkans 87, 476
  and Bavarian Succession 159
  and Bukovina 159
  and Convention of London 405
  and Crimean War 436
  diplomacy 86–87, 125*n*65, 127*n*102
  geopolitical ambitions 86–87
  and Hungary 102
  mediation 103–6
  militarisation of society 120*n*17, 137
  and Orsova 138, 166
  and Ottomans 107–10, 224, 226
  Prussian challenge 34
  1737–39 War 136, 104–5*plate*
  1768–92 136
  1787–92 War 136, 166, 178*n*89
  tax revenues 89–90
  Third Coalition 236
  Treaty of London 298
  and Varadin fortress 101–2
Austrian army 130, 478
  Catholic proselytising 59–60
  commanders 113–14
  ethnic regiments 58–59
  expenses 435
  federative arrangements 86
  financing 119*n*5
  hayduks 58
  Hofkriegsrat (Vienna) 59, 432
  Inner Austrian administration (Vienna) 59
  Landwehr 432, 436
  losses at Belgrade 127*n*99
  mobilisation in Balkans 106–7, 112
  officers 125*n*65
  peacetime army 120*n*17
  rebellions 59–60
  reform 86–87, 89–90, 137, 169
  in 1700 45–46
  in 1850 432–36
  strengths 120*n*17
  supply difficulties 112
  universal conscription 169
  weaknesses 119*n*5
Austrian Council 106

Austrian Netherlands
resistance 166
Austrian Succession, War of 137
Austro-Hungarian Empire:
ethnic regiments 62
peasants 426
Austro-Prussian relations 142–43
and Crimean War 456
military alliance 433–36
Austro-Prussian Treaty (1791) 138
Austro-Russian relations:
cooperation 426–27, 431, 434
deterioration 435–36
Avlonya 189, 287
ayan see local elites *(ayan, derebeys)*
Aydın:
anti-reform movement 366
and *Redif* 383
Aydos 351, 352
and 1828–29 War 354–55
Aynalıkavak, Convention of (1779) 160
'Azaban 232
azabs (infantry) 63
Azadlı 195
Azeris 462
al-Azhar mosque (Cairo) 241
al-'Azm family 232
Azov fortress 18, 24, 68, 83, 90–91, 93, 98, 158
capture 29
surrender 103
Azov-Mozdok line 463

Babadaği 106, 143, 235
and Crimean War 453
Bade, Louis Guillaume de, General Prince 67
Baghdad 33, 62, 85, 223, 261, 286
anti-reform movement 366
frontier 174*n*32
siege of (1638–39) 63, 69–70
Bagration, General: and Crimean War 474
Bahçesaray 159
sack 103
bakshish (bribery) 127*n*102
Balaklava, Battle of (1854): Ottoman forces 458–59
Balkans:
and Austria 87
balance of power 215
campaign routes 68–69
cavalrymen 232
command centres 273
early 18th century 84–86
historiography 58
movements of populations 84
Muslims and 60
Ottoman policy 285
private armies 244
provincial elites 288, 289
refugees 442
religious ideologies 29
revival of traditional armies 336
Russian encroachments 343
sectarian rivalries 22
socio-political setting 1768 129
Balta Limani Convention (1849) 427
Baltacı Mehmed Pasha 75, 97–98
and Prut 95–96, 117
Baltic provinces 266
Baltic Sea 90
Banat/Temeşvar 24, 65–68, 86, 90, 101–2, 112–13, 163
fall 87
Banjaluka:
losses 126*n*84
siege 110–11
Bar Confederation (1768) 140, 142
Barker, Thomas Jones: *General Williams and His Staff Leaving Kars* 474–75
Başgedikler: and Crimean War 444, 467
Bashir II (Shihab) 370, 405, 416
Bashir Qāsim 405
başibozuks (irregular cavalry) 304*n*69, 410–11*plate*, 417
and Crimean War 456–57, 473
discipline 457
and pacification of Aleppo 418
social impact 478
Başmet Giray 156
Basra 286
bastinado 145
Bathurst, Captain 452
Battal Hüseyin Pasha 218
1787–92 War 463

Batum 462–63, 466
and Crimean War 473–75
1828–29 War 465
Bavaria 23
War of Succession 159
Bayburt 466
1828–29 War 465
Bayezid 463
Bazarcık 349
Beatson, General: and Crimean War 457
Bebutov, General 471
Bedouins 62–63, 232
rebellions 486
Bedr Khan 386
Behiç Efendi (Mehmed Emin) 204–5
Behram Pasha (General Cannon) 386
and Crimean War 456
Beirut 370, 405
Bekir Agha 465
Bektashis 215, 340*n*45, 358
elimination 322–25
beldars (trench diggers) 63
Belgrade 20, 24, 83, 112, 220–21, 284
and Austrians 166
Janissary dominance 222
Belgrade, Siege of (1739) 68, 85–86, 101–2, 118, 126*n*87, 130, 137, 265
numbers of troops 64, 124*n*60
Belgrade, Treaty of (1739) 28, 83–84, 86–87, 112, 127*n*102, 138, 166, 278, 427, 438, 485
impact 116–17
negotiations 115–18, 119*n*1, 127*n*102
terms 84, 115–17
Belgrade-Azov fortress line 28, 58, 90–91
maintaining 68–72
Belgrade-Danube fortress line: 1806–12 War 273
Belgrade-İsmail fortress line 131
Belorussia 168
Bem, General 426, 427
Bender fortress 68, 85, 90, 107–10, 158
numbers of troops 126*n*83
provisioning 176*n*59
and Prut 123*n*40
Russians and 147, 166, 245
1768–74 War 143
Bérenger 120*n*17
Berlin: Ottoman embassy 226–27
Berlin, Treaty of (1856) 449
Beshir III 416
Bessarabia 137–38, 160–61, 242, 292
colonisation 289
expulsion of Tatars 282
Greeks in 289
and Russia 246, 392
Beylan 371
Bibescu, George 425–26
binbaşı (colonel, later major) 196
Birecik 389–90
biscuit *(peksimed)* 148
Black Sea 87, 461, 462
border skirmishes 134
defence 32
displacement of coastal populations 160
1828–29 War 353
fortress line 64, 84, 90–91, 194
imperial confrontations 75
neutralisation 475
Russian access 129, 135, 140, 155, 158, 259, 269, 278–81, 297–98
1768–74 War 143
strategic importance 466
and Treaty of Adrianople 363
Blacque, Alexandre:
Le Moniteur Ottoman 377
Spectateur de l'Orient 377
Blaquière, Edward 303*n*58
Bohemia 23
bombardiers 324
Bon, Alexandre 124*n*59
border raids *(razzia)* 27
Borsha 116
Boşnak Agha 272–74
Bosnia 19, 22, 85, 242, 284
autonomy 392
Christian populations 92–93, 430
and Crimean conflict 110–13
and Crimean War 431
1828–29 War 345
ethnic forces 480
militia 73
Muslim nobility 429
rebellions 358–59, 477
reorganisation of 1834 429–32

Bosnia (*continued*)
resistance to reforms 324, 340*n*49, 352, 366
subdivided 431
and Tanzimat 429–32
Bosnia-Herzegovina:
census 431
riots 477
Bosnian militias 113, 221, 355
and Crimea 106
and Hekimoğlu Ali Pasha 110–13
loyalty to Ottomans 84
1736–39 War 84
Bosphorus 194, 237, 407
and Crimean War 444
fortifications 64, 246–47, 315, 316
rebellion 248–49
Russia and 298, 353
and Treaty of Adrianople 363
Bostancı guard (palace guard) 196, 258*n*77
replaced 330
Bostancı Musket Regiment *(Bostancı Tüfenkci Oçaği)* 196
Bourbon-Habsburg interests 87
boyars 140
Brancoveanu, Constantine 29, 91–92, 121*n*20
Brandenburg 23
Brasov 290
Britain 130 *see also* Anglo-French conflicts, *etc*
advisors to Ottomans 185
and Alexandria 307
and Ali Pasha of Iannina 287
and Arabs 239
colonial expansion 239, 400
and Convention of London 405
and Crimean War 444–45, 453
Eastern Question 487
and Egypt 308
1828–29 War 343–44
and European diplomacy 93
and Greek Question 352
and Habsburgs 229
impressions of Ottomans 237–42
and mamluks 238–39
mediation in Ottoman affairs 84, 161, 260, 463
missions to Ottoman court 375–76
and Napoleon 237
Ottoman territorial integrity 363, 440, 443
routes to India 375, 389, 407, 463, 487
and Russia 139
and Greek Revolt 297–98
and Selim III 228
and Syria 384–85
technical experts 195
Third Coalition 236
Treaty of Adrianople 355–56
Treaty of Hünkâr İskelesi 374–76
Treaty of London 298
Triple Alliance 165
British army 199
officer class 239–40
British-French-Ottoman alliance
and Crimean War 444–45
British India 119*n*8, 130, 199, 227, 236, 239, 281
militarisation 239
model for Mehmed Ali 312
British Military Mission 239
Brønsted, Peter 287, 303*n*51
Brue, Benjamin 100–1, 123*n*50
Bucharest 153, 271, 278, 280, 353
battle 151
and Crimean War 452, 456
French consulate 223
partition 456
rebellion of 1848 426–27, 447*n*27
and Russia 246, 285, 347
Russian consul 290
and Russo-Ottoman Wars 131
Bucharest, Treaty of (1812) 228, 259, 291, 302*n*37; *n*38, 315, 346, 362, 422
Article VIII 281, 283
and Canning 279
implementation 298
mediators 279
negotiations 277–82
terms 463
Buckley, Lieutenant 452
Buda 20
Budapest: rebellion of 1848 426
Bug River 107, 161, 164

Bukovina: 163
and Austria 137, 159
Bulgaria 22, 259, 261, 284
border defences 63–64
and Crimean War 454
1877–78 War 476
ethno-religious movements 251–52
historiography 58
nationalism 220
secular schools 376
social transformation 422
timariots 424
troops 221
Bulgarian-Wallachian border: movements of populations 84
Burgas: 1828–29 War 354
Burgos depot 148
Bursa 315
and *Redif* 383
Buşatli Mustafa Pasha 288, 394*n*23, 355
Abanian revolt 366–59, 364
at Filbibe 355–56
Butler, Captain James A: and Crimean War 454
Butrinto 237
Byron, Lord 56, 293, 304*n*66
Byzantium: restoration 290

Cairo 261, 364
factionalism 327
and French 306
rebellions 236, 241
regiments 234
Caliph of Islam 214
Calosso, Captain 327, 341*n*58
Campoformio, Treaty of (1797) 138, 229
Canad 66–67
Canakkale 293
Canib Mehmed Efendi 288
Canikli Ali Pasha 84, 144, 216–20, 225
Nasayih al-Muluk 190–92, 215–16, 218
nepotism 217–18, 252*n*5
and reform agenda 223–24
Canning, Stratford 270, 280, 301*n*21, 302*n*37; *n*38, 317, 403, 427, 437–44, 460–61
Christian agenda 408, 417–18
and Crimean War 444, 457, 459, 480*n*30
and Missolonghi 295
Ottoman forces 446*n*11
and Treaty of Bucharest 279, 281
Cantemir, Dimitrius 29, 61, 96–97, 121*n*20
and Russians 91–92
Çapanoğlu (Cebbarzade) family 144, 261
Capodistrias, Ioannis 290
Capriola, Duc Sierra 279
Cardigan, Lord 458
Çarhacı Ali Pasha 183, 296
Carpathian Mountains 137
Catherine II, 'the Great,' Empress of Russia 29, 34, 60, 88, 97, 154, 160, 167, 250, 265 *see also* Romanov Empire; Russian army
and 1768–74 War 142–43
annexation of Crimea 204, 284
challenges 224
diplomacy 254*n*28
expansion of empire 135–36
and Joseph II 137–38
military reform 267
and Selim III 227
southern policy 129, 138, 140, 155–61, 217
Catholic Church 22–23, 34
Caucasus 18, 33, 218, 346
exiles 79*n*50, 251
frontier 174*n*32
Ottoman-Persian struggles 103
ports 462
and Russia 60, 135, 242, 259, 269, 278, 297, 343, 351–52, 392, 461–62
Russo-Ottoman disputes 401–2
slaves 232
Cebbarzade (Çapanoğlu) 182–84, 353*n*10
Cebbarzade (Çapanoğlu) Mustafa 217, 221
cebeci (armourers) 144
Celali revolts 54–55
Celaluddin Mehmed Agha 316

Çelebi Mustafa Pasha 247–49, 270
Central Asia: Russia and 135–36
Çerkes Hasan Pasha 144, 156
Çerkes Mehmed Pasha 96
Çeşme 154
  Russians at 199
Cevat Erbakan 125*n*76
Cezzar Ahmed Pasha 62, 230–32, 234, 255*n*38; *n*39, 257*n*62
Cezzarzade family 261
Chandler, David 94–95
Charles VI, Emperor 107, 113
Charles XII, King of Sweden 29, 93, 122*n*34
  and debts 123*n*45
  and Prut 90–95, 97, 123*n*40
Charter of the Nobility (1785) 34, 135, 171*n*10
Chernyshev, War Minister 436
Chesney, Captain 375
Chichagov, Paul 281, 302*n*42
China 87
Chios 292–93
Choiseul-Gouffier, Duc de 140, 162, 227–28
Christian armies: reform exemplars 187
Christians 285–86 *see also* Ottoman Empire
  attacks on 386, 419–21
  Chios massacre 292–93
  commanding non-Christian troops 457
  displacement 160
  medical school doctors 329
  in Mount Lebanon 416–17
  and Ottoman navy 316
  and Ottoman rule 92–93
  proselytising 408
  Protestants 23
  and Sekban 262–63
  in Syria 365
Chrzanowski (military advisor) 375
Church, Sir Richard 297
Circassia/Circassians 33, 136, 444, 462
  and Crimean War 457
  1806–12 War 463
  exiles 79*n*50, 251
  Muslim rebels 466
  and Ottomans 466
  and Russia 281
  settlement 204
Citate 452
Clarendon, Foreign Minister 471
Clausewitz, Karl Marie von 46
Cochrane, Lord 297
Cochrane, Thomas 295
Coehorn (seven-pounder) 349
Collège de France 426
Çoluk River 471
Concert of Europe *see* Europe
Considine, Lieutenant Colonel 375
Convention for the Pacification of the Levant *see* London, Treaty of (1827)
Corinth 99, 124*n*55
  Greek control 297
Cornea 113
Cossacks 28, 33, 46, 90, 146, 266, 269, 348
  in *Asakir-i Mansure* 331–32
  brutality 126*n*79
  control of 136
  and Crimea 103
  and Crimean War 457
  Don 103
  1828–29 War 464, 465
  haidamuk revolt 140
  military settlements 58–59, 168–69
  and peasants 140
  and Pughachev Rebellion 156, 224
  Ukrainian 103
  Zaporozhian 62, 93, 103, 139–40, 331
Council of Judicial Ordinances (1838) 404
Court Battalion of the Aghas of the Inner Service 327
Craiova (Little Wallachia) 137, 141, 152, 221, 273–76, 285
  and Crimean War 452
  plague 271
Crete 309, 313, 364, 370, 404
  and Ali Pasha of Iannina 294
  riots 477
  unrest 392
Crimea 24, 28, 103–9*plate*, 142, 146–47, 157, 160, 167–68, 218
  annexation by Russia 83, 87, 135–36, 138, 204, 217, 284

Crimea (*continued*)
  and Crimean War 458, 470
  defence 95–96
  independence 160
  Lacy in 114
  Ottoman loss 485
  refugees 79*n*50, 477
  relationship with Empire 487
  1736–39 War 103–10, 125*n*65
  1768–74 War 143, 155–58
  1787–92 War 166–67
  and Tatars 282
  Tsarist colonistaion 60
Crimean Khanate 139
  independence 159–60
Crimean War (1853–56) 240, 362, 400
  aims 458
  auxiliary systems 458
  British forces 444, 456
  British-French-Ottoman alliance 444–45
  causes 401, 435, 448*n*48
  cholera 456
  Christian officers 457
  communications 458
  diseases 453, 456, 472
  efforts to prevent 440–45
  Florence Nightingale 453
  French forces 444, 456, 458
  Habsburgs 426
  human cost 476
  losses 474
  medical provision 453, 458–59, 473
  military high command 446*n*8
  origins 437–45
  Ottoman forces 131, 156, 444, 446*n*11, 450, 456–61, 467, 473, 475
  and *redif* system 478
  refugees 482*n*42
  Russian forces 265, 449, 452–53, 457, 470–71, 474
  and Trans-Caucasus 134
Croatia 18, 22, 24, 163
  army 111
  military service 59
Croats 29
Cvetkova, Bistra 58
Cyprus 313, 405
Dacia (Romania and Bessarabia) 137–38, 160–61
Dagistan 103, 444, 462
  Russian occupation 463
Dagistanlı Ali Pasha 144, 156
Dalmatia 18, 19, 24, 242
Damad Ali Pasha: at Nauplion 99–101
  and Varadin 101–2
Damascus 62, 223, 230–34, 261, 325
  annual pilgrimage caravan 309
  conscription 421
  and İbrahim Pasha 384
  Janissaries in 340*n*51
  massacres of Christians 421
  and Mehmed Ali Pasha 259
  Ottoman 5th Army 418
  post-1841 416
  rebellions 255*n*39, 477
  reform of fiefs 336
  and reforms 365
  town council 383
Damat Mehmed Ali Pasha 409
Damietta 238, 307
Daniel, Norman 239
Danube frontier 28 *see also* Ottoman borderlands
  closing 83
  conflicts 178*n*89
  and Crimean War 450
  logistics 131
  1768–74 War 143
Danube region 21*plate*
  and Crimean War 457
  depopulation 141
  encroachments 141–42, 194
  estuary 132–33*plate*, 166–67
  fortifications 316
  garrison line 334
  geopolitical importance 284–85
  geopolitics 19–23
  imperial confrontations 75
  stabilisation 141–42
  topography and environment 22
  warfare 65–72
Danube River 59
Dardanelles Straits 407, 440
  Anglo-Russian-Ottoman confrontation 181
  fortifications 316

Dardanelles Straits (*continued*)
Ottoman closure 299
and Russia 243, 352–53
Russian blockade 147, 247
and Treaty of Adrianople 363
Dardenelles, Treaty of (1809) 279
Davut Pasha 270
Decembrist Revolt (1825) 265
Dede Mustafa Agha 316
Deed of Agreement/Sened-i-Ittifak 261–65
Deli Abdullah 315
Delvine 189
Derbent 462
derebeys see local elites *(ayan, derebeys)*
Derviş Kuşmanı: *Zebire-i Kuşmani fi Ta'rif Nizâm-i Ilhami* 204
Derviş Mustafa Pasha 290
Deryal Pass 462
Déscription d'Egypte 230
Devlet Giray, Tatar Khan 95, 159–60
devşirme (Janissary recruits) 49–50, 77*n*15
impact of system 77*n*22
Diebitsch, Count: and Treaty of Adrianople 352–56
Diriyah 309
Diyarbakir: ethno-religious unrest 385–88
dizdar (local fortress commander) 63
Dnieper River 28, 157, 161
borderlands 140
Dniester River 96, 114, 149, 269
1768–74 War 116
Dobruja 348
raising cavalry 330–31
D'Ohsson, Ignatius Mouradgea 253*n*21
Tableau général de l'Empire Othoman 213*n*36
Dolmabahçe Palace 409
Don Cossacks 103
Don River 28, 158
Doxat, General 111–12
Dragatsani, Battle of 291
dragoman (*tercüman*/translator) 25–26, 139, 227–28, 292
Dramalı Mahmud Pasha 296
Druze 62, 231, 255*n*39, 370
conscription 421
and İbrahim Pasha 385
in Mount Lebanon 416–17
rebellions 486
Dubayet, Albert 202
Duckworth, Admiral 243
DuPlat, Captain 375
Durmaş Pasha 93
Dutch Republic:
and European diplomacy 93
mediation 84
Triple Alliance 165

East India Company 236, 239
Eastern Crisis *see* Syrian Crisis (1839–41)
Eastern Question 88, 129, 137–39, 170, 250, 299, 375, 407, 417, 436, 438, 441, 487
diplomacy 215
and revolutions of 1848 427
Ebu Bekir Pasha 220
Ebubekir Ratib Efendi 225–26
Edirne 36, 68–69, 95–98, 100, 117, 248, 355–56, 370
conscription 244
1828–29 War 354
plague 352
and *Redif* 383
resistance to reforms 324
Edirne, Treaty of 422
Edirne Incident 244
Egypt 62 63
agricultural reform 307–8, 366–67
Albanian forces 306–7
and Ali Bey el-Kebir 339*n*30
autonomy 235
beylicate 235
blockade of ports 370
and British 236, 243, 484
civil war 237
economy 215
fleet 310, 364, 384
and French 195–96, 222, 235–37, 240
and global trade 367
hereditary rule 405–6
household system 231–35
and Mahmud's reforms 282–83
and Mehmed Ali 243, 306–8

Egypt (*continued*)
and Napoleon 287
New Army 308–13
Ottoman forces 232–33
Ottoman restoration 237
pacification 234–35, 401
pre-Napoleonic 231–35
private armies 231–32
reform 364–65
revolts 312, 339*n*30, 392
social transformation 239, 307–12, 312
trade 308
Egyptian army:
absorbed by Ottomans 406–7
and Battle of Nizib 390–91
conscription 268, 310–12
and Crimean War 480*n*10
desertion 372
1828–29 War 345
ethno-linguistic distinctions 365–66
French advisors 312
and Greek Revolt 294, 363
and Mehmed Ali 357, 366–67
officers 310, 312
reform 251, 308–13, 364
and Syria 371, 383–84
uniforms 310*plate*
Egyptian-Levantine coast 259
Elbasan 271, 288
Enlightenment spirit 293
Enver Pasha 484
Enverî, Sadullah 150, 182–83, 200
1768–74 War 174*n*34
and military reform 192
Epirus 293
revolt 453
Erivan 33, 466
and Crimean War 471
Russian siege 463
Erzincan 462
Erzurum 95, 134, 216–17, 261, 296, 460–66
and Crimean War 473
and Ottoman-Russian confrontations 449–50
and *Redif* 383
resistance to reforms 324
Esad Efendi (Mehmed)
Üss-ü Zafer (The Foundation of Victory) 317
and Janissaries 321–22, 339*n*36
Eşkinci Corps: formation 317–22
Essen, Count 273
Esztergom 41*n*2
Esztergom-Istanbul: diplomatic passage 20
Et Meydani 320
Eugene, Prince of Savoy 18, 24, 36, 46, 84, 113
and Belgrade 103
and military reform 89–90
triumphs 86
and Varadin 102
Eupatoria 461
and Crimean War 459
Euphrates-Basra 375
Europe:
and Chios massacre 292
Concert 436, 476
Eastern Question 88
1848 407, 424–27, 476
future of Ottoman Empire 260
and global balance of power 130
and Middle East 417–18
military bureaucracy 89
military revolution 171*n*9
and Napoleon 169, 282
philhellenes 293
reform 476–77
relations with Ottomans 225, 399, 401
and Russian expansionism 93
territorial realignments 86–87
European armies 300*n*19
artillery corps 198–99
concept of honour 121*n*18
constitutional challenges 206*n*1
drill and discipline 78*n*31, 144–45
modernisation 134, 206
organisation and recruitment 86
and Prussian model 477–78
rate of advance 81*n*69
European diplomacy 224
and Ottomans 112, 161, 165–66, 225–29

European diplomacy (*continued*)
  and participation 26
  radical 224–26
  revolution (1756) 86–87
Evlad-ı Fatihan 358
Eyup Bey 386
Eyyubî Efendi Kanûnnâmesi (military regulations) 53–54

Fahmy 309–10
Farshut 312
Fazil Ahmed Pasha, Grand Vizier 65
Fethülislam 113
Feyzullah, *Seyhülislam* 36–37
Filibe 355–56
Filibe-Sofia-Belgrade axis 63
First Russo-Ottoman War (1768–74) *see* Russo-Ottoman War (1768–74)
Fischer, Lieutenant 67
Florence Nightingale: and Crimean War 453
Foksani 152
France 130
  and Ali Pasha of Iannina 287
  alliances 139
  citizens' army 168
  commerce 215
  and Convention of London 407
  and Crimean War 444–45, 453
  Eastern Question 487
  and Egypt 235–37, 236, 306
  1828–29 War 343–44
  and Greek Question 352
  and Greek Revolt 297–98
  mediating with Ottomans 84, 112–15, 161, 260
  and Mediterranean 237
  and Mehmed Ali 404, 407
  missions to Ottomans 163, 178*n*86, 185, 195, 201*plate*, 202, 225
  under Napoleon 169
  and Ottomans 139, 224
  protection of Christian populations 438
  1768–74 War 138–39
  and Sweden 139
  and Syria 385
  total war 169
  Treaty of Adrianople 355–56
  Treaty of Hünkâr Iskelesi 374–76
  Treaty of Kütahya 374
  Treaty of London 298
  weapons 199
Francis II Rákóczi 87
Franco-Austrian alliance 86–87
Franco-British relations 161
  colonial wars 139
  rivalries 229
Franco-Ottoman relations 227–28
  alliance 257*n*63
Franco-Prussian War (1870) 476
Franco-Russian-British relations 215
Franco-Russian relations 181, 246, 265, 271
  deterioration 273
Franz Joseph, Habsburg Emperor 426, 442
  absolutism 435–36
  coronation 426, 434
  and Crimean War 456
  suppression of Ottoman Ultimatum 441
Franz Stephan of Lorraine 113, 115
Frazier, Mackenzie General 243
Frederick II, 'the Great' 86, 87, 90, 118, 143, 172*n*20, 189, 198–99, 253*n*20
  expansionism 138–39
  military discipline 145, 268
  and Mustafa III 139
  and Russia 88
  and Silesia 137
French army 302*n*42, 478
French Revolution 170, 225, 228
  impact 293–94, 299, 433
Friedland 243
Fuad Pasha 402, 426, 438, 439, 477
  and Gülhane/Tanzimat 377

Gaillard 330
  and *Asakir-i Mansure* 326
Galatasaray (medical school) 376
Galatz 152
Galib Efendi (also Pasha) 228, 247, 277, 279–80, 315, 324, 330, 338*n*22, 340*n*49
  and Treaty of Bucharest 281, 463

Galicia: 163
  and Austria 137
  and Crimean War 456
Gallipoli 95
  and Crimean War 453
galloper guns 198
Gazi Hasan Pasha 161, 178*n*85
Gelibolu (prison) 145, 293
Georgia 160, 461
  and Crimean War 457
  Russian annexation 278, 281, 463
Georgians 33, 136, 462
  settlement 204
Ghazlay, Mahmud 33
Ghica, Alexander, Prince of Wallachia 425
Giurgevo 153, 247, 271, 276–78
  battle 151
  and Crimean War 453, 454
  and Russians 245
global balance of power 130
Golden Horde 30
Goltz, Colmar von der 484
Golitsyn, General:
  at Hotin 149–51
  Russian First Army 146
  troops under command 174*n*37
Gorchakov, Prince MD 440
  and Crimean War 450–52
Gordon, Ambassador 355
Grach, Colonel 454
Gradeva, Rossitsa 64
Gradiska 111
Grain Administration 194, 209*n*17
Grand Council 389
grand viziers 35–37, 46
  reformers 52
  role 39
  1736–39 War 85
Grant, Jonathan 195
Great Britain *see* Britain
Great Game 32, 281, 487
Great Northern War (1700–21) 29, 87
  and Cossacks 93
  mediation 93
Greater Kabarda 158
Greece 442
  Adriatic Islands 289
  and Ali Pasha of Iannina 287
  allied occupation 453
  French influence 240
  French occupation 363
  independence 392
  Ionian Islands 291–94
  and Mahmud II 282–83
  militias *(klefts)* 292
  nationalism 240
Greek Committees 293–94
Greek Empire 160–61
Greek Project 137–38
Greek Question 298, 352
  and Treaty of Adrianople 363
Greek Revolt (1821–27) 154, 238, 259, 286–92, 313, 402, 453
  allied naval blockade 299
  diplomatic interventions 297–98
  eyewitness accounts 295–96
  Ionian Islands 229, 291–94
  losses 294–95
  militias *(klefts)* 292–93
  Morea 307, 316
  Ottoman forces 293, 295–96
  Ottoman perception 299
  outcomes 297–98
  Peloponessus 292–94
  recruitment 358
Greeks 29, 288 *see also* Ottoman Empire
  in Morea 153, 235
Gregory V, Patriarch 292
Grenville, PM 226, 254*n*27
Gribeauval 198
Grieg, Admiral 350
Grocka 85, 115
  numbers of troops 127*n*105
Grünne, Field Marshal 434–35
Guilleminot, Ambassador 355
Gülhane Edict 322, 323, 361, 485 *see also* Tanzimat (1839–76)
Gümüşhane 216
  mines 464
Gustavus Adolphus 198

Habsburg-Bourbon relations 93
Habsburg Empire 19, 85 *see also* Austria; Austrian army
  assimilation 35
  buffer role 22

Habsburg Empire (*continued*)
- and Catholic faith 60
- clamour for autonomy 488
- competing loyalties 61
- and Crimean War 426
- ethnic diversity 433
- federative imperialism 23–24
- and French 229
- global conflict 119
- Hungarian problem 425–26
- imperial aims 34–36
- Long War (1593–1606) 54
- *Militärgrenze* 59–60, 79*n*47, 90, 168
- Military Border System 25, 33–34, 59–61, 446*n*13, 487
- and Poland 88
- position in Europe 86–87
- re-ordering of territories 137–38

Habsburg-Ottoman line 163
Habsburg-Ottoman relations 99
Hacı Ali 248–49
Hacıoğlu Pazari 182–83, 271
Hacıveli 471
hadiths 251
Hafiz İsmail 244
Hafiz Mehmed Pasha 387–88
- Battle of Nizib 388–91

haidamuks: and Cossacks 140
Haifa 370
Hajji Bektash 323
Hakkari 386
Hakkı Pasha of Sivas 464
Halet Efendi 228, 286–89, 303*n*53, 314–15, 316
Halil Hamid 218
Halil Rifa'at Pasha 102, 154, 355, 379
- and Abdülmecid's reform agenda 403
- 1828–29 War 354
- military reform 409
- and Treaty of Kütahya 374

Halvetis 323
Hamdi Bey 378
Hamdullah Pasha 315
Hamid Efendi 274
Hammer, von 122*n*38, 124*n*61
Hamza Pasha 142
Hanafi school 309
Hanbalis 308
Hantepesi (supply depot) 147–48, 150
- provisioning 176*n*59

Hanya 221
harçlik (supplies) 54
Harput headquarters 410, 466
Hasan Agha 317, 325
- on Janissaries 340*n*51

Hasan Kurdi 117, 121*n*25; *n*26, 122*n*39
- Prut campaign 92–98

Hasan Pasha 166, 235
Hasan Rıza Pasha 461
- military reform 409

Hasankale 464, 465
- and Crimean War 473

Hasköy:
- arsenal 264
- foundry 195

Hassa (palace guard) 330, 378–79
Hatibzade Ahmed Agha 148
haydut (bandits) 41*n*3, 58
- Serbian 116

Hawran 231
Haynau, Field Marshal 434
Heideck, Colonel 297
Hekimoğlu Ali Pasha 84, 115, 117
- and Bosnian Army 110–13
- and Serbia 113

Heliopolis 236
Henry, Prince (Frederick's brother) 88
Hertz, AZ 114
Herzegovina 429
- Christians 92–93
- subdivided 431

Hetman Mazeppa 93
Hezargrad 221, 245
High Council of Judicial Ordinances 379
Hijaz 309
- Wahhabi resurgence 312

Hildburghausen, Joseph Friedrich 106–7
- at Banjaluka 110–11

Hilmi Pasha 246, 247, 273
Hirka-yi Şerif (Mantle of the Prophet) 258*n*77
Hirsova 156, 349
historiography 45–48, 58
historiography: nationalist 58
Hochedlinger, 86, 120*n*17, 127*n*99, 102
Hofskriegsrat (Vienna) 59
Hohenzollerns 87

Holy League 18, 22–28, 30, 49
Holy Roman Empire 23, 90
Holy War Duties 380, 383
Homs 370
Honvéd 434
Hotin 28–29, 68, 159
  and Ottomans 150–51
  and Russians 147, 164, 245
  1768–74 War 115–16, 143
Howe, Samuel 296
Hudaybiya, Treaty of (628 AD) 27
Humbaraci 198
  barracks 264
Hungarian campaigns (1684–99) 54
Hungary 24, 425–26
  army service 432
  and Austria 102
  border defences 63–64
  competing loyalties within 61
  and Crimean War 456
  fortresses 64
  independence 87
  Muslim exodus 140
  refugees 447*n*29
  revolt 166
  revolution 434
  struggle with Austria 35
  and Transylvania 426
  universal conscription 169
Hünkâr Iskelesi, Treaty of 374–76, 399
  secret article 375
Hurewitz, JC 26
Hurşid Pasha (General Guyon) 270, 273, 283, 288, 471
Husayn Gradasevic 359
Hüseyin Agha Pasha 315, 320, 326, 330, 347
  at Aydos 352–54
  and Mahmud's confiscation of wealth 341*n*54
  and *Redif* 382
  at Şumnu 348–51
Hüseyin Avni Pasha: military reforms 478–79, 483–84
Hüseyin Daim 483
Hüseyin Pasha 370–71
Hüsrev Pasha 238, 306–7, 310, 323, 326, 352, 393*n*22, 396*n*52, 397*n*68, 439
  and Danube flotilla 316
  Husrev regulations 412
  and Mahmud's reforms 237, 377–79
  and Mehmed Ali Pasha 337*n*1, 338*n*9, 383
  and Mehmed Selim 348
  and *Redif* 382
  reform 330–32, 409, 486
  replaced 405
Hussey, Sir William 20, 41*n*4
Hutchinson, General Hely 236

Iannina 84
  siege 289
Ibn Saud family 309
İbrahim Afif Efendi 226, 229
İbrahim Bey 238
İbrahim Hilmi Pasha 244
İbrahim Müteferrika 118
  and military reform 187–89
İbrahim (Ottoman commissioner) 22, 24
İbrahim Pasha 238, 259, 309, 330, 352, 355, 359–60, 396*n*52, 405
  and Adana 371
  and Battle of Nizib 390–91
  and Damascus 364–65
  Greek campaigns 368*plate*
  and Greek Revolt 294
  and Missolonghi 294–95
  and religious diversity 365
  revolts against 464
  and Siege of Acre 367–71
  and Syria 312, 371–75, 406, 416
  and Treaty of Kütahya 375
İbrahim Sari Mehmed Pasha 118
İbrail 68, 96, 97, 271, 316
  and Crimean War 456
  defence 152–53
  disrepair 334
  1828–29 War 346–49
  fall 131–34
  1768–74 War 143
  siege 270
Ilbasan 271
Illustrated London News: and Crimean War 452
Imam Shamil 466
imdadiye (campaign tax) 70, 73
Imeretia 462

Imperial Ottoman Bank 479
Imperial War Council 89
imperialism:
  federative arrangements 86
  global 119*n*8
  multi-ethnic communities 62
  rhetoric of identity 47–48
India *see* British India
Indian Mutiny (1857) 458
İnebahtı 99
Ingur River 475
Inner Austrian administration (Vienna) 59
Inner Treasury (Privy Purse) 72
international law 400
Ionian Islands 229
  revolt 292–94
İrad-ı Cedid (New Revenue/Treasury) 185, 193
Iran 286
  frontier unrest 130
  and Russia 281
Iraq:
  ethnic forces 480
  and Mamluks 286
  resistance to Mahmud 282–83
  Wahhabi incursions 286
İsakçı 95–96, 146, 167, 349
  and Crimean War 453
  warehouse 147–49
Isfahan 33
İskodra 288
Islam:
  Caliph of 214
  heterodoxy 215
İslimiye 271
İsmail Bey 273, 386
  at Vidin 276
İsmail fortress 68, 71, 149, 159, 270
  battle 167
  Russian occupation 147
İsmail Pasha (General Kméty) 95–96, 471, 474
  and Crimean War 452
İsmail Tirsiniklioğlu Agha 220, 221, 223, 244–45
Istanbul 195, 347 *see also* Ottoman Empire
  discontent 245
  elites 185
  factionalism 84, 327
  French influence 240
  Greek Orthodox families 25–26, 289, 324
    reprisals against 292
  Janissary rampage 316–17
  under Mahmud 269–70
  and Mehmed Ali Pasha 363
  need for reform 191–92
  Patriarch 297
  plague 175*n*46
  police force 205
  provincial relations 197, 214, 217–19, 223, 250, 261–65
  purge 324
  revolts 35–38, 46, 117, 165–66, 483
  under Selim 243–44, 250
  and Serbian revolts 284–86
  vulnerability 346
Italinsky, Andrew Count 274, 277, 280
Italy 86
Ivan IV 31
İvaz Mehmed Pasha 85, 111, 115, 127*n*100
  at Orsova 113–14
Izmir (Smyrna) 227–28
İzmit 315
İzvornik 110
İzzet Mehmed Pasha 320, 351–52
  and *Redif* 382
Izzet Molla 347

Jaffa 230, 235, 370
  Egyptians abandon 406
  and İbrahim Pasha 384
Jalili family (Mosul) 85
James I, King of England 32
Janissaries 19, 35, 92, 177*n*67, 238, 241, 248–49, 260, 265, 301*n*24, 394*n*27
  accession gifts 50, 52
  admittance 39–40
  Agha (commander) 36; becomes *Serasker* 53
  Alemdar Incident 262–64
  allegiance to new sultans 37–38
  artillery corps 209*n*19

Janissaries (*continued*)
and arts of war 193
auxiliary forces 51*plate*, 76*n*11
in Balkans 85
in Belgrade 222
challenge to Mahmud 315
command and discipline 53–54, 117, 134
continuation of system 74–75
corruption 184
in Damascus 340*n*51
demands 55–56
deployment 64
discontent 50
disloyalty 221
Egyptian 20
1806–12 War 276
elimination (1826) 52, 203, 306, 313, 320–22, 339*n*36, 403, 485
planning 315–17
entitlements 57–58
and Eşkinci Corps 317–20
expansion 40
failure 45
financing 267–68
fortress troops 73
ideal 38
imperial 76*n*11
internal reorganisation 117–18
in Istanbul 270
at Kartal 154
and local revolts 57
loyalties and ethos 38–40
manumission 49
mobilisation against Ypsilanti 316–17
at Morea 124*n*55
at Nauplion 99–101
at Ochakov 161–63
payment of salaries 66, 100, 124*n*51, 317
privileges 50–52
provincial 189–90
at Prut 95, 122*n*39
rations 69
recruits 49–50, 134
and reform 130, 196, 205–6, 247, 286–87
regulations 53–54
relations with provincial forces 145
reorganisation of loyal 324
replacement 185
revolts 33, 36, 222, 268, 301*n*22
Selim's reforms 184, 194
1768–74 War 116, 143
Sipahis (cavalry) 76*n*11
structure 48–53, 339*n*36
süratçis (rapid-fire artillerymen) 162
and Varadin 102
at Vidin 220
weapons 195
Janissary Revolt (1730) 33
Jassy 96
Austrians and 164
and Crimean War 456
French consulate 223
Russian consul 290
and Russians 151, 153, 246, 289
and Russo-Ottoman Wars 131
Jassy, Peace of (1792) 134, 184, 222
terms 167
Jelavich, Barbara 430
Jena 242
Jerusalem 370
access to sacred places 400–1
Egyptians abandon 406
European involvement 417–18
Protestant church 408
Jiddah 370
jihad (holy war):
continuation 26–27, 41*n*19
1828–29 War 343
Jochmus, General 405, 406
Joseph II of Austria 34–35, 120*n*17, 137, 159, 160, 165–66, 178*n*86, 432
and Catherine the Great 137–38
military reorganisation 169
1787–92 War 163–64
Juchereau de St. Denys 195, 202, 250, 257*n*65
judges *(kadi)* 57

Kabakçı Mustafa 247, 248, 250
Kabarda 462
Kadi Abdurrahman Pasha 244
Kadiköy 274
Kadri Efendi 183

Kaehler, General 484
Kagul River 154
Kahraman Pasha (Arnavut) 143, 144, 174*n*34
Kalafat:
  and Crimean War 450, 452
Kalmuks 146
Kamanke 98
Kamenice 18
Kamensky, Mikhail Count 271
Kandia (Crete) 45–46
Kanizsa 65
Kaplan Giray 153
kapu halkı (household folk; provincial governer's forces) 38, 57
kapu kulu (household-slave; Janissary) 38
Kapudan Pasha 176*n*49
Kara George 219, 242, 273, 281
  1804 revolt 282–83
Kara Mustafa 35
Karabağ 463
Karahisar 216
Karaiskakis, Greek Commander 297
Karamanli family 385
Karaosmanoğlu family 261
Karaosmanoğlu 182–84
Karaosmanoğullari 144
Karasu 156
Karayal Hill 471
Karinabad 271
Karlowitz, Treaty of (1699) 20, 36, 67, 98, 102, 118, 129, 225, 485
  frontiers 24–25, 28
  as *jihad* 26–27, 41*n*19
  negotiations 18, 22, 24–26
  Ottoman violations 101–2
  and *sipahis* 56–59
  and Tatars 91
  terms 26–28
Karlstadt fortress 59
Kars 134, 217, 261, 460–63, 466
  and Crimean War 449
  fortress 468–69*plate*
  Russian occupation 464
  Russo-Ottoman conflicts 449–50
  strategic importance 462
Kars, Siege of 466–76
  commemorated 474–75
Kartal, Battle of (1770) 145, 149, 153–54, 170, 272, 301*n*35
  numbers of troops 154
  ordnance 174*n*33
  Ottoman losses 154
Kartlis 462
Kastamonu 216
Kaunitz, Austrian Chancellor 143
Kayseri 370
  anti-reform movement 366
Kazakhs 88
kaziaskers (military judges) 53
Kerch Straits 462
  fortress 155, 158, 166
Kevenhüller, 111
Khedives 407
Kherson 160
Kievan Rus 28
Kilburun fortress 68, 157–58, 160–62
Kilya (Kili, Killi, Kilya) 68, 158, 167, 280
  Russian occupation 147
  1768–74 War 143
Kırim Giray 143
Kırkkilise (Thrace) 71
Kiselev, Pavel D 424
Kladovo 271
Kléber, General 236
Klyuchevsky, VO 28
Koca Ragıb Pasha 84, 117–18, 130, 139, 189
Koca Sekbanbaşı 207*n*2
Koca Yusuf Pasha 164, 165, 166, 182–83
  and military reform 192
Koehler, General 239
  in Acre 257*n*55
Königsegg, Lothar 113, 114
Konya 370, 373, 378, 396*n*58
  and Battle of Nizib 388
  Ottoman-Egyptian confronation 372–74
  and *Redif* 383
Köprülü, Ahmed 111
Köprülü, Fazıl Mustafa 20–22, 41*n*4
Köprülü family 35, 46
Kör Yusuf Ziya Pasha 231, 236, 238, 270–73
Koron 99
Köse Mehmed Pasha 464

Kosovo 359
Kossuth, Lajos 426, 427, 434
Köstence 271, 349
Köstendil 271
Kozluca 271, 354
Kraina 222
Kriashens (baptised Tatars) 59
Kuban 28, 160, 278
Kuban River 462
Küçük Hüseyin Pasha 221–22, 231, 238
Küçük Kaynara Treaty (1774) 61, 188, 225, 278, 438, 441, 462
  negotiations 155–58
  terms 159–60
Küçük Seyyid Mustafa: *Diatribe de l'ingenieur Seid Moustafa* 203–4
Kuleli Vakası (revolt) 483
Kulewtscha, Battle of 353–54
Kunersdorf, Battle of (1759) 120*n*13
Kurdistan: pacification 427
Kurds 33, 60, 62, 136, 231, 241, 286, 462
  Battle of Nizib 388–90
  discipline 221
  exiles 251
  impressment 386
  and Ottomans 385–88
  rebellions 477, 486
Kurt-Tepe, Battle of 351
Kuruc rebellion 24, 87
Kurudere, Battle of 471
Küfahyalı Rıvdan Agha 144
Kütahya 286
Kütahya, Convention of (1833) 374
Kutaisi 462, 470
Kutuzov, General Mikhail I 164, 269, 466
  at İsmail 167
  replaced 281
  at Rusçuk 272–77
  and Treaty of Bucharest 277–78

Lacy, Field Marshal Franz Moritz 163
Lacy, Peter, General 83
  and Azov 103
  and Crimea 110, 114
Lafitte-Clavé, André
  at Ochakov 161–62
  and Ottomans 178*n*86
Lake, Colonel Atwell:
  and Crimean War 471–73
  Narratve of the Defence of Kars 471
Landwehr 382
Langeron, Louis Alexandre Andrault de 269, 271, 273
Larissa 100
Latakia 370
Laudon, Field Marshall 166, 167
Laz 33, 246, 251
Le Moniteur Ottoman 377
Leake, William Martin 303*n*51
Lebanon:
  and Europe 417–18
  and İbrahim Pasha 384
  riots 477
Leopold I, Holy Roman Emperor 20, 22–23, 65
Leopold II, Holy Roman Emperor 138
Lesczynski, Stanislas 97
Lesser Kabarda 158
Levant Company 227
Levend Çiftliği 195–97, 262–64
levend (irregulars) 56–57, 151
Lewis, Bernard 228
Light Brigade: and Crimean War 458, 460
Limnos 247
Lipova 66–67
Lippa 66
Liston, Robert 281, 302*n*37
Livonia 90
local elites *(ayan, derebeys)* 185, 301*n*24, 397*n*66
  consolidation 84
  discontent 416–18
  growth 130–31
  Osman Pasvanoğlu 219–24
  Ottoman 140, 171*n*3, 181, 214–16
  power 218–19
  rise in importance 84–86
  supply and mobilisation 182, 271–72
local militias:
  importance 126*n*83
  reliability 127*n*105
  role 63–65
Lofça 222, 272
Lom Palanka 276

London:
  Ottoman embassy 226–27, 377
  philhellenes 291–92
London, Convention of (1840) 400, 436
  and France 407
  terms 405
London, Treaty of (1827) 298, 355, 423
  terms 399
Long War (1593–1606) 54
  Janissaries 49
Louis Philippe, King of France 426
Louis XIII, King of France 57
Louis XIV, King of France 20, 24, 57
  wars 94–95
Lynn, John 57–58

Macedonia 242, 261, 293, 336
  anti-reform movement 366
  Christians 92–93
  revolt 453
Macedonians 29
Macfarlane, Charles 313, 327, 344, 347, 352
Maçin 182, 271
  and Crimean War 453
Ma'den 370
Maghrebins 232
Mahmud I 52, 115
  and Janissaries 317
  reformed army 251
Mahmud II 85, 131, 136, 169–70, 180, 185, 242, 247, 259, 264–65, 293, 301*n*24, 314*plate*, 464 *see also* Ottoman military
  and Abdullah Pasha 367
  absolutism 223, 299, 314–15, 322–25
  achievements 252
  and administrative/military class 251
  and Akkirman Convention 297, 299
  and Ali Pasha of Iannina 287
  and Battle of Kulewtscha 354
  and Battle of Nizib 388–91
  and British 305*n*74
  civil service 376
  clarification of Muslimness 285–86, 385
  confiscations 325, 341*n*53
  death 391
  and Deed of Agreement 262
  and Diyarbakir 385–88
  and economy 325–26
  enthronement 249
  foreign advisors 210*n*28, 326, 376
  governance 376, 379–80
  and Greek Revolt 298–99, 343–44
  and Hüsrev Pasha 377–79
  and İbrahim Pasha 372
  imperial consolidation 198, 260
  and Janissaries 297, 316–17
  jihad against Russia 299–300
  and Mehmed Ali Pasha 237, 312–13, 365, 383–85
  modernising dynastic image 403–4
  and Mustafa Pasha 359
  and Obrenovich 422
  and Ottoman sovereignty 284
  Ottomanism 485
  and provincial elites 286–87, 388
  provincial relations 282–83
  recentralisation 184, 285
  reforms 216, 317–20, 320, 377, 379, 391–92, 396*n*65; agenda 1834–39 376–82
  religious toleration 323, 360–61
  revenues 341*n*53
  and Russian imperialism 269–70
  Russian military assistance 374–75, 388
  and Selim's reforms 205
  steadfastness 272
  and subject populations 134
  and Syria 370, 374
  and Treaty of Adrianople 362–63
  and Treaty of Bucharest 278–81, 463
  visit to Tutrekan 362*plate*
  and von Moltke 240
  and Wahhabis 308–9
Mahmud Raif 246–47
  Tableau des nouveaux règlemens de l'Empire ottoman 226–27
Makarov, Ivan 266–67
Malatya 336
malikane 185
Malvasie 99
Mamluk armies 240–42
  and Napoleon 229–30
  sources 256*n*44
  training 233*plate*
  wars 235

Mamluk Beys 307
Mamluks 31, 62, 215, 238–39, 243, 286, 337*n*3
 civil war 237
 control 308
 elimination 310, 337*n*3
 factionalism 327
 households 215, 232–35, 307
 and Mehmed Ali 307
Manolaki *Ban* of Craiova 152
Marani: and Crimean War 473, 475
Maraş 370, 371
 and İbrahim Pasha 371
 reform of fiefs 336
Mardin 286
Maria Theresa, Habsburg Empress 34–35, 86, 137
Marie-Antoinette 178*n*86
Markov, Commander 276
Marmont, Marshal 371, 373, 396*n*52; *n*58, 397*n*68
Maros River 59, 65
Marsigli, Luigi Ferdinando 19–22, 24, 40*n*1, 49, 55, 62, 64, 77*n*15
 Ottoman military transport 70*plate*
Martineshti 166
martolos (cavalry) 63–64
Maurocordato, Lord Alexander 25
Mavrocordato, Nicholas 91, 140
McNeill, 116–17
Mecca *see* Sacred cities of Mecca and Medina
Mecidiye, 451*plate*, 455*plate*
 and Crimean War 454
Medina *see* Sacred Cities of Mecca and Medina
Mehadia 113
Mehmed Ali Pasha (of Egypt) 215–16, 235, 242, 316, 337*n*4, 401–7
 see also Egypt; Egyptian army; Mamluks; Syria
 Albanian troops 310
 and Battle of Nizib 388–91
 consolidation of power 307–8
 1828–29 War 345
 exiled 461
 expansionist agenda 366–67
 and French 364
 and Greek Revolt 294, 297, 304*n*66, 312–13
 hereditary rule of Egypt 405–6
 and Hüsrev Pasha 337*n*1
 independence 385
 and Mahmud II 299, 306, 364–65, 384
 and Morea 352
 Napoleonic Wars 308
 and Ottomans 237, 366
 reforms 259
 religious orthodoxy 366
 rise 238–39
 and routes to India 375
 taming 404–7
Mehmed Ali Pasha (minister of war) 439–42
 exiled 461
Mehmed Bey Lazoğlu 312
Mehmed Celaluddin 316, 317–18
Mehmed Emin Agha 202
 Sevânihü'l-Levâyih'i 204–5
Mehmed IV 53
Mehmed Pasha:
 at Aleppo 370
 and İbrahim Pasha 371
Mehmed Raşid Efendi (*Reis*) 226, 286
Mehmed Reşid Pasha
 and Missolonghi 294–95
Mehmed Rüşdü 461
Mehmed Selim Pasha 317–18, 331, 350–52, 352, 405
 and Hüsrev Pasha 348
Mehmed Tahir 148
Memiş Agha 218–19
Memoirs of Saint Remy 213*n*34
Menekşe 221
Menou, General 236
Menshikov, Prince (Russian envoy) 350, 438, 440
 impact in Istanbul 437–38
menzils (way-stations) 69–70
Mes'ud Agha 197
Metternich, Clemens von 425–26, 432
Mevlevis 303*n*53, 323, 340*n*45
Michael, Grand Duke 348
Michailovsky-Danielevsky, 277–78
Michelson, General 245
 and Principalities 271

Middle East:
access to sacred places 400–1
British diplomatic presence 487
colonialism 229, 236–37, 400–1
European involvement 417–18
international intervention 306
religious proselytising 400–1
Mikdad Ahmed 218
Military College 89
Military Commission (1762) 168
Military Council 379
Mingrelia: and Crimean War 473
Ministry of War (Ottoman) 409
Mir Mehmed Pasha 386
miri system 194–95
Misistre 99
Misivri 354
Missolonghi 293, 303*n*58, 304*n*66
response to fall 297–98
retaken by Ottomans 294–95
siege 316
survivors 304*n*67
Modon 99, 124*n*55
siege 100
Moldavia 22, 25, 28–89, 91, 137, 140, 265, 278, 287, 444
autonomy 392, 423–25, 475
border dispute 278, 280
and Crimean War 456
depopulation 141, 146–47
famine 92, 95–96
and George Bibescu 425
and Greek Revolt 288
impositions and obligations 209*n*17, 302*n*43
native rule 292
and Ottomans 61–62
peasant conscription 281–82
and Phanariots 140–41
rebellions 98, 288–89; 1848 426–27
and Russia 142, 456
Russian occupation 243, 269–70, 282, 426–27, 435–37, 440
Russification 289
Russo-Ottoman disputes 401–2
1768–74 War 143
Treaty of Adrianople 361–62
Treaty of Bucharest 278, 280
and Ypsilanti 290–91, 316–17
Moldavian army 90
and Kutuzov 272
and siege of Vienna 62
Moldovancı Ali Pasha 150
Molière: 'Bourgeois Gentilhomme' 228
Moltke, Helmut von (the Elder) 203, 213*n*49, 240, 387–88
and Battle of Nizib 388–91
on Mahmud 360–61
and Ottomans 376
on Russian troops 346–47
Monastir 359
Moniteur 230
Montecuccoli, Count Raimondo 46
Monteith, William 464, 465
Montenegro 435
rebels 99
secularisation 430–31
and Tanzimat 429–32
Montesquieu, Charles: *Lettres persanes* 211*n*36
Morava River 106
Morea 18, 24, 221, 287–88, 309, 313
Egyptian troops 304*n*66
flight of populations 153
French forces 352
Greek Revolt 296–97, 316
and Ottoman fleet 123*n*47
and philhellenes 295
rebellions 153–54, 288
Morea campaign (1715) 83, 90, 101
losses 100
numbers of troops 99–100, 123*n*50
Ottoman recapture 98–101
Morier, D R 303*n*51
Mortar Corps *(Humbaracı)* 193, 202
Moscow 28, 164, 432
and Napoleon 248, 265, 271–72
Mosul 261
Mount Lebanon 370
partition 416–17
post-1841 416
Mourouzis, Constantin 292
mubaayacı (state purchaser) 73
Müftizade Selim Efendi 277
Muhammad Bey Abu al-Dhabab 215, 234
Muhammad ibn Su'ūd: and Wahhabis 337*n*7

Muhammad (prophet) 27
  authority 251
  military innovations 188
Muhammed ibn 'Abd al-Wahhāb 308–9, 337*n*7
Muhib Efendi 228, 242, 243, 247, 257*n*62; 63
Muhsinzade Abdullah Pasha 85, 106–10, 125*n*76
Muhsinzade Mehmed Pasha 43*n*39, 125*n*76, 142, 153, 157–58, 177*n*62, 188–89, 201, 234, 287, 349
Mukhtar Pasha 287
Mulla Pasha: treachery 273–74, 276
Münchengrätz Convention (1833) 375
Münnich, General 83, 85, 103, 165
  and Bender 114
  brutality 126*n*79
  Crimean campaigns 107–10, 125*n*65
  at Ochakov 120*n*13
  1768–74 War 116
  Wallachia and Moldavia 106–7
Murad IV 33
Muraviev, General 460–61
  and Crimean War 473, 474
Murids 466
Murphey, Rhoads 49, 63, 65, 72
  Ottoman Warfare 187
Musa Pasha (*Kaymakam*) 246–47, 248
Muscovy 31, 58
Muslims *see also* Ottoman Empire; Syria
  in Albania 359
  atrocities against 291
  and conscription 372
  constitutionalism 477
  demographic displacement 160, 269
  in Mount Lebanon 416–17
  and non-Muslims 299
  and Organic Regulations 424
  orthodoxy and reform
  Ottoman navy 316
  and reform 477
  reformist literature 188
  rivalries among landowners 359
  in Syria 365
Mustafa Alemdar Pasha 84, 259–61
  and Serbia 283
Mustafa Asım Efendi 323
  and Abdülmecid's reform agenda 403
Mustafa II 35–37
Mustafa III 72, 82*n*80, 135, 142, 183, 189, 193, 202, 213*n*34
  expenditure on war 149
  and Frederick the Great 139
  military reform 212*n*40; 42
Mustafa IV 218, 247, 248, 286
  execution 263
  supporters 262–64
Mustafa Kesbi 143, 173*n*29, 174*n*31
Mustafa Naili 441
Mustafa Nuri Pasha 409
Mustafa Pasha (Belgrade commander) 394*n*23
  and Serbia 222
Mustafa Pasha (commander, Crimean War) 450
Mustafa Reşid (*çavuş*, d. 1819) 185, 207*n*2, 287
Mustahfizan 232
müstahfizes (local fortress guards) 64
Mutasarrıfs 171*n*4

Nablus 231, 370
  violence against Christians 422
Nadir Shah 85, 106
Najd 309
Namık Pasha 327, 374
  and Sixth Army 418–19
  pacification of Aleppo 418–20
Napoleon 34, 63, 214, 236–37, 243, 257*n*62; 63, 269, 310
  at Acre 232
  Adriatic conquests 287
  citizens army 168
  coronation 242
  and Egypt 195–96, 216, 225, 229–31, 241, 287
  fall 432
  ideological propaganda 240–42
  and Istanbul 272
  and Mahmud II 280
  Moscow campaign 164, 248, 265, 271, 302*n*42
  and Osman Pasvantoğlu 222
  and Ottomans 198

Napoleon (*continued*)
  structure of regiments 330
  and Third Coalition 242
  and Venice 99
Napoleon III 435
  and Crimean War 457–58, 476
  and Levant 438
Napoleonic era 136, 170, 181
  armies 477–78
  close 458
Napoleonic Wars 138, 169, 264–65, 308, 346
  and military reform 206
Naqshbandi Sufis 204, 215, 251, 340*n*45, 366, 403
  teachings 323–24
nasihatnames (advice manuals) 187–92
Nasmyth, Lieutenant Charles: and Crimean War 454
National Army Museum 474–75
nationalism 299
  South Slavs 439
Nauplion (Napoli) 124*n*55
  defenders 124*n*59
  Greek control 297
  siege 99–101
Navarino 123*n*47, 294, 313, 363–64
  and Ottoman fleet 299, 333, 343–44, 463
Necati Salim 122*n*38
Necib Efendi 312, 320, 338*n*20
Neipperg, Count Reinhard Wilhelm 85, 127*n*102
  1768–74 War 115–16
Nelson, Admiral Horatio 230
  and Egypt 236
Nemirov Peace Congress (1737) 106, 112, 119*n*1, 125*n*65
  Austrians at 125*n*65
Nesselrode, Chancellor 407
New Order *see Nizâm-ı Cedid*
New Orsova *see* Adakale (New Orsova)
New Russia 139–40
  civilian populations 59
Newcastle, Duke of: and Crimean War 457
Nicholas I 265, 434–42
  death 460
  1828–29 War 346, 348, 350–53, 352–53
  future of Ottoman territories 363, 399–400, 407
  and Münchengrätz Convention 375
  and Ottomans 297–98
Niğbolu 157, 220, 271–72, 274
  reorganisation 336
Nile, Battle of (1798) 230
Nine Years War 24
Niş 103, 106–7, 273
  Ottomans retake 110–12
  rebellion 423
  and Serbia 447*n*23
Nizâm-ı Cedid 129, 181, 194, 203–4, 238, 257*n*65, 310
  Albanians 246, 255*n*39
  Anatolians 145
  creation 196
  definition 206
  discipline 205–6
  elimination 247
  financing 185
  foreigners 198
  and Janissaries 260, 321
  medical school 327, 329, 376
  and Napoleon 231
  provincial 197–98, 214, 219, 244
  recruitment and training 185, 196, 197–98
  supporters 223
  Talimli Askerler (Trained Soldiers) 185
Nizamiye 410, 478–79
  in Bosnia 431–32
  composition 412–13
  divisional armies 415–16
  and Egyptian army 410
  enlisted men 413–14
  irregular cavalry (*see başıbozuk*)
  miliary preparatory schools 413
  structure and deployment 410–14
  training and drill 412
Nizib 386
  battle 388–92
North America 130
North Anatolia *see also* Anatolia
  and Canikli Ali 216–19
Novoştanilisti 97

Numan Pasha 156–57
Nurullah 386

Obrenovic, Milos 283–84, 289
  Prince of Serbia 422
Obreskov 142, 155
ocaklık (group of tax farms) 73, 75
Ochakov 64, 68, 83, 95, 107, 138, 159–60, 160, 268
  barbarity 165
  crisis 170
  disease and dysentery 165
  fall 131–34, 167
  losses 110, 120*n*13, 126*n*79, 164–65, 179*n*93
  numbers of troops 164–65
  Ottoman forces 164
  preparations for defence 103–6
  Russian blockade 164
  Russian losses 107–10, 112
  1768–74 War 114, 143
  1787–92 War 161–65
  siege 110, 256*n*47, 272
  supplies 148–49
odabaşı (head of cavalry) 66
Odessa 87
  Greeks in 289
Okhri 288
Olmütz 442
Olt estuary 274
Oltenia 102, 290 *see also* Wallachia
Oltenitsa 353
  and Crimean War 444, 452–54
Ömer Pasha (Michael Lattas of Croatia) 296, 427–28, 460–61, 461, 473, 480
  and Crimean War 450, 452, 454, 458–60, 470, 474–75
  Third Army 424
  ultimatum to Russia 443
Ömer Pasvanoğlu 220
Ömer Vrione Pasha 345
  and Albanian troops 349–51
ordu akçesi (guild contibutions) 71–72
Organic Regulations (1829) 424–25
Orlov, Admiral Count Alexis 154, 234
Orsova (Old Orsova) 102, 164, 166 *see also* Adakale (New Orsova)
  and Austria 138
  Ottoman victory 112–17
  raids 111
Orthodox Church 23
  politics of resistance 153
Orthodox Patriarch: and Ypsilanti 290
Ösijek 65, 67
Osman, House of 215
Osman Agha (of Temeşvar) 65–67
Osman Agha Şatiroğlu 218
Osman Halisa Efendi 106
Osman III 189
Osman Pasha (at Plevne) 139
Osman Pasha (*serasker* 1828–29) 465
  War 465
Osman Pasha, Governor of Anatolia 271
Osman Pasha of Tripoli 370
Osman Pasha (Tirhala *Mutasarrif*) 221
Osman Pasha *(Tüfenkçi Başi)* 222
Osman Pasvanoğlu 56, 84, 215–16, 229–30, 237, 245, 287
  and Selim III 241
  and Vidin 219–24
Ossetians 462
Otto I, King 453
Ottoman Bank 409, 440
Ottoman borderlands 65, 140, 174*n*32, 191, 466
  eastern 31–4, 60–61
  fortress system 63–65, 91, 102–3, 112, 131, 135, 153, 158, 166–67, 333
  Iranian 130
  northern 28–29, 61–62
  Persian 32
  Russian 361–63
  Safavid 85–86, 130
  southern 62–63
  stabilisation 141–42
  Sunni-Shiite 32–33
  western zone 28
Ottoman court 83–84
  and British interests 388
  British missions to 375–76
  criticism 185
  household patronage 409
  influence and clientage 61
  pasha households 36
  pro-war party 161

Ottoman court (*continued*)
promotion of dynastic power 47
and Prussians 240
reform 186
religious innovation 322–25
rivalries 161, 166, 217–18
under Selim 259–60
Ottoman diplomacy 253*n*19, 254*n*27
ambassadors 227, 228–29
and Concert of Europe 399, 401, 476
diplomatic corps 25–26, 41*n*15
embassies from the West 228
embassies to the West 84, 121*n*19, 139, 253*n*20, 253*n*21
re-opening of 377
European initiatives 224–29
Europeanisation 25–26, 84, 170
foreign affairs bureaucracy 139
foreign mediation 284
Greeks and 227
late eighteenth century 139–40
under Mahmud II 228–29, 376–77
Russian overtures 222
Ottoman-Egyptian army 295
Ottoman-Egyptian elites 308
Ottoman-Egyptian fleet 299
Ottoman Empire 90 *see also* Istanbul; Russo-Ottoman Wars
administrative systems 29–30
aims 34–36
anti-Russian feeling 246
and Balkans 84–86
bandits *(eşkıya)* 191
besieged 251–52
and Caucasus 106
centralisation of control 61–62
Christians 77*n*15, 242, 251–52, 285–86, 289, 360, 394*n*34, 437–39, 438–341, 475
and conscription 479–780
protection 440
relations with Muslims 224
civilising mission 485–86
clamour for autonomy 487–88
colonialism 215
communications across 250
disintegration 343
economy:
commercial code 415
financial crisis 479
fiscal innovation 170*n*1
paper money 414
reform 414–16
treasury reorganisation 184–85
education 415 (*see also* Ottoman military)
and Egypt 232–33
equality of subjects 476
ethnic and religious diversity 35, 134, 224, 286, 417–18
freedom of worship 476
and loyalty 23
militant Islam 366
non-Muslim rights 360
perceptions of 285–86
Sunnis 31
tensions 32–33, 154, 205, 251–52, 343, 358–59, 408
European intervention 399–400
foreign missionaries 251–52
and French 115, 139, 198
future 259–60
geopolitical realities 33
governance 85, 130–31, 169, 251, 285
Greeks 98, 134, 289, 358
impact of military reform 223 (*see also* Ottoman military)
intelligence sources 139
and Islam 60, 485–86
Long War (1593–1606) 54
middle class 183
military needs 214–15
modern age 170
Muslims 215, 343, 394*n*34
and conscription 479–80
relations with non-Muslims 477
and Nadir Shah 85
Napoleonic threat 222
native mercantile class 205
new allegiances to 85
newspapers 377
orthodoxy 244, 260
at peace 118–19, 130
police force 415, 479
political crisis 218–19
population 394*n*34
and Portuguese 31–32

Ottoman Empire (*continued*)
and Principalities 91
provincial relations 35, 38–39, 56–57, 73–75, 416–18 (*see also* local elites *(ayan, derebeys)*)
elites 215–18, 260–65
and Qazdağlıs 235
reform 476–77
public support 316
social context 187–92
social impact 206
refugees 427–28, 477, 481*n*40
Hungarian 427–28
Polish 427–28
rise and eclipse 34–35
sedition 324–25
social identities 366
social transformation 185–86
sovereignty challenged 85
strategic imperatives 139
and Tatars 30–31, 155
taxes and tax collecting 130–31, 341*n*55
groups *(ocaklıks)* 73
imdadiye (campaign tax) 71, 74
non-Muslim 73
provincial 73–74
reform 72–73
revenues *(mukataa)* 38–39, 43–4*n*42, 55, 73, 85
tax farms *(malikane)* 39, 72–73, 130–31, 185, 261
technology 195
territorial maintenance 399–400
trade 375–76, 388
Treaty of Adrianople 361–63
Turkification 264
universal citizenship 401
unregulated populations 264
war measures 171*n*8
and Western historians 205
westernisation 366
Ottoman fleet 463
Ottoman-French relations 227–28
Ottoman-Habsburg conflicts
1593–1606 41*n*12
1787–91 220
and Osman Pasvanoğlu 220
peace treaty 24
Ottoman-local elites *see* local elites *(ayan, derebeys)*
Ottoman military 48, 415–16 *see also Asakir-i Mansure; Nizamiye; Redif*; war and warfare
administrative reform 193
Albanians 189–90, 288, 306–7
Anatolians 188
army's rate of advance 123*n*50
arsenal 195
artillery corps 162–63, 185, 193, 202–3, 315–16, 324
discipline and training 201
field 135, 162, 202
foreign advisors and 198–206, 210*n*28
gunners 198, 201
Janissary 209*n*19
numbers under Selim III 202–3
privileging 213*n*49
and arts of war 193
auxiliaries 62, 119*n*5, 207*n*3, 480
Abkhazians 463
Albanian 84
Anatolians 216–17
Bosnian 84, 126*n*83
Christian 63–65
financing 72
Kurdish cavalry 478
local commanders 156
local militias 126*n*83, 127*n*105, 130–31
and Mahmud's reforms 334–36
Muslim 63–65
religious factors 63–65
in battle 150–51
at Belgrade 124*n*60
camp followers 188
campaign financing 71–75, 134
campaign routes 68–69
cavalry *(sipahis)* 54–56, 150–51 (*see also başıbozuks*)
Anatolian 232
evolution 56–58
mutiny 182–84
at Ochakov 161
and reform 206
change to voluntary 134
Christian/Muslim relations 480

Ottoman military (*continued*)
- collapse 449
- communications 100
- conscription 57–58, 256*n*44, 349*n*29, 479–80, 482*n*42
  - exemptions 479–80
- corruption 176*n*59, 190–91, 472–73, 481*n*22
- desertions 365–66
- discipline and training 100, 118, 134, 144–45, 168, 174*n*34, 201, 221, 238, 268, 292–93
- education 376, 483–84
- ethnic and religious diversity 65, 251, 268, 360–61
- ethno-religious distinctions 197
- European view 186
- financing 82*n*80, 146, 193
  - Jews' estates seized 380
  - *miri levend* 56
- foreign advisors 135, 163, 178*n*86, 185, 198, 484
- French contingent 202
- general staff 378, 409
- guilds' campaign contributions 71–72, 82*n*77
- *hayduts* 58
- hunger 268
- informants 123*n*41
- irregulars 56–57, 191
  - Muslim 465
- justice 53
- leadership 177*n*62, 179*n*95, 221, 292–93; local 85, 156
- logistics 75
- looting 268
- losses 1826–37 396*n*55
- Mahmud's reforms 129, 203, 210*n*28, 238, 320, 376
- mobilisation 68–69, 126*n*90, 130–31
- mobility 250–51
- Muslim orthodoxy 205
- and Muslim social mobility 483
- night attacks 350
- officers 118, 144, 239–40, 377–79
  - training 409, 472
- organisation 83, 88
- peasant soldiers 145
- post-Crimea 483–84
- prices and wages 82*n*82, 188–89
- and prisoners of war 268–69
- and provinces 54–58, 191
- and Prussians 240
- Prussification 240, 286, 325–28
- rations 69
- rebuilding after wars 169
- recruits/recruitment 134, 189–90
- reform 49, 118, 128*n*108, 129, 169–70, 180, 181–93 (*see also Nizâm-ı Cedid*)
  - financing 212*n*40
  - post-Crimea 408–14
- regulations 53–54
- Selim's reforms 129, 185–86, 191–95, 203, 214
- in 1700 45–46
- slavery 77*n*22, 338*n*13
- strengths and weaknesses 117–19, 158
- supply systems 69–71, 81*n*75, 100, 130–31, 176*n*59, 189
  - collapse 129, 147–51
  - reform 194–95
- technology 118, 486
  - gap 88, 134–35
- transport 70*plate*
- warfare administration 90
- Western impressions 144–45, 200–1, 237–42

Ottoman/Naqshbandi synthesis 323
Ottoman navy 100–1, 123*n*47, 330, 404
- Black Sea fleet 163, 316
- at Çeşme 154, 199
- and Crimean War 444
- Danube fleet 147, 176*n*49, 276
- and Greek sailors 316
- Mediterranean fleet 147
- at Navarino 299, 333, 343–44, 463
- at Ochakov 164
- 1768–74 War 147

Ottoman-Persian War (1821) 463
Ottoman-Prussian alliance 165
Ottoman-Russian confrontations 449–50 *see also* Russo-Ottoman Wars
- and Tatars 91

Ottoman Ultimatum of 1853 441
Ottomanism 487
- evolution 484–85

Ozurgeti: and Crimean War 471

Pahlen, Count 347, 354
Palestine 234
  European involvement 417–18
  resistance to İbrahim Pasha 384
Palmerston, Lord 374–76, 427
  and Crimean War 444, 457–58
  diplomacy 401
Pamuk, Sevket 325
Pangiotis Codrika 228
Paris:
  Ottoman embassy 226–28, 242, 377
  philhellenes 291–92
Paris, Treaty of (1802) 228
Paris, Treaty of (1856) 475
pasha households 36
Paskevich, General 346, 434
  in the Caucasus 463–66
Passarowitz, Treaty of (1718) 138
  terms 102
Pasvanoğlu, Osman *see* Osman Pasvanoğlu
Patras 294
patriarchs (Orthodox) 292, 297
Patrona Halil Revolt (1730):
  and guilds 33, 71
Paul I, Tsar 136, 167, 265, 267 *see also* Russian army
  diplomacy 254*n*28
Pecs 65
Pehlivan Hüseyin 247, 301*n*27
Pehlivan İbrahim Pasha 272
Peloponessus 100, 290, 326, 330
  Greek Revolt 238, 291–94
  Greeks in 289
  and İbrahim Pasha 294
  persecution of Muslims 292
  restoration of Ottoman authority 293
Perekop 52, 103
Persia 85
  conflict with Ottomans 71–72
  and Ottomans 106, 125*n*65
  peace with 117
  Safavids 31, 33
Pertev Pasha 323
  and Gülhane/Tanzimat 377
Peter the Great 18, 24, 29, 48, 75 *see also* Russian army
  and arts of war 193
  and Prut 90–98
  and Russian navy 91
  social restructuring 186
Peyssonnel, Louis Charles de 76*n*12, 200–1, 212*n*37
Phanariots 29, 61, 80*n*53, 140–41, 289
philhellenes/philhellenism 94–205, 240, 291, 297, 303*n*58
Philike Hetairia 289, 292
Piraeus: blockade 453
pişkeş 91
plagues 112, 146, 271, 301*n*26, 346, 352, 393*n*16 *see also* war and warfare
Plevne, Battle of 131
Podolia 18, 46
Poland 18, 24, 30, 46, 146, 280
  borderlands 140
  break-up 88
  Catholicism 140
  civil conflict 142
  partitions 29, 137, 139, 143, 155, 170
  refugees 447*n*29
  and Russia 135, 172*n*20
  1768–74 War 138–39
  territorial integrity 137
  troops 150
Poles 90
Polish-Lithuanian Commonwealth 46
Polish Question 155
Pölöske, siege 65
Poltava 110
Poltava, Battle of (1709) 90, 93
Pomaks 272
Poniatowski, Stanislaw, King of Poland 140
Ponsonby, John 403
Porter, Sir James 76*n*12
Portuguese 31–32
Potemkin, Grigorii 160–61, 164–68, 268
  at Ochakov 163–65, 179*n*93
Poti 463, 466
  1806–12 War 463
  and Crimean War 466, 473
  Russo-Ottoman War (1768–74) 462–63
Potoski 142
Potsdam 145

Pozzo di Borgo, Count 356
Pravadi 221
  1828–29 War 353–54
pre-Napoleonic warfare 46–48
Pressburg, Battle of (1805) 242
Preveza 237
Princes' Islands 243
Principalities 28–30, 87, 167, 276, 290, 346
  borderlands 140
  Christian populations 159, 423–24
  Convention of Akkirman 297
  and Crimean War 456
  dual protectorate 424–25
  ethno-religious movements 251–52
  flight of populations 291
  governors 98
  and Greeks 289
  importance to Ottomans 61–62
  Muslim populations 291, 423–24
  Ottoman-Russian disagreements 292, 297–98, 313, 401–2
  peasant unrest 141–42
  Phanariot rulers 140–41
  relationship with Empire 487
  religious diversity 360
  religious propaganda 242
  and Russia 92–93, 163, 265, 426–27, 440, 456
  Russian occupation 158, 222–23, 242, 246, 271, 282, 443
  supply obligations 194, 209*n*17, 285, 393*n*10; 1828–29 347
  and Tanzimat 423–24
  timariots 423–24
  and Treaty of Adrianople 361–63, 363
  and Treaty of Bucharest 280–81
  and Ypsilanti 290–91
private armies 39, 142, 182–84, 244
  Balkans 244
  competing loyalties 61–62
  Egypt 231–32
  Osman Pasvanoğlu's 220–21
  Syria 231–32
Prozorovsky, Prince 271, 273
Prussia 34, 145
  and Austria 35, 166
  and Bavarian Succession 159
  and Convention of London 405
  and Eastern Europe 130
  and global warfare 87
  Landwehr 382
  mediation 84
  Ottoman embassy 226–27
  and Ottomans 165
  rise 87, 119
  Russian drive into 120*n*15
  Triple Alliance 165
Prussian army 86, 477–79
  discipline 168
Prusso-Ottoman relations 118, 376
  alliance (1790) 161
  defensive alliance 138
Prut, Battle of (1711) 29, 48, 75, 83, 95–98, 101, 122*n*38
  and Baltacı Mehmed's army 117
  Egyptians 95
  mobilisation 95–97, 122*n*39
  origins of campaign 90–95
  Ottoman forces 93
  peace terms 29, 97–98
  Russian forces 94–95, 122*n*39
Prut River 28–29, 149, 272
Pugachev Rebellion (1773–74) 136, 156, 224
Pyramids, Battle of the (1798) 229

Qadiris 323
Qajar Turco-Persian dynasty 463
Qazdağlı households 215, 234–35, 238, 256*n*44
Quraysh 251
Qusayr 236

Radetzky, General 433
Radhujevac 111
Raglan, General:
  and Crimean War 454, 459, 471
  and Turkish Contingent 460
Raguzinsky 92, 97
Rami Mehmed Effendi 26
Ramiz Pasha 263, 264
Rauf Pasha (Mehmed Emin) 142, 150, 378, 405
  military reform 409
Raymond, André 52

Redif (Reserve Army) 436
 and Crimean War 478
 desertion 389–90
 establishment 380–83
 expansion of system 410–14
 financing 383
 and the Kurds 385–89
 muster roll 381*plate*
Redut-Kale: and Crimean War 474–75
Reichenbach, Convention of (1790) 166
Reisülküttab (Chief of the Scribes) 26
religion:
 and international relations 23
 proselytising 400–1
 rhetoric 37
 rivalries 84
Repnin, Nikolai 158
Rescript *see* Gülhane Edict; Tanzimat (1839–76)
Reşid Mehmed Pasha 352–54, 359, 379, 386, 388, 396*n*52, 441, 477
 at Konya 373
 and Syria 384–85
Reşid, Mustafa Pasha (foreign minister and grand vizier)
 and Tanzimat reforms 323
 and Crimean War 439–44
 and Abdülmecid's reform agenda 403
 and Gülhane Edict 377, 402–3
 lecture to Janissaries 204
 mission to London 389
 resignation 408
 and Treaty of Hünkâr Iskelesi 375
Resmi Ahmed *see* Ahmed Resmi Efendi
River Olt 152
Riyadh 309
Ridvanoğlu, Ali Pasha 359, 429–30
Robertson, James: *başıbozuk* 411*plate*
Romania 137–38, 160–61
 development 140–41
 historiography 92
 independence 362
 rebellion of 1848 426–27
 revolt 290
Romanians 159
Romanov Empire 281 *see also* Russia; Russian army
 aims 297
 assimilation 35
 borderlands 33–34, 59–61, 87–88, 140, 215, 346
 Christians 136
 clamour for autonomy 488
 colonial wars 269
 colonisation measures 135–36, 172*n*14, 215, 487
 expansion 34–36, 87–88, 135–36, 139–40, 142
 financial challenges 136
 and Georgia 462
 global conflict 119
 Muslims 136
 and Orthodox faith 28, 60
 and Orthodox populations 159
 Orthodox proslytising 172*n*14, 215
 and Poland 88
 provincial administration 175*n*40
 provincial estates 266
 religious/ethnic diversity 135–36; intolerance 32–33
 serfdom 135
 social impact of reform 135–36, 206
 support for co-religionists 29, 438–41
Romanovs 19, 88–89
romanticism 299
Ronne, General 97
Rosetta 236, 243, 307
Ruffin, Pierre 228
Rumeli 220
 cavalrymen 182
 centralised control 287
 and recruitment 198
 reform of fiefs 336
 revolt 288–89
Rumeli Kavağı 246–47
Rumeli Mehmed Pasha 150
Rumelia:
 military judges 53
 and Prut 95
 sipahis 55
Rumi 323
Rumiantscv, P A, Field Marshal 46, 58, 89, 130, 140, 145, 160–61, 164–65, 176*n*48, 198, 266, 269, 272, 301*n*35, 466
 Azov-Mozdok line 463
 Kartal ordnance 174*n*33

Rumiantsev, P A, Field Marshal (*continued*)
  and military reform 168
  Russian Second Army 146
  1768–74 War 149, 151, 153–55, 156–58
  troops under command 174*n*37
Rusçuk 20, 68, 84, 131, 156–57, 223, 244–45, 275*plate*, 284, 316, 429
  disrepair 334
  1828–29 War 349, 354
  supplies 147, 149
Rusçuk, Battle of 270–77
Rusçuk Committee 247–52
Russia 18, 23 *see also* Romanov Empire
  and Britain 139
  control of Tatars 30–31
  and Convention of London 405
  and the Crimean peninsula 28
  Decembrist Revolt (1825) 265
  diplomacy 254*n*28, 290–91
  and Eastern Europe 130
  and Greek Revolt 297–98
  and Habsburgs 229
  and Holy League 24
  and İsmail fortress 167
  and Napoleon 237
  and Ottoman affairs 260
  and Ottoman court 160, 222, 224
  Ottoman policy 440
  and Poland 172*n*20
  and Principalities 435–36
  and Selim III 228
  and Serbia 242
  1768–92 135–36
  strategic problems 87–88
  Third Coalition 236
  and Transcaucasus 134
  Treaty of Adrianople 361–63
  Treaty of Hünkâr İskelesi 374–76
  Treaty of Kütahya 374
Russian army 146 *see also* Cossacks
  Albanian troops 150
  border garrisons 168
  conscription 168, 265–66
    serfs 436–37
    substitution 266
  Cossacks 103, 168–69, 267
  defence 58
  discipline 266–67
  disease 268
  in 1850 436–37
  elimination of *streltsy* 88–89
  ethnic regiments 58–59
  European officers 198
  Europeanisation 145
  expenditure and financing 145, 172*n*13, 478
  General Staff 168
  hunger 268
  infantry 122*n*30, 168
  irregulars 146
  leadership 130, 265–66
  logistics 265–66
  looting 268
  military intelligence 94
  modernisation 93–94, 206
  in 19th century 265–69
  officers 125*n*65, 168, 269
  Paul's reforms 167, 267
  pay and rations 267
  in peacetime 175*n*40
  Peter's reforms 34, 88–89, 118, 135, 170, 269
  Polish troops 150
  and prisoners of war 268–69
  Prussification 269
  recruits/recruitment 86, 88–89, 93–94, 168, 265–66
  reform 49, 88–89, 168–69, 170, 180
  religious propaganda 268
  and Russia 135–36, 147
  and serfdom 135, 265–66
  in 1700 45–46
  1768–74 War 145, 168
  supply system 89, 94–95, 282; difficulties 112, 120*n*15, 146–47
  Tatar Raids 103, 143
Russian navy 147, 153
  1768–74 War 153
  development 91
  European officers 198
  fleet 237
  at Varna 167
Russo-Austrian alliance (1726) 87, 137–38
  1768–74 War 114–15

Russo-Austrian-Ottoman War (1736–39) 90, 225
Russo-Ottoman conflicts 28–29, 401–2
border 93
Caucasus 462–66
Danube region 422–24
Erzurum 449–50
Kars 449–50
Wallachia 401–2
Russo-Ottoman cooperation
alliance 230
Balta Limanı Convention (1849) 427
diplomacy 290–91
fleet 237
supplies 194
Russo-Ottoman War (1677–81)
and Cossacks 58
Russo-Ottoman War (1736–39) 102
and Crimea 94, 103
numbers of troops 126*n*83, 137
Ottoman forces 117–18, 156
supply 94
Russo-Ottoman War (1768–74) 30, 129, 136, 462
Anatolian troops 216
Bucharest 151, 153
causes 138–40
centre of conflict 140–42
Çeşme 154
and civilian populations 116–17
and Cossacks 58–59
cost 146, 149
and Danube fortresses 151–53
eye-witness accounts 174*n*34
financing 143
Giurgevo 151, 153
Janissaries 49, 174*n*33
Jassy 151
Kartal 149, 153–54
leadership 125*n*76
losses 147, 151, 158
and military reform 485
mobilisation 145–46
and Muhsinzade Mehmed Pasha 235
numbers of troops 63, 145, 175*n*38
Ochakov (*see* Ochakov)
Ottoman forces 143–44; mobilisation 143–45, 151
Ottoman navy 147
peace negotiations 115–17, 149, 154–60, 235
preparations 142–47
Russian forces 146–47
Russian headquarters 131
siege of Hotin 149–51
supply systems 146–47, 156–57
and Tatars 61, 174*n*31
timariots 55
truce 201
Ukrainian militia 175*n*38
Russo-Ottoman War (1787–92) 129, 136, 160–67, 184, 462–63
and Austria 137–38, 178*n*89
Crimean front 166–67
Danube estuary 166–67
disease 164–65
human and financial costs 163–65, 163–66
Ochakov (*see* Ochakov)
Ottoman forces 156, 162–63, 168
peace 167–70
refugees 164
supply systems 164–65, 168
Russo-Ottoman War (1806–12) 223, 225, 271, 463
and Ali Pasha of Iannina 287
Ottoman forces 156, 269–70, 271–72
Russo-Ottoman War (1828–29) 306, 313, 343–48, 434, 436
challenges 346–47
and Christian populations 355
Cossacks and 348
filth and disease 351–55
foreign intervention 343–44
as jihad 343
Muslim irregulars 465
Ottoman forces 156, 344–48, 349
Russian forces 345–47, 353
Russian gains 351–52
Russian objectives 346
Russo-Ottoman War (1877–78) 131, 476
Russo-Ottoman Wars 129, 251
and Catherine's reform agenda 136
competition for supplies 194
contested borders 215

Russo-Ottoman Wars (*continued*)
 demographic upheavals 204
 and geopolitics 259
 Ottoman strategic objectives 157–58
Russo-Turkish Treaty of Alliance (1805) 242
Rüstem 386
Ruthenians 159

Sabaneev, Major 277
Sabri Mustafa Pasha 423
Sacred Band 291
Sacred Cities of Mecca and Medina 31
 annual pilgrimage 62, 232–34, 308–9; revenues 185
 recapture 309
Sadık Rifat Pasha 439
Safavids 85–86, 130
 Persian 462
 Shiites 31
Şahin Giray 159
Said, Khedive 407
Said Bey 386
Said Mehmed Efendi 327, 379
Salih Bey 230
Salih Pasha 292–93, 464
Salonika 68
Sami Pasha 450
Samokov 142
San Stefano, Treaty of (1878) 283
sancaks 171*n*3
Sandwich, Captain Humphrey: and Crimean War 471–73
Sapper *(Lağımcı)* 193
Sarajevo 431
Sarı Mehmed Pasha 52
sarıca/sekban (musketeers) 56–57
Sarıkamış 464
Sarım İbrahim Pasha 143, 152
Satunovo 348
Sava River 111, 163
Saxony 87
Schmettau 115–16
School for Military Science 327, 376, 379
School of Mathematics 327
Sea of Azov 155, 158, 462
 and Crimean War 457
Sébastiani, French ambassador 243
Seckendorf, Friedrich Heinrich von 106–7, 110–12
Second Russo-Ottoman War (1787–92) *see* Russo-Ottoman War (1787–92)
Second Serbian Uprising (1815) 283–84, 302*n*46
Sekban:
 and former Janissaries 260
 recruits 262–63
Şekib Efendi 429
Selim III 26, 64, 136, 163, 166 *see also* Ottoman army
 and Ali Pasha of Iannina 229
 alliance with France 243
 Caliph of Islam 214
 and Canikli Ali 217
 centralisation 184, 223
 challenges to 214–15, 224, 242–48
 consolidation 198
 death 249–50, 252, 258*n*77
 diplomacy 224–26
 economic reforms 184–85
 and Egypt 231, 307
 finances 225
  bankruptcy 72–73
  reorganisation of treasury 184–85
 Francophilia 242–44
 and French 227–28, 237
 impact of reign 250–52
 and Janissaries 317–18, 321
 legitimacy 214–15, 224
 and Napoleon 241
 and Naqshbandis 323
 and Osman Pasvanoğlu 229
 and Prussia 138
 purges 315
 reform 169–70, 180, 193–95, 205–6, 214
 revolt against 246–47
 vacillation 245
 Wahhabi revolts 308–9
Selim Pasha (governor) 365
Selim Pasha (Crimean War) 470, 473, 474
Selimiye Barracks (Üsküdar) 204, 264
 and Crimean War 453
Semendre 115

Sened-i-İttifak/Deed of Agreement 261–65
Septinsular Republic 237
Serbia 22, 90, 442, 444
  autonomy 224, 272, 278, 280–81, 283, 392, 422
  Christians 92–93
  and Crimean War 450
  1804 revolt 282–83
  1815 uprising 283–84
  geopolitical importance 284–85
  and Janissaries 222
  and Muslims 422–23
  nationalism 220
  resistance to Mahmud 282–83
  revolts 242, 259, 282–88, 302*n*43, 424
  and Treaty of Adrianople 363
Serbians (Rascians) 29, 134, 290
  at Grocka 127*n*105
  Orthodox communities 224
  1768–74 War 116–17
  at Varadin 102
Seret River: and Moldavian border 278, 280
serhad kulu (frontier troops) 56–58
serhad strategy 65
Sevastopol
  British ultimatum 444
  and Crimean War 444, 449, 457–60, 474
Seven Years War (1756–63) 88–89, 130, 137, 198–99
  Grenzers 59
  and Ottomans 134–35
Sèves, Colonel *see* Süleyman Pasha
Şeyhülislam (chief religious officer) 27, 36
Seyyid (Moralı) Ali Efendi 228, 249
Shahrizur 286
shari'a (Islamic law) 27, 53
Shaykh Khalid 323
Shaykh Zahir al-'Umar 230, 234–35
Sheremetev, Field Marshal 95–6
Shihab family 416–17
Shiites 31, 286, 308, 385
  Safavids 31
Shorofi (Soroki) 96
Sibiu 291
Sidon 230, 370
  post-1841 416
Silahdar Ali Pasha 315
Silesia 137
  Prussian invasion 86
Silhadar Mehmed Pasha 103, 106, 107
Silistre 62, 68, 156, 244, 245, 277, 316, 451*plate*
  and Crimean War 449, 453–56, 454
  disrepair 334
  1828–29 War 347, 353
  reorganisation 336
  siege 271
  and Treaty of Adrianople 363
Simmons, Colonel: and Crimean War 474
Sinai 235
Sinop 68, 221
  and Crimean War 444, 467
sipahis (holders of *timars*) 76*n*11
Sipahis (Janissary cavalry) 76*n*11
Sistova 271
  and Crimean War 450
  Russian reoccupation 280
Sistova, Treaty of (1791) 138, 166, 220, 225
Sivas 216–17, 370, 387, 462
  and *Redif* 383
  Reşid Mehmed Pasha 384
Siverek 386
Sizeboli 353
Skene, James Henry 412–13
Slade, Adolphus 440
Slatina 152
slavery *see also* Ottoman military
  Mamluks 232
Slavonia 18
  Christians 92–93
  revolts 60
Slivno 355
Slobozia 273, 276
  armistice 247
  armistice (1807) 271
Smith, Sydney 231, 236, 239
  in Acre 257*n*55
Sobieski, Jan, King of Poland 24
Sofia 95, 142, 164, 270, 273, 283, 288
Soğanli Dağ 464
Sokhum 463
Sophie, Archduchess 434–35

Spahis d'Orient: and Crimean War 456–57
Spanish Succession, War of (to 1714) 93, 99
Spectateur de l'Orient 377
St Arnaud, Commander: and Crimean War 454, 456
St Gotthard, Battle of (1664) 45, 65
St Nicholas 466
  and Crimean War 444, 466
St Petersburg:
  Ottoman ambassador 227
St Petersburg, Convention of (1826) 298
St Petersburg, Convention of (1834) 425
St Priest, Ambassador 159
state provisioners *(mubayaacı)* 148
Stoffeln, General: 1768–74 War 151–52
Stoianovich 141
Stopford, Admiral 405–6
Stoye, John 40
Straits Convention (1841) 399
  terms 407
streltsy 93–94
Sublime Porte/*Bab-i Ali* 263, 298–99, 320, 377, 379, 409
  and Abdülmecid 414
  and Treaty of Adrianople 361–62
Sudan 309
  conscription 311
Suez-Red Sea 375
sufis 31, 308, 322–23 *see also* Naqshbandi Sufis
Sugar, Peter 40, 47
Şükrü Pasha 472
Süleyman Agha 221, 238, 248–49
Süleyman II 20, 41*n*4
Süleyman Pasha (Colonel Sèves) 197, 312, 326, 348–49, 365, 405, 426
Süleyman Penah Efendi 191
  Mora İhtilâli Tarihi 189–90
Süleyman the Magnificent 36
  ideal 187
  model for reform 251
Süleymanic age 55
sultanic ordinances *(kanun)* 27
sultans 19
  accession gifts 50, 52
  image of 445*n*5
Şumnu 156, 166–67, 182–83, 221, 271, 273
  and Crimean War 450, 453–54
  1828–29 War 345–50
  Ottoman headquarters 134
  1768–74 War 157–58
Sunnis 31, 286
  Mughal house 31
  Shaybanids 31
Supreme Council 429
Suvorov, Field Marshal AV 46, 130, 136, 160–62, 268, 269, 272, 466
  brutality 300*n*18
  and Crimea 156
  at İsmail 167
  and military reform 168
  at Ochakov 164–65
Sweden 88, 139
Syria 216, 230, 231
  centralised control 287
  Egyptian campaign 369*plate*
  ethnic forces 480
  ethno-religious movements 251–52
  European mediation 385
  financing occupation 383–84
  and İbrahim Pasha 312, 371–75, 384–85
  internal tensions 416–17
  invasion 367
  and Mehmed Ali Pasha 259, 306, 308, 359, 364
  Muslims 365
  pacification 401
  private armies 231–32
  resistance to conscription 345
  rout of Mehmed Ali 404–6
  and Shaykh Zahir al-'Umar 234–35
  unrest 392
Syrian Crisis (1839–41) 401, 436 *see also* Mehmed Ali
  European intervention 399–400
  Ottoman forces 396*n*52
Syrian Question (1831–33) 363–67

Table of Ranks (1722) 34, 88, 135
Taganrog (naval base) 28, 91, 158
Tahir Agha 149
Tahir Efendi 323
Tahir Pasha 306–7, 430

Takvim-i Vekayi 382
Talimli Askerler (Trained Soldiers) 185
Talleyrand 228
Talman 106
Taman 160
Tanzimat (1839–76) 401, 417–18, 485–86
  agenda 404
  in Arab provinces 418–22
  ideology 486–87
  leadership 377
  provincial implementation 414–16, 462
  regulations 194–95
  resistance of Aleppo 418–22
  resistance of Vidin 427–28
  revolts against 464
Tarsus 371
  Egyptians abandon 406
  and İbrahim Pasha 371
Tarsuslu Koca Agha 144
Tatar Khan 30, 97, 103
Tatar Khanate 61, 95, 159–60
Tatar Question 155
Tatarcikzade Abdullah Efendi:
  and military reform 192
  reform agenda 193–95
  reform recommendations 226
Tatariçe 272
Tatars 28, 30–31, 33, 90, 136, 269
  and Crimea 106
  Crimean 91
  exiles 60, 79*n*50, 139, 251, 282
  horsemen 331
  independence 155, 159
  military role 119*n*5, 122*n*39, 174*n*31
  settlement 204
  1768–74 War 143
  and siege of Vienna 62, 43*n*26
Taurus Mountains 366
Tavuk çu basi, Mustafa 117–18
Tayyar Mahmud 218, 248
Teesdale, Captain CC
  and Crimean War 471, 474
Temeşvar *see* Banat/Temeşvar
Thessaly 293
Third Coalition 236
  and Napoleon 242
Thirty Years War (1618–48) 23, 33, 198
Thököly, Imre, Prince of Hungary 24
Thompson, Captain Henry Langhorne 471
Thornton, Thomas 48–49, 52, 267–68
Thrace 259
  secular schools 376
Tiflis 462, 463
Tilsit, Treaty of (1807) 181, 243, 246–47, 257*n*63, 269, 271, 280
timar (land grant) 54–56, 78*n*29
  abolition of system 197
timariots 185, 223, 423–24, 486
  entitlements 57–58
Times: and Crimean War 460
Timok River 111
Tirana 288
Tisza River 59, 65–66
Tkat 216
Tokat 216, 217
  anti-reform movement 366
Tolstoy (Russian representative) 95
Topal Osman Pasha 430–31
Topçus (gunners/artillerymen) 144, 198
Topçular (artillery corps): numbers under Selim III 202–3
Tophane 332
  arsenal 263, 264
  foundry 195, 200–1
Topkapı palace 250, 295
Topkapı Sarayı 263
Tott, François, Baron de 30, 43*n*26, 52, 76*n*12, 135, 149, 195, 212*n*40, 211*n*33
  Memoirs 175*n*46, 200–1, 211*n*36, 212*n*42, 213*n*36, 255*n*32
  military advisor 199–201
  and Ottoman armies 193
  1768–74 War 176*n*59
Trabzon 84, 190, 216–18, 462, 465
  and Crimean War 473
  governors 95
  and *Nizâm-ı Cedid* 219
  recruits 246
Trabzon-Anapa line 463
Transcaucasus 462, 466
  frontier 466
  Russian encroachment 463
Translation Bureau (Istanbul) 376

Transylvania 18, 22, 28, 114, 163, 442
and General Bem 426
Muslim exodus 141
rebellion 1848 426–27
Russian invasion 427
and siege of Vienna 62
Travnik 110
Triple Alliance: mediating with Ottomans 165–66
Tripoli 364, 370
autonomy 392
Ottoman forces 384
Tulcea 68, 156, 167, 349
and Crimean War 453
Tunis 364
Tunisia: autonomy 392
Turk-Arab rivalry 366
Turkey (Republican) 88
historiography 58
Turkish Contingent: and Crimean War 459–61
Turkish Spy, The 228
Türkmen 62–63, 232, 462
horsemen 331
Türkmencay, Treaty of (1828) 463
Turks 221
nationalism 41*n*9
use of term 23, 41–9
Tusun Pasha 309
Tutrekan:
and Crimean War 452–53
Mahmud's visit 362*plate*
Tyre 370

Ukraine 18, 83, 88–90, 107, 110
contest over 46
ulema (religious class) 27, 36–37
Umar Makram 307
United Provinces *see* Dutch Republic
United States:
army 169
and Middle East 400
trade with Egypt 308
Upper Egypt 236
conscription 311–12
Urfa 371
and İbrahim Pasha 371
and Mehmed Ali 384–85
Urquhart, David 357, 443
Üsküdar 196, 197, 231, 244, 262–64, 372
and Crimean War 453
First Army 410
Selimiye Barracks 204
Utrecht, Treaty of 93
Utrecht-Rastatt, Treaties of (1713–14) 89
Uyvar 65–66
Uzunçarşılı, I Hakkı 52

Varadin 59–60, 101
and Eugene of Savoy 86
numbers of troops 124*n*60
Varna 221, 345, 346, 347
and Crimean War 454, 456, 458
1828–29 War 349–54
and Russia 160, 167
Vasfı Ahmed *see* Ahmed Vasıf Efendi
Vasfı Mustafa Efendi 152, 154, 158, 296–97
need for *Nizâm-ı Cedid* 181–84
1768–74 War 176*n*59
Vasif Pasha 429, 472–73
Vede Khan Bey 386–87
Veli Pasha 287
and Crimean War 473, 474
Venetian-Ottoman relations 99
Venice 18, 19, 24, 83, 229
empire 123*n*49
forces at Nauplion 124*n*59
and Habsburgs 99, 101–2
loss of Morea 98–101
and Peloponessus 100
siege of Kandia 45–46
Vergennes 227
Verinac, Raymond 227
Vidin 25, 58, 68, 73, 84–85, 106–7, 110–11, 141, 163, 219–23, 230, 273, 291, 316, 461
and Crimean War 444, 449
demographic changes 423–24
disrepair 334
and Janissaries 220, 222
numbers of troops 64
and Osman Pasvanoğlu 219–24
rebellion 423
and reforms 324, 427–28

Vidin (*continued*)
reorganisation 336
separation by religion 65
and Serbia 447*n*23
1736–39 War 111–12
siege 220–22
Vidin-Rusçuk line 347
Vienna 432
Ottoman embassy 226–27, 377
rebellion of 1848 426
Vienna, Siege of (1683) 20, 24, 36
Janissaries 49
numbers of troops 62
timariots 55
Vienna Conference
and Crimean War 475
Vienna Note 440–41, 442
Villeneuve, Marquis de 115–16, 117, 138
Vladimirescu, Tudor 290
Voinov 274
Volga River 28
Volney, Constantin 230, 235
on Mamluks 240–41
Voltaire 122*n*39
Voronezh shipyards 95
Vorontsov, Prince 466
voynuk corps 63–64

Wahhabis 215, 232, 241–42, 252, 366
challenge of 308–13
resurgence 312
spread of influence 286
Wallachia 22, 28–29, 137, 140, 245, 265, 271, 276–78, 287, 425–26, 444 *see also* Craiova (Little Wallachia); Oltenia
and Alexander Ghica 425
annual tribute 91
Austrians and 166
autonomy 392, 423–24, 475
and Crimean War 450, 453, 456
depopulation 141, 146–47
Greeks 289
Hellenisation 289
impositions and supply obligations 209*n*17, 302*n*43, 393*n*10
and Ömer Pasha 427
and Osman Pasvanoğlu 220
and Ottomans 61–62
peasant conscription 281–82
and Phanariots 140–41
rebellions 98; 1848 436–37
Russian occupation 83, 152, 246, 269–70, 282, 426–27, 435–37, 440
and Treaty of Adrianople 361–62
Wallis, George Olivier 106, 111, 127*n*100
1739 campaign 115
Walsh, Reverend 333
war and warfare *see also* plagues
captives 81*n*67
and civilian populations 59, 116–17, 141
disease/poor hygiene 112, 114, 146, 164–65, 268, 351–55, 393*n*20, 453, 456, 472
financing 134
global 87, 119, 230
ideological propaganda 241–42
refugees 79*n*50, 164, 251, 427–28, 442, 447*n*29, 481*n*40, 482*n*42
and state development 47
weapons 195, 198–99
War College 484
War of Austrian Succession (1740–48) 88
Warsaw 454
Weissman, General 156
Wellington, Duke of 297
Western Wallachia *see* Oltenia
Whitcombe, Thomas 295–96
William III, Prince of Orange and King of England 20
Williams, Colonel Fenwick: and rimean War 471–73
Williams, General WF 460
and Crimean War 467–70
Wilson, Herbert: and Crimean War 452
Wittgenstein, Field Marshall Count 346, 352
World War I 240, 458, 484
Worsley, General Lord: and Crimean War 452

Yahya Pasha 103–7
yamaks (recruits/garrison guards) 64, 80*n*61
Yanbolu 271, 355

Yedikule fortress 142
Yedisan (Later Bessarabia) 158
Yeğen Mehmed 110
  at Vidin 113–14
Yemen: coffee trade 234
Yenibazar 349
Yenikale fortress 68, 155, 158
yerli (local forces) 64
yerliye (Jerusalem) 62
Yıllıkoğlu Süleyman 272
Young Turks 480
Ypsilanti, Alexander 290
  in Moldavia 316–17
Ypsilanti, Constantin 281, 290
Yusuf, General: and Crimean War 456–57
Yusuf Agah Efendi 226, 254*n*27
  and Selim III 227
Yusuf Pasha 98, 161, 179*n*95, 350
  betrayal 352
  at Prut 95–96
Yusuf Pasha (Adana *Mutasarrif*) 221
Zahir al-'Umar *see* Shaykh Zahir al-'Umar
Zaporozhian Cossacks 62, 103 *see also* Cossacks
  horsemen 330
  and New Russia 139–40
  split 93
Zarif Mustafa Pasha, 429
  dismissal 470
  in Aleppo 419–21
  in Erzurum 460–61
  in Urfa 368–87
  on Konya 372–73
  training 378
Zenta, Battle of (1697) 24, 36, 69
zimmi pact 360
  Muslim/non-Muslim relations 299
Zivin 464
Zriny, Nicholas, *Ban* of Croatia 65
Zsitva-Torok, Peace of 41*n*12
Zuhab, Treaty of (1639)
  Ottoman-Persian border 32